寰宇文献 Universal Library | SINOLOGY 系列

# SELECTED WORKS OF BERTHOLD LAUFER

# 劳费尔著作集

## 第三卷

[美] 劳费尔 著

黄曙辉 编

中西书局 ZHONGXI BOOK COMPANY

**图书在版编目(CIP)数据**

劳费尔著作集 / (美) 劳费尔著;黄曙辉编. —上海:中西书局,2022

(寰宇文献)

ISBN 978-7-5475-2015-4

Ⅰ. ①劳… Ⅱ. ①劳… ②黄… Ⅲ. ①劳费尔 – 人类学 – 文集 Ⅳ. ①Q98-53

中国版本图书馆CIP数据核字(2022)第207067号

# 第 3 卷

# 056

死亡舞蹈的起源

# The Open Court

## A MONTHLY MAGAZINE

### Devoted to the Science of Religion, the Religion of Science, and the Extension of the Religious Parliament Idea

*Editor:* Dr. Paul Carus.    *Associates:* { E. C. Hegeler. Mary Carus.

VOL. XXII. (No. 10.)   OCTOBER, 1908.   NO. 629.

## CONTENTS:

### CHICAGO

## The Open Court Publishing Company

LONDON: Kegan Paul, Trench, Trübner & Co., Ltd.

Per copy, 10 cents (sixpence).   Yearly, $1.00 (in the U. P. U., 5s. 6d.).

# ORIGIN OF OUR DANCES OF DEATH.

BY BERTHOLD LAUFER.

THE so-called Dances of Death, the *trionfi della morte* of Italy, came into being during the fifteenth century and reached their culmination in the sixteenth when Hans Holbein the Younger created his famous pictures in 1530. Various explanations have been given to account for the origin of this motive of art in Europe. The great epidemic plagues of the fourteenth century have been made responsible for its rise,[1] an argument which has little convincing force, as such natural phenomena may sufficiently explain a certain propensity of the time for reflections on death, but not the material foundation of an artistic conception of a motive of very peculiar and individual character. Deserving of more consideration is the suggestion that it presents the reproduction of a real dance of Death,[2] such as was performed in 1424 in the Cemetery of the Innocent at Paris, and in 1499 at the Castle of Bruges, and as it is represented in the shape of really dancing skeletons in Hartmann Schedel's *Weltchronik* of 1493. But the idea of Kraus[3] that we must descend into classical antiquity where we find Greek and Latin inscriptions repeatedly referring to death as the one who seizes all mankind, in order to explain the origin of our Dances of Death, is altogether too far-fetched, and such general reflections on the power of death as may occur at all times and almost any-

---

[1] F. Xaver Kraus, *Geschichte der christlichen Kunst,* Vol. II, 1. Freiburg. 1897, pp. 448-451, where also the entire former literature on the subject is quoted.

[2] "Ein ins Bild übersetztes Spiel des Totentanzes," as Kraus expresses it.

[3] *Loc. cit.,* p. 450. There is, altogether, no greater contrast imaginable than that existing between the idea of Death as a dancing skeleton and the Greek representation of Death as a beautiful serious youth. If Didron (*Christian Iconography,* Vol. II, London, 1891, p. 156) refers to the antique larva as the early Christian model for Death in the form of a skeleton, this may be right; but this representation has no direct connection with the Dances of Death in which the essential point is that the skeletons are represented *dancing,* and not merely skeletons.

where are too vague and general to be admitted as arguments in the present question. Certain it is that classical art did not possess this motive, that it was likewise unknown to early Christian art and sprang up in Europe at a late date, not before the fifteenth century. As regards all motives of art, we are justified in searching for their historical foundation and for their occurrence in other spheres of art, from which they may have eventually been derived.

Such a province of art, in which the motive under consideration is widely made use of, indeed exists, and it is found in Buddhism, more particularly in the Buddhism of Tibet or Lamaism. As is well known, in Tibet and the other countries where Lamaism prevails, certain kinds of mystery plays are performed, in which masks of very elaborate make are used. Up to the present time but little has become known of the plots of these masquerade moralities, though in Tibetan literature a special class of books is devoted to the subject of their performance and rites. Some represent the advent of Buddhist monks from India into Tibet, and their struggle with, and final triumph over the native adherents to shamanism; and others give scenes from the life of Buddha, while still others relate the so-called birth-stories or Jatakas, the deeds and miracles of remarkable saints, or the horrors and torments of the Inferno. The subjects of other pantomimes are taken from Tibetan history, like the assassination of King Glang-dar-ma by a Lama, because of his hostility to Buddhism. Others are emblematic, one for instance being symbolical of the departure of the old year and the ushering-in of good luck with the new. Many of these lamaistic dances suggest the exorcism of devils or survivals of ancient shamanistic rites. The principal deities and demons represented by the masks of these plays in a series which I obtained from the great Lama Temple Yung-huo-kung at Peking, are the four Great Kings of Heaven (*maharaja*), distinguished, according to their colors, as the Yellow, Red, Blue, and Black King, each a guardian on one of the four sides of the world-mountain Sumeru, where they command hosts of demons; further two men-devouring ogres or Rākshasa, painted yellow and red, with protruding tusks and brute-like ears and snout; and two aerial demons or Yaksha,—one red, the other blue,—with elephant's trunk, tusks, and ears, and with a wreath of five skulls around the forehead. Four masks represented ghosts of small-pox, others animals—the stag, spotted black and white, the monkey, the blue and the red ox, these being helpers to the god of Death, Mahākāla, whose masks are made in four different colors. Contrasting with these fierce-looking demoniacal faces are the masks of the jovial and humorous monks. The largest one signifies the great or chief Huo-shang, who in some plays is intended to represent an historical Chinese monk who appeared in Tibet at the end of the eighth century, while in others he symbolizes Maitreya, the Buddha of the future. He is bald-headed, has a big wart on his forehead, an almost scythe-like mouth, and his face is convulsed with laughter. A peach of colossal size, expressive of longevity, is his attribute. Six other heads belong to young monks, his disciples, who play the parts of buffoons with him. They have youthful red lips, and their skulls are painted with a small tuft of hair with a short cue attached.

A special group of these masks is formed by that of the graveyard ghouls (*çmaçānapati* or *citipati*) which are intended to represent skulls. They are pale-faced, have circular eyes with red rings around them and flames over them, a flattened nose, and compressed mouth. They wear clothes to represent skeletons. In one of the sacred dances they scare away with their sticks a raven who is about to steal the strewn offering of sacred meal.[4]

Here, accordingly, we meet with a real Dance of Death, and further, the same ghosts performing their weird dance in the mystery plays find their counterpart also in a pictorial representation. The most common one of these is a pair of skeletons dancing over a human corpse.[5] They brandish staves made into the form of a skeleton. This is a favorite household picture of all Lamaists and easily procurable at Peking of all dealers in Lamaistic objects.

[4] An illustration of such a performance, in which nine skeletons take part, is inserted in the book of E. F. Knight, *Where Three Empires Meet,* London, 1897, on the plate opposite p. 216. The author's description of them (p. 219) runs thus: "A small black image representing a human corpse was placed within a magic triangle designed upon the pavement of the quadrangle. Figures painted black and white to simulate skeletons, some in chains, others bearing sickles or swords, engaged in a frantic dance around the corpse. They were apparently attempting to snatch it away or inflict some injury upon it, but were deterred by the magic of the surrounding triangle, and by the chanting and censer-swinging of several holy men in mitres and purple copes, who stood beneath the temple porch." . A single skeleton-dancer is figured in Waddell's *Buddhism of Tibet,* p. 525. See also Chandra Das, *Journey to Lhasa* (2d ed.), pp. 115-116 and p. 263. The same dance occurs among the Mongols and is described and figured by A. Posdnäyev in *Sketches from the Life of Buddhist Monasteries in Mongolia* (in Russian), St. Petersburg, 1887, pp. 396, 397.

[5] Illustrations in Pander-Grünwedel, *Das Pantheon des Tschangtscha Hutuktu,* p. 98, No. 253; Grünwedel, *Mythologie des Buddhismus in Tibet und der Mongolei,* p. 170. These ghosts are called "Lords of the graveyard" or "protectors of the cemetery" (Tibetan, *zhing skyong*); they belong to the retinue of Yama, the god of the nether world, and are accordingly real personifications of Death.—As to the development of the history of art, it is interesting to note that a mask intended for a death's-head is already represented on one of the sculptures of Gandhara in the demons of Mara's army (see Grünwedel, *Buddhist Art in India,* p. 99, German edition, pp. 94, 106.).

In this and kindred representations we have doubtless to see prototypes of our Dances of Death. In Buddhism the Dance of Death has its suitable well-founded place and is connected with many other phenomena of Indian religious lore. We must here call attention to a peculiar class of spirits called Vetāla, who have become so familiar to us from the entertaining collection of stories, the Vetālapañcaviṃçati.

The Vetāla is a ghost who haunts graveyards and is possessed with the ability of passing into a corpse which is thus resuscitated and begins to move and to dance.[6] Miraculous powers may be obtained from such a ghost (vetālasiddhi), and to conjure them, a special method is employed which plays an important part in the Yoga and Tantra schools of later Buddhism. It is especially conspicuous in the legends of Padmasambhava (eighth century), who meditates and conjures ghosts for five years in one cemetery. It seems to me probable that it was Padmasambhava himself who introduced the dance of the skeletons in Tibet while actively engaged there in the suppression of demons, for just at his time we find a description of masked pantomimes in celebration of the completion of the temples of bSam-yas built after his plans.[7]

No example of this subject in Chinese art is known to me; it seems to have had its special connection with Tantrism which penetrated into Tibet. The Japanese, however, must have had a certain tradition relating to it, for Kyosai has taken up the theme with some eagerness, and incorporated two Dances of Death in his Mangwa,[8] in which the skeletons perform wonderful acrobatic feats.

The analogy between the two phenomena in the East and the West is most striking. On both sides we encounter skeleton-dancers in mystery plays and pictorial representations of the same subject. The similarity goes still further in a very peculiar point. "The spectacles or performances of the Dance of Death, so common in the Middle Ages, were often relieved of their gloom by the introduction of interludes in which the Fool took a prominent part. Such scenes when illustrated formed part of the series of subjects of engravings of the 'Danse macabre.' The Fool is seen at strife with

[6] See the illustration in Grünwedel's *Mythologie*, p. 192.

[7] See my paper "Die Bru-ža Sprache und die historische Stellung des Padmasambhara," in *T'oung Pao*, 1907. Also Chandra Das (*Journey to Lhasa*, p. 155) joins in the opinion that Padmasambhara is the reputed originator of religious dances in Tibet.

[8] Reproduced by C. H. Stratz, *Die Körperformen in Kunst und Leben der Japaner*, Stuttgart, 1904, pp. 126, 127. On other representations of skeletons in Japanese art see A. Brockhaus, *Netsuke, Versuch einer Geschichte der japanischen Schnitzkunst*, Leipsic, 1905, p. 367.

his adversary Death, and hitting him with a bladder full of peas or pebbles. We frequently meet with allusions to Death's fool in Shakespeare."[9]  In the Tibetan moralities also, the fool or mime is of utmost importance and appears under a great variety of forms and masks;[10] whether the skeleton masqueraders figure also in this rôle, I am not prepared now to assert positively, as we know too little about the plots of these plays, but it certainly seems that their actions and dances are better calculated to bring about a humorous and comical effect than a serious one, in the same way as the Vetāla is a rather jolly and jovial creature in the Indian stories. If the whole subject could be scrutinized more closely in Tibet, both in the mystery plays and in the line of iconography, and in the texts relative to the subject, the points of coincidences between East and West would probably increase to a considerable extent.

But the reason I am inclined to believe that we are compelled to admit an historical connection between the two phenomena, lies still deeper. That the personification of Death in the shape of a human skeleton may have arisen independently in various quarters is obvious, although the idea is by no means of frequent occurrence among mankind, and if this were all, the whole matter would not be a case of great significance. But the idea of a temporary rising of the dead conceived as skeletons, and of their ability to move freely around and commit extraordinary actions, is very specific and does not find any explanation from the thoughts of Christianity, to the whole spirit of which it seems to be entirely alien. It has certainly nothing to do with the idea of resurrection, which is eternal, while here it is the question of a merely transitory rising with a final return to the grave, so it has nothing to do in Buddhism with the doctrine of transmigration. It is the idea which we find expressed in numerous German folk-tales and songs, of the midnight dance of the dead over the graveyard, of the man rising from the grave to punish his faithless sweetheart, of the dead man climbing a steeple—the dead always appearing as moving skeletons in popular imagination. No doubt, this conception is foreign to the early periods of Christianity and probably may not be older than the times when the Dances of Death began to come into more general vogue. On the contrary, the orthodox Jewish-Christian notion is that the corpse does not continue a material existence, but that it will decay and crumble away into earth and dust. This notion is strongly contra-

[9] Didron, *Christian Iconography*, Vol. II, p. 169.

[10] The mask of one type of buffoon is figured in *Globus*, Vol. LXXIII, p. 6. As everywhere, he is armed with a large stick, as already G. Boyle (1774) emphasizes (L. R. Markham, *Narratives of the Mission etc. to Tibet*, p. 93).

dicted by the whole conception of the Dances of Death and of the dead, in which the *moving power of the skeletons is implicitly pre-supposed,* but not by any means accounted for. This shows that it must be a foreign, a borrowed idea in European Christianity, and we find this idea fully developed and rationally accounted for in Buddhism. In India, it seems to me, the idea must ultimately be traced back to the system of Yoga, the practice of which was considered the safest way of acquiring many kinds of supernatural, miraculous powers (*siddhi*),—among others, the ability of causing deceased persons to appear and communicating with them, of passing into another body and returning into one's own.[11] The notion of the Vetāla penetrating into a corpse and filling it with life is perhaps connected therewith. At all events, Indian tradition offers an interpretation for the moving power of the dead or skeletons derived from and consistent with indigenous religious beliefs.

We have heretofore considered only those lamaistic representations in which solely dancing skeletons figure. As is well known, in the European Dances of Death, the figure of the latter is usually associated with several or even a whole company of human beings whom he leads away into the realm of shadows. This was a moral point specially emphasized by the Church which availed itself of this motive for educational religious purposes. Certainly we have here a peculiar Christian development of it, and the great artists of the Renaissance treated the theme with the spirit of their individuality. On the other hand, however, it must not be passed over in silence that also in lamaistic art human life is brought into close connection with the powers of death with a utilitarian viewpoint in mind for impressing the masses. I think, in this connection, of the numerous representations of the punishments and tortures of Hell, as they particularly appear on the so-called "Wheels of Life." On many of these, the demons inflicting castigation are drawn in dancing postures, and the dance of the Preta signifies a regular Dance of Death.[12]

Quite recently, R. Pischel[13] justly remarked that "without doubt much has migrated from Lamaism into the Catholic Church."

[11] See R. Garbe, *Samkhya und Yoga,* Strassburg, 1896, p. 46. Compare also the two interesting legends contributed by Sarat Chandra Das in his article "On the Translation of the Soul from One Body to Another," *Journal of the Buddhist Text Society of India,* Vol. V, Part III, 1897, pp. (1)-(7).

[12] See especially Posdnäyev, *Sketches from the Life of Buddhist Monasteries in Mongolia,* Plate opposite p. 80, where the ribs in the emaciated bodies of the dancing Preta are clearly outlined, and Plate VII, in Waddell's "Lamaism in Sikhim" (in Risley's *Gazetteer of Sikhim,* Calcutta, 1894.).

[13] In his excellent book *Leben und Lehre des Buddha,* Leipsic, 1906, p. 124.

Among these migrations, we may now count the artistic motive of the Dance of Death. Several ways of how the transmission took place may be indicated. First of all, I wish to call attention to the fact that a representation very similar to the Tibetan conception has been discovered in Turkistan. I refer to Plate XII in Grünwedel's Report.[14] This illustration represents an ink-drawing derived from an Uighur inscribed roll and showing, as Grünwedel remarks,[15] "the boldly painted figure of a demon, a sketch which might be called Japanese, if it should fall into our hands without knowledge of where it was found."

The most striking features about this demon are first, that he is represented in a dancing posture with crossed legs and outstretched arms, and secondly, his general skeleton-like appearance. This has been brought out by the artist by the ghastly thinness and leanness of the limbs and bones, and by clearly outlining his ribs on one side. Whether the figure must be conceived of as wholly nude or as being clad with a tightly fitting linen robe, such as the lamaistic skeleton dancers wear, with the ribs painted on, may be a debatable question. Arms, breast, abdomen, and legs at all events convey the impression of being uncovered, while a sort of breechcloth seems to be present. Arms, hands and fingers, legs and feet, have upon the whole a skeleton-like character. The head is very curious: the tremendous eye-sockets and the big skull with the large tuft of hair are intentionally made quite out of proportion with the smallness of the face. I think we need not hesitate to look upon this figure as an offshoot of the "Dances of Death" series. It would hardly be a matter of great surprise, if more and still more impressive representations of the same subject were to come to light in Turkistan, and then we might be able to establish a similar case, as R. Pischel did in regard to the penetration of the fish-symbol into Christianity.[16]

As the ideas concerning the Death Dances did not crystallize in Europe before the fifteenth century and may extend back as far as into the fourteenth, it may be well to suppose that it was the Mongols who brought a knowledge of the subject to Europe; to them it was familiar from Lamaism, and it may not be superfluous to recall the fact that they possessed in their Siddhi-kür a version of the

[14] A. Grünwedel, "Bericht über archäologische Arbeiten in Idikutschari und Umgebung im Winter 1902-1903." Aus *Abhandlungen der Bayer. Akademie*, Munich, 1906.

[15] *Loc. cit.*, p. 71.

[16] "Der Ursprung des christlichen Fischsymbols," *Sitzungsberichte der Berliner Akademie*, 1905, pp. 506-532.

Indian Vetālapañcavimçati, many stories of which are found in European folk-lore.

It has been asserted that Dances of Death are represented also on works of art coming down from classical antiquity; this opinion, however, seems to be in general erroneous, as the skeletons found on such representations are by no means in a dancing posture, but reclining, standing, or walking. The best known subject of this kind is that occurring on the silver goblets from the Treasure of Boscoreale ascribed to the time of the reign of the Emperor Augustus.[17] De Villefosse[18] remarks regarding this motive: "It has been pretended that the reliefs of our goblets were to represent a dance of skeletons (*Totentanz*). The expression is far from being correct; no detail justifies us to entertain such an idea. The principal actors of these scenes are evidently not given to a dance." Besides, it is shown by the profound investigation of the same author that these reliefs are not at all connected with the idea of death; they do not recall to mind the briefness of existence; their object is to demonstrate the uselessness of philosophy and the hypocrisy of morals.[19] This is sufficient to exclude any relation of these Greek subjects to the Christian and Buddhistic Dances of Death. The only known Greek example of really dancing skeletons remains one moulded in a relief on a terracotta goblet, apparently of Alexandrine art, now in the Musée du Louvre.[20] It does not seem improbable that this exceptional case may have received a certain impetus from traditions derived from India, for it is noteworthy that according to de Villefosse[21] the antique lamps adorned with figures of skeletons appear to have been turned out by one special factory which, in all probability, had its seat in Alexandria. And, in the judgment of the same scholar, it was just after the epoch of Alexander the Great that ancient art has increased the representations of skeletons and *larvae*, which, I am inclined to think, can hardly be a case of mere chance, but is possibly traceable to an incentive from Buddhist India.

[17] Described and figured by Ant. Héron de Villefosse, "Le Trésor de Boscoreale (Fondation Eugène Piot, *Monument et Mémoires*, Vol. V, Paris, 1899, pp. 223-245, and plates VII and VIII).

[18] *Loc. cit.*, p. 240.

[19] *Ibid.*, p. 245.

[20] Illustrated and described by E. Pottier, *Revue archéologique*, 1903, I, pp. 12-16.

[21] *Loc. cit.*, p. 225.

# 057

## 勃律语和莲花生大士的历史地位

# T'OUNG PAO

# 通報

OU

# ARCHIVES

CONCERNANT L'HISTOIRE, LES LANGUES,
LA GÉOGRAPHIE ET L'ETHNOGRAPHIE
DE
L'ASIE ORIENTALE

Revue dirigée par

**Henri CORDIER**
Membre de l'Institut
Professeur à l'Ecole spéciale des Langues orientales vivantes
ET
**Edouard CHAVANNES**
Membre de l'Institut, Professeur au Collège de France.

SÉRIE II. VOL. IX.

LIBRAIRIE ET IMPRIMERIE
CI-DEVANT
E. J. BRILL
LEIDE — 1908

# DIE BRU-ŽA SPRACHE UND DIE HISTORISCHE STELLUNG DES PADMASAMBHAVA

VON

## BERTHOLD LAUFER.

Unter dem Namen Bru-ža, auch ḁBru-ža, Gru-ža, Gru-ša, auch Bru-šal, ḁBru-šal wird in der tibetischen Litteratur häufig eine uns noch unbekannte Sprache erwähnt. Die verschiedenen Formen der Schreibung zeigen, dass es sich um den Versuch der Fixirung eines fremden Namens handelt. Der berühmte Sa-skya Paṇḍita (1181—1252) soll schon in seiner frühen Jugend Sanskrit und Bru-ža gelernt haben [1]). Diese Sprache spielt nach GRÜNWEDEL besonders in den Legenden des Padmasambhava [2]) eine Rolle, des Stifters des Lamaismus im achten Jahrhundert, der seine Bücher von den Ḍākinī's

---

1) GRÜNWEDEL, Mythologie des Buddhismus, S. 61, 66.

2) Die zeitliche Ansetzung des Buches über Padmasambhava ist noch sehr schwierig. Einen wunderlichen Versuch zu seiner Fixirung hat E. SCHLAGINTWEIT (Abhandlungen der Bayer. Akademie, 1903, S. 522) gemacht, indem er sich darauf beruht, dass *kLa-kLo* zur Bezeichnung der Mohammedaner in dem Buche noch nicht vorkomme und zum ersten Male im Zamatog erscheine, dessen Abfassung, unter Berufung auf mich, in das elfte Jahrhundert falle. Dagegen habe ich an der angeführten Stelle (Sitzungsberichte der Bayer. Akademie, 1898, S. 524) das Datum der Abfassungszeit des Zamatog auf das Jahr 1513 berechnet. Auch abgesehen davon wird die ganze Argumentation hinfällig, da das Wort *kLa-kLo* im Sinne von Mohammedaner bereits in der Kālacakra Litteratur auftritt, die zum grössten Teil vor dem elften Jahrhundert verfasst wurde. Die Ansicht GRÜNWEDEL's (T'oung Pao, Vol. VII, 1896, p. 529), dass das *Padma t'añ-yig* während oder nach der Yüan Dynastie seine gegenwärtige Fassung erhalten habe, hat die grösste Wahrscheinlichkeit für sich, obwohl die einzelnen Bestandteile natürlich älteren Datums sind.

1

in einer fremden, unbekannten Sprache erhielt und sie in Höhlen
für eine spätere Zeit verbarg, wo man sie verstehen würde. In den
Titeln seiner Legendenbücher kommen merkwürdige unverständliche
Titel vor, die sich mit den als Übersetzung gegebenen tibetischen
Titeln nicht vereinigen lassen, und ein von seiner Schule herrühren-
des grosses Werk trägt einen Titel in Bru-ža Sprache [1]). Im Folgenden soll ein Versuch zur Identifikation des Landes Bru-ža gemacht werden.

Die erste Erwähnung des Landes Bru-ža in der Geschichte ge-
schieht unter dem König Kʻri-sroṅ-lde-btsan (zweite Hälfte des
achten Jahrhunderts), der die beiden Gebiete des Westens, sBal-ti
und ḥBru-šal seiner Herrschaft unterwarf [2]). Ebenso dehnte König
Ral-pa-can seine Macht auf dasselbe Gebiet aus, das an dieser Stelle
als an Persien grenzend bezeichnet wird [3]). Damit ist die geogra-
phische Lage als die eines Nachbargebietes von Baltistan genügend
charakterisirt. Die nähere Bestimmung ergibt sich aus einer historischen Erwägung. Über die in Rede stehenden Territorien besitzen
wir auf dieselbe Zeit bezügliche Berichte in den Annalen der Tʻang-
Dynastie, die von CHAVANNES [4]) übersetzt und erläutert worden sind.
Die chinesischen Nachrichten sprechen von dem grossen Pu-lü oder
Pu-lu, das Baltistan entspricht, und dem kleinen Pu-lü, das CHA-
VANNES [5]) mit guten Gründen mit dem Distrikt von Gilgit identi-
ficirt. Beide Gebiete, das grosse und kleine Pu-lü oder Baltistan
und Gilgit standen nach dem *Tʻang-shu* im achten Jahrhundert
unter tibetischer Herrschaft, wozu denn die oben mitgeteilte Notiz

---

1) GRÜNWEDEL, l. c., S. 46.

2) E. SCHLAGINTWEIT, Die Könige von Tibet, Abhandlungen der Bayerischen Akademie, X, Bd. III, S. 847.

3) Ibidem, S. 850.

4) Documents sur les Tou-kiue (Turcs) occidentaux, St. Petersburg 1903, pp. 149—
154. S auch BUSHELL, The Early History of Tibet, p. 96, Note 36.

5) L. c, p. 150, Note 1. M. A. STEIN (Kalhaṇa's Rājataraṅginī, Vol. II, p. 435)
ist geneigt, die beiden Pu-lü mit Baltistan und Ladākh zu identificiren.

aus der officiellen tibetischen Geschichte die bestätigende Parallele liefert, und in derselben dürfen wir nun sBal-ti und ạBru-šal mit Baltistan und Gilgit identificiren, so dass wir nun die Gleichung ạBru-šal oder Bru-ža = Gilgit hätten [1]). Die Bedeutung dieses ersten Ergebnisses tritt sofort in ein helles Licht, wenn wir bedenken, dass sich Udyāna, die Stätte der ersten Wirksamkeit des Padma-sambhava, unmittelbar im Südwesten an Gilgit anschliesst [2]). Es klingt wie eine fern verhallende Reminiscenz an seine und seiner Schule Tätigkeit, wenn wir im *T'ang-shu* lesen, dass sich die Bewohner von Udyāna in den magischen Künsten auszeichnen [3]).

Die vermittelst tibetisch-chinesischer Quellen erschlossene Identifikation wird nun von indischer Seite her bestätigt. Im zehnten Kapitel des Lalitavistara nämlich wird in der bekannten Aufzählung der Alphabete als neunzehntes die Schrift der Darada genannt. In der tibetischen Übersetzung [4]) entspricht dem der Ausdruck Bru-šai yi-ge „die Schrift der Bru-ša" [5]). Somit können wir Bru-ša oder Bru-ža vom indischen Standpunkt aus mit Dardistan identificiren und dürfen wohl darunter das ganze von den Stämmen der Darden bewohnte Territorium mit Einschluss von Gilgit verstehen.

---

1) Das chinesische *Pu-lü* wird wohl auf den alten Namen von Baltistan *Bolor* (STEIN, l c., Vol. II, p. 363; YULE's Marco Polo, 3. Aufl., Vol. I, pp. 172, 178, 179) zurückgehen, und mit *Bolor* dürfte vielleicht auch der tibetische Name *Bru-ža* zusammenhängen, wenn man nicht etwa an *Burishki* denken will, womit die Hunza ihre Sprache bezeichnen. Der Name *Gru-ža* erinnert ein wenig an *Guraizi*, einen Dialekt der Darden (G. W. LEITNER, The Languages and Races of Dardistan, Lahore 1877, Part II, p. 41). Doch ich will mit diesen lediglich auf Vermutungen beruhenden Andeutungen nichts Abschliessendes gesagt haben.

2) S. die Karte in dem citirten Werke von CHAVANNES.

3) CHAVANNES, l. c., p. 128.

4) FOUCAUX, Rgya Tch'er Rol pa, Vol. I, Paris 1847, p. 113, 7.

5) Die darauf folgende Schrift der Cīna ist im tibetischen Text fälschlich mit *rgyai yi-ge* „chinesische Schrift" wiedergegeben, obwohl natürlich die dardischen Shina gemeint sind. Über die Wohnsitze der Darada s. LASSEN, Indische Altertumskunde, Bd. I, S. 39, 418 ff.; STEIN, l. c., Vol. I, p. 47.

In den im 94. Bande des Tanjur enthaltenen Annalen von Khotan lesen wir in der Übersetzung Rockhill's [1]):

Now the Bhikshus of An-tse, of Gus-tik, of Par-mk'an-pa and of Shu-lig [2]) were also greatly afflicted; so they set out for the Bru-sha country, and there also repaired the Bhikshus of Tokara and of K'a-c'e (Kashmir), who were persecuted by unbelievers. When they had all come to Bru-sha, they heard of the vihāras which were being built in Bod-yul [Tibet], and that the king was a Bodhisattva who honoured the Triratna and made much of the images; so they started out for Bod rejoicing, and they all lived there for three years in peace and plenty.

Der hier genannte tibetische König ist Ral-pa-can oder K'ri-lde-sroṅ-btsan, der von 816 bis 838 regirte. Es scheint, dass die erwähnten Bhikṣu von Mohammedanern vertrieben wurden oder vor ihnen flohen, und es stimmt ganz gut zu der Annahme der Identität von Bru-ža und Gilgit, wenn sie sowohl von Turkistan als von Tokharestan und Kashmir aus ihre Zuflucht in dem wohlgeschützten Bergland Gilgit suchten. Da das oben citirte Werk, das diese Nachricht enthält, nach einer jedenfalls correkten Vermutung von Rockhill aus der Sprache von Khotan ins Tibetische übersetzt worden ist [3]),

---

1) The Early History of Li-yul (Khoten), in seinem Buche The Life of the Buddha, London 1884, p. 243.

2) Rockhill hat diese Namen nicht erklärt Ein auf p. 240 im Texte vorkommendes An-se (in Verbindung mit Li-yul und Shu-lig) deutet er als möglicher Weise identisch mit dem chinesischen An-hsi, was im höchsten Grade unwahrscheinlich ist. Der einzige definirbare der vier Namen ist Shu-lig, der offenbar nichts anderes als chin. Su-le (contonesisch lék) 疎勒, also Kashgar ist. Es ist daher zu vermuten, dass die vier genannten Gebiete die tibetischen Äquivalente der Namen der „Vier Garnisonen" 四鎮 repräsentiren, die in den Beziehungen der Tibeter mit den westlichen Türken eine so grosse Rolle spielen, nämlich Kucha, Khotan, Kashgar und Harashar (s. Chavannes, Documents etc., p. 45, Note 4 und Index). An-tse könnte etwa für Yen-k'i (Harashar), Gus-tik für K'iu-tse, K'üt-ts'e (Kucha, s. Hirth, Sinologische Beiträge zur Geschichte der Türkvölker, S. 245) stehen; Par-mk'an-pa ist tibetische Lautgebung.

3) Das Gocṛṅga-vyākaraṇa (Kanjur, Vol. 30) wurde aus der Sprache von Khotan übersetzt. Zu Kanjur, N° 242 (Kanjur-Index, ed. Schmidt, p. 40) wird als Mitübersetzer

so ist es interessant, auf gleichzeitige Beziehungen Tibets zu Khotan nach einer originaltibetischen Quelle hinzuweisen. Zufolge der tibetischen Geschichte des *rGyal-rabs gsal-bai me-loṅ* (fol. 94, 8), herausgegeben im Jahre 1327, liess derselbe König K'ri-lde-sroṅ-btsan für den Bau des Klosters dPe-med bkra-šis dge-ạp'el erfahrene Kunsthandwerker aus Khotan berufen [1]).

Im Legendenbuch des Padmasambhava wird Bru-ža einmal im 21. Kapitel unter den Ländern aufgezählt, die als Verbannungsort für den jugendlichen Padmasambhava vorgeschlagen werden; im 28. Kapitel, in dem die Bergung buddhistischer Schriften in den verschiedensten Ländern erzählt wird, wird das Lalitavistara in Bru-ža versteckt; und im 52. Kapitel zieht Padmasambhava selbst zur Bekehrung von Bru-ža aus [2]). Ausserdem wird in dem Buche, wie bereits erwähnt, auf die Schrift und Sprache von Bru-ža häufig angespielt.

Rajendralala Mitra in seiner Übersetzung des Lalitavistara (p. 189) ist der Ansicht, dass die Darada keine Schrift besessen hätten. Doch ist in diesem Zusammenhang darauf hinzuweisen, dass es eine ornamentale tibetische Schrift, ạBru-ts'a oder ạBru-ts'ag genannt, gibt. Die Form ạBru-ts'a oder Bru-ts'a [3]) kommt nach CHANDRA DAS für Bru-ža vor, und so könnte die Bezeichnung jenes Alphabets von dem Namen dieses Landes herstammen, oder gar die Schrift selbst in Bru-ža entstanden sein [4]). Eine willkommene Be-

---

ein Bhikṣu von Khotan, namens Çrīdharma (mit dem Beinamen sDe-snod-gsum-daṅ-ldan-pa „Tripiṭaka-besitzer") im Colophon genannt.

1) Tibetisch: *dPe-med bkra-šis dge-ạp'el-gyi gtsug-lag-k'aṅ bžeṅs-par ạdod-nas, Lii-yul-nas rig-byed-la mk'as-pai bzo-bo bos.*

2) E. SCHLAGINTWEIT, Lebensbeschreibung von Padma Sambhava, II. Teil, S. 535, 539, 570 (Abhandlungen der Bayer. Akademie, 1903).

3) Nach ROCKHILL, The Life of the Buddha, p. 243, Note 1, auch Bru-dza.

4) Sie ist dargestellt in CSOMA, A Grammar of the Tibetan Language, Tafel 36 und 37, und bei SARAT CHANDRA DAS, The Sacred and Ornamental Characters of Tibet, Plate III, d, e- (Journal Asiatic Society of Bengal, Part I, N° II, 1888).

stätigung für diese Vermutung liefert die Bemerkung von CHANDRA DAS [1]), dass fast alle Schriften der Bon Religion in diesem Alphabet abgefasst seien, wie ihm scheine, aus Antagonismus gegen das Lāñca-Alphabet der Buddhisten. Der Grund liegt aber wohl tiefer, denn in dem historischen Werke *Grub-mt'a šel-kyi me-loñ* wird ein Bon Priester aus ạBru-ža erwähnt [2]), so dass die weite Verbreitung der Bru-ts'a Schrift unter den Bon sehr wohl auf ihre Beziehungen zu dem Lande Bru-ža zurückgehen kann. Eine Bru-ža Schrift hat es sicher gegeben, sonst wäre der Passus in der Biographie des Sa-skya Paṇḍita, dass er die Sanskrit, Lāñca, Vartu und Bru-ža „Sprachen” [3]) erlernt habe, kaum verständlich; analog der hier genannten Lāñca und Vartu Schrift muss auch eine Bru-ža Schrift existirt haben, — wie hätte Sa-skya bei seiner Entfernung von dem Lande selbst auch die Sprache erlernen sollen?

Das vermutliche Vorhandensein einer Bru-ža Schrift bringt uns nun zu dem greifbareren Faktum der ehemaligen Existenz einer Litteratur in Bru-ža Sprache. Dass eine solche einst vorhanden war, ersehen wir vor allem aus dem tibetischen Kanjur, wo sich im zwanzigsten Bande der Abteilung rGyud oder Tantra ein Werk befindet, das der in demselben enthaltenen Angabe zufolge in dem Lande Bru-ža aus der Bru-ža Sprache von Dharmabodhidānarakṣita und dem tibetischen Übersetzer C'e-tsan-skyes ins Tibetische übersetzt worden ist [4]). Der Titel desselben ist auf Sanskrit und Tibetisch, ausserdem auch in der Bru-ža Sprache selbst mitgeteilt, letzterer aber in tibetischen Schriftzeichen transcribirt. Derselbe lautet in der nach dem Tibetischen gegebenen Umschrift:

---

1) L. c., p. 48.

2) Journal Asiatic Society of Bengal, 1881, p. 198.

3) So in der Übersetzung von CHANDRA DAS, Journal Asiatic Society of Bengal, 1882, Part I, N° I, p. 19.

4) ALEXANDER CSOMA in Asiatic Researches, Vol. XX, Calcutta 1836, p. 547.

Ho na pan ril til pi bu bi ti la ti ta siṅ un ub haṅ paṅ ril ub pi su baṅ ri že hal pai ma kyaṅ kui daṅ rad ti [1]).

Es wäre nun ein gänzlich voreiliges und verfehltes Beginnen, sich an eine sprachliche Analyse dieses Titels zu wagen, sei es auf Grund der gegebenen Identifikation in Anlehnung an die Sprachen von Gilgit [2]) oder an irgend eine andere Sprachengruppe. Denn vor allem ist es recht zweifelhaft, ob der Bru-ža Titel dem Sanskrit oder tibetischen Titel wirklich genau entspricht [3]), da wir viele Beispiele dafür haben, worauf ich bereits früher an verschiedenen Stellen hinwies, dass solche in tibetischen Büchern gegebene fremdsprachliche Titel häufig genug Fiktionen sind, die von späteren unverständigen Copisten an Stelle der jedenfalls ursprünglich echt vorhandenen Titel gesetzt worden sind [4]). Ferner wissen wir ja gar nichts von der oder den alten Sprachen von Dardistan und ihrer buddhistischen Terminologie, auf die ja beim Lesen des obigen Titels alles ankäme, und die modernen, auch recht dürftigen Vokabulare

1) Der entsprechende Sanskrittitel lautet: Sarva-tathāgata-citta-jñāna-guhyārtha-garbha-vyūha-vajra-tantra-siddhi-yogāgama-samāja-sarva-vidyā-sūtra-mahā-yāna-sābhisamaya-dharma-paryāya-vivṛ ṇ'a-nāma-sūtra.

2) Materialien über die Gilgit-Sprachen findet man bei ALEXANDER CUNNINGHAM, Ladák, pp. 397—419, in dem vorher citirten Werke von LEITNER, und einige Texte bei GHULAM MUHAMMAD, Festivals und Folklore of Gilgit (Memoirs of the Asiatic Society of Bengal, Vol. I, N° 7, Calcutta, 1905).

3) Man beachte z. B., dass der Bru-ža Titel wesentlich kürzer ist als die beiden anderen; er besteht aus 32 Silben, während der Sanskrit Titel 26 Wörter von 59 Silben, und der tibetische Titel ebenso viele Wörter von im ganzen 50 Silben enthält.

4) S. Sitzungsberichte der Bayer. Akademie, 1898, S. 590—2. Hier einige weitere Beispiele. Die grosse Ausgabe des Legendenbuches von Padmasambhava beginnt mit Anführung eines Titels in der Sprache von Udyāna, der *Ru ,akṣa-ša-ka-ri-ni* lautet. Darin mögen ja immerhin einige wirkliche Udyāna Wörter stecken. Dagegen gibt es einen kurzen sowohl in Tibet als in Peking gedruckten Auszug des grösseren Werkes von 31 Blättern, der gleichfalls mit den Worten „In der Sprache von Udyāna" beginnt, worauf aber *Buddha-dharma-saṅgha-dhaya* folgt! Ein tibetisches aus dem Chinesischen übersetztes Sūtra mit dem Titel *Dag-pa gser-gyi mdo-t'ig* beginnt: *rgya-nag skad-du* „auf Chinesisch", worauf im reinsten Tibetisch folgt: *gtsug-lag ap'rul-gyi agyur-rtsis;* weiter: „in der Sprache der Götter der Welt" (*srid-pa lhai-skad-du*) mit dem folgenden Titel: *koṅ tse liṅ tse mer-ma rol-ma'žes-bya-ba,* was wohl halb chinesisch und halb tibetisch sein soll.

können dafür keinen Ersatz bieten. Der Gang der Untersuchung,
die in den alten Klosterbibliotheken von Tibet beginnen müsste,
wird vielmehr der sein, nach ausgedehnteren Materialien zur Bru-
ža Litteratur, die sicher in der tibetischen Litteratur enthalten sind,
zu forschen, um danach den Charakter des Idioms zu bestimmen.

Aber das oben genannte Werk gibt zu einer weiteren wichtigen Schlussfolgerung Anlass. Zu dem Bande des Kanjur, in welchem
dasselbe enthalten ist, bemerkt nämlich Csoma am Schlusse: "This
whole volume is old-fashioned, and of little authority, except to the
rÑiṅ-ma-pa sect (the most ancient among the Buddhistic sects in
Tibet)." In der Tat scheint dieses wie die beiden anderen von
Csoma für diesen Band aufgeführten Werke in anderen Recensionen
des Kanjur zu fehlen, wie sie denn nicht in dem von Schmidt
herausgegebenen Kanjur-Index erscheinen. Wenn wir es also in
dem aus der Bru-ža Sprache übersetzten Werke mit einem Erzeugnis
der rÑiṅ-ma-pa [1]) Sekte zu tun haben, die den Padmasambhava
zum Stifter hat und ihn als ihren Heiligen und Schutzpatron ver-
ehrt, so haben wir hier einen bedeutsamen Fingerzeig für die Tatsache gegeben, warum in den von Padmasambhava handelnden
Legenden die Bru-ža Sprache eine solche Rolle spielt. Und noch
mehr. Das erste der in demselben Bande des Kanjur erwähnten
Werke ist nach der Angabe von Csoma im achten oder neunten

---

1) Csoma befolgt die Schreibung *sñig-ma*, das mit *rñig-ma* oder *rñiṅ-ma* identisch ist.
Direkt als Stifter der Sekte wird Padmasambhava von Csoma, A Grammar of the Tibetan
language, p. 197 und A. Cunningham, Ladák, p. 367, bezeichnet. Dagegen ist Grünwedel
(Mythologie des Buddhismus, S. 56) in Übereinstimmung mit Köppen (Die Religion des
Buddha, Bd. II, S. 71) geneigt, die rÑig-ma-pa mit dem primitiven Buddhismus aus der
Zeit des Königs Sroṅ-btsan-sgam-po zu identificiren. Wie dem auch sein mag, die Tatsache
ist jedenfalls unleugbar, dass Padmasambhava oder die Schule, die von ihm ausging, enge
Beziehungen zu jener Sekte gehabt hat und noch hat (s. besonders Sarat Chandra Das,
Journey to Lhasa, p. 305, und Carlo Puini, Il Tibet secondo la relazione del viaggio del
P. Ippolito Desideri, Roma 1904, p. 334). Desideri, der sich von 1715 bis 1721 in Tibet
aufhielt, gibt einen sehr interessanten und lesenswerten Bericht über das Leben und Wirken des Padmasambhava und seinen Kultus.

Jahrhundert von Çrī-Siṁhaprabha und Vairocana übersetzt worden. Beide aber gehörten zu den hervorragendsten Schülern des Padmasambhava. Ersterer erscheint als solcher unter dem mongolischen Namen Tsoktu Arsalan bei SANANG SETSEN [1]), während Vairocana zu den sieben sogenannten »Probeschülern" des Çāntirakṣita, des Freundes und Gefährten von Padmasambhava, gehörte [2]).

Ferner wird im Kanjur unter N° 409 (Kanjur-Index, ed. SCHMIDT, p. 61) eine Schrift Sarvatathāgata-citta-garbhārtha-tantra »Tantra des Sinnes der Quintessenz der Gedanken aller Tathāgata" aufgeführt, die, dem Titel nach zu urteilen, über denselben Gegenstand,

---

1) I. J. SCHMIDT, Geschichte der Ost-Mongolen, S. 43.

2) Im rGyal-rabs (fol. 85b, 6—86a, 3) wird dieses Ereignis so erzählt: „Darauf versammelte der König (K'ri-sroṅ-lde-btsan) das Volk von Tibet, und da es früher in Tibet noch keinen Präcedenzfall für den Eintritt in den Priesterstand gegeben hatte, wollte er sehen, ob sich in Tibet ein geistlicher Stand verwirklichen lasse. Daher empfingen sieben geistig begabte Söhne aus dem Beamtenstande und dem Volke bei dem Bodhisattva (d. i. Çāntirakṣita) die geistlichen Weihen. Diese waren Ratna, der Sohn des sBa-san-ši; Çākyaprabha, der Sohn des mC'ims ,A-nu (mC'ims ist der Familienname eines tibetischen Geschlechts, ,A-nu der persönliche Name); Vairocana, der Sohn des Pa-gor-ratna; rGyal-ba mc'og-dbyaṅs von Ñan-lam; der rMa ācārya (d. i. Lehrer von rMa) Rin-c'en-mc'og; aK'on-klui-dbaṅ-po-bsruṅs-pa (Note im Text: In einem Buche oder Berichte heisst es, dass letzterer nicht zu den sieben Probeschülern gehört habe); und bTsan-legs-grub. Diese sieben wurden also Geistliche und sind als die sieben Probeschüler (sad-mi mi-bdun) bekannt." Nach Tāranātha (SCHIEFNER's Übersetzung, S. 213) hätte jeder dieser sieben Schüler das -rakṣita vom Namen Çāntirakṣita in seinem Namen geführt. Der obige Name sBa-san-ši (vergl. HUTH, Geschichte des Buddhismus, Bd. II, S. 9: sPas [für sBas] -ratna) ist sicher ein Schreibfehler für sBas-ži, einer tibetischen Übersetzung von Guptaçānti (vergl. Ži-ba sbas-pa = Çāntigupta, bei Tāranātha, S. 263, 265), sicher in Anlehnung an den Namen Çāntirakṣita (tib. Ži-ba ạts'o). Beachtenswert ist, dass Padmasambhava in einer Legende als Rākṣasa Rakṣiçānta bezeichnet wird und ebenda von einer Ḍākinī Ži-ba mts'o die Weihe Abhiṣeka erhält (GRÜNWEDEL, Tä-še-suṅ, S. 12). Letzterer Name könnte für Ži-ba ạts'o stehen, was gleich Çāntirakṣita ist. Vgl. auch SCHLAGINTWEIT, Abhandlungen der bayer. Akademie, 1899, S. 433.

Vairocana, der vom König einst nach Indien gesandt wurde (ROCKHILL, The Life of the Buddha, p. 221), erscheint als Verfasser von Tanjur, Vol. 124, N° 8, und mit Padmasambhava zusammen als Übersetzer eines Werkes im 22. Bande, N° 14, der rGyud oder Tantra des Kanjur (s. CSOMA, Asiatic Researches, Vol. XX, p. 552), das gleichfalls in dem Kanjur-Index von SCHMIDT fehlt, somit von den Orthodoxen als unkanonisch angesehen werden muss.

wenn auch weit kürzer, zu handeln scheint als das aus der Bru-ža
Sprache übersetzte obengenannte Werk. Nach Angabe des bei
SCHMIDT mitgeteilten Colophons ist die Schrift von Seiner Ehrwür-
den rJe-btsun Gu-ru und dem Übersetzer Seṅ-kar Çākyaprabha (tib.
*Šā-kya od*) in P'a-mt'iṅ [1]) übersetzt und redigirt worden. Ich ver-
mute, dass der hier genannte Çākyaprabha mit dem zweiten der
»sieben Probeschüler" identisch ist (s. den Text aus dem *rGyal-
rabs* in der vorher mitgeteilten Anmerkung), dass somit auch jenes
Werk aus der Schule des Padmasambhava stammt. In dem Beinamen Seṅ-kar, offenbar gleich *Seṅ-dkar* »weisser Löwe" könnte vielleicht eine Anspielung auf einen in den Padmasambhava-Legenden
vorkommenden weissen Löwen gesucht werden [2]).

Ist somit die Annahme der Existenz einer buddhistischen Litteratur in Bru-ža Sprache gesichert, so hören wir auch von Paṇḍita's
aus Bru-ža. So wird in einer Schrift über Chronologie, verfasst von
dem Tibeter Sureçamatibhadra im Jahre 1591, ein Paṇḍita Dhar-
mamati (tib. *C'os-kyi blo-gros*) von Bru-ža erwähnt [3]), dessen Zeit
auf 836 berechnet wird.

Die bunte Vermengung von Ideen in Padmasambhava's Legendenbuch, die aus den verschiedensten Religionen geschöpft sind,

---

1) Ist ein Ort in Nepal, wie ich aus TĀRANĀTHA (Text, p. 188, 18) schliesse. SCHIEF-
NER (p. 249) und WASSILJEW (p. 239) in ihren Übersetzungen des Tāranātha behandeln
den Namen irrtümlich als Personennamen, was aber ganz unmöglich ist, da im Text nicht
*P'a-mt'iṅ*, sondern *P'a-mt'iṅ-pa* steht. Die Stelle kann nur so verstanden werden: der aus
P'a-mt'iṅ in Nepal und sein Bruder, oder noch wahrscheinlicher ist *sku-mc'ed* als sein
Name aufzufassen und demgemäss zu übersetzen: der aus P'a-mt'iṅ in Nepal, namens
sKu-mc'ed.

2) E. SCHLAGINTWEIT, Die Lebensbeschreibung von Padma Sambhava, II. Teil, S. 527,
552. Çākyaprabha erscheint sonst nur noch an einer Stelle im Kanjur, und zwar als Übersetzer zu N° 356 (Kanjur-Index, ed. SCHMIDT, p. 54) zusammen mit dem indischen Gelehrten Ajitaçrībhadra; hier wird er als *žu-c'en-gyi lo-tsts'a-ba Siṅ* (wohl für *Seṅ*)-kar dge-
sloṅ *Šā-kya od* bezeichnet, wonach *Seṅ-kar* („der Bhikṣu von Seṅ-kar") auch Name eines
Ortes oder Klosters sein könnte. Aber es geht aus dieser Stelle unzweideutig hervor, dass
*Seṅ-kar* ein Beiname und kein ursprünglicher Bestandteil des Personennamens ist.

3) E. SCHLAGINTWEIT in Abhandlungen der Bayer. Akademie, 1896, S. 624.

Buddhismus, Parsismus, Christentum, Islam — wahrscheinlich wird auch der Manichäismus hinzuzuzählen sein — ist das getreue Spiegelbild einer Epoche, in welcher alle diese Religionen einen ungewöhnlichen Grad von Expansion erreicht hatten. Nur wenn wir verstehen, wie im siebenten und achten Jahrhundert durch die Vermittlung des Reiches der westlichen Türken diese verschiedenen Glaubenssysteme sich nach Central- und Ostasien verbreitet haben [1]), können wir den Gehalt jenes Buches, die Bedeutung seines Helden und die Fortentwicklung der von ihm verpflanzten Ideen in Tibet verstehen lernen. In Padmasambhava oder in der unter seinem Namen gehenden Litteratur spiegelt sich wie in einem Brennpunkt ein grosser Teil der religiösen Anschauungen wieder, welche die damalige Welt Asiens beherrschten, und da er bestimmend auf die Geschicke des Buddhismus in Tibet eingewirkt hat, ist eine Analyse dieser fremden Elemente für die Geschichte des Buddhismus von denkbar grösstem Interesse [2]).

---

1) Wie treffend von Chavannes, Documents etc., pp. 301—303, auseinandergesetzt.

2) Charakteristisch für das Eindringen vorderasiatischer Ideen in Indien, worauf Grünwedel bei seinen Studien der Padmasambhava-Legenden wiederholt hingewiesen hat, sind zwei Varianten der Jonah-Legende, die in der Geschichte der Siddha's vorkommen und in Tāranātha's Werk *bKa-babs-bdun-ldan*, verfasst 1600 (ed. by Chandra Das, Darjeeling 1895) erzählt werden. Die eine betrifft den Siddha Na-ro-pa, der einst in Meditation versunken von einer Strömung fortgerissen und von einem Fisch verschlungen wurde. Im Innern des Fischbauches wurde er das Maṇḍala des Heruka gewahr, so dass er keinen Schaden erlitt und wieder ausgespieen wurde (p. 37, 7). Die andere handelt von dem Siddha Mīnapa (d. i. Fischer), dem Schüler von Kakkuṭipa (p. 58). „Dieser war ein Fischer in Kāmarūpa im Osten Indiens und pflegte zuweilen Meditationen nach der Atmungsmethode des Yoga zu betreiben. Einstmals warf er seinen Angelhaken nach einem Fisch aus, und dieweil er die Leine anzog, wurde er von dem Fisch geschnappt und verschluckt. Kraft seiner Werke und Meditationsübungen starb er aber nicht, sondern trieb auf dem Flusse Rohita nach Kāmarūpa. Dort predigte Maheçvara auf dem kleinen Hügel Umagiri der Göttin Umā Instruktionen über den Wind-Yoga. Da der Fisch in jenen Fluss kam, hörte der im Bauch des Fisches befindliche Fischer die Predigt, meditirte und erlangte viele Vorzüge. Einstmals wurde dieser Fisch von Fischern gefangen und getötet, als ein Mann daraus hervorkam. Der frühere König war indessen verstorben; er fand seinen Sohn, seit dessen Geburt dreizehn Jahre verflossen waren: zwölf Jahre hatte er im Magen des Fisches zugebracht. Darauf gingen sie zu dem Meister Carpaṭipa, Vater und

GRÜNWEDEL [1]) hat auf den merkwürdigen Zusammenhang hingewiesen, der darin besteht, dass Padmasambhava nach seinem Abhiniṣkramaṇa auf *acht Leichenstätten* in Bannungen verharrt und dort
Geheimlehren von den Ḍākinī's erhält, und dass auch die Bon-
Religion *acht Leichenstätten-Bannungen* kennt. Es erscheint ihm
daher nicht unmöglich, dass ein alter Connex zwischen Padmasambhava's Wirken und der Bon-Religion besteht. Nun ist es beachtenswert, dass eine der früheren Sekten der Bon, mit Namen Dur-
Bon, d. i. Bon der Leichenstätten, in Bru-ža und Žaṅ-žuṅ (Guge)
entstanden ist und sich aus diesen Gebieten unter dem König Gri-
gum btsan-po in die Centralprovinzen Tibets verbreitet hat [2]). Der
Connex zwischen beiden Systemen ist also wohl so zu deuten, dass
beide aus derselben gemeinsamen Quelle, bezeichnet durch den Namen Bru-ža, geschöpft haben [3]). Vielleicht stammt aus diesem Lande
die Idee der auf Leichenstätten vollzogenen Bannungen; sicher hat
sich Padmasambhava einen Teil seiner Weisheit von da hergeholt.

Auch das Legendenbuch bringt Padmasambhava in einer Geschichte mit Bon-po's zusammen [4]), und die Geschichte der Bon behauptet, dass Padmasambhava die Lehre des gŠen-rabs, des Stifters
ihrer Religion, angenommen habe, betrachtet ihn also als einen
der ihrigen [5]).

---

Sohn ersuchten ihn um eine Predigt, meditirten und erlangten beide die Siddhi. Der Vater
ist als der Siddha Mīnapa bekannt, der Sohn heisst der Siddha Ma-tsʻin-dra-pa". Letzterer
Name ist offenbar gleich Matsyendra „der Fürst der Fische" (s. R. PISCHEL, Der Ursprung
des christlichen Fischsymbols, Sitzungsberichte d. Preuss. Akademie, 1905, S. 521). Vermutlich ist die Jonah-Legende, die sich bekanntlich im Koran (Sūrah XXXVII, 139—148)
findet, durch die Mohammedaner in Indien verbreitet worden.

1) Zeitschrift der deutschen Morgenl. Ges., Bd. 52, S. 449.
2) CHANDRA DAS, A Tibetan-English Dictionary, p. 631a.
3) Vergl. was bereits vorher über die Bru-tsʻa Schrift bei den Bon und den Bon-
Priester aus Bru-ža bemerkt wurde.
4) SCHLAGINTWEIT, Die Lebensbeschreibung von Padma Sambhava, II. Teil, S. 542.
5) *rGyal-rabs Bon-gyi abyuṅ-gnas* (ed. CHANDRA DAS), p. 23, 8. Vergl. über dieses
Werk Tʻoung Pao, 1901, pp. 24 et seq. In einer kleinen Schrift im British Museum (Or.
5373) heisst es von Padmasambhava, dass er als ein den Buddhisten und den Bon glei-
chender Sohn geboren wurde.

Wenn nun, wie GRÜNWEDEL constatirt hat, sich auch persische Elemente in dem Legendenbuche befinden, so erhält diese Erscheinung ihr Gegenstück in der Bon-Religion, die nicht nur stark mit persischen Ideen durchsetzt ist, sondern in deren Geschichte es auch sehr weitläufig auseinandergesetzt wird, dass eine Reihe ihrer heiligen Bücher aus dem Persischen in die Sprache von Žaṅ-žuṅ und aus letzterer ins Tibetische übersetzt worden sind [1]). Ausserdem werden drei persische Weise als geistige Nachkommen des gSenrabs genannt [2]).

Es ist irrtümlich, wie bisher zuweilen geschehen, die Bon-Religion ohne weiteres mit der einheimischen tibetischen Volksreligion zu identificiren. Sie hat reichliche Anleihen aus diesem Gebiete gemacht, ebenso wie der Buddhismus, ob mehr oder weniger als letzterer, lässt sich noch nicht entscheiden und ist zunächst für die Beurteilung unserer Frage gleichgültig, sie ist aber wie der Buddhismus keine tibetische, sondern eine fremde Religion, die auf persischer Grundlage basirt ist, mit allen möglichen fremden Elementen vermischt in Dardistan entwickelt wurde, von da zunächst nach Guge in den westlichen Teil Tibets und später in das centrale Tibet gelangte. Als System ist die Bon-Religion alles eher denn eine Volksreligion; in ihrem letzten Entwicklungsstadium hat sie sich eine complicirte kosmologische Geographie, ein Göttersystem, eine wilde Mystik und eine Geschichte, alles nach buddhistischem Vorbild, zurechtgezimmert. Ihre ältesten Traditionen weisen aber, wie gesagt, deutlich genug auf Persien hin. Charakteristisch ist die Tatsache, dass in ihrer Geschichte [3]) Ol-mo luṅ-riṅ wiederholt als ein Ort in Persien bezeichnet wird, und dass ihre heilige Stätte in

---

1) L. c, p. 36.

2) Journal Asiatic Soc. of Bengal, 1881, p. 196. Über andere iranische Einflüsse in Tibet s. Sitzungsberichte der Bayer. Akademie, 1898, S. 591, und GRÜNWEDEL, Mythologie des Buddhismus, S. 114, 209.

3) L. c., pp. 7, 19; 11, 5 etc.

Guge, wo ihr Stifter gŠen-rabs mi-bo geboren sein soll, gleichfalls das »persische Ol-mo luṅ-riṅ" (*sTag-gzig Ol-mo luṅ-riṅ*) genannt wird, und zwar nicht nur in ihren eigenen Schriften, sondern auch so in der orthodoxen buddhistischen Litteratur; z. B. im grossen *rGyal-rabs,* Kap. 8, fol. 23, 11, wo eine kurze Geschichte der Bon skizzirt wird. *Luṅ-riṅ* ist tibetisch und bedeutet »langes Tal." *Ol-mo* dürfte vermutlich nichts anderes als eine durch viele trübe Quellen erfolgte Entstellung des persischen Namens Ormuzd (Ormazd, Ahura-Mazda) sein.

Nachdem uns Grünwedel's Forschungen die Legenden Padma-sambhava's erschlossen und in den Geist seines Systems eingeführt haben, dürfte es nunmehr am Platze sein, einmal zu untersuchen, was wir über Padmasambhava aus der historischen Litteratur er-fahren, und wie sich der historische zu dem legendären Padma-sambhava verhält.

Padmasambhava wird in Tāranātha's Geschichte des Buddhismus erwähnt, und es ist sonderbar, dass ihn weder SCHIEFNER noch WASSILJEW an dieser Stelle erkannt haben. Beide (p. 264 der Über-setzung von SCHIEFNER, p. 253 der russischen von WASSILJEW) reden vom Ācārya Padmākara, während im Text (p. 201, 1) deutlich *slob-dpon Pad-ma-ɑbyuṅ-gnas* dasteht, was einem Sanskrit *Padma-sambhava* entspricht. Dass an der betreffenden Stelle wirklich von Padmasambhava die Rede ist, der den Buddhismus auf der Insel Dramila (tib. *ɑGro-ldiṅ*) eingeführt haben soll [1]), geht aus der im Folgenden zweimal vorkommenden Erwähnung von Udyāna hervor, wobei die Bemerkung, dass auch Tantra aus Udyāna nach Indien gebracht worden seien, die zuvor dort nicht gewesen wären, beson-ders interessant ist.

---

1) Dies fällt mit der in dem tibetischen Legendenbuche von Padmasambhava erzählten Bekehrung von Tāmradvīpa und Suvarṇadvīpa zusammen (s. SCHLAGINTWEIT, Lebensbeschrei-bung von Padma Sambhava, II. Teil, S. 569).

Nun wäre ferner zu erwägen, ob der in SCHIEFNER's Übersetzung, p. 219, von Tāranātha genannte Ācārya Padmākaraghoṣa in Kashmir, wo gleichfalls wohl Padma-sambhava-ghoṣa zu lesen ist (tib. *Pad-ma-abyuṅ-gnas-dbyaṅs*) mit unserem Padmasambhava identisch ist. Dafür würde sprechen, dass er dort in einer Reihe von Männern aufgezählt wird, die als Zeitgenossen des Königs K'ri-sroṅ-lde-btsan genannt werden; dagegen aber vielleicht die auf p. 220 gemachte Angabe, dass er der Paṇḍita von Lo-dri zu sein scheine, ein Name, den weder SCHIEFNER noch WASSILJEW erklären.

Einige weitere Nachrichten über Padmasambhava, die vielleicht seine Eingliederung in uns bekannte Schulen des Buddhismus gestatten, erfahren wir aus Tāranātha's anderem Geschichtswerke *bKa-babs bdun-ldan*, das er laut seiner eigenen Angabe am Schlusse im 26. Jahre seines Lebens, also im Jahre 1600, acht Jahre früher als sein Buch *rGya-gar c'os-byuṅ* geschrieben hat. Jenes ist offenbar auch das Werk, auf welches er in letzterem so oft für die Biographieen von Persönlichkeiten mit den Worten verweist, dass man sie an anderer Stelle zu suchen habe [1]). Das *bKa-babs-bdun-ldan* ist bedeutend kürzer und behandelt in neun Kapiteln hauptsächlich die Lebensbeschreibungen der Siddha's oder Zauberer, unter denen dem Çāntigupta der weitaus grösste Raum, fast das ganze Schlusskapitel, zugewiesen ist.

Padmasambhava wird in diesem Buche nur an zwei Stellen erwähnt, und zwar einmal (p. 47, 5) als Schüler des Mahācārya Buddhajñānapāda (tib. *Saṅs-rgyas ye-šes žabs*), dessen Leben wir aus Tāranātha's Geschichte des Buddhismus in Indien (S. 220 der Über-

---

1) Die Schrift wird in der „Geschichte des Buddhismus in Indien" (Text, p. 58, 7; Übersetzung, p. 73) für die Biographie des Nāgārjuna citirt, der Titel ist aber von SCHIEFNER „Geschichte der sieben Übergaben des Buddha-Wortes oder die Erzählung, welche der Fundgrube von Edelsteinen gleicht" in durchaus irrtümlicher Weise übersetzt worden. Es muss heissen: Kostbare Quelle der Biographieen der Tantra-Vertreter der sieben Schulen der bKa-babs (d. h. Religiose, die den Auftrag ihres Lehrers ausgeführt haben).

setzung von SCHIEFNER) kennen. An jener Stelle wird dem Padmasambhava nun auch der Name *Padma-vajra-p'yi-ma* oder *-c'un̐-ba*
(d. h. der spätere oder jüngste Sohn) zugeschrieben, was vielleicht
sein eigentlicher Name war, und bemerkt, dass es bekannt sei, dass
eine ausführliche separate Lebensbeschreibung von ihm existire.
Damit ist offenbar das bekannte Legendenbuch gemeint, für dessen
Vorhandensein das Jahr 1600 somit die untere Abgrenzung bildet.
An einer anderen Stelle derselben Schrift (p. 52, 4) lässt Tāranātha
den Padmasambhava bei dem Meister Kukurarāja, dem zweiten
Sohne des Königs Indrabhūti [1]), studiren und Erklärungen zu vielen
Tantra abfassen, für deren Lektüre aber gegenwärtig (d. h. zur
Zeit des Autors) keine Verbindlichkeit bestehe, und die auch in
der Geschichte des Atīça nicht aufgezählt würden. Aus dem ganzen
Bericht des Tāranātha geht nun hervor, dass Padmasambhava zu
einer Schule gerechnet wird, an deren Spitze Naro-pa und Maitri-
pa (p. 48, 26), Hauptvertreter der Mahāsiddha's [2]), stehen. Weitere
hervorragende Mitglieder derselben sind nach Tāranātha's *bKa-babs-
bdun-ldan* Avadhūtipa [3]), Ratnākaraçānti [4]), Mahāvajrāsana [5]), Ku-
sa-li-pa [6]), Bhaiṣajyapāda (tib. *sMan-pa žabs*), und der Meister von
Udyāna Buddhaçrīçānti [7]). Die interessanteste dieser Persönlichkeiten
ist unstreitig Ratnākaraçānti, der in der »Geschichte des Buddhismus in Indien" als der »östliche Torwart" (*šar sgo bsrun̐-ba-po*)
bezeichnet, und für dessen Biographie auf eine andere Stelle verwiesen wird. Tatsächlich findet sie sich in unserem Werke auf p. 50.

---

1) ‹Darin liegt vielleicht eingeschlossen, dass Padmasambhava als Sohn des Königs
Indrabhūti dem Bereich der Legende angehört.

2) GRÜNWEDEL, Mythologie des Buddhismus, S. 40.

3) S. SCHIEFNER, Tāranātha, S. 237.

4) Ibid. S. 234—235.

5) Ibid. S. 244.

6) Bei GRÜNWEDEL, l. c., Kasoripa.

7) Tib. *Sañs-rgyas dpal ži-ba*, wahrscheinlich mit dem bei SCHIEFNER, l. c, S. 219
als Zeitgenossen des K'ri-sron̐-lde-btsan genannten Buddhaçānti identisch.

Danach — ich gebe nur einen kurzen Abriss — war er im Lande
Magadha aus einem Brāhmaṇa-Geschlecht, nach anderen aus einem
Vaiçya-Geschlecht geboren, früh in den Wissenschaften bewandert,
lernte das ganze Tripiṭaka auswendig, studirte in Otantapurī und
Vikramaçīla und hörte bei Mahājetari und vielen anderen Gelehrten
die Sūtra und Çāstra des Mahāyāna. Später ging er nach Malava,
wo er sieben Jahre in Meditation verweilte, das Antlitz des Mañjughoṣa,
der Tārā und des Ajitanātha (tib. *Mi-p'am mgon-po*) schauend und
den Nektar der heiligen Religion erflehend, wobei er schliesslich
bei der Methode des Ārya-Asaṅga angelangte. Da erschien ihm einst
im Traume die Tārā mit der Aufforderung, sich nach Ceylon zu
begeben, und gleichzeitig vernahm der König von Ceylon im Traume
eine Stimme, die ihn ermahnte, den Ācārya Ratnākaraçānti in
Jambudvīpa zu berufen, um in seinem Lande das Mahāyāna zu
verbreiten. Der Meister reiste nach Bengalen, wo ihn ein bereits
von Ceylon abgesandter Bote traf. Zweihundert Sūtra des Mahāyāna
nahm er mit sich und predigte auf Ceylon sieben Jahre die Lehre.
Dort erstanden fünfhundert Geistliche des Mahāyāna, dessen Sūtra
weite Verbreitung fanden. Später, als er nach Süd-Indien zurückgekehrt war, fasste er den Plan, den Wu-tai-shan in China zu
besuchen, wurde aber vom König, der nach einigen Mahāpāla, nach
andern Canaka sein soll (Tāranātha entscheidet sich für letzteren),
an diesem Vorhaben gehindert und erhielt einen Posten in Vikramaçīla. Später finden wir ihn in Udyāna wieder, dem Wirkungskreis
des Padmasambhava. Seine Verbindung mit Ratnākaraçānti und
Çāntirakṣita erklärt nun wohl auch die Tatsache, warum in so vielen der mit Padmasambhava in Beziehung stehenden mythischen
und wirklichen Wesen die Bestandteile ratna, çānti und rakṣita in
ihren Namen erscheinen.

Die in den tibetischen Annalen enthaltenen Berichte über
Padmasambhava sehen nun ganz anders aus als die Erzählungen

2

über ihn in dem von ihm handelnden Legendenbuche. Die grosse
Ausgabe der Königsannalen (*rGyal-rabs*, herausgegeben 1327) berichtet über ihn in kühlen und nüchternen Worten und schildert
seine Unbeliebtheit am Hofe und die Abneigung der Minister gegen
ihn. Diese discreditiren das von ihm dem König gereichte Lebens-
wasser, und der König schliesst sich ihren Zweifeln an¹). »Die
schlechten [d. h. nicht-buddhistischen] Minister", heisst es nun im
*rGyal-rabs* (fol. 85, 10) weiter, »sagten dem Könige, dass der
Meister, da er ein Zauberer sei, nicht lange verweilen solle und
eine gefährliche Person sei, doch sie vermochten nicht, den König
dahin zu bringen. Ferner erhielt der Meister eine königliche Gemahlin, namens mK'ar-c'en-bza ats'o-rgyal zum Geschenk und nahm
sie zur Frau, worüber sämtliche [also auch die buddhistischen]
Minister ungehalten waren und dem Könige Verleumdungen über
ihn zutrugen. Doch der König schenkte ihnen dem Meister zuliebe
kein Gehör. Darauf sagten sie ihm, dass er einem dreifach grossen
Irrtum verfallen sei, und stellten das Gesuch, dass der Meister, da
er ganz gewiss mit seiner Zauberkraft dem Reiche schaden würde,
in sein eigenes Land zurückkehren solle. In dem Buche *bSam-yas-
kyi ka-ts'igs*²) heisst es, dass der Meister daraufhin erklärt habe,
er wolle in das Land Udyāna zurückreisen, dass er sich aber dann
mit dem Könige beraten, eine Zeit lang in Tsa-ri Aufenthalt genommen habe, und nachdem der Minister Ma-žaň K'ron-pa-skyes

---

1) Ich enthalte mich der Übersetzung dieser Stelle, da sie bereits SCHMIDT (Geschichte
der Ostmongolen, S. 354) nach dem Bodhimör gegeben hat. Der dort erwähnte „sehr
mächtige Drachenkönig" ist der Nāga Manasvin. Die oben folgenden nach dem *rGyal-rabs*
übersetzten Stellen finden sich nicht bei Schmidt. — Die Art und Weise, wie SANANG
SETSEN (ibid. S. 39 ff.) die Berufung und weitere Geschichte des Padmasambhava erzählt,
lässt darauf schliessen, dass er von den rein legendären Berichten starken Gebrauch gemacht hat.

2) Wahrscheinlich eine alte Chronik des Klosters bSam-yas, die nach *rGyal-rabs*,
fol. 92, 12 eine ausführliche Darstellung des Lebens des Königs K'ri-sroň-lde-btsan enthal-
en soll.

aus dem Wege geräumt war [1]), wieder nach bSam-yas zurückgekehrt sei". Darauf wird die oben S. 9 bereits mitgeteilte Wahl der sieben Probeschüler erzählt, und es ist im Zusammenhang der Stelle beachtenswert, dass dieselben nicht von Padmasambhava, sondern von Çāntirakṣita ordinirt werden.

Padmasambhava beginnt erst beim Bau des Klosters bSam-yas seine Rolle zu spielen, dem eine kleine Episode vorausgeht. »Der König fragte: Grosser Meister, wo befinden sich jetzt meine beiden Eltern? Der Meister antwortete: König, dein Vater ist in Indien als ein grosser Paṇḍita wiedergeboren worden und wird zur Zeit deines Enkels (Anmerkung im Text: diese Weissagung bezieht sich auf den Paṇḍita Dānaçīla) nach Tibet kommen; deine Mutter ist in Zuṅ-mkʻar als Tochter eines armen Ehepaars wiedergeboren worden, und ihr Name und ihr Geschlecht ist so und so, also weissagte er. Darauf nahm der König, der schon vier Frauen besass, nämlich Tsʻe-spoṅ-bza me-tog sgron [2]), mKʻar-cʻen-bza ạtsʻo-rgyal (Anmerkung im Text: dies ist die dem Meister zum Geschenk gegebene Gemahlin), ạBro-bza byaṅ-cʻub sgron [3]) und ạCʻims-bza lhamo btsan, aus Dankbarkeit gegen seine Mutter die Tochter eines armen Ehepaares zur fünften Frau, die als Pʻo-yoṅ-bza rgyal-mo btsun [4]) bekannt ist. In einem anderen Buche heisst es, dass er dieses Mädchen, das eine Wiedergeburt der Mutter des Königs gewesen sei, aus Zur-cʻu in ạPʻan-yul erlangt habe, was, wie man die Sache auch ansehen möge, denselben Sinn hat, dass es eben Pʻo-yoṅ-bza war. Darauf prüfte der Meister die Ortsverhältnisse für

---

1) Also wohl um den Widerstand der Oppositionspartei im Adel zu brechen. Er wurde nach Bu-ston lebendig in einer Höhle begraben (s. Journal of the Buddhist Text Soc., Vol. I, p. 3, und Journal Asiatic Soc. Beng., 1881, pp. 225—226). Die ganze Geschichte ist romantisch-sagenhaft ausgeschmückt und erinnert sehr stark an EDGAR ALLAN POE's Novelle The Cask.

2) D. i. Frau Tsʻe-spoⁿ, die mit Blumen geschmückte.

3) D. i. Frau ạBro, die mit der Bodhi geschmückte.

4) D. i. Frau Pʻo-yoṅ, die königliche Gemahlin.

den Bau des Klosters [1]) auf die Vorzeichen hin und fand, dass der

---

[1]) Ich lasse eine Übersetzung des Berichts über den Tempelbau in extenso folgen, da er grosses kunsthistorisches und religionsgeschichtliches Interesse bietet. SANANG SETSEN hat bekanntlich einen ganz kurzen Abriss des Bauberichts geliefert, der an Unklarheit nichts zu wünschen übrig lässt, und auf den sich KÖPPEN (Religion des Buddha, Bd. II, S. 68) bezieht. Da ich im Folgenden einige Male auf diesen Text zurückgreifen muss, teile ich zunächst den Versuch einer neuen Übersetzung desselben mit, da die von SCHMIDT (Geschichte der Ostmongolen, S. 41, 43) gegebene unverständlich und unzulänglich ist, was zum Teil am Original liegt, da Sanang Setsen offenbar selbst keine richtige Vorstellung von der Sache hatte; ich selbst habe den Text erst nach Durcharbeitung der tibetischen Beschreibung entwirren und verstehen können. Glücklicher Weise findet sich ein Abdruck des Textes in KOWALEWSKI's Mongolischer Chrestomathie, Bd. II, S. 58—60, in dem die Druckfehler bei Schmidt berichtigt sind. Der bequemeren Übersicht halber numerire ich die von Sanang Setsen gelieferten Fakten. „1) Entsprechend der Form des grossen Maṇḍala im Stil (oder nach den Angaben) der mystischen Dhāraṇī (mongolisch *nighutsa tarni* = Skr. *guhyadhāraṇī*), entsprechend den Angaben des Abhidharma, wie es im Tripiṭaka enthalten ist, entsprechend den Ursachen und Folgen nach der Angabe der Grundlagen der Sūtra, erbaute er (der König) den Bima (s. KÖPPEN, l. c.) Tempel, den aus natürlichen Ursachen von selbst entstandenen unvergleichlichen (mongolisch *učir činar-un ülisi-üghei* = tib. *rgyu raṅ-bžin-gyi gžal-yas*, offenbar Versuch zu einer Wiedergabe der officiellen Bezeichnung des Klosters *mi-aġyur lhun-grub*, s. weiter unten am Schlusse des Textes). 2) Die Mehrzahl der Menge der Buddha im Innern des Tempels machte er nach den Angaben der mystischen Dhāraṇī. (Dieser Satz fehlt in der Übersetzung von SCHMIDT.) 3) Die (nämlich die Buddhafiguren) im unteren Stockwerk waren im tibetischen Stil, die im mittleren im chinesischen und die im oberen im indischen Stil. 4) Erfüllt von den *vier* grossen Erhabenen, besass das untere Stockwerk *drei* Tore. (SCHMIDT übersetzt, wie um einen Ausgleich anzubahnen, *vier* Tore, obwohl sein und auch KOWALEWSKI's Text *ghurban* schreiben. Dass es *drei* Tore waren, ergibt sich aus dem tibetischen Text, in welchem die Symbolik der Dreizahl auf die drei Tore des Vimokṣa gedeutet wird. Obwohl dem Sanang Setsen viel zuzutrauen und in Sachen von Zahlen und anderen kleine Denkoperationen erfordernden Dingen nicht mit ihm zu rechten ist, wie er denn selbst dafür am Schlusse des Werkes an die Nachsicht des Lesers appellirt, so ist doch kaum anzunehmen, dass er solchen Unsinn geschrieben habe; man wird vielmehr *dürbän* des Textes in *ghurban* verbessern und bei *ghurban yäkä däghädü-s* (tib. *mc'og-gsum*) an das Triratna denken müssen.) Als Erinnerungszeichen des Vijaya (mongolisch *tein büyhät ilaghuksan* = tib. *rnam-par rgyal-ba*) hatte das mittlere Stockwerk *ein* Tor, und als Denkmal des *einzigen* Dharmakāya (mongolisch *nom-un bäyä* = tib. *c'os-sku*) hatte das obere Stockwerk *vier* Tore. (Dies Rätsel ist vorläufig unlösbar, da im tibetischen Text die Tore dieses und des mittleren Stockwerks nicht erwähnt sind.) 5) Als Erinnerungszeichen der vier Unermesslichen (mongolisch: *dürbän tsaghlaši üghei* = tib. *ts'ad-med bži* = Skr. *catvāro brahmavihārāḥ*, s. Dharmasaṁgraha, Sect. XVI; Triglotte, fol. 12) und der vier vollendeten Taten hat man in dPal bSam-yas (mongolisch: *Tsoktu Sätkiši-üghei*; SCHMIDT hat den Namen nicht darin erkannt), jenem Kloster des Dharmacakra, um den mittleren Haupttempel (mongolisch *tsomtsak* entspricht dem *dbu-rtse* des tibetischen Textes), welcher drei Stockwerke, vier Seiten und vier Ecken

Ostberg, einem auf dem Tron sitzenden König gleichend, gut war,
dass der Ri-c'uṅ (kleiner Berg), einem von dem Muttervogel be-
deckten Vöglein gleichend, gut war, dass der sMan-ri (Berg der
Heilkräuter), einem Haufen von Edelsteinen gleichend, gut war,
dass der Has-po Ri, einer mit weissseidenem Gewande bekleideten
königlichen Gemahlin gleichend, gut war, dass der Ri-nag (Schwarz-
berg), einem in die Erde geschlagenen Eisennagel gleichend, gut
war, dass der Me-yar, einem wassertrinkenden Maultier gleichend,
gut war, dass die Ebene von Del (*Del-t'aṅ*), einem vorgezogenen
Seidenvorhang gleichend, gut war, dass der Grund und Boden,
einem mit Safran gefüllten Goldgefässe gleichend, gut war; da befahl der Herr: an dieser Stelle bauet! und zeichnete einen Grundriss.
Darauf bannte der Meister die schädlichen Dämonen und machte
sie unschädlich. Unter Begleitung von fünfzig aufs beste geschmück-
ter Knaben und Mädchen aus sechs Familien des höchsten Adels,
die mit Segenswasser gefüllte kostbare Flaschen (*bum-pa*) hielten,

---

hat, vier grosse Dvīpa und acht kleine Dvīpa errichtet, 6) ferner einen Tempel der Deva
(*tägri*) Sūrya und Candra (*Naran Saran tägri*) innerhalb des Torpfostens und der Türschwelle
der Yakṣa, 7) ferner einen Tempel der kraftvollen vier grossen Yama und der acht grossen Mahākāla, 8) ferner vier grosse Stūpa und den Licht ausstrahlenden (*bataragghui
ghäräl-tü* = tib. *od ạbar-ba*) Stūpa, insgesamt *dreissig* Bauten. 9) Den Tempel, der von
einer einzigen grossen Mauer umschlossen wurde, hat er (der König) nach dem Vorbild des
im Meere von Indien verborgenen, Ottaburi (so!) genannten, Klosters erbaut." („Mauer"
ist im letzten Satze durch *tämür-ün küriyä* „eiserne Mauer" ausgedrückt; SCHMIDT und
KOWALEWSKI, vielleicht auch SANANG SETSEN selbst, scheinen dabei an eine wirkliche
Mauer von Eisen zu denken; doch die Phrase ist offenbar von tibetisch *lcags-ri* „Eisen-
berg", d. i. einfach „Mauer", suggerirt.) — Von diesen Angaben erweisen sich 2, 3, 4
teilweise, 5, 6, 8 und 9 in Übereinstimmung mit den tibetischen Angaben als ziemlich
correct. Dagegen ist von dem unter 7 erwähnten Tempel der Yama und Mahākāla im
tibetischen Text keine Rede. Unter den in 5 erwähnten vier vollendeten Taten ist eine
Auswahl aus der bekannten Reihe von Buddha's zwölf Taten zu verstehen, nämlich das
Abhiniṣkramaṇa, Buddha als Māra-Bezwinger, Besuch im Tuṣita, und Nirvāṇa, die nach
dem tibetischen Bericht in den Dvīpa-Tempeln dargestellt waren. — Die Summe von dreissig Bauten lässt sich natürlich nicht aus den fragmentarischen Angaben unseres Autors
herausrechnen, ergibt sich aber als correct, wenn man die Zahl der im tibetischen Text
genannten Bauten addirt.

schlug er einen Nagel in die Erde und weihte den Boden. Darauf trieb der König die Beamten und das Volk zur Arbeit an, und als er nach dem Vorbild des Sumeru, des Königs der Berge, in der Mitte den Grundstein zu dem grossen mittleren Tempelgebäude gelegt hatte [1]), baute er vor allem zuerst nach der Weissagung der ehrwürdigen Tārā die Halle für Āryapalo [2]) und stellte als Hauptgottheit des Tempels den Ārya Avalokiteçvara auf, zu seiner Rechten die Tārā, zu seiner Linken die Göttin Marīcī, weiter zu seiner Rechten die Sechs-Silben Gebetsformel (d. i. Oṁ maṇi padme hūṁ), zur Linken den Çrī-Hayagrīva, im ganzen also den Herrn und vier Begleiter. Darauf gab der Meister dem König Belehrung über die Bannung des Hayagrīva, und als der König die Bannung begann, und der Gott dadurch herbeigezogen wurde, ertönte ein dreimaliges Pferdegewieher, das man über zwei Drittel von Jambudvīpa hörte.

---

1) Im allgemeinen ist die Beschreibung der Tempelanlage ziemlich durchsichtig, und man kann sich danach ein ungefähres Bild des Bauplanes skizziren. Die Grundidee des Gesamtplanes war, das buddhistische Weltall in der Anordnung der einzelnen Tempel symbolisch darzustellen. Der Haupttempel (*dbu-rtse c'en-po*), mit dem besonderen Tempel des Avalokiteçvara, lag im Centrum und versinnbildlichte den Sumeru; um ihn gruppirten sich zwölf Tempel, je drei auf einer Seite, als Repräsentanten der zwölf Dvīpa. Westlich von dieser Gruppe erstreckte sich der grosse centrale Tempelhof, auf dessen Nord- und Südseite je ein Schatzhaus, und auf dessen Westseite die Bibliothek lag. Die Lage der übrigen Bauten — der Nāgatempel, Sonne- und Mondtempel, das Badehaus, Tsaṅ-maṅ-ken, die fünf Stūpa, und die drei Tempel der königlichen Gemahlinnen — lässt sich nach den Angaben des Textes nicht genau bestimmen, sondern es lässt sich nur im allgemeinen sagen, dass sie sich in irgendeiner symmetrischen Form um die Haupttempel herum gelagert haben müssen. Auffallend ist, dass über die Wohnräume für die Geistlichkeit nichts bemerkt ist, die aber der alte Desideri an zwei Stellen ausdrücklich hervorhebt (s. C. PUINI, Il Tibet, Roma 1904, pp. 68, 331).

2) Die Form Āryapalo ist nicht, wie bisher angenommen, lediglich mongolische Entstellung des Namens Avalokiteçvara, sondern gehört bereits der tibetischen Volkssprache an, wie aus *Li-šii-gur-k'aṅ*, fol. 20*b*, hervorgeht: *Āryā-wa-lo-ki-ta ap'ags-pa sPyan-ras-gzigs žes-pa zur-c'ag-pas Ārya-pa-lo*. Ebenda findet sich A-mi-de-wa als volkstümliche Form für Amitābha verzeichnet. — Der genannte Tempel des Avalokiteçvara ist ein besonders von dem Sumeru-Haupttempel verschiedenes Gebäude und ergibt mit den übrigen Tempeln, Stūpas und anderen Bauten die von Sanang Setsen richtig angegebene Zahl von dreissig Bauten.

Der Meister sagte, dass auch des Königs Herrschaft sich über zwei Drittel von Jambudvīpa ausdehnen werde. Darauf erbaute er im Hasenjahre [1]) das untere Stockwerk des grossen mittleren Tempelgebäudes, das er dem vom Has-po Berg herbeigeholten, von selbst entstandenen steinernen Çākya (*T'ub-pa*) weihte, eine Statue, die mit einer aus zerriebenen Edelsteinen bestehenden Erde bedeckt und mit allen Merkmalen des Buddha versehen war. In der rechten Reihe stellte er Maitreya auf, Avalokiteçvara, Kṣitigarbha [2]), Nandaçrī (*dGa-bai dpal*), Krōdha-trilokya-vijaya (*K'ro-bo k'ams-gsum rnam-rgyal*); in der linken Reihe Samantabhadra, Vajrapāṇi, Mañjuçrī, Sarvanivaraṇaviṣkambhī [3]), Upāsakavimala (*dGe-bsñen drima-med*) und Krodha-Acala [4]) (*dK'ro-bo mi-γyo-ba*), im ganzen dreizehn, den Herrn mit zwölf Begleitern, in tibetischem Stil gearbeitet. Ferner hatte er dort Malereien, welche die zwölf Taten Buddhas darstellten, Deckenmalereien, und Wandmalereien, welche den Schutzgott des Herrn der Mantra-Halle (*sñags-k'añ rjei t'ugsdam*) darstellten, anfertigen lassen. Dieses untere Stockwerk stellte er unter den Schutz der Ḍākinī Siṁhaçirā (*Señ-gei mgo-can*, die Löwenköpfige). Darauf erbaute er den Mittelstock, den er dem Buddha Vairocana als der Hauptgottheit weihte; auf seiner rechten Seite stellte er Dīpaṅkara auf, auf seiner linken Maitreya, und vorne Çākyamuni, Bhaiṣajyadeva [5]) und Amitābha (*sNañ-ba mt'ayas*); zur rechten und linken die Bodhisattva-aṣṭa-upaputra [6]), Upāsaka-vimala, den Bodhisattva Nandaçrī, den Krodha Kañ-dañ-kiñ [7]),

---

1) Über dieses Datum s. weiter unten am Schlusse des Textes.

2) GRÜNWEDEL, Mythologie des Buddhismus, S. 141.

3) GRÜNWEDEL, l. c., S. 141.        4) Ibid., S. 162.

5) Tib. *sMan-gyi lha*, gewöhnlich *sMan-gyi bla* (s. GRÜNWEDEL, Mythologie, S. 118).

6) Tib. *Byañ-c'ub-sems-dpa ñe-bai sras brgyad*. Ihre Namen s. bei CHANDRA DAS, Tib.-Engl. Dict. p. 485b.

7) Wie schon JÄSCHKE bemerkte, ist *kañ-dañ-kiñ* mit *kiñ-kañ* gleichzusetzen; offenbar ist damit das chinesische *kin-kang* 金剛 beabsichtigt (s. PANDER und GRÜNWEDEL, Pantheon, S. 59).

diese Figuren in chinesischem Stil verfertigt. An den Wänden befanden sich ein Kapitel aus den Scenen des Yum-rgyas-pa [1]) und
die vier Mahārāja dargestellt; ferner rückwärts schauend nach dem
der heiligen Umwandlung dienenden Pfade (*bskor-k'an*) waren an
der Wand acht Caitya und Wandmalereien, die das Nirvāṇa darstellten. In der Mantra-Halle (*snags-k'an*) waren die Buddha der
zehn Gegenden, eine Reihe von Bildern des Yamāntaka und die
Masken (*ḥbag*) der Dharmapāla's von bSam-yas. Dieses Stockwerk
war unter den Schutz des Dharmapālarāja Sin-bya-can [2]) gestellt.

Der Hauptgott des oberen Stockwerks war der Buddha Vairocana
mit vier Gesichtern und zwei Begleitern [3]), ferner waren dort die
Bodhisattva-aṣṭa-upaputra, und unter den Gottheiten im Innenraum
der Bodhisattva Vajradhvaja (*rDo-rje rgyal-mts'an*) und die übrigen
Buddha und Bodhisattva der zehn Gegenden, der Krodha Acala und
Vajrapāṇi, diese Figuren in indischem Stil verfertigt [4]). Auf den

---

1) Die ausführlichere Version (*ḥbum-rgyas*) der das Abhidharma enthaltenden Abteilung (*yum*) des Kanons.

2) Wörtlich: der mit dem Holzvogel. Über diese Schutzgottheit ist sonst nichts bekannt.

3) Eine solche Darstellung scheint aus der lamaischen Kunst nicht bekannt zu sein,
doch der Wortlaut des Textes (*Sans-rgyas rNam-par-snan-mdzad kun-tu žal re-re-la ak'or
gñis-re*) lässt wohl kaum eine andere Deutung zu.

4) SARAT CHANDRA DAS (Journey to Lhasa, ed. by ROCKHILL, London 1904, p. 295)
behauptet von dem Tempel bSam-yas: "Its three stories were each in a different style of
architecture, one Tibetan, another Indian, and the third Chinese." Und dabei war, wie er
selbst berichtet, der ganze Tempel nach dem Vorbild von Otantapurī in Magadha erbaut.
Wie sollte solch eine architektonische Chimäre auch möglich sein! Es geht aus dem tibetischen Text ganz deutlich hervor, dass sich die drei genannten Stilarten auf nichts anderes
als auf die Statuen beziehen (*bzo rgya-gar-gyi lugs-su bžens* etc.), und hier haben wir die
kunstgeschichtlich sehr beachtenswerte Tatsache, dass ein tibetischer, chinesischer und indischer Stil in der buddhistischen Plastik existirt hat und in Tibet im achten Jahrhundert
bewusst angewandt wurde.

Auch KÖPPEN (l. c., S. 68) ist durch SCHMIDT's Übersetzung des Berichts von Sanang
Setsen verleitet worden, von einem „wahren Wunder von Architektur, wenn auch ein sehr
buntscheckig-geschmackloses, ein Gemisch tibetanischen, indischen und chinesischen Stiles
und mit der complicirtesten Symbolik überladen" zu sprechen. Freilich kann man den Satz
des Sanang Setsen (vergl. die Übersetzung S. 20 unter 3), aus dem Zusammenhang heraus-
gerissen, so verstehen: „Was das untere Stockwerk betrifft, so war es tibetischer Stil etc."

Wänden war das Daçabhūmika der Sautrāntika [1]) geschrieben. Dieses Stockwerk war dem Schutz des Dharmapālarāja Zaṅs-mai ạbelog-can unterstellt. Die Decken waren mit schweren decorirten Seidenstoffen bekleidet. In den vier Dachtürmchen stellte er den Buddha Naudaçrī umgeben von Bodhisattva's auf. Dieses Stockwerk wurde unter den Schutz der Dharmapāla, der vier mit blauen Gewändern versehenen Vajrapāṇi [2]), gestellt.

Darauf erbaute er den mittleren Tempelhof (ạk'or-sa bar-pa). [3]) Das Nāga-Schatzhaus auf dessen Südseite füllte er mit Musikinstrumenten an und stellte es unter den Schutz der Yakṣa, nämlich der einen Stock in den Händen haltenden drei Brüder. [4]) Auf der Westseite befand sich das Schatzhaus der Sūtra, der Tantra und des Dharma, 300 Schriften im ganzen, mit tibetischen und Sanskrit Büchern angefüllt, und unter den Schutz der drei ein Messer in den Händen haltenden A-tsa-ra [5]) gestellt. Auf der Nordseite waren die drei Schatzhäuser der kostbaren Kleinodien gelegen, mit Gold, Silber, Kupfer u. s. w. angefüllt und unter den Schutz der drei eine Keule in den Händen haltenden Yamāntaka [6]) gestellt.

---

Aber der Zusammenhang der Stelle zeigt klar, dass unser Autor sich auf die im vorhergehenden Satze genannten Burchane oder Buddhafiguren bezieht und darin mit der Auffassung des tibetischen Textes übereinstimmt.

1) Tib. *mdo-sde sa-bcu-pa*. Vergl. SCHIEFNER's Übersetzung des Tāranātha, S. 119, 122.

2) Tib. *C'os-skyoṅ Lag-na-rdo-rje gos-sṅon-can bži*; *gos-sṅon-can* = Sanskrit *nīlāmbaradhara*.

3) Der, wie aus den folgenden Angaben hervorgeht, westlich von dem Sumeru-Haupttempel gelegen haben muss.

4) Tib. *gnod-sbyin lag-na dbyug-t'o t'ogs-pa mc'ed gsum.*

5) Tib. „*A-tsa-ra lag-na spu-gri t'ogs-pa gsum.* -- DESIDERI (C. PUINI, Il Tibet, p. 67) spricht bei der Beschreibung von bSam-yas von einer „grandissima libreria, con immensità di libri; ma specialmente sono considerabili gli originali de' libri appartenenti alla setta di quelle genti; e che furono le prime versioni, fatte in quella lingua, di moltissimi libri fatti venire con grande spesa, anticamente del regno dell'Indostano".

6) Tib. *gŠin-rje-gšed lag-na be-con-can gsum.* Nach NAIN SINGH befindet sich der tibetische Staatsschatz noch jetzt in bSam-yas (s. CHANDRA DAS, Journey to Lhasa, p. 297, Note 2).

Auf den Wänden war das Lalitavistara geschrieben, und zwischen den Zeilen waren die tausend und zehntausend Buddha gemalt.

Darauf errichtete er, den grossen Tempelhof umwandelnd, ein die Verdammnis beseitigendes Maṇḍala des Çrī-Vairocana. Auf den drei Seiten der Brustwehr waren Bilder der fünf Dhyānibuddha, die, wenn man zurückschaute, einen gefälligen Schmuck darboten. Sie waren unter den Schutz des Maṇibhadra [1]) und des Dharmapāla Nāgarāja Ānanda gestellt. Im Hintergrunde errichtete er eine Steinsäule, unter den Schutz des Dharmapāla Siṁhamukha [2]) gestellt.

Die drei Tore des Haupttempels stellten die Tore des Vimokṣa (rnam-par t'ar-pai sgo) symbolisch dar, die sechs Treppen in den Ecken die sechs Prajñāpāramitā.

Das untere Stockwerk war aus Steinen, das mittlere Stockwerk aus Ziegeln errichtet; das obere Stockwerk war aus unvergleichlich kostbarem Holze erbaut. Sämtliche Statuen aus Metall (bzo-rnams) waren mit den Angaben des Vinaya in Übereinstimmung, sämtliche Tonfiguren (lder-so) mit den Worten der Geheimzaubersprüche [3]), so heisst es.

Darauf erbaute er nach dem Vorbild der drei halbkreisförmigen Pūrvavidehadvīpa des Ostens auf der Ostseite [4]) drei Tempel: in dem Tempel rNam-dag k'rims-k'aṅ [5]) machte er Çākyamuni mit vier Begleitern zur Hauptgottheit; die dort befindlichen Wandmalereien stellten das Abhiniṣkramaṇa [6]) dar; der Tempel stand unter

---

1) Tib. *Nor-bu bzaṅ-po*, Beiname des Kubera.

2) Der Name ist im Text in Transcription des Sanskrit wiedergegeben.

3) Tib. *gsaṅ-sṅags-ṅag*. Hier ist wohl in erster Reihe an die Mantra-Kenntnis des Padmasambhava zu denken. Vergl. den obigen Bericht des Sanang Setsen, S. 20.

4) Nämlich auf der Ostseite des Sumeru-Haupttempels.

5) Wörtlich: „das reine Gesetzeshaus". Das Wort *gliṅ* ist hier jedesmal doppelsinnig in dem Sinne von Land (dvīpa) und Tempel gebraucht. Jeder der vier bekannten Dvīpa ist als aus drei Dvīpa bestehend gedacht, denen drei Tempel entsprechen, und deren Namen mit den Gottheiten des Tempels in Beziehung gesetzt, resp. von ihnen abgeleitet sind.

6) Tib. *mṅon-par abyuṅ-ba*. Oder: das A. war auf den Wänden geschrieben.

dem Schutze des Dharmapāla Brahma mit dem muschelförmigen Haarknoten [1]). In dem Tempel des wissenden Mañjuçrī war als Hauptgott Ārya-Mañjuçrī mit sechs Begleitern; auf den Wänden waren zwei torbewachende (*sgo-sruṅ*, dvārapāla) Yamāntaka, das Mañjuçrīmūlatantra und das Durgatiçodhana [2]); der Tempel war unter den Schutz des Dharmapāla Yama mit dem Rade (*ṅkor-lo-can*, cakradhara) gestellt. In dem Tempel brDa-sbyor Tsʿaṅs-pa [3]) war die Hauptgottheit Çākyamuni mit sechs Begleitern; die Wände waren mit einer Darstellung des Nirvāṇa bemalt; der Tempel war unter den Schutz des Dharmapāla »der blitzende Drache" gestellt.

Nach dem Vorbild der drei schulterblattförmigen Jambudvīpa des Südens errichtete er auf der Südseite drei Tempel. In dem Tempel bDud-ạdul sṅags-pa (»der Māra bezwingende Zauberer") war Çākya im Stile des Māra-Bezwingers die Hauptgottheit mit vier Begleitern; auf den Wänden war in der Mitte das Daçabhūmika der Sautrāntika und ringsherum Ākāçagarbha [4]); der Tempel war unter den Schutz der 28 Içvarī [5]) gestellt. In dem Tempel des Āryapalo war der Hauptgott Kharsapaṇa mit vier Begleitern; im oberen Stock waren Amitābha mit vier Begleitern und unter den Skulpturen sechs mit Silber bedeckte Statuen der Könige aus Saudelholz; auf den Wänden war ein Caitya gemalt. Ferner war da

---

1) Tib. *Tsʿaṅs-pa duṅ-gi tʿor-tsʿugs-can*, letzteres vielleicht einem Sanskrit *çaṅkhaçi-khābandhana* entsprechend.

2) Tib. *ṅan-soṅ sbyoṅ-ba*. S. Kanjur-Index, ed. SCHMIDT, N° 596, p. 86; N° 777, p. 105; N° 957, p 124; N° 986, p. 127.

3) Es ist schwer, eine Übersetzung zu wagen. Die Bedeutung könnte »gereinigte (oder ist *Tsʿaṅs-pa = Brahma?*) Orthographie oder Sprachwissenschaft" sein. Die Begriffe »Tempel, Kloster, Schule" fallen ja im Buddhismus zusammen, und das besagte Gebäude könnte gleichzeitig als Schule für sprachlichen Unterricht gedient haben. Diese Vermutung ist um so wahrscheinlicher, als zufolge Bu-ston nach Berufung der Paṇḍita Jinamitra und Dānaçīla die Arbeit des Schreibens und grammatisches Studium in eben jenem Tempel betrieben wurde (s. Journal Buddhist Text Soc. of India, Vol. I, p. 5).

4) Wohl ein von ihm handelndes Sūtra, s. Kanjur-Index, ed. SCHMIDT, No 259, p. 43.

5) Tib. *dBaṅ-pʿyug-ma ñi-šu rtsa brgyad.*

ein Za-ma-tog. [1]) Der Tempel war unter den Schutz der 1002
Devī und des Dharmapāla Rṣi Dharmarāja gestellt. In dem Tempel
sGra-sgyur rGya-gar (»das Indien der Übersetzer") war Çākya in
indischem Stil mit vier Begleitern die Hauptgottheit; an den Wänden waren Amitāyus und Figuren der Locāva-Paṇḍita Übersetzer
dargestellt; der Tempel war unter den Schutz des Dharmapāla
Yakṣa Rāhula gestellt.

Nach dem Vorbild der drei runden Godānīyadvīpa des Westens
erbaute er auf der Westseite drei Tempel. In dem Tempel dPa-bo
Be-tsa (»der Held Be-tsa") errichtete er eine Statue des Vairocana
aus Kupfer und gab ihm als Begleiter seine mystische Çakti [2]); auf
den Wänden war die Abhisaṁbodhi des Vairocana [3]); der Tempel
war unter den Schutz des Dharmapāla Yakṣa mit dem Ochsenkopf
gestellt. In dem Tempel Tuṣita-Maitreya war der Beschützer (nātha)
Maitreya die Hauptgottheit mit sechs Begleitern, ferner zwei torbewachende Yamāntaka. Auf den Wänden waren die sechzehn
Sthavira, die Erbauung von bSam-yas und die Anordnung der Welt
gemalt. Der Tempel war unter den Schutz des Dharmapāla Krodha
mit dem blauen Gewand gestellt. In dem Tempel Acaladhyāna [4])
waren die fünf Dhyānibuddha und die sechzehn Çrāvaka; auf der
Wand befand sich ein Gemälde des Wasser herabsendenden Knaben
nach dem Suvarṇaprabhāsosattama [5]); der Tempel war unter den
Schutz des Dharmapāla Rāja mit dem Silberkopf gestellt.

Nach dem Vorbild der drei viereckigen Uttarakurudvīpa des
Nordens errichtete er auf der Nordseite drei Tempel. In dem Tem-

---

1) Ohne weitere Angaben, so dass sich nicht entscheiden lässt, ob es sich um einen
realen Gegenstand oder ein literarisches Werk handelt.

2) GRÜNWEDEL, Mythologie des Buddhismus, S. 98.

3) S. Kanjur-Index, ed. SCHMIDT, N° 487, p. 73.

4) Tib. *Mi-ɣyo-ba bsam-gtan.*

5) Tib. *gser-od dam-pa-nas kʰyeu cʰu ạbebs-kyi ri-mo.* S. Kanjur-Index, ed. SCHMIDT,
N° 555—7, p. 81; CSOMA in Asiatic Researches, Vol. XX, pp. 514—6. Es handelt sich
wohl um die malerische Darstellung eines Jātaka.

pel Rin-c'en sna-ts'ogs (»mannigfache Kostbarkeiten") war die Hauptgottheit Çakyamuni mit vier Begleitern; auf den Wänden waren die Dankbezeigungen des Bhagavat gemalt, als er im Tuṣita-Himmel seiner Mutter die Lehre predigte; der Tempel war unter den Schutz des Dharmapāla Rāja mit eisernen Klauen gestellt. In dem Tempel Cittotpādabodhi [1]) waren der Buddha Padmapāṇi, Āvaraṇaviṣkambhī [2]), Vajrapāṇi und bDud-kyi [3]) ạk'yil-ba; auf den Wänden waren das Ratnamegha der Sautrāntika [4]) und Auszüge (?) aus der Prajñāpāramitā [5]); der Tempel war unter den Schutz des Dharmapāla, der Ḍākinī Simhaçirā, gestellt. Im Tempel der Schatz-kammer des Pe-dkar war Çakyamuni der Hauptgott mit acht Be-gleitern; an den Wänden war das Pitāputrasamāgama [6]). Hier wurde auch der Rest des beim Bau von bSam-yas auf die Kunstwerke verausgabten Geldes aufbewahrt, und eine Rechnungsablage war beigelegt. Dieser Tempel war unter den Schutz des Beschirmers K'raṅ-k'aṅ-gi rgyal-po [7]) gestellt.

Der ganze Tempel aber war dem Schutze anvertraut des allge-meinen Hüters der Lehre, des von dem Meister (d. i. Padmasam-bhava) aus dem Lande von Za-hor [8]) mitgebrachten Deva-mahā-

---

1) Tib. *Sems-bskyed byaṅ-c'ub.*

2) Tib. *sGrib-pa rnam-sel.*

3) Eine andere Handschrift liest *bdud-rtsi.*

4) Tib. *mdo-sde dkon-mc'og sprin,* s. Kanjur-Index, ed. SCHMIDT, N° 231, p. 38.

5) Tib. *rtag-tu dus* (?) *šes-rab-kyi p'a-rol-tu p'yin-pa stsal-pa-rnams.*

6) Tib. *yab sras mjal-ba,* Begegnung Buddhas mit seinem Vater Çuddhodana. S. Kanjur-Index, ed. SCHMIDT, N° 60, p. 9; Csoma in Asiatic Researches, Vol. XX, p. 409.

7) Vielleicht „König des starken Hauses oder Tempels".

8) Über die Lage dieses Landes ist seit den Tagen des Schmidt-Klaproth'schen Uigu-renstreites viel Tinte vergossen worden (s. SCHMIDT, Geschichte der Ost-Mongolen, S. 351; Philologisch-kritische Zugabe etc., S. 20—21; ABEL-RÉMUSAT, Observations sur l'histoire des Mongols orientaux, pp. 39—41; KÖPPEN, Religion des Buddha, Bd. II, S. 68), ohne dass nennenswertes documentarisches Material beigebracht worden wäre. Die von JÄSCHKE und neuerdings von SCHLAGINTWEIT auf Grund des *Padma-t'aṅ-yig* gegebene Identifikation ist übrigens schon lange vorher von KOWALEWSKI (Mongolische Chrestomathie, Vol. II, p. 356) erkannt worden. Beachtenswert ist der folgende Passus im *Li-šii-gur-k'aṅ,* fol. 20*a*:

upāsaka [1]), des grossen Generals gegen die Heere des Māra, des Herrschers über die acht Klassen der Deva, Asura, Rākṣasa u. s. w. des Saṁsāra, dessen, der allen Wesen den Atem benimmt, des grossen Herrn des Lebens, Pe-har, dem sein Heiligtum im Pe-dkar-gyi gliṅ (»Tempel des Pe-dkar") errichtet wurde [2]).

Der Tempel der Sonne war in dem Tempel des oberen Yakṣa Pūrṇabhadra [3]), wo Çākyamuni mit vier Begleitern die Hauptgottheit war; an den Wänden waren die 1002 Buddha. Der Tempel des Mondes war in dem Tempel des unteren Yakṣa Maṇibhadra [4]), wo dieselben Wandmalereien waren wie in dem vorigen.

Im Tempelgebäude Ts'aṅ-maṅ-keu befand sich oben der Tempel des Punarvasu-nakṣatra [5]). Im Tempelgebäude des Badehauses für die Geistlichkeit war Quellwasser, voll von Sandelholz. In dem

---

*Sa-hor žes-pa rGya-gar-gyi rgyal-rigs yan-gar-ba žig-gi miṅ, zur-c'ag-pas Za-hor žes-pa*, d. i. „Das Wort Sa-hor ist der Name einer besonderen Dynastie von Indien, in der Vulgärsprache Za-hor genannt".

1) Tib. *Lhai dge-bsñen c'en-po.*

2) S. besonders CHANDRA DAS, Tibetan-English Dictionary, p. 785a, und WADDELL, The Buddhism of Tibet, pp. 371, 478. Dem Einwand des letzteren, dass die Ableitung des Namens des Gottes von Bihar (*vihāra*) unwahrscheinlich sei, muss entgegengehalten werden, dass nach *Li-šii-gur-k'aṅ*, fol. 20a, die beiden Formen *dpe-har* und *dpe-dkar* (beachte die verschiedenen Schreibungen im obigen Texte) als vulgäre oder volkstümliche Bezeichnungen von Sanskrit *vihāra* erklärt werden. Es gab jedenfalls zu jener Zeit einen indischen Schutzgott der Klöster, genannt Vihāra, der, in Tibet von Padmasambhava eingeführt, einen Ausgleichsprocess durchmachte, wie das in so vielen anderen Fällen geschah, und einer schon vorhandenen einheimischen Gottheit assimilirt wurde.

3) Tib. *yag-ša goṅ-ma Gaṅ-ba bzaṅ-po.*

4) Tib. *yag-ša og-ma Nor-bu bzaṅ-po.* — Es handelt sich bei dem Tempel der Sonne und des Mondes um die indischen Götter Sūrya und Candra, wie auch aus dem Bericht des Sanang Setsen (6) hervorgeht, der ie als Deva (*tägri*) bezeichnet. Aus dem obigen Text wird auch seine etwas unklare Definition von der Lage der beiden Tempel „innerhalb des Torpfostens und der Türschwelle der Yakṣa" deutlich, womit wohl nur gesagt sein soll, dass sie sich in den beiden Nebentempeln der oben erwähnten Yakṣa befanden. Es ist interessant, Sūrya und Candra als besondere Götter hier aufgeführt zu sehen, da sie z. B. in dem von GRÜNWEDEL herausgegebenen Pantheon des lCaṅ-skya Hutuktu nicht aufgezählt werden, während sie als Nebenfiguren in lamaistischen Kunstdarstellungen ziemlich häufig sind (GRÜNWEDEL, Mythologie, S. 105, 120, 141).

5) Tib. *nabs-so lha-k'aṅ.*

Tempel des »Nâga-Hauses der Zauberei" [1]) befanden sich an einem aus acht Zweigen bestehenden Rosenbusch die acht grossen Nāga; in der Mitte war Vajrapāṇi; an den Wänden war das Ži-ba ạdul-byed [2]); ausserdem waren Schildkröten, Fische und Makara (c'u-srin) da.

Darauf wurde ein weisser Stūpa, ein Stūpa der Mahābodhi, im Stile der der Çrāvaka, mit Löwen geschmückt, von dem Baumeister pud-pu dpal-gyi seṅ-ge erbaut. Er wurde dem Schutz des Dharma-Šala Yakṣa mit den Sternschnuppen (skar-mda-can) anvertraut. Ein roter Stūpa, im Stile der der Dharmacakravartin's, mit Lotusblumen geschmückt, wurde von dem Baumeister rGyal-ts'a lha-gnaṅ aus sNa-nam [3]) errichtet. Er wurde dem Schutz des Dharmapāla Planeten Mars (mig-dmar) anvertraut. Ein schwarzer Stūpa, im Stile der der Pratyeka-Buddha, wurde von dem Baumeister sTag-sgra klu-goṅ aus Li-ṅam errichtet. Er wurde dem Schutz des Dharmapāla Yakṣa mit der Eisenlippe (lcaẏs-kyi mc'u-can) anvertraut. Ein blauer Stūpa, im Stile der der Tathāgata, die aus der Götterregion herabgestiegen sind, die Kapelle mit sechzehn Toren geschmückt, wurde von dem Baumeister mC'ims rDo-rje sprel-c'uṅ errichtet. Er wurde unter den Schutz des Dharmapāla Yakṣa Sūryavaktra [4]) gestellt.

Als zu dieser Zeit die Dämonin Prabhāvatī (Od-can-ma) allen Menschen Harm zufügte, erbaute der Meister im Tempel des Maitreya einen Licht ausstrahlenden Stūpa [5]), wodurch er die Dämonin überwand und ihre Anschläge zunichtemachte.

Darauf wurde eine vieleckige Mauer erbaut, deren heilige Um-

---

1) Tib. mt'u-rtsal klu-k'aṅ.

2) Eine andere Handschrift schreibt žib statt ži-ba. Ich weiss nicht, um welches Werk es sich hier handelt.

3) Bezeichnet wohl in manchen Texten Samarkaud, ist aber auch und das trifft an der obigen Stelle zu, ein Ort im Distrikte Žaṅ in der Provinz gTsaṅ in Tibet.

4) Tib. Ñi-ma gdoṅ.

5) Tib. mc'od-rten ọd ạbar-ba (etwa Skr. prabhajvala). Vergl. den S. 20 gegebenen Bericht des Sanang Setsen unter 8, der den Ausdruck wörtlich ins Mongolische übersetzt.

wandlung über 108 Stūpas erfolgte [1]): in jedem derselben war je
eine Reliquie Buddhas enthalten.

Was nun die drei Tempel der königlichen Gemahlinnen betrifft,
so errichtete ạBro-bza byaṅ-c'ub sgron den Tempel dGe-rgyas bye-
ma. Als Hauptgottheit liess sie den Buddha Amitābha mit sechs
Begleitern in Gussbronze herstellen, zu seiner Rechten die Beschützer der drei Klassen (d. i. Götter, Menschen und Nāga), zu seiner
Linken Bhaiṣajyaguru, Samantabhadra und Acala. Da sie keinen
Sohn hatte, und ihre väterlichen Verwandten weit entfernt lebten,
hegte sie die Besorgnis, dass, wenn einmal in Zukunft in ihrem
Tempel ein Riss entstehen sollte, niemand da wäre, der ihn repa-
riren würde; daher verwendete sie beim Bau nur Ziegel und liess
die Fugen derselben mit geschmolzenem Blei ausfüllen; das Dach
liess sie aus Kupfer verfertigen. Auch stiftete sie Musikinstrumente
und liess an Stelle von Lampen leuchtende Edelsteine an den Pfei-
lern befestigen; für das Weihwasser liess sie Brunnen herstellen.
Die Königin Ts'e-spoṅ-bza me-tog sgron liess, da sie drei Söhne
geboren hatte, nämlich Mu-ne btsan-po, Mu-rug btsan-po und Mu-
tig btsan-po Sad-na-legs, nach dem Vorbild der drei Haupttempel
des Vaters den Tempel K'ams-gsum zaṅs-k'aṅ gliṅ (»Tempel des
Kupferhauses der drei Welten") errichten. P'o-yoṅ-bza rgyal-mo
btsun liess den Tempel Bu-ts'al gser-k'aṅ und ein Maṇḍala des
Vajradhātu errichten. Da sie den Bildhauern und übrigen Kunst-
handwerkern jedesmal je dreizehn verschiedene Speisen vorsetzte,
verfertigten diese zum Lohn für ihre Güte dreizehn Meisterwerke:
die Aussenseite, obwohl ohne Mauer, war ein Meisterwerk stark
wie ein Donnerkeil; das Innere, obwohl ohne Säulen, war ein

---

[1]) Die Mauer wird auch von SANANG SETSEN (s. S. 20 unter 9) und NAIN SINGH
(Journal Roy. Geogr. Soc., Vol. 47, p. 114) erwähnt, der bSam-yas 1873 besuchte. Er sagt,
dass die Mauer sehr hoch, kreisförmig, $1\frac{1}{2}$ Meilen im Umfang sei und vier Tore habe.
Oben auf der Mauer zählte er 1030 Chorten's aus gebrannten Ziegeln, so dass die Zahl
von 108 schon bei der Gründung bestanden haben mag.

Meisterwerk schön wie ein Zelt; der Boden war aus Messing gearbeitet; an den Balken von Türkis waren gallopirende Pferde aus
Gold, an den goldenen Balken waren Drachen aus Türkis ange-
bracht; das Dach im chinesischen Stil war nach oben zurückgebo-
gen; die im Innern aufgestellten Götterbilder waren von einem
allgemeinen Schirm, und jedes einzelne von einem eigenen Schirm bedeckt; beim Öffnen der Tür liess ein goldnes Vöglein seine Stimme
erschallen; die zwölf Taten des Buddha waren in Hochreliefs [1]) dargestellt u. s. w.: so schufen sie dreizehn staunenswerte Meisterwerke.

So wurde der schöne Bau des ganzen Umkreises, des »herrlichen bSam-yas, des unvergänglichen durch Zauberkraft geschaffenen
Klosters" [2]), vollendet. Von den Erstlingen des Bauholzes von
Lhasa wurden die grossen Pforten vor der Mitte des btsan-kʿan [3])
mit Bogen geschmückt. Von der Erde, die beim Hausbau als Opfer
dargebracht wird, verfertigte man die vier Mahārāja. Nachdem so
in einem Hasen-Jahre der Grundstein zu bSam-yas gelegt worden
war, war der Bau in einem Hasenjahre ganz und gar vollendet, so
dass er einen Cyklus von zwölf Jahren gedauert hatte [4]).

---

1) Tib. *aḅur-du dod-pa.*

2) Dies ist der officielle Titel, den zuerst GRÜNWEDEL (Mythologie, S. 54) richtig
übersetzt hat. CHANDRA DAS (Journey to Lhasa, p. 296) übersetzt the temple of the
unalterable mass of perfection, und WADDELL (Buddhism in Tibet, p. 266) leistet sich die
Übersetzung the academy for obtaining the heap of unchanging meditation!

3) Nach JÄSCHKE der innerste dunkle Raum in einem Tempel, in welchem die Götter
wohnen.

4) Nach der Chronologie des Historikers BU-STON begann der Bau des Tempels im
Feuer-Hasen Jahre und endete im Erde-Hasen Jahre; das ergäbe die Zeit von 786 bis
798. Auch SANANG SETSEN nimmt zwölf Jahre für den Bau des Tempels in Anspruch,
und zwar 811—823, begeht aber gleichzeitig die Torheit, das Jahr 811 mit dem 22.
Lebensjahr des Königs Kʿri-sroṅ und das Jahr 823 mit seinem 24. Lebensjahr zu identificiren, einen Widerspruch, den weder sein Übersetzer I. J. SCHMIDT noch sein Kritiker ABEL-
RÉMUSAT (Observations sur l'histoire des Mongols orientaux de Sanang-Setsen, Paris 1832)
beobachtet haben. Wegen des grossen Umfanges der Bauten ist die Annahme einer bloss

3

Darauf wurden für die Feier der Einweihung unermessliche Genüsse an Speise und Trank herbeigeschafft. Das ganze Volk versammelte sich und veranstaltete ein grosses Fest. Der König gab dann dem Meister (Padmasambhava) und dem Bodhisattva (Çāntirakṣita) Geschenke an Gold. Der König legte ein mit vielen Edelsteinen besetztes Gewand an und machte im ganzen Umkreis des herrlichen bSam-yas, des unvergänglichen durch Zauberkraft geschaffenen Klosters, siebenmal vollzählig die wunderwirkenden Weihen. Er zeigte sich in einem Nirmāṇakāya [1]). Darstellungen dieses Ereignisses wurden auf die Rückseite des Tores des Haupttempels und auf die beiden Flügel des Tores zu dem Tempelhof gemalt."

Es kann keinem Zweifel unterliegen, dass der Anteil Padmasambhava's an dem Bau des Tempels nicht gering gewesen ist; auf ihn geht jedenfalls die Grundidee, den Tempel Otantapurī in

---

zweijährigen Bauperiode ausgeschlossen. Dem *rGyal-rabs* zufolge starb König K'ri-sroṅ im Holz-Rind Jahre, d. i. 784, im Alter von 56 Jahren, wonach er also 728 geboren worden wäre, und da er nach demselben Werke im 8. Lebensjahre den Tron bestiegen hätte, wäre 736 das Jahr seines Regirungsantritts, und er hätte 48 Jahre regirt. Nach dem *T'ang-shu*, in dem sich übrigens viele unrichtige Angaben über die Könige von Tibet befinden (s. z. B. ROCKHILL, The Life of the Buddha, p. 219, Note) hätte er von 755 bis 796 regirt, also 41 Jahre. Weder mit der tibetischen noch mit der chinesischen Berechnung seiner Regirung lässt sich Bu-ston's Ansetzung des Tempelbaus vereinigen. Nimmt man an, dass Bu-ston statt des vierten den fünften Cyklus, der 726 beginnt, gemeint habe, so erhält man die Jahre 726—738, was gleichfalls unmöglich anzunehmen ist. Der Fehler Bu-ston's ist um so greller, da nach seiner Berechnung K'ri-sroṅ von 730 bis 789 gelebt hat (Regirungsantritt 743). Seine Datirung des Tempelbaus ist daher für die Ansetzung desselben ganz auszuschalten. CSOMA (A Grammar of the Tibetan Language, p. 183) hat nach dem Vaiḍūrya-dkar-po die Zeit der Erbauung von bSam-yas auf 749 bestimmt, was der Wahrheit schon näher kommen mag; GRÜNWEDEL (Mythologie des Buddhismus, S. 55) gibt das Jahr als um 770. Mit der zwölfjährigen Dauer des Baues wird es wohl seine Richtigkeit haben, und aus der Reihenfolge der Ereignisse unter der Regirung des Königs zu schliessen, scheint der Bau ungefähr in die Mitte seiner Regirungszeit zu fallen.

1) Das Subjekt ist nicht ausgedrückt, es kann aber wohl nur Padmasambhava sein.

Magadha als Vorbild zu benutzen, zurück [1]). Aber damit ist auch nach dem *rGyal-rabs* seine Rolle völlig ausgespielt. Sein Name wird nicht mehr erwähnt, ausgenommen noch ein einziges Mal in einer von Çāntirakṣita kurz vor seinem (Çāntirakṣita's) Tode ausgesprochenen Prophezeiung, in der er als »der zwölf Dämoninnen bannende" gerühmt wird. Unter den Namen der indischen und tibetischen Übersetzer, die unter König K'ri-sroṅ mit der Übertragung des Tripiṭaka beschäftigt waren, wird er im *rGyal-rabs* nicht genannt [2]). Dagegen erzählt der Historiker BU-STON (geboren 1288), dass sowohl Çāntirakṣita als Padmasambhava dreizehn Jahre lang in bSam-yas erfolgreich für die Verbreitung des Buddhismus gewirkt hätten [3]). Auch nach dem *Grub-mt'a šel-kyi me-loṅ* (verfasst 1740) hätten beide viele Sūtra und Tantra übersetzt [4]). SANANG SETSEN [5]) weiss viel von Padmasambhava's Übersetzertätigkeit zu berichten, doch hat er, wie wir bereits gesehen, von den legendären Traditionen über ihn reichlich Gebrauch gemacht und kann kaum als lautere Quelle für diese Zeitperiode angesehen werden. Für die Beurteilung dieser Frage müssen wir in Erwägung ziehen, dass die Mehrzahl der tibetischen historischen Werke der orthodoxen Schule entstammt, und dass Padmasambhava, der nicht in ihr System hineingehört, darin schlecht fortkommt. Eine Anzahl von Übersetzungen hat er sicher geliefert, dafür bürgt die Tatsache, dass er, wie schon erwähnt, im Kanjur als Übersetzer erscheint. Jedenfalls

---

1) So sagt auch der alte DESIDERI (ed. C. PUINI, Il Tibet, Roma 1904, p. 330): Ne prescrisse Urghien il sito, e ne formò il disegno; e il re offrì largamente, pronti ad ogni bisogno, gli erarii.

2) Ebensowenig im Bodhimör (s. SCHMIDT, Geschichte der Ost-Mongolen, S. 356).

3) Journal Buddhist Text Soc. of India, Vol. I, 1893, p. 4.

4) Journal Asiatic Soc. of Bengal, 1881, p. 226.

5) SCHMIDT, l. c., S. 45—47.

war die Zahl seiner und seiner Schule Werke viel grösser als wir
nach dem Kanjur annehmen dürfen, da diese nur in einer Recension desselben vorhanden sind, während sie in einer anderen, der
anscheinend rechtgläubigen Ausgabe, herausredigirt worden sind [1]).
Merkwürdig ist auch die Tatsache, dass weder das *rGyal-rabs*, noch
wie es scheint, eines der übrigen Geschichtswerke, auch nur mit
einer Silbe etwas von seinem Ende erzählen [2]), während der Tod
des Çāntirakṣita ausdrücklich berichtet wird. Nach allem erhält man
den Eindruck, dass die grosse Bedeutung, die Padmasambhava im
Laufe der Zeit im System des Lamaismus erhalten hat, erst das
Produkt einer lange nach seinem Tode liegenden Periode gewesen
ist. Nach den officiellen Annalen zu urteilen, war der historische
Padmasambhava lediglich der persönliche Freund und Berater des
Königs K'ri-sron, wie auch aus der nicht-officiellen Geschichte der
Bon-Sekte [3]) hervorgeht, und von der Adelspartei, die auf die Vor-
rechte des neuen Günstlings eifersüchtig war, weidlich gehasst. Der
einfache Bericht des *rGyal-rabs* trägt hier durchaus den Stempel
historischer Glaubwürdigkeit. Seine Rolle beschränkte sich auf die
eines gefürchteten Dämonenbezwingers oder Hexenmeisters, worin
hauptsächlich der Keim für seine spätere immense Popularität erblickt werden muss. Doch vergeblich suchen wir in ihm das bele-
bende, fruchtbare geistige Element, als dessen Vertreter vielmehr

---

1) So konnte ROCKHILL (The Life of the Buddha, p. 220) bemerken: "I have not
met with any works of his in the Tibetan Tripiṭaka."

2) Den episch ausgeschmückten Bericht der Legende über seinen Abschied von Tibet
hat GRÜNWEDEL (Mythologie des Buddhismus, S. 55) übersetzt. — Die Dauer seines
Aufenthalts in Tibet wird verschieden angegeben. Nach CHANDRA DAS (Journey to Lhasa,
p. 296) soll er sechs, nach einer anderen Tradition achtzehn Jahre lang in Tibet geblieben
und dann nach Indien zurückgekehrt sein. Nach dem Werke *rGyal-rabs Bon-gyi ąbyuṅ-
gnas*, p. 54, 24, waren es elf Jahre.

3) T'oung Pao, 1901, p. 44.

Çāntirakṣita erscheint. Ich glaube daher, dass man wohl zu viel behauptet hat, wenn man Padmasambhava schlechtweg als Gründer des Lamaismus ins Treffen führte; er hat einer gewissen Richtung innerhalb des Lamaismus den Anstoss gegeben, die schliesslich in einem formelhaften Zauberwesen und in Teufelsaustreibung gipfelte, er darf daher als einer der vielen Faktoren betrachtet werden, die sich zur Bildung des Lamaismus vereinigt haben [1]). Der geistige Vater des Lamaismus aber war zweifelsohne Çāntirakṣita, mit dem sich ihm anschliessenden Heer von Übersetzern, die auf dem wirklichen Boden des Buddhismus standen.

# ANHANG.

## Über den epischen Charakter der tibetischen Annalen.

H. C. v. D. GABELENTZ war, soviel ich weiss, der erste, welcher in einem historischen Werke einer ostasiatischen Litteratur das Vorhandensein von Bruchstücken einer ehemaligen epischen Dichtung nachwies. In seiner trefflichen Abhandlung »Einiges über mongolische Poesie" [2]) zeigte er, dass »Sanang Setsen an vielen Stellen seines Geschichtswerks grössere oder kleinere Bruchstücke von Gedichten eingeflochten hat, welche offenbar einem epischen Cyclus der Mongolen angehören und vielleicht noch jetzt im Munde

---

1) Mit vollem Recht bemerkt Desideri (l. c , p. 333) von ihm: A tutto ciò mi piace aggiunger qui alcuna cosa, che stimo utile ricordare. Quantunque Urghien fondasse nel Thibet la Religione, non è però da quelle genti tenuto per legislatore; perchè la legge che egli colà insegnò e diffuse, non era da lui instituita, ma da Çākya-t'ub-pa. Vero è però, che Urghien vi aggiunse del suo; che fu l'insegnamento di certe arti magiche, le quali pretendeva dovessero servire a liberare i viventi da' mali e dalle sventure, non già per recar danno altrui.

2) Zeitschrift für die Kunde des Morgenlandes, Bd. I, Göttingen 1837, S. 20—37.

des Volkes fortleben, wie dies wahrscheinlich auch zu Sanang Setsen's Zeit (1662) der Fall war".

Neuerdings hat W. GRUBE [1]) mit vollem Recht den dichterischen Charakter des Shu-king beleuchtet, in welchem vielfach gereimte Verse völlig unvermittelt und zwanglos in den Gang der Erzählung oder Rede eingeschoben sind. Einzelne Partieen im Shu-king, wie vor allem die von Grube ausgewählten zeigen, tragen durchaus epischen Charakter.

Derselbe Zug lässt sich in den tibetischen Annalen des *rGyal-rabs gsal-bai me-lon* (»Der die Genealogie der Könige klar darstellende Spiegel") verfolgen. Dieses Werk wurde im Jahre 1327 von dem Sa-skya-pa bSod-nams rgyal-mts'an im Kloster bSam-yas nach alten Quellen compilirt, die er wiederholt in seinem Buche citirt. Dasselbe besteht in seinem ersten Teil vorwiegend aus alten Sagen und religiös-buddhistischen Erzählungen, im zweiten Teil aus einem chronikartigen Bericht der Geschichte der Könige von Tibet mit besonderer Berücksichtigung dessen, was sie für die Einführung des Buddhismus getan haben. Obwohl im allgemeinen in Prosa geschrieben, finden sich darin eine Reihe in Versen abgefasster Bestandteile von teilweise epischem Gepräge. Ein gewisser epischer Ton waltet fast über der ganzen Erzählung und charakterisirt sich in der Kürze und Prägnanz des Stils, in häufiger Anwendung der Dialogform und volkstümlicher sprüchwörtlicher Redensarten, und wie gesagt, in Unterbrechung der ungebundenen durch gebundene Rede in längeren Abschnitten, die in der Regel nach einer älteren Quelle angeführt werden. Einer der interessantesten derselben, der für die Geschichte der tibetischen Volkspoesie von einiger Bedeutung ist,

---

1) Geschichte der chinesischen Litteratur, Leipzig 1902, S. 41, 46.

schliesst sich unmittelbar an die oben mitgeteilte Geschichte der Erbauung von bSam-yas an uud dürfte daher hier eine passende Stelle finden. Er beschreibt das Fest der Einweihung der beendeten Tempelbauten und ist einem Werke *bKai-t'an-yig c'en-mo* entlehnt, von dem mir sonst nichts bekannt ist.


Als auf diese Art im herrlichen bSam-yas
Das Fest der Einweihung gefeiert wurde,
Sass da, wo im Seepavillon
Der Lotusstengel hervorspriesst,
Der König auf goldenem Tron;
Sassen anmutige königliche Frauen fünf,
Prangend in schönem Schmuck.
Die Priester, Übersetzer und Paṇḍits alle
Führten in freudiger Stimmung religiöse Gespräche.
Die frommen Beamten und Herren all
Standen um den Fuss des Königstrones herum,
Des Reiches Untertanen
Aus allen Gebieten von dBus und gTsaṅ
Wurden von den vielerlei Genüssen an Speise und Trank —
Es gab nichts, was sich dort nicht gefunden hätte —
Alle nach Herzenslust satt.
Der Freudentänze und des frohen Liederklangs
War Tag für Tag kein Ende.
Von der Menge der Schirme, Standarten und Flaggen
Wurde am Mittag die Sonne verfinstert,
Und die befiederten Vögel hatten zum Fliegen keinen Raum.

Von schwarzköpfigen [1]) Männern war der ganze Platz erfüllt.

Vom Cymbelklang erdröhnte die Luft wie vom Donner.

Kein Platz war da zum Lauf der edlen Renner.

Knaben und Mädchen alle

In vollem Schmuck schwangen Trommeln in den Händen;

Trommelschlagend und singend drehten sie sich in Tanzbewegungen.

Mimen mit den Masken des Yak-Löwen, Seiden-Löwen und Tiger-
Löwen,

Der den jungen Löwen darstellende Mime,

Und schöngestaltige, trommelschwingende Tsam-Tänzer [2])

---

1) Die Bezeichnung des Volkes als „schwarzköpfig" ist im *rGyal-rabs* häufig. Der
Ausdruck, der sich auch in Denkmälern der türkischen und mongolischen Litteratur findet,
dürfte wohl aus der chinesischen Litteratur entlehnt sein. Man weiss, zu welchen phantasti-
schen Combinationen er Veranlassung geboten hat. CHAVANNES (Les mémoires historiques
de Se-ma Ts'ien, Vol. II, pp. 133—134) hat diese Theorieen hinreichend gewürdigt. Der
Umstand, dass die Chinesen selbst keine Definition des Terminus geben, scheint mir dafür
zu sprechen, dass es sich dabei um ein recht einfaches Faktum handeln dürfte. Eine, wie
mir scheint, ganz plausible Lösung der Frage kam mir ganz zufällig, als ich einst am
Westsee (Hsi-hu) von Hang-chou einem grossen Tempelfest beiwohnte. Ich sass mit meinen
chinesischen Freunden auf dem Balkon eines Theehauses und schaute auf die riesige Volks-
menge herab, die sich unten auf dem Vorplatz des Tempels angesammelt hatte, eine einzige zusammenhängende Masse von Schwarz, die schwarzen Köpfe der barhäuptigen Leute.
In jenem Moment fiel es mir ein, dass sich der Ausdruck „Schwarzköpfe" oder „schwarz-
haariges Volk" ausschliesslich auf den Mangel jeglicher Kopfbedeckung bei den Männern
des Volkes beziehen muss. Fast überall hat der Hut ursprünglich eine rein ceremonielle
und officielle Bedeutung und ist anfänglich das Vorrecht des Königs und seiner Granden.
Man denke an die mannigfaltige Reihe ceremonieller Kopfbedeckungen im alten China,
Korea und Japan, von denen sich noch viele Überreste bis jetzt bewahrt haben. Noch
heute sind die Hüte der chinesischen Beamten im wesentlichen Ceremonialobjekte, während
die Volksklassen im Sommer überhaupt keine Kopfbedeckungen tragen (ausgenommen die
breitkrämpigen Strohhüte der Landleute und Schiffer zum Schutz gegen die Sonne). Es war
daher ganz natürlich, wenn im alten China der Kaiser, der mit seinen Granden durch
rangbezeichnende Ceremonialhüte ausgezeichnet war, von dem „schwarzhaarigen Volke"
sprechen konnte, das keine Hüte trug und sein schwarzes Haar dem Anblick der stets auf
erhöhtem Standpunkt befindlichen Grossen offen aussetzte. Das schwarzhaarige Volk ist
nichts anderes als das hutlose Volk.

2) Im 10. Kapitel des *rGyal-rabs* wird ein ähnlicher Maskentanz zur Zeit des Königs
Sron-btsan-sgam-po beschrieben; dort werden Löwe, Yak-Löwe und Tiger-Löwe als Masken

## Brachten des Königs Majestät ihre Huldigung dar.

erwähnt. Die Löwentänze, die noch jetzt in fast allen buddhistischen Ländern, besonders in Birma, Siam, Tibet, China und Japan (ich selbst habe sie in Peking gesehen) aufgeführt werden, sind von grösster Bedeutung für die Geschichte der Wanderungen des Mimus. Es kann schon jetzt gezeigt werden, dass der indische Mimus ziemlich frühe, sicher bereits zur T'ang-Zeit, nach Tibet, Turkistan und China gewandert ist. Sehr beachtenswert in dieser Hinsicht sind die steinernen Löwensäulen in der Provinz Shensi, die sich nach meinen Beobachtungen auf der ganzen Route von T'ung-kuan bis Hsi-an fu und in grosser Menge in letzterer Stadt befinden, nach mündlichen Angaben von Chinesen auch in der Provinz Szech'uan vorkommen sollen. FORKE (Mitteilungen des Seminars für oriental. Sprachen, Bd. I, Abt. 1, 1898, S. 72) erwähnt sie für den Ort Chih-shui. Diese Steinsäulen, von viereckigem Querschnitt mit abgeflachten Ecken, gewöhnlich 1.5 m hoch, sind, zuweilen in Gruppen von zwei bis vier, vor den Fassaden der Häuser, in einer Entfernung von etwa 1 m aufgestellt. Stehen zwei Säulen vor einem Hause, so sind sie gerade vor den Ecken desselben aufgepflanzt, während eine einzelne Säule vor der Mitte desselben errichtet ist. Man begegnet ihnen allenthalben zerstreut in den Strassen von Hsi-an, Lin-tung, Wei-nan, und besonders in den beiden Vorstädten auf der Ost- und Westseite des letzteren Ortes. Einige dieser Säulen sind oben nur mit einem runden Knopf oder einem birnenförmigen Gegenstand versehen, dessen Spitze nach unten gerichtet ist. Die Mehrzahl aber ist mit dem Löwen von dem ausgesprochen indisch-buddhistischen Typus, mit Figuren von Affen und dem „Narren" der indischen Bühne geschmückt. Der Affe kommt entweder allein für sich oder auf dem Rücken eines Löwen reitend oder kauernd vor. Im ersteren Falle sitzt er aufrecht da und hält zwischen seinen Vorderpfoten einen nicht näher definirbaren Gegenstand, vielleicht eine Frucht, an der er kaut. Ein anderer Typus hält die Vorderpfoten über die Brust gekreuzt. Der löwenreitende Affe hält ein Objekt in der rechten Pfote, während die Linke sich am Löwen festhält (vergl. die in Turkistan gefundenen Affenfiguren: HOERNLE, A Report on the British Collection of Central Asian Antiquities, Part II, pp. 47, 48, 55; M. A. STEIN, Preliminary Report etc., pp. 30, 31) Da, wie wir sehen werden, der Löwe den Mimus symbolisirt oder der Mimus in der Maske des Löwen auftritt, so haben wir es hier unzweifelhaft mit dem Affen des Gauklers zu tun, den man so oft unter den Volksbelustigungen von China und Japan auf den Strassen bewundern kann (s. bes. ALBERT BROCKHAUS, Netsuke, Versuch einer Geschichte der japanischen Schnitzkunst, Leipzig 1905, S. 407). Der Mimus tritt entweder allein auf der Spitze der Steinsäule auf oder auf dem Rücken des Löwen reitend, zuweilen beide Hände in seinen Rachen streckend. Sein charakteristisches Attribut ist die weite, kegelförmige, spitze Narrenkappe, deren Zipfel nach vorne herunterfällt. Einige dieser Kappen zeigen über die Ohren herabgezogene Seitenklappen; auf mehreren Darstellungen läuft die Mütze in spiralförmigen Windungen aus, beinahe von der Form eines Schneckenhauses. Sein Gesicht trägt den melancholischen Ausdruck eines gutmütigen Humors. Nach den traditionellen Angaben der Ortsbevölkerung sollen diese Säulen aus der Zeit der T'ang-Dynastie stammen, was eine durchaus annehmbare Ansicht ist. Sicher sind sie nichts anderes als Repliken indischer Stambha's, und insbesondere der indischen Löwensäulen (siṁhastambha). Sind sie aber indisch, so folgt daraus, dass auch die Figur des Mimus ihren Ursprung aus Indien herleiten muss, wie mir auch durch die Darstellung desselben Mimus bestätigt wird, den ich

So waren der Herrscher und das ganze Volk

Ausgelassenem Jubel hingegeben;

Des Tanzes Stampfen war gleich dem Klatschen des Himmelregens;

In ihrer Ekstase waren sie wie von Sinnen,

Und jeder stimmte für sich ein Freudenlied an.

Da nun erhob sich der Gebieter der Menschen, der von den Göttern
eingesetzte

Schirmer der Lehre, er, K'ri-sroṅ-lde-btsan,

Von seinem edelsteingeschmückten Tron,

Und der König stimmte einen Freudensang an:

»Dieser mein Haupttempel mit den drei Stockwerken

Ist aus den fünf Kleinodien erzaubert,

Ungleich einem Bau von Menschenhand, wie von selbst emporge-
wachsen.

Mein Tempel ist höchst wundervoll,

---

wiederholt auf Reliefs steinerner Pai-lou in der Provinz Honan gesehen habe, Pai-lou,
deren architektonischer und decorativer Stil entschieden und ausgesprochen indisch ist. In
Turkistan wurde ein keramisches Fragment entdeckt, das nach HOERNLE (A Note on the
British Collection etc., p. 2) einen Mann darstellt, bekleidet mit einem Costüm, das auf-
fallend dem des mittelalterlichen Hofnarren gleicht. Dies ist dieselbe indische Darstellung
des Mimus wie die von mir oben aus China beschriebene. Der indische Mimus ist also
nach Turkistan und China gewandert, wie uns denn auch in der Geschichte der T'ang-
Dynastie ausdrücklich von der Ankunft indischer Gaukler in China erzählt wird (HIRTH,
Chinesische Ansichten über Bronzetrommeln, S. 56). Der gegenwärtige Löwentänzer ist
Gaukler und Spassmacher zugleich (GRUBE, Zur Pekinger Volkskunde, S. 111), ist daher
in Peking auch eine beliebte Kinderunterhaltung (HEADLAND, The Chinese Boy and Girl,
p. 143; Abbildung p. 144), ebenso in Japan (BROCKHAUS, l. c., S. 408; CHAMBERLAIN,
Things Japanese, p. 103), und in Tibet wird der Löwentänzer von einem „harlequin
mummer" eingeführt (WADDELL, The Buddhism of Tibet, p. 539). Das alles deutet darauf
hin, dass der Löwentänzer eine Form des indischen Mimus ist. Der indische Mimus als
Löwentänzer hat nun sicher auch den Anstoss zu dem künstlerischen Motiv des auf dem
Löwen reitenden Narren der erwähnten Löwensäulen gegeben. Warum erscheint nun auch
der europäische Harlekin so häufig mit der Maske des Löwen (O. DRIESEN, Der Ursprung
des Harlekin, Berlin 1904, S. 144, 169—170)? Sollte das blosser Zufall sein oder eher
eine von Indien entlehnte Idee?

Sein blosser Anblick ist ein Entzücken,

Jedes Herz ist daher von Glück erfüllt.

Nach dem Vorbild des Landes Pūrvavideha

Sind jene drei Tempel des Ostens

Aus den fünf Kleinodien erzaubert,

Ungleich ... (u. s. w., wie oben).

Nach dem Vorbild des Landes Godānīya

Sind jene drei Tempel des Westens

Aus den fünf Kleinodien erzaubert (u. s. w., wie oben).

Nach dem Vorbild des Landes Uttarakuru

Sind jene drei Tempel des Nordens (u. s. w., wie oben).

Nach dem Vorbild des Landes Jambudvīpa

Sind jene drei Tempel des Südens (u. s. w., wie oben).

Meine Tempel des oberen und unteren Yakṣa

Gleichen dem Aufgang von Sonne und Mond am Himmel.

Die drei Tempel meiner Gemahlinnen

Gleichen dem Bau eines Maṇḍala von Türkis.

Dieser mein weisser Stūpa

Gleicht der nach rechts gewundenen weissen Muschel;

Dieser mein roter Stūpa

Gleicht der zum Himmel emporlodernden Flamme;

Dieser mein blauer Stūpa

Gleicht einer Säule von Türkis;

Dieser mein schwarzer Stūpa

Gleicht einem in die Erde geschlagenen Eisenkeil.

Meine Stūpa's sind höchst wundervoll,

Ihr blosser Anblick ist ein Entzücken,

Jedes Herz ist daher von Glück erfüllt.''

Diesen und andere Freudengesänge

Sang er nach dem Muster der von den 33 Göttern

Gelehrten Lieder; und dann

Stimmte er die Weise vom »Türkispalast und dem Goldtron" an [1]).

Der Göttersohn Mu-ne btsan-po [2])

Sang die Weise von »der Leuchte der Welt";

Der Göttersohn Mu-tig btsan-po [2])

Sang die Weise von »dem trotzigen Löwen";

Die Königinnen stimmten ihre Lieder an,

Sangen die Weise von »den Wirbeln des türkisfarbenen Sees",

Sangen die Weise von »den türkisfarbenen Blättern und Zweigen".

Nun stimmten die Priester ihre Lieder an:

Der Gelehrte Bodhisattva [3])

Sang die Weise von »dem weissen Rosenkranz der Meditation" [4]);

Der Meister Padmasambhava

Sang die Weise von »der Überwindung der Dämonen" [5]);

Der Gelehrte Vairocana

Sang die Weise von »dem Hervorlocken der Musiktöne";

Nam-mkʻa sñiṅ-po von gCubs-pan

Sang die Weise von »dem am Himmel schwebenden Garuḍa";

bTsan-pa rgyal-mcʻog van Ńan-lam

Sang die Weise von »dem trotzigen Hayagrīva";

Der Gelehrte dPal-bzaṅ von Cog-ro

---

1) Dieses und die im folgenden Katalog aufgezählten Lieder sind gegenwärtig noch gangbare Volksweisen.

2) Sohn des Königs Kʻri-sroṅ.

3) Çāntirakṣita.

4) Es scheint also damals schon Volkslieder buddhistischen Inhalts gegeben zu haben.

5) Ganz in Übereinstimmung mit seinem traditionellen Beruf als Dämonenbezwinger.

Sang die Weise von »den neun Herrlichkeiten der Seligkeit".

Dann stimmten die Beamten ihre Lieder an:

Der grosse Herr der Lehre, der alte ạGos,

Sang die Weise von »dem stämmigen weissen Baume";

Sein Sohn Yab-lhag kʻri-bzaṅs

Sang die Einleitung zur »Magie der weisen Schrift";

Šud-pu dpal-gyi seṅ-ge [1])

Sang die Weise »dur cʻuṅ γyu riṅs";

bsTan-bzaṅ dpal-legs

Sang die Weise von »den acht Wünschen";

Kʻyuṅ-po dum-tsʻugs von mGar

Sang die Weise von »dem schwebenden Geier";

rDo-rje sprel-cʻuṅ aus dem Geschlechte mCʻims [2])

Sang die Weise »Ehrensang auf das Geschlecht mCʻims";

rGyal-tsʻa lha-gnaṅ aus sNa-nam [3])

Sang die Weise von »der erhabenen Götter-Ceder";

Kʻri-bzaṅ yaṅ-bon vom Geschlechte gÑags

Sang die Weise von »dem alles bescheinenden Mondlicht";

sTag-sgra klu-goṅ aus Li-ṅam [4])

Sang die Weise von »den sechs Quellen des Gangges".

Darauf sangen die adligen Herren ihre Lieder,

Sangen von »den wunderbaren Goldblumen".

Die Jünglinge sangen ihre Lieder,

Sangen von »den gelben Blumen des Tigerberges".

Die Nonnen sangen ihre Lieder,

---

[1]) Der vorher erwähnte Baumeister des weissen Stūpa.

[2]) Baumeister des blauen Stūpa.

[3]) Baumeister des roten Stūpa.

[4]) Baumeister des schwarzen Stūpa.

Sangen von »dem Blumen-Rosenkranz".

Die Ärzte sangen ihre Lieder,

Sangen die Weise »gron zer rin-mo".

Es ist unmöglich, jedes Lied zu nennen,

Das im Laufe eines ganzen Jahres

Jeglicher Mann gesungen hat.

In jener frohen Zeit eine Armspanne des Segens umfassend,

Schwelgten sie in der Freude und des Glückes Vollgenuss;

Diese Wohltat erfüllte wie das Licht von Sonne und Mond

Während eines ganzen Jahres alle Weltgegenden.

# *058*

掘金蚁的传说

# T'OUNG PAO

# 通報

OU

## ARCHIVES

*CONCERNANT L'HISTOIRE, LES LANGUES,*
*LA GÉOGRAPHIE ET L'ETHNOGRAPHIE*
*DE*
*L'ASIE ORIENTALE*

Revue dirigée par

**Henri CORDIER**
Membre de l'Institut
Professeur à l'Ecole spéciale des Langues orientales vivantes
ET
**Edouard CHAVANNES**
Membre de l'Institut, Professeur au Collège de France.

**SÉRIE II. VOL. IX.**

LIBRAIRIE ET IMPRIMERIE
CI-DEVANT
E. J. BRILL
LEIDE — 1908

# DIE SAGE VON DEN GOLDGRABENDEN AMEISEN

## BERTHOLD LAUFER.

———◦∞◦———

Das im Mahābhārata erwähnte Ameisengold und Herodots (III,
102—105) Erzählung von den goldgrabenden Ameisen haben eine
grosse Reihe scharfsinniger Untersuchungen hervorgerufen, die als
bekannt vorausgesetzt werden dürfen. Man kann nicht behaupten,
dass auch nur einer der vielen Deutungsversuche wirklich befriedi-
gend ausgefallen wäre. Nimmt man einfach, wie geschehen, eine
Verwechslung mit einem grabenden oder Goldsand aufwerfenden
Tiere an, so bleibt doch gänzlich unerklärt, warum man dieses Tier
nicht mit seinem eigentlichen Namen, sondern gerade als »Ameise”
bezeichnet hat, und geht man von der zoologischen zur ethnographischen Erklärungsmethode über, indem man in den goldgrabenden
Ameisen ein goldgrabendes Volk sieht, wie SCHIERN in scharfsinniger Weise dargelegt hat.[1]), so darf man nicht verabsäumen, den
Zusammenhang aufzudecken, der zwischen diesem Volke und der
Bezeichnung »Ameisengold” besteht oder bestanden hat, ein Versuch, den meines Wissens niemand zuvor unternommen. Dass das
Ameisengold in letzter Instanz auf irgendein Volk zurückgehen

---

1) FREDERIK SCHIERN, Über den Ursprung der Sage von den goldgrabenden Ameisen.
Kopenhagen und Leipzig, 1873.

29

muss, liegt auf der Hand und bedarf kaum eines Beweises. Dieses
Volk aber können nicht, wie Schiern geschlossen hat, tibetische
Goldgräber gewesen sein, wie sich aus dieser Untersuchung ergeben
wird. Es kommt bei der ganzen Frage überhaupt nicht so sehr auf
die Identifikation mit einem bestimmten goldgrabenden centralasiatischen Volke an als vielmehr darauf, zu zeigen, wie entweder
von diesem oder einem benachbarten Volke der eigentümliche Name
»Ameisengold" hat ausgehen können, und der Gradmesser für die
Richtigkeit der Identifikation wird in dem mehr oder minder glück-
lichen Erfolg gerade dieser Erklärung gesucht werden müssen, d. h.
da, wo die Möglichkeit einer plausiblen Interpretation der Entstehung
des Terminus »Ameisengold" gegeben ist, werden wir mit grösster
Berechtigung auch das Volk erblicken dürfen, welches in der Han-
delsgeschichte des Altertums der Träger des Ameisengoldes gewesen
ist. Die Idee des nach den Ameisen benannten Goldes muss im
Mittelpunkt des Problems stehen, darauf kommt alles an: vom
Ameisengold erzählen uns nicht nur Griechen und Inder, sondern,
wie wir sehen werden, auch Mongolen und Tibeter. Es kann daher
kein blosser Zufall sein, dass dieses Gold an den verschiedensten
Stellen der Welt zum »Ameisengold" gemacht worden ist, und die
Frage, warum es gerade »Ameisengold" genannt wurde, ist die
Achse, um die sich alles andere dreht.

Ohne mich vorläufig in eine Erörterung der früheren Ansichten
einzulassen, gehe ich zunächst in medias res, um neues Material
vorzulegen, das jene von selbst wesentlich modificiren wird.

Zuerst sei aus einer Stelle der Sage von Geser Chan gezeigt,
dass das Ameisengold den Mongolen bekannt gewesen ist. Der Text
ist der Ausgabe von I. J. Schmidt, p. 63, entnommen.

(1) *dung tsaghan* [1]) *aghūla bui. dung tsaghan aghūla du dung tsaghan churagha übär-yēn mailaju baichu buyu.*

(2) *altan aghūla buyu. altan aghūla du altan täghärmä übär-yēn ärghijü baichu buyu.*

(3) *tämür aghūla buyu. tämür aghūla du xükä xiris* [2]) *ükär übär-yēn toghlaju baichu buyu.*

(4) *altan aghūla buyu. altan aghūla du altan sabagha übär-yēn sabaju baichu buyu.*

(5) *dsäs aghūla buyu. dsäs aghūla du dsäs nochai übär-yēn chutsaju baichu buyu.*

(6) *altan aghūla buyu. altan aghūla du altan sono übär-yēn dürghiräjü baichu buyu.*

(7) *širgholjin-u chān üilä-dü churiyāksan xürmäk altan buyu.*

(8) *naran barichu altan tsalma.*

(1) Es gibt einen muschelweissen Berg. Auf dem muschelweissen Berge lebt ein von selbst blökendes muschelweisses Schaf.

(2) Es gibt einen goldenen Berg. Auf dem goldenen Berge ist eine sich von selbst drehende goldene Mühle.

(3) Es gibt einen eisernen Berg. Auf dem cisernen Berge lest ein von selbst umherspringendes türkisblaues Rind.

(4) Es gibt einen goldenen Berg. Auf dem goldenen Berge befindet sich ein von selbst schlagender goldener Stock.

(5) Es gibt einen Kupferberg. Auf dem Kupferberg befindet sich ein von selbst bellender kupferner Hund.

(6) Es gibt einen goldenen Berg. Auf dem goldenen Berge befindet sich eine von selbst summende goldene Bremse.

(7) Es gibt *Gold* in Klumpen, welches der *König der Ameisen* in seiner Tätigkeit angesammelt hat.

(8) Es gibt eine goldene

---

1) Übersetzung von tib. *duṅ dkar-po.*

2) SCHMIDT übersetzt *xükä xiris* durch „eisenblau", was ihm wohl der Parallelismus zu den übrigen Beispielen nahe gelegt hat. KOWALEWSKI und GOLSTUNSKI jedoch weisen in ihren Wörterbüchern dem Worte *xiris* für diese Stelle, an der allein dasselbe vorzukommen scheint, die Bedeutung „Türkis" zu.

*saran barichu müngghün tsalma.*

(9) *širgholjin-u chabar-in čisün nighän chaldsa.*

(10) *büghäsün-ü širbusun nighän atchu.*

(11) *ärä charabtur šibaghūn-u chabar-un čisün nighän chaldsa.*

(12) *ämä charabtur šibaghūn-u kükän-i sün nighän chaldsa.*

(13) *dsultsagha charabtur ši-baghūn-u nidün-ü nilbusun nighän chaldsa.*

(14) *yäkä dalai dotoraki bulu čilaghūn-i činäghän šiküsü-tü molor ärdäni.*

---

Schlinge, die Sonne zu fangen, eine silberne Schlinge, den Mond zu fangen.

(9) Es gibt ein Fläschchen mit dem Nasenblut von Ameisen.

(10) Es gibt eine Handvoll Sehnen von Läusen.

(11) Es gibt ein Fläschchen mit dem Nasenblute des männlichen schwarzen Adlers.

(12) Es gibt ein Fläschchen mit der Milch aus den Brüsten des weiblichen schwarzen Adlers.

(13) Es gibt ein Fläschchen mit den Tränen aus den Augen der jungen schwarzen Adler.

(14) Im Innern des grossen Meeres gibt es einen Saft enthaltenden Krystalledelstein von der Grösse einer steinernen Dreschwalze.

---

Diese vierzehn Schätze werden dem Geser angeboten für die Freilassung der Seele des Čoridong Lama, die er in Gestalt eines Insekts in seiner Hand halt. Es wird aber im weiteren Verlaufe dieses Buches nicht berichtet, ob Geser die Kleinodien tatsächlich erhält. Unter diesen wird also als siebentes das vom Ameisenkönig angesammelte Gold aufgezählt. [1]) Im dritten Buche unserer Helden-

---

1) Meine Übersetzung weicht in zwei Punkten von der SCHMIDT's (S. 54) ab, welcher diesen Passus mit den Worten übertragen hat: „Es gibt ferner Goldstaub, den der Ameisenkönig sich zum Bedarf gesammelt hat". Das Wort *üilä* bedeutet indessen niemals „Bedarf", sondern nur „Werk, Arbeit, Beschäftigung", dann insbesondere die gewöhnliche,

sage werden alle hier aufgezählten Kostbarkeitem nochmals in einem anderen Zusammenhange genannt. Geser wird zur Aufheiterung eines sagenhaften Herrschers Kümä Chaghan, der über den Tod seiner Gemahlin in einen bedenklichen Zustand von Melancholie verfallen ist, nach China berufen und stellt zur Bedingung, dass ihm zunächst jene Schätze herbeigeschafft werden. Nun muss man sich die Tatsache vorgegenwärtigen, dass die uns vorliegende Recension der mongolischen Sage durchaus kein einheitliches, aus einem Gusse geschaffenes Werk vorstellt. Es zeigt sich vielmehr deutlich, dass das Buch aus verschiedenen, oft nur lose unter einander verbundenen Teilen zusammengesetzt ist, die auf verschiedene Überlieferungen oder Handschriften zurückgehen. Leider verfügen wir noch nicht über eine eingehende kritische Analyse dieser einzelnen Bestandteile. Um nur ein Beispiel für die Tatsache anzuführen, dass diese zum Teil ihren Ursprung in abweichenden Quellen haben, sei darauf hingewiesen, dass Geser am Schlusse des ersten Buches seiner Gattin Rogmo Gō einen kurz zusammenfassenden Abriss aller seiner im Vorhergehenden beschriebenen Taten vorträgt, aber in dieser Inhaltsübersicht ein Abenteuer mit der Tochter des Drachenfürsten, Adju Märghän, beschreibt, von dem vorher gar nicht die Rede gewesen ist; als handelnde Person tritt diese Amazonin erst im fünften und sechsten Buche auf. [1]) So liegen auch, was die oben

---

regelmässige Berufstätigkeit; *üilä-dü* entspricht vollständig der tibetischen Verbindung *lassu* d. h. in Ausübung seiner gewohnten Tätigkeit, berufsmässig, kraft seines Amtes, ex officio. Das Goldsammeln wird demnach als Geschäft des Ameisenherrschers angesehen. SCHMIDT übersetzt *kürmäk altan* mit „Goldstaub", was die Lexika keineswegs bestätigen. KOWALEWSKI, Dictionnair mongol-russe-français, p. 2651 gibt zu *kürmek* nichts anderes an als „ein Haufen verwelkter Blätter"; GOLSTUNSKI dagegen weist in seinem lithographirten mongolisch-russischen Wörterbuch, vol. III, p. 486, diesem Ausdruck für die betreffende Stelle die Bedeutung „ganzes Stück Gold" zu, die denn auch das Richtige zu treffen scheint. „Goldstaub" heisst übrigens gewöhnlich *chumaki altan* von *chumak* „feiner Sand". Anstatt des im Texte von SCHMIDT, p. 35, Zeile 9, dastehenden intransitiven Verbums *churaksan* hsbe ich die hier erforderliche transitive Form *churiyäksan* nach der Parallelstelle p. 64, Zeile 5, wiederhergestellt.

1) Vergl. SCHOTT, Abhandl. der Berliner Akademie, 1851, S. 571.

citirten Stelle betrifft, Widersprüche des dritten zum ersten Buche vor. Nach diesem sollte man eigentlich bereits annehmen, dass Geser in den Besitz jener Schätze gelangt sein müsste. Noch auffallender aber ist folgender Umstand. Da die chinesischen Minister ausserstande sind, Geser's Wünsche zu erfüllen, steigt er auf einer von oben herabgelassenen Kettenleiter zu seiner Grossmutter Absa Gürtsä in den Himmel und berichtet ihr, dass ihm sein Weib Rogmo Gō erzählt habe, ihr Vater Sängäslü Chaghan sei im Besitz der genannten Kostbarkeiten. Im ersten Buche ist dagegen nur gesagt, dass dieselben das Lösegeld des Čoridong Lama darstellen; von einer Beziehung der Schätze zu Rogmo oder deren Vater ist im Vorausgehenden nichts erwähnt. So dürfte daher wohl anzunehmen sein, dass diese Variante des dritten Buches auf einen anderen Bericht, oder wenn man sich so ausdrücken darf, auf einen anderen Autor zurückzuführen ist. Dafür lassen sich im einzelnen noch folgende Gründe aufführen. Bei der Aufzählung der Schätze gegenüber dem Abgesandten der chinesischen Minister kommen eine Reihe von der parallelen Stelle im ersten Buche abweichender Lesarten vor, bei deren Anführung ich mich auf die in diesem Falle allein in Betracht kommende Lectio *üli-dü* statt *üilä-dü* in (7) beschränken kann; *üli* bedeutet »Nest, Ameisenhaufen". Ferner ist hier die Reihenfolge insofern verändert, als das Ameisengold seine Stelle zwischen der goldenen Sonnenfangschlinge einerseits und der silbernen Mondfangschlinge anderseits findet. Trotzdem lässt sich der Gedanke nicht abweisen, dass diese Partie bei der vorhandenen Übereinstimmung in den Hauptsachen dieselbe Grundlage hat wie die entsprechende im ersten Buch. Dies kann jedoch nicht von der folgenden Episode gelten, die Geser's Besuch bei seiner himmlischen Grossmutter erzählt. Denn bei der hier gegebenen Aufzählung der Schätze wird des Ameisengoldes nicht gedacht, und zwar, wie aus dem weiteren Verlaufe dieser Erzählung deutlich wird, aus einem

ganz bestimmten Grunde. Der Verfasser dieser Episode lässt nämlich seinen Helden von jenen Kostbarkeiten zu einem wohl beab-
sichtigten Zweck Besitz ergreifen. Der Herrscher von China will
ihn zur Strafe dafür, dass er ihm den Leichnam seiner Gemahlin
entrissen, von dem er sich nicht trennen konnte, auf grausame Art
sterben lassen indem er ihn in eine Schlangengrube, eine Wespen-
höhle, ein finsteres Loch u. s. w. werfen lässt; doch durch die Zauberkraft seiner Schätze besteht Geser siegreich alle diese Gefahren.
Da nun das Ameisengold mangels solcher magischen Eigenschaften
in diesem Zusammenhange nicht verwertbar war, so konnte jenes
bei der Anführung der Schätze gegenüber Absa Gürtsä als überflüssig übergangen werden. Aus dieser Erörterung ergibt sich nun
ein für unsere Betrachtung sehr wichtiger Schluss. Das Ameisengold ist nicht wunderwirkend, kein Talisman, kein Retter in Be-
drängnis wie alle übrigen Schätze, deren durchaus märchenhafter
Charakter sich ohne weiteres aufdrängt; es gehört also nicht ohne
weiteres in das Bereich der frei erfundenen, aus der Phantasie ge-
schöpften Fabel, es muss ihm vielmehr eine realere Vorstellung
anhaften. Wir begegnen also zwei Variationen desselben Themas
in der Gesersage. Die berühmten Kleinodien werden einmal mit
Einschluss des Ameisengoldes als Sühn- und Lösegeld, sodann mit
Ausschluss desselben als märchenhafte Rettungsmittel in Lebensge-
fahr genannt. Was wir aus dem mongolischen Geserepos erfahren,
ist die Vorstellung, dass der König der Ameisen berufsmässig oder
(nach einer anderen Lesart) in seinem Neste Stücke oder Klumpen
Gold anhäuft, und dass dieses Gold zu den hervorragenden Schät-
zen (*ärdäni*) gerechnet wird, aber keineswegs zu den fabelhaften.
Beachtenswert ist ferner, dass irgendwelche Erzählung darüber, welche
nähere Bewandtnis es mit diesem Ameisengolde hat, im ganzen
Buche nicht gegeben wird, dass dasselbe vielmehr nur in jener
descriptiven Reihe erscheint, gleichsam als das Facit eines bekann-

ten Berichts. Es ist also daraus zu schliessen, dass es sich um eine der Volksvorstellung wohl vertraute Erscheinung handelt, die einer Erläuterung ebensowenig bedurfte als die goldene Mühle, welche von selbst mahlt, oder die Schlingen, welche Sonne und Mond fangen. Die blosse Erwähnung des Ameisengoldes als eines gegebenen Faktums setzt das Vorhandensein einer ausführlicheren geläufigen Tradition voraus, deren allgemeine Kenntnis von vornherein angenommen wird. Schon daraus ist zu vermuten, dass diese Überlieferung keine entlehnte sein dürfte. Sicherlich ist sie nicht aus indischen Quellen geschöpft. Denn indische Stoffe sind in den Sagen von Geser überhaupt nicht vorhanden, wenn auch in manchen der uns bisher zugänglichen Redaktionen eine buddhistische Färbung verwaltet. Das Grundmotiv wie die einzelnen Gegenstände der Sage müssen als ein echtes und ursprüngliches Erzeugnis centralasiatischer Stämme angesehen werden, denn sie schildern in einem getreuen Spiegel die alten Kulturzustände dieser Völker, ihr Leben, ihre Sitten, ihre Kämpfe. Mit Recht behauptete bereits SCHOTT (l. c. S. 286): »Die Gesersage kann nur entweder aus Tibet stammen oder aus der Mongolei; allein das Verhältnis des tibetischen Textes zum mongolischen ist noch nicht aufgeklärt." Leider trifft dieser letzte Satz auch noch heute zu, nachdem ein halbes Jahrhundert seit jener Abhandlung verflossen ist. Man kann sich daher einstweilen mit allem Vorbehalt nur so ausdrücken, dass der Bericht vom Ameisengolde dem weit verzweigten tibetisch-mongolischen Sagenkreise von Geser angehört. Noch schwieriger als die Frage nach der Heimat dieser Stoffe ist die Frage nach ihrer zeitlichen Entstehung. Die uns vorliegende mongolische Redaktion ist verhältnismässig spät abgeschlossen worden, wie z. B. aus der Erwähnung des Kanjur und Tanjur hervorgeht, dagegen tragen die einzelnen Sagen, wenigstens teilweise, die Speeren eines hohen Altertums an sich, wie vor allem aus dem Stande der Kultur geschlossen werden kann, den sie widerspiegeln. Wenn sich vor der Hand auch noch keine bestimmte Epoche weder für die Bildung der Sagen noch für das Sagenzeitalter selbst bestimmen lässt, so scheint doch das eine wenigstens gewiss zu sein, dass die Entstehung jener in eine Periode zurückreicht, die weit vor den Anfängen geschichtlicher Aufzeichnungen in Tibet und damit auch in der Mongolei fällt.

Der Ameisenkönig spielt auch noch gegenwärtig in den Volkssagen der Mongolen eine Rolle. Nach der Vorstellung der Burjaten gibt es sechs über das Tierreich herrschende Könige[1]: 1) den König der Menschen, 2) den König der Vögel, den Vogel Khanchärägdä, 3) den König der Vierfüssler Argalan-zon, 4) den König der Schlangen, die Schlange Abarga, 5) den König der Ameisen Šoro-tämän-šurgaldjin-chan, und 6) den König der Fische, den grossen Fisch Abarga. Der Ameisenkönig erscheint ferner in der burjatischen Sage von Garyulai-Morgon[2]), in der die verschiedensten Motive unter einander verwebt worden sind. Bei der hier in Betracht kommenden Partie handelt es sich um das Motiv der »Proben des Schwiegersohns". Den drei Bewerbern wird die Aufgabe gestellt, in einem Topf vermischte rote und weisse Hirse reinlich auszuscheiden und jede Sorte in einen Topf für sich zu sammeln. Auf den Rat seines Rotfuchses ruft der Held den Ameisenkönig Biltagar-kara-šurgaldjan zu Hülfe, der mit seinen Untertanen erscheint und den anderen unsichtbar die Arbeit ausführt.[3])

Die Sage von den goldgrabenden Ameisen muss in Tibet in verhältnismässig früher Zeit bekannt gewesen sein. Dafür lässt sich als Beleg eine merkwürdige Erwähnung des Ameisengoldes im

---

1) CHANGALOV und SATOPLAJEV, Burjatische Erzählungen (russisch), in Denkschriften der Ostsibirischen Abteilung der Kaiserl. Russischen Geographischen Gesellschaft, Band I, Heft 1, Irkutsk, 1889, p. 119.

2) Eine Variante zu der altaischen Erzählung von Altaïn Sain Salam bei W. RADLOFF, Proben der Volkslitteratur der türkischen Stämme Süd-Sibiriens, Band I, S. 12—28.

3) CHANGALOV und SATOPLAJEV, l. c., p. 42.

Munde des Königs K'ri-sroṅ-lde-btsan (zweite Hälffte des achten Jahrhunderts) anführen, die in der officiellen Geschichte der tibetischen Königsdynastie (*rGyal-rabs gsal-bai me-loṅ* verfasst im Jahre 1327, fol. 86, 2—8) enthalten ist. Ich lasse die bisher unveröffentlichte Stelle in Text und Übersetzung folgen. Es handelt sich um die Zeit nach der Ankunft des Padmasambhava, als der König den Plan fasste, das später so berühmte Kloster bSam-yas zu gründen.

(1) *De-nas rgyal-poi t'ugs la dpal bSam-yas mi-ạgyur lhun-gyis grub-pai gtsug-lag-k'aṅ bžeṅs-par ạdod-nas, blon-po ạGos la sogs-pai c'os blon rnams daṅ gros mdzad-de,*

*dka-bai las-rnams ma bsten-na,*

*sla-bai las-la grub-pa med,*

*gsuṅs-te,*

(1) Darauf entstand im Herzen des Königs der Wunsch, das herrliche bSam-yas, das unvergängliche durch Zauberkraft geschaffene Kloster [1]) zu gründen, und bei einer Beratung mit dem Minister ạGos und den übrigen buddhistischen Ministern, sagte er:

»Wer nicht nach schweren Werken trachtet,

Vollendet auch die leichten nicht." [2])

(2) *mṅa-og-gi blon-po Bod ạbaṅs-rnams bsags-te, rgyal-poi žal-nas sṅon-c'ad Bod-kyi rgyal-po byon-pa-las ṅa c'e, ṅas lag-rjes c'en-po cig ạjog dgos.*

(2) Da versammelte der König die Minister des Reiches und das Volk von Tibet und sprak: »Ich bin grösser als die früheren Könige von Tibet, die dahingegangen sind, und muss daher ein grosses Denkmal hinterlassen.

(3) *rgyu šel-las grub-pai mc'od-*

(3) Soll ich einen Stūpa, ganz

---

1) Dies ist der officielle Titel des Klosters (s. GRÜNWEDEL, Mythologie des Buddhismus, S. 54).

2) Die Anwendung volkstümlicher Sprüchwörter in Versen ist in den tibetischen Annalen sehr häufig. Vergl. T'oung Pao, 1901, p. 36.

rten šar ri dań mñam-pa cig
ạc'os-sam.

(4) Žań-poi-rgyai-yul mt'oń-
bai mk'ar-cig brtsig-gam.

(5) Has-po ri zańs-kyis ạt'um-
mam.

(6) dKa-bcu t'ań-la k'ron-pa
ạdom dgu brgya dgu bcu yod-pa
bcu yod-pa ạbru-am.

(7) Wa-luń grogs-mo gser
p'yes ạgeńs-sam.

(8) gTsań-po sbubs-su ạjug-
gam:

(9) dKon-mc'og gsum-gyi rten
bžugs-pai lha-k'ań bre ts'ad tsam
cig bžeńs-sam ạdoms šig gsuńs-pas.

(10) der t'ams-cad-kyis rgyal-
poi bka de gser-gyi p'a-boń bžin
sems t'og-tu lji-bar byuń-nas, žu
ma šes-par mig hu-re lus-so.

aus Krystall bestehend, dem Ostberg an Grösse gleich, errichten?

(4) Soll ich ein Schloss erbauen, von dem man bis nach
China, dem Lande meines Oheims,[1])
schauen kann?

(5) Soll ich den Has-po Berg[2])
mit Kupfer überziehen?

(6) Soll ich in der Ebene von
dKa-bcu[3]) einen 990 Klafter tie-
fen Brunnen graben?

(7) Soll ich Wa-luń[4]) mit
dem *Ameisen-Goldstaub* anfüllen?

(8) Soll ich den Brahmaputra
in eine Höhle stecken?

(9) Oder soll ich einen Tempel als Sitz der drei Kleinodien
(*triratna*), nicht grösser als 1 *bre*,
errichten? Wählet!" So sprach er.

(10) Auf dieses Wort des
Königs hin legte es sich allen
schwer wie ein Goldklumpen auf
die Seele, und unfähig zu ant-
worten, sassen sie da mit starren-
den Augen.

---

1) Der Oheim des Königs war der Kaiser Chung Tsung (684—709) der T'ang Dynastie.

2) Nicht weit vom Kloster bSam-yas.

3) Wörtlich: Ebene der zehn Askesen, eine Sandfläche nahe beim Kloster bSam-yas. Nach KOWALEWSKI (Dict. mongol, p. 2430b) bezeichnet *dKa-bcu* einen wissenschaftlichen Grad.

4) *Wa-luń* bedeutet wörtlich „Fuchs-Tal". Sollten die „Ameisen des Fuchstales" irgendwie zu den fuchsgrossen Ameisen des Herodot (μύρμηκες μεγάθεα ἔχοντες κυνῶν μὲν ἐλάσσονα, ἀλωπέκων δὲ μέζονα, Herodotus III, 102) in Beziehung stehen? CHANDRA DAS (Tibetan-English Dictionary, p. 1062) erklärt *Wa-luń* als „a district in East Nepal inha-

(11) *der blon-po aGos dan, Žan-ñan-bzan dan, gÑer-stag-bstan-adon gzigs-la sogs-pai c'os blon-rnams bžans-te, rje gcig lags,*

(11) Endlich dämmerte es dem Minister aGos, Žan-ñan-bzan und gÑer-stag-btsan-adon, und indem auch die übrigen buddhistischen Minister aufstanden, sprachen sie: »Unser lieber Herr!"

(12) *šel-gyi mc'od-rten šar ri dan mt'o dman mñam-pa ni skye-ba adi-la grub-pa mi srid.*

(12) Ein Krystall-Stūpa, an Höhe dem Ostberg gleich, ist in dieser einen Geburtsperiode zu vollenden unmöglich.

(13) *rGya yul mt'on-bai mk'ar bsams-pas agrub-pa mi srid.*

(13) Die Ausführung des Planes, ein Schloss zu erbauen, von dem man bis nach China schauen kann, ist unmöglich.

(14) *Bod mña-rigs-kyi zans t'ams-cad bsdus-kyan, Has-po ri at'ums-pa med.*

(14) Wenn man auch alles Kupfer des Reiches von Tibet ansammelte, so bedeckt es nicht den Has-po Berg.

(15) *Wa-lun grog-mo gser bye-mas kyan mi k'ens.*

(15) Wa-lun wird nicht von dem *Ameisen-Goldsand* angefüllt.

(16) *k'ron-pa adom dgu brgya bas brgya yan abru mi t'ub.*

(16) Einen Brunnen kann man nicht hundert, geschweige denn neunhundert Klafter tief, graben.

(17) *gTsan-po sbubs-su dgun ts'ud kyan dbyar mi ts'ud.*

(17) Was das Einschliessen des Brahmaputra in eine Höhle betrifft, so geht dies vielleicht im Winter, im Sommer geht es aber nicht.

---

bited mainly by Tibetans lying just where the river Arun coming from Tibet enters the Himalayan gorges to join the Kosi rivir". Es wird sich aber an unserer Stelle wohl um eine in Tibet gedachte Lokalität handeln.

(18) *de-bas rjei t'ugs-dam,* | (18) Daher, o heiliger Herr,
*ạbaṅs-kyi skyabs-gnas, p'an-bde* | Hort des Volkes, Urquell jegli-
*t'ams-cad ạbyuṅ-bai gnas, gtsug-* | chen Heils,
*lag-k'aṅ bre ts'ad tsam cig bžeṅs-* | soll ein Kloster, nicht grösser
*pa lags žus-pas.* | als 1 *bre*, erbaut werden. So ent-
 | worteten sie.

(19) *der t'ams-cad k'a-ạc'am-* | (19) Da stimmten alle zu und
*par de-bžin legs-so žes gdao.* | sagten, dass es so gut sei.

Wir haben also hier denselben Fall vor Augen wie in der mongolischen Sage, nämlich dass auch den Tibetern das Ameisengold als eine geläufige Sache bekannt war, denn der König setzt bei blosser Nennung des Namens desselben seine Bekanntschaft bei den Hörern voraus, aus deren Antwort hervorgeht, dass sie verstehen, was er meint.

Schliesslich kann ich noch ein drittes Dokument zur Kenntnis der Sage bei den Centralasiaten vorlegen, das ich der Freundlichkeit des Missionars A. H. FRANCKE, des besten Kenners der Volkssagen und der einheimischen Religion von Ladākh, verdanke. Auf eine diesbezügliche Anfrage hin teilte mir Herr Francke in einem Briefe vom 27. April 1901 aus Leh die folgende von ihm dort aufgezeichnete Sage mit:

»Ein König, namens K'ri-t'ob, wollte seine Tochter Ñi-dar-mk'an an den Minister dKar-ri-t'ob verheiraten und verlangte von ihm einen hohen Kaufpreis. Dieser sagte, er würde den hohen Preis nur zahlen, wenn alles Hausgerät, das die Tochter mitbringen musste, von Gold wäre. Der König kam in Verlegenheit, weil er gar kein Gold hatte. Man riet ihm, er solle den Ameisenkönig fangen und das Gold von den Ameisen aus dem Meere holen lassen. Ein Lama machte Regen, und die Ameisen kamen zum Vorschein. Man fing einige grosse, unter welchen sich wirklich der König befand. Als diesem mit dem Leben gedroht wurde, befahl er allen Ameisen, das Gold aus dem Meer zu holen. Diesen wurde ein Draht um den Leib gebunden (daher der tiefe Einschnitt), und so wurden sie hinabgelassen. Es kam viel Gold zum Vorschein, und die Hochzeit wurde fröhlich gefeiert."

Man sieht, die Sage van goldsammelnden Ameisen ist in Ladākh bekannt; der Zug, dass sie das Gold aus dem Meere holen, ist freilich auffallend und vielleicht eine spätere Umdeutung, zum Teil durch das Bestreben veranlasst, die Form des Ameisenkörpers zu erklären.

Die Chinesen wissen viel des Interessanten über Ameisen, auch über streitbare Ameisen, zu erzählen [1]), doch sie besitzen keine Nachrichten über goldgrabende Ameisen, was jedenfalls darauf hinweisst, dass das ursprüngliche Verbreitungsgebiet der Sage auf Centralasien beschränkt gewesen ist. Dagegen findet sich eine Stelle im Commentar zu den Elegieen von Ts'u, der zufolge »es in der Wildnis der westlichen Gegenden rote Ameisen gab, die so gross waren wie Elephanten und Menschen zu töten vermochten." [2]) Vielleicht dürfte es nicht zu gewagt erscheinen, bei dieser Anspielung auf die westlichen Gegenden an einen fernen Nachhall einer centralasiatischen Sage von einem goldgrabenden Ameisen-Volke zu denken, das, nach den Berichten der Alten zu schliessen, aus recht streitbaren Männern bestanden haben muss.

Wir sehen also, dass das Ameisengold den Mongolen sowohl

---

1) S. *T'u shu tsi ch'éng*, Sect. 19, Kap. 189. Einige Notizen, wenn auch nicht in zuverlässiger Übersetzung, bei A. PFIZMAIER, Denkwürdigkeiten von den Insecten China's, Wien 1874, S. 41—45.

2) Da mir der Originaltext augenblicklich nicht zur Hand ist, kann ich die Stelle leider nur nach 格致鏡原, Kap. 98, p. 5*b*, citiren, wo es heisst: 楚詞。赤蟻如象。注。西方曠野有赤蟻其大如象能殺人。

als Tibetern, und anscheinend seit alter Zeit, bekannt gewesen ist
und in ihren Traditionen eine Rolle spielt. Aus dieser Tatsache ist
eine wichtige Schlussfolgerung zu ziehen, nämlich die, dass der
Ausdruck *pipīlika* (Ameisengold) nicht erst in Indien entstanden
sein kann, sondern als Äquivalent, als Übersetzung eines entsprechenden aus einer centralasiatischen Sprache (welcher? — lässt sich
naturgemäss nicht mit Bestimmtheit sagen) stammenden Ausdrucks
herübergenommen worden ist. Das will ferner besagen, dass die
Inder auch die ganze Sage von den goldgrabenden Ameisen aus
Centralasien entlehnt haben, und dass alle früheren Erklärungsver-
suche, die eine Verwechslung von Ameisen mit Murmel- oder anderen Tieren [1]) annahmen, nicht correct sind. Die Inder erzählten
vom Ameisengolde, und Herodot berichtete von goldgrabenden Ameisen, eben aus dem und keinem anderen Grunde, weil das Ameisengold aus Centralasien unter dem dort gebräuchlichen Namen »Ameisengold" gebracht wurde. Hier hätten wir denn wieder eine glän-
zende Ehrenrettung des grossen Herodot zu verzeichnen, der durchaus
nichts missverstanden, nichts verwechselt, sondern vollkommen richtig gehört und wiedergegeben hat. Von einer Substitution von Wörtern, von einer Vertauschung von Tiernamen kann in diesem Falle
ganz und gar keine Rede sein, dafür bürgt — und darauf ist grosser Nachdruck zu legen — die in den tibetischen Wörterbüchern
verzeichnete Gleichung Sanskrit *pipīlaka* = tibetisch *grog-ma* (oder
*grog-mo*), die sich schon in der Mahāvyutpatti findet [2]), und dieselbe
Gleichung mit Hinzufügung des mongolischen *širgholjin* im *Li-šii-
gur-n'ai* [3]) fol. 22*a*). Diese drei Wörter bedeuten »Ameise" und

---

1) So z. B. ALEXANDER CUNNINGHAM, Ladák, London 1854, pp. 232—33; LASSEN,
Indische Altertumskunde, Bd. I, S. 850; RAWLINSON, History of Herodotus, Vol. II, p.
492; O. SCHRADER, Sprachvergleichung und Urgeschichte, II. Teil, Die Metalle, 3. Aufl., S. 34.

2) S. auch L. FEER, Le Karma-Çataka (Extrait du Journal Asiatique, Paris, 1901),
pp. 81, 82.

3) S. Sitzungsberichte der Bayer. Akademie, 1898, S. 523.

sind eben diejenigen Wörter, die in jeder dieser drei Sprachen zur Bildung als Begriffs »Ameisengold" verwendet werden.

Ich verhehle mir nicht, dass man meiner Theorie gegenüber einen gewichtigen Einwand machen kann, nämlich den, dass die von mir citirten mongolischen und tibetischen Quellen verhältnismässig jung sind und keinen Vergleich mit der weit älteren griechischen und indischen Berichten aushalten, und dass es daher vom rein historischen Standpunkt berechtigter erscheinen möchte, die centralasiatischen Traditionen von der indischen, nicht aber umgekehrt, abzuleiten [1]). Ein solcher Einwand ist völlig berechtigt und wird für den bestehen bleiben, der auf dem extremen chronologisch-historischen Standpunkt verharrt. Es lassen sich aber zwei Erwägungen anstellen, die meine Ansicht, wenn nicht zu einem endgültig festgelegten historischen Faktum, so doch zu einem gewissen Grade von Wahrscheinlichkeit erheben.

Einmal ist doch zu erwägen, dass wir eben nicht ältere historisch definirbare Berichte aus centralasiatischen Litteraturen beibringen können als die Periode schriftlicher Fixirung reicht, und es käme nun alles darauf an zu zeigen, in wieweit in den ältesten erreichbaren Denkmälern dieser Litteraturen Traditionen erhalten sind, die weit über die Zeit ihrer Abfassung hinausreichen. Mit anderen Worten, wir stehen dann vor der Principienfrage, wie lange sich überhaupt mündliche Überlieferungen in einem Volke erhalten, und mit welchem Grade von Treue und Genauigkeit sie sich durch das Gedächtnis von Generationen fortpflanzen. Diese

---

1) Das wäre indessen schwer zu erweisen, da die in Tibet und der Mongolei vorhandenen indischen Vorstellungen sämtlich aus der indischen-buddhistischen Litteratur stammen, und in dieser meines Wissens das Ameisengold nicht erwähnt wird. Obwohl ich T'oung Pao, 1901, S. 27 und 32, gezeigt habe, dass sich fragmentarische Traditionen aus dem Rāmāyana und Mahābhārata in der tibetischen Litteratur finden, so ist es doch nicht erweislich, ja in hohem Grade unwahrscheinlich, dass das ganze Mahābhārata je nach Tibet gelangt sei.

Frage lässt sich freilich nicht mit einem Schwertstreich entscheiden; es werden sich in dieser Hinsicht auch nur wenige allgemeine Principien aufstellen lassen. Die Verhältnisse liegen bei den einzelnen Volkern verschieden; Gedächtniskraft, Güte, Umfang und Tiefe der Tradition sind nicht überall gleich. Auch dem persönlichen Eindruck steht dabei ein weites Tor offen. Aber nach Analogieen und Erfahrungen zu schliessen, hat es nichts so unwahrscheinliches an sich, dass diese Tradition bei Mongolen und Tibetern aus uralter Zeit herabgekommen sein mag und, als sie Schrift und Litteratur erhielten, schriftlich aufgezeichnet wurde. Bessere Bekanntschaft mit ihren zum grössten Teil noch verschlossenen Litteraturen wird uns ja wohl auch ausfürlichere Versionen der Sage bringen. Zweitens scheint mir eine Erwägung des gesunden Menschenverstandes hinzuzukommen. Wenn die Berichte der Inder und des Herodot in gleicher Weiser auf Centralasien als das Ursprungsland des Ameisengoldes hindeuten, wäre es beinahe absurd, annehmen zu wollen, dass die Sage von den goldgrabenden Ameisen von Indien nach Centralasien gleichsam zurückgewandert sei. Wenn das Ameisengold aus Centralasien kam, dann ist mit denkbar grösster Wahrscheinlichkeit vorauszusetzen, dass dann auch die damit verknüpfte Tradition gleichfalls aus derselben Gegend kam, d. h. also in Centralasien entstanden ist. Ich muss gestehen, dass selbst ohne das Vorhandensein diesbezüglicher centralasiatischer Traditionen der centralasiatische Ursprung der Sage mir wahrscheinlich wäre, und dass ich stets, lange bevor ich jene kannte, diese Ansicht gehabt habe. Ich gebe gern zu, dass meine Ansicht infolge des Mangels einer concreten chronologischen Basis hypothetisch bleibt, aber eine Hypothese, die sich einem nicht geringen Grade von Wahrscheinlichkeit und Gewissheit nähert.

SCHIERN hat seine Untersuchung mit den Worten geschlossen: »Uns klingt die Erzählung nicht mehr als ein Wunder: 'Die gold-

30

grabenden Ameisen' haben ursprünglich nicht Tiere bezeichnet, weder wirkliche Ameisen — wie das naive Altertum annahm — noch andere grössere Tiere, mit denen sie wegen des Grabens oder Aussehens derselben verwechselt sein sollen — wie so viele berühmte Gelehrte bis auf unsare Tage gemeint haben —, sondern vollblütige Menschen: tibetanische Goldgräber, die im fernen Altertum lebten und sich zeigten, wie sie noch jetzt leben und sich zeigen". Das klingt ganz gut, wenn nicht der eine wunde Punkt ware, dass wir über das »ferne Altertum" der tibetischen Goldgräber einfach gar nichts wissen. Schiern hat nur die modernen Verhältnisse des gegenwärtigen Tibet, hauptsächlich nach den Berichten der reisenden Pandits der englischen Regirung, seiner Darstellung zu Grunde gelegt und von der Gegenwart unmittelbar auf das Altertum geschlossen, ohne sich im mindesten die Frage vorzulegen, ob es wirklich so in den Zeiten des Altertums gewesen sein kann, und was denn eigentlich das Altertum von Tibet sei. Von einem Tibet kann indessen zur Zeit des Herodot und des Mahābhārata ganz und gar keine Rede sein, und wir wissen nicht einmal, ob das Hochplateau des centralen Tibet damals bevölkert gewesen ist, was mir aus manchen Gründen recht zweifelhaft erscheinen will. Von einem politischen und nationalen Gemeinwesen Tibet in historischer Zeit können wir nicht vor dem Anfang des siebenten Jahrhunderts n. Chr. sprechen, und von einer alten Kultur auf dem Boden des eigentlichen Tibet ist uns nicht das allergeringste bekannt. Vor jener Zeit haben wir eine Menge einzelner Stämme vor uns, die zum Teil zur Bildung der späteren tibetischen Nation beigetragen haben mögen, deren Ursitze aber nicht in Tibet, sondern weiter östlich im westlichen China gelegen waren. Die Traditionen der historischen Tibeter, die indochinesische Sprachwissenschaft und die Geschichte der Wanderungen der indochinesischen Stämme befinden sich über diesen Punkt in Übereinstimmung, dass die Ausbreitung

der Tibeter von Osten nach Westen erfolgt, und dass das heutige
Tibet der Endpunkt dieser Wanderungen gewesen ist [1]). Die tibetische Tradition weiss nichts von Kämpfen zu berichten, die mit
einer eingesessenen Urbevölkerung stattgefunden hätten, so dass wir
annehmen können, dass weite Strecken von Tibet damals herrenlo-
ses unbesetztes Land gewesen sind. Dagegen dürfen wir wohl den
heutigen zahlreichen mit der tibeto-barmanischen Familie verwandten Stammen des Himalaya eine ehemalige grössere Ausdehnung
nach Norden hin und spätere Zurückdrängung nach Süden durch
die Tibeter zuschreiben. Die Zeit der Einwanderung der Tibeter
lässt sich noch nicht genau fixiren, dürfte aber kaum vor dem
vierten oder fünften Jahrhundert n. Chr. stattgefunden haben. Wie
dem auch sein mag, es bleibt vorläufig, so lange wir nicht durch
neue Tatsachen eines Besseren belehrt werden, höchst problematisch,
von einem Altertum auf dem Boden des eigentlichen Tibet zu
reden, da sich die älteste Geschichte der tibetischen Stämme, die
wir aus den chinesischen Annalen kennen, durchaus nicht auf diesem Boden abgespielt hat. Es ist daher unmöglich, moderne Zustände
Tibets, wie es das von Schiern angeführte Goldwaschen ist, —
über dessen Geschichte wir zudem gar nichts wissen, — auf eine
Zeit zu übertragen, die doch mindestens vor der Epoche des Herodot liegen muss; die Theorie von Schiern kann darum nicht angenommen werden.

Überdies stimmt ja diese Theorie gar nicht zu der altindischen
Tradition. Im indischen Epos sind es die Khaça und andere ihrer
Nachbarvölker, welche das Ameisengold an den Hof des Königs
Yudhiṣṭhira brachten [2]). Und Tibeter waren diese Stämme ganz ge-

---

1) Besonders ERNST KUHN, Über Herkunft und Sprache der transgangetischen Völker,
München 1883, S. 5—6.

2) LASSEN, Indische Altertumskunde, Bd. I, S. 848. Die Nachrichten über diese Stämme
findet man jetzt übersichtlich zusammengestellt in dem ausgezeichneten Werke von SYLVAIN
LÉVI, Le Népal (Vol. I, Paris 1905, pp. 259 et seq.).

wiss nicht. Das Ameisengold ist niemals auf dem Wege über Tibet nach Indien gekommen, und Tibet ist meines Erachtens für die Beurteilung der ganzen Frage auszuschalten. Es hat niemals etwas mit dem Ameisengolde zu tun gehabt, ausgenommen dass es die Vorstellung und die Sage davon den Mongolen entlehnt hat.

Bei Strabo [1]) (XV, I, 57) ist eine klare Andeutung über den Weg enthalten, den das Ameisengold nach Indien genommen hat. Nach ihm sind es die Darden oder Darada, welche das den Ameisen gehörende Gold raubten. Diese Tradition ist verschieden von der des Herodot, welcher die goldraubenden Inder, ohne ihnen einen bestimmten Namen beizulegen, an Afghanistan oder Kabul (Paktyike) angrenzen lässt. Herodot's Bericht stammt nach seiner eigenen Angabe aus persischer Quelle, die Perser haben also die Sage gekannt und in verschiedenen Gegenden des nordwestlichen Indiens lokalisirt. Vielleicht stammt auch die Form, in welcher die Perser die Sage erhielten, gar nicht aus Indien selbst, sondern aus einer direkteren centralasiatischen Quelle, denn es lässt sich doch keine exakte Parallele zu Herodot's Version aus der indischen Litteratur nachweissen. Nun mögen ja die Darden seit alten Zeiten Goldwäschereien betrieben haben [2]), aber es ist deshalb nicht anzunehmen, dass die Sage von den goldgrabenden Ameisen bei ihnen entstanden ist, oder das wirkliche Ameisengold unmittelbar von ihnen kam. Wir haben vielmehr in der Augabe von Dardistan nur eine Etappe in dem Handel mit dem Ameisengolde und eine Lokalisirung der viel weiter aus dem Norden stammenden Sage zu erblicken.

Das Ameisengold hätte nie seinen sensationellen Anstrich und seine weite Verbreitung erhalten können, wenn es sich dabei nicht um etwas Besonderes gehandelt hatte. Sein sagenumwohner Charak-

---

1) Der in diesem Falle nach Megasthenes berichtet, s. SCHIERN, l. c., S. 23. McCRINDLE, The Invasion of India by Alexander the Great, pp. 187, 341.

2) M. A. STEIN, Kalhaṇa's Rājataraṅgiṇī, Vol. II, p. 280.

ter zeigt, dass es aus weiter Entfernung kam. Eine der ältesten Goldproduktionsquellen Asiens aber waren die Gebirge des Altai, von denen wir mit Sicherheit wissen, dass sie seit den ältesten Zeiten ganz Sibirien mit Gold erfüllt haben, wovon die vielen herrlichen, aus der sibirischen Bronzezeit stammenden Goldaltertümer der Eremitage in St. Petersburg beredtes Zeugnis ablegen. Aristeas von Prokonnesos, der Verfasser des epischen Gedichts Arimaspeia, hat bekanntlich die Kämpfe der Arimaspen mit den goldhütenden Greifen geschildert, und bei der Goldgewinnung der Arimaspen kann es sich wohl nur um die Altai-Region handeln, wie auch schon A. v. HUMBOLDT [1]) und TOMASCHEK [2]) vermutet haben.

Die bergbautreibenden Stämme des Altai waren zweifellos türkische Völker, vielleicht waren auch solche mongolischer Abkunft [3]) darunter. Aus dem Mongolischen lässt sich nun eine befriedigende Erklärung des Namens »Ameisengold" gewinnen. Eine der grossen Abzweigungen der mongolischen Familie, die sich südlich von den Hängen des Altai bis in die Gebiete des Kuku-nōr und des Oberlaufes des Huang-ho hinzieht, führt den Namen *Shiraighol* (d. i. »gelber Fluss"), und der Name der Ameise lautet im Mongolischen *shirgholjin* oder *shirgholji*, in welchem *-jin* oder *-ji* jedenfalls als Suffix aufzufassen ist [4]). Die Ähnlichkeit zwischen dem Volksnamen

---

1) L'Asie centrale, Vol. I, p. 402.

2) Kritik der ältesten Nachrichten über den skythischen Norden, I. Sitzungsberichte der Wiener Akademie, 1888, S. 763. Dieser Ansicht neigt sich auch O. M. DALTON (The Treasure of the Oxus, London, 1905, p. 8) zu.

3) Die Geschichte der Anfänge der Mongolen bleibt noch zu untersuchen, obwohl russische Forscher namhafte Beiträge zu dieser Frage geliefert haben. Sicher haben türkische und tungusische Stämme und jedenfalls auch ein alteinheimisches Element zu ihrer Bildung beigetragen. Der Name der Mongolen ist wohl viel älter als allgemein angenommen, worauf auch jüngst A. FORKE (Lun Hêng, Part I, p. 338) hinweist, gelegentlich des Vorkommes des Namens *Mong-ku* im Lun Hêng des Wang Ch'ung.

4) Vergl. die verschiedenen burjatischen Formen *šorgol-d'eng, šorgol-jeng, šorgol-je* (*šorgol* durch Attraktion aus *širgol* entstanden) bei CASTRÉN, Versuch einer burjätischen Sprachlehre, S. 140 b.

*Shiraighol* [1]) und dem Namen der Ameise *shirghol* dürfte sicher die

1) Mit den Königen der Shiraighol kämpft Geser, der Held der mongolischen und tibetischen Sage. Ein alter Zweig dieses Volkes waren die Tuluhun (tib. *T*u-lu-hun*), die in der Gegend des Kuku-nōr ansässig waren, und gegen die König Sroṅ-btsan-sgam-po Krieg führte. Ihre Zugehörigkeit zu den Shiraighol geht hervor aus ihrer mongolischen Benennung Shira Shiraighol (SCHMIDT, Forschungen im Gebiete der älteren religiösen, etc. Bildungsgeschichte der Völker Mittel-Asiens, S. 228) und aus ihrem tibetischen Namen *Hor-ser T*u-lu-hun*, d.i. die gelben Mongolen T'u-lu-hun oder die T'u-lu-hun, die zu den gelben Mongolen gehören. Diese T'u-lu-hun sind nun ferner identisch mit den im *T*ang-shu* genannten T'u-yü-hun (wie nach CHAVANNES richtiger statt T'u-ku-hun zu lesen ist); diese Gleichung ergibt sich aus der Tatsache, dass der im *T*ang-shu* berichtete Feldzug des Sroṅ-btsan-sgam-po gegen die T'u-yü-hun in völliger Übereinstimmung in den tibetischen Annalen erzählt wird, wo die T'u-yü-hun eben T'u-lu-hun genannt werden. Es seien die wichtigeren Punkte aus dem *T*ang-shu* (nach der Übersetzung von BUSHELL, The Early History of Tibet, pp. 9—10) hervorgehoben und in Parallele mit dem tibetischen Bericht des *rGyal-rabs*, fol. 95, gesetzt:

|  |  |
|---|---|
| *T*ang-shu.* | *rGyal-rabs.* |

**Erste** Gesandtschaft des K'i-tsung-lung-tsan (= Sroṅ-btsan-sgam-po) an den Kaiser von China im Jahre 634. Da er gehört hatte, dass die T'u-küeh und *T*u-yü-hun* kaiserliche Prinzessinnen zur Ehe erhalten hatten, schickte er einen Gesandten mit Geschenken und einem Brief und bat um eine Prinzessin. Der Kaiser verweigerte es. Als der Gesandte zurückkehrte, berichtete er Lung-tsan: „Als wir anfänglich am Hofe anlangten, empfing man uns höchst ehrenvoll und versprach eine Prinzessin zur Ehe, doch da erschien gerade der *T*u-yü-hun* Fürst am Hofe und mischte sich ein, so dass die Verhandlungen abgebrochen wurden. Darauf wurden wir mit geringer Höflichkeit behandelt, und das Bündnis wurde abgelehnt".

Darauf führte Lung-tsan zusammen mit den Yang-t'ung die vereinigten Heere, um die *T*u-yü-hun* anzugreifen.

Die *T*u-yü-hun* vermochten ihm nicht zu widerstehen und flohen an die Ufer des Ts'ing-hai, um der Schneide des Schwertes zu entgehen. Die Einwohner und ihre Herden wurden alle von den T'u-fan fortgeschafft.

**Der** tibetische König Sroṅ-btsan-sgam-po war ein Zeitgenosse des Kaisers T'ai Tsung. Der König von Tibet entsandte einen Boten, der um die Tochter des Herrschers von China bitten sollte; doch da der Herrscher von China die Tochter nicht gab, kehrte der Gesandte nach Tibet zurück und berichtete dem König: „Der Herrscher von China hat zuerst grosse Freude an mir gehabt und war im Begriff, seine Tochter zu geben, als die *Hor-ser T*u-lu-hun* mich bei dem Herrscher von China verleumdeten".

Der König von Tibet geriet in Zorn und marschirte an der Spitze eines Heeres von hundert tausend Mann in das Gebiet von T'uṅ-ciu. Darauf übertrug er das Commando dem Beamten gYa-t'oṅ und entsandet ihn zur Unterwerfung des Landes der *T*u-lu-hun*.

Die *T*u-lu-hun* flohen in der Richtung nach dem Kuku-nōr (tib. *mTs'o-sñon*) im Gebiet von gTsaṅ-к'a. Die von ihnen zurückgelassenen Menschen und Vieh wurden sämtlich von den Tibetern fortgeschafft.

Entstehung der Sage von einem goldgrabenden Ameisen-Volke veranlasst haben. Wie bereits Tomaschek ganz richtig dargelegt hat, waren die am Goldhandel beteiligten Völker sehr interessirt und darauf bedacht, den wahren Ursprungsort des Goldes ihren entfernter wohnenden Abnehmern zu verheimlichen, um etwaige direkte Anknüpfungen von Handelsbeziehungen der letzteren mit den Bergbau treibenden Stämmen zu verhindern; so mochte die ursprüngliche Bezeichnung »Gold der Shiraighol" durch den Gleichklang der Wörter die Handelsmarke »Gold der Ameisen" erzeugt haben. Diese Annahme ist *möglich*, aber es mag sich mit der Beziehung der Shiraighol zu den Ameisen auch anders verhalten haben, es mag in einer Volkssage begründet gewesen sein. Da uns aber alle tatsächliche Kenntnis devon mangelt, so ist es völlig zwecklos, sich darüber Spekulationen hinzugeben; ich will auch nur so viel behaupten, dass mir in den Namen Shiraighol und shirghol ein Fingerzeig zu einer vernünftigen Lösung der Frage gegeben zu sein scheint, die jedenfalls nicht unvernünftiger als die bisher gegebenen Erklärungen ist, in welchen die Entstehung des Namens

---

Wie oben gezeigt, waren die T'u-lu-hun nach tibetischer und mongolischer Anschauung ein Zweig der Shiraighol, also ein mongolischer Stamm. Dagegen CHAVANNES (Documents sur les T'ou-kiue (Turcs) occidentaux, p. 179, Note 1) und O. FRANKE (Beiträge zur Kenntnis der Türkvölker und Skythen Zentralasiens, S. 17) bezeichnen sie als ein Volk tungusischen Stammen (BUSHELL, l. c., p. 93, Note 11, nennt sie „an Eastern Tartar race", was natürlich nicht viel besagen will). Für diese Anschauung sind allerdings in der chinesischen Litteratur Anhaltspunkte gegeben, die wahrscheinlich auf *Sui-shu*, Kap. 83, basirt sind, wo die T'u-yü-hun von einem Manne gleichen Namens aus Liao-tung abgeleitet werden, aber ich stimme hier mit ROCKHILL (The Land of the Lamas, p. 335) in der Ansicht überein, dass dieser östliche Ursprung der T'u-yü-hun unwahrscheinlich ist. Es wird sich hier wohl um einen analogen Fall wie bei den Si-hia handeln (CHAVANNES, Dix Inscriptions de l'Asie centrale, p. 13, Note). Das Auftreten eines tungusischen Stammes in der Gegend des Kuku-nōr ist auch aus ethnographischen Gründen wenig einleuchtend, während das betreffende Gebiet stets von einem Gewimmel mongolischer und tibetischer Stämme besetzt war und den Zankapfel zwischen beiden bildete. Die Schilderung der Lebensgewohnheiten der T'u-yü-hun, die im *Sui-shu* entworfen ist, besonders die Angabe, dass sie sehr viele Yaks hielten, passt auch gar nicht auf die Kultur und Wirtschaftsverhältnisse eines tungusischen Volkes. ber die Geschichte der T'u-yü-hun s. auch P. S. POPOV, Nachrichten über die mongolischen Nomaden (russische Übersetzung des 蒙古游牧記), St. Pet., 1895, p. 117.

»Ameisengold" gar keine Berücksichtigung gefunden hat. Dazu ist natürlich in Rechnung zu ziehen, dass, wie bereits gezeigt, das Ameisengold der mongolischen Volkstradition bekannt ist.

Ich trage ferner kein Bedenken, die Arimaspen [1]) des Aristeas und Herodot mit einer Gruppe mongolischer Stämme in Verbindung zu bringen. Nach Herodot ist der Name skythisch und bedeutet »einäugig" (ἄριμα γὰρ ἕν καλέουσι Σκύθαι, σποῦ δὲ ὀφθαλμόν, IV, 27). Es müsste nun ein grosser Zufall sein, wenn wir im Mongolischen das Wort *äräm-däk* [2]) in dem Sinne von »einäugig" finden und dasselbe nicht mit dem Namen der Arimaspen irgendwie in Zusammenhang stehen sollte [3]).

---

1) Alle Nachrichten über die Arimaspen findet man in PAULY's Realencyklopädie (Artikel Arimaspoi von WERNICKE), Bd. II, S. 826—827, zusammengestellt.

2) KOWALEWSKI, Dictionnaire mongol-russe-français, p. 250 a. Das mongolisch-chinesische Wörterbuch (Vol. II, p. 36 a) erklärt das Wort durch 眇 一 目 und manju *gakda*. Der Stamm des Wortes ist zweifellos *äräm*, und *-däk* ist das zur Bildung des Participium perfecti dienende bekannte Suffix (BOBROWNIKOV, Grammatik der mongolisch-kalmükischen Sprache, § 244). — Auch das *Shan hai king* berichtet bekanntlich von einäugigen Menschen (*Tu shu tsi ch'êng*, Sect. 8, Buch 139, 1, p. 4 und F. DE MÉLY, Le „de Monstris" chinois et les bestiaires occidentaux, Revue archéologique, Vol. III, 1897, p. 367).

3) TOMASCHEK's Erklärung des Namens (bei PAULY, l.c., S. 826) aus dem Skythischen in dem Sinne von ,Besitzer wilder Steppenrosse' ist mir unverständlich. Ich sehe keinen Grund, warum Herodot's Überlieferung, der Name bedeute einäugig, nicht correct sein soll. — Die ganze Theorie von der iranischen Abstammung der Skythen ist sehr verdächtig und bedarf einer gründlichen Revision. Die geringen skythischen Sprachreste sind doch recht wenig beweiskräftig. Die Hauptsache ist, dass die skythische Kultur, vor allem im Wirtschaftsleben, den türkisch-mongolischen, oder allgemeiner gesagt, den altsibirischen Kulturtypus repräsentirt. Wenn man Herodot's Skythenberichte und die altchinesischen Texte über die Türken zusammenhält, dazu Pallas' Sammlungen historischer Nachrichten und B. Bergmann's klassisches Buch (Nomadische Streifereien unter den Kalmüken) liest, fragt man sich vergeblich, was denn eigentlich an der Kultur der Skythen indogermanisch sein soll. Ausschlaggebend ist ferner das archäologische Material aus dem südrussischen Boden, das unzweideutig die Verbindung skythischer mit sibirischer Kunst und Kultur erweist, wie S. REINACH (Antiquités de la Russie méridionale) richtig gezeigt hat, und worauf neuerdings wieder O. M. DALTON (The Treasure of the Oxus, p. 14) aufmerksam macht. J. PEISKER in seiner interessanten Untersuchung Die älteren Beziehungen der Slawen zu Turko-Tataren und Germanen (Stuttgart, 1905, S. 48) erklärt die Skythen für iranisirte Uralaltaier. — Die Arimaspeia und Herodot's Skythenberichte bedürfen sehr der Revision, besonders im Lichte sinologischer Forschung. Die Issedonen mögen immerhin ein mit Tibetern verwandtes Volk oder ein Volk tibetischen Stammes gewesen sein, aber ihre Wohnsitze sind gewiss nicht im heutigen Tibet, sondern viel weiter nach Norden und Osten zu suchen. Die von Lassen und Tomaschek aufgestellte Identifikation der K'iang 羌 mit den Kanka des Mahābhārata, die doch nur auf einem rein äusserlichen Wortgleichklang beruht, ist aus historischen und geographischen Gründen völlig unhaltbar.

# 059

满文文献概说

IX. évfolyam.          1908.          1—2. szám.

# KELETI SZEMLE.

KÖZLEMÉNYEK AZ URAL-ALTAJI NÉP- ÉS NYELVTUDOMÁNY KÖRÉBŐL

A M. TUD. AKADÉMIA TÁMOGATÁSÁVAL

A NEMZETKÖZI KÖZÉP- ÉS KELETÁZSIAI TÁRSASÁG MAGYAR
BIZOTTSÁGÁNAK ÉS A KELETI KERESKEDELMI AKADÉMIÁNAK
ÉRTESITŐJE.

# REVUE ORIENTALE

POUR LES ÉTUDES OURALO-ALTAÏQUES.

SUBVENTIONNÉE PAR L'ACADÉMIE HONGROISE DES SCIENCES.

*

JOURNAL DU COMITÉ HONGROIS DE L'ASSOCIATION INTER-
NATIONALE POUR L'EXPLORATION DE L'ASIE CENTRALE ET
DE L'EXTRÊME-ORIENT.

SZERKESZTIK ÉS KIADJÁK
*Rédigée par*
Dꞧ KÚNOS IGNÁCZ ⋆ Dꞧ MUNKÁCSI BERNÁT.

BUDAPEST.

*En commission chez Otto Harrassowitz*
*Leipsic.*

# SKIZZE DER MANJURISCHEN LITERATUR.

— Von Berthold Laufer. —

Die folgende Skizze beruht teilweise auf den früheren Forschungen über diesen Gegenstand, teilweise auf eigenen Studien,
die durch eine von mir in Peking 1901 und 1902 erworbene
ziemlich vollständige Bibliothek manjurischer Bücher ermöglicht
wurden; dieselben befinden sich jetzt teils im Chinese Department der Columbia University, teils im American Museum of
Natural History, New York. Es sind darunter manche in europäischen Bibliotheken nicht vorhandene, manche, die Möllendorff und anderen Bibliographen nicht bekannt waren; merkwürdig ist, dass die grösste Zahl der Manju-Werke in der Sammlung Möllendorff's unvollständig, selbst von leicht zu erlangen-
den Werken nur einzelne Bände vorhanden sind; auch in den
Bibliotheken Europas existiren viele der umfangreicheren Werke
nur fragmentarisch, während die unserer Sammlung alle vollständig sind. Auf Grund des mir zur Verfügung stehenden Materials habe ich die Daten und Übersetzer der Werke festgestellt
und die bibliographischen Angaben meiner Vorgänger vielfach
berichtigt. In der Transkription des Manju bin ich der von
H. C. v. d. Gabelentz gefolgt. nur dass ich unter *h* den von
ihm angewandten Punkt weglasse.

Eine Geschichte der manjurischen Literatur, die ja im
wesentlichen nur ein Abglanz der chinesischen ist, muss naturgemäss anders behandelt werden als die eines selbständigen, in
sich abgeschlossenen Literaturgebietes. Es kann nicht die Aufgabe ihres Historikers sein, die einzelnen Übersetzungen auf
ihren Inhalt zu prüfen und zu analysiren, eine Aufgabe, die
nur im Zusammenhang mit der Entwicklung der chinesischen

Literatur verständlich würde und füglich einer Darstellung dieser
überlassen bleiben muss. Der Leser, der über die hier berührten
Werke nähere Auskunft zu erhalten wünscht, sei vor allem auf
das feinsinnige Buch von WILHELM GRUBE, Geschichte der chinesischen Literatur (Die Literaturen des Ostens in Einzeldarstel-
lungen, Band VIII, Leipzig, 1902) verwiesen.[1]) Der hauptsäch-
liche Gesichtspunkt bei einer Übersicht der manjurischen Literatur sollte vielmehr die Frage sein, was wir aus derselben in-
betreff der Manju selbst, ihrer Geschichte, Geistesentwicklung
und ihres Charakters lernen können, inwieweit sie, obwohl ein
erborgtes Gut, der Spiegel ihrer geistigen Verfassung zu sein
vermag, und von diesem Gesichtspunkt aus habe ich den Gegenstand zu behandeln versucht.[2])

---

[1]) Ich habe daher auch von vollständigen Literaturangaben zu den
einzelnen Werken Abstand genommen, da man diese in H. CORDIER's
Bibliotheca Sinica zusammengestellt findet, dagegen öfters auf letztere
verwiesen.

[2]) Grammatiken des Manju gibt es fast mehr als es je Lente gegeben hat, welche die Sprache und Literatur wirklich kannten. Sie alle
aufzuzählen verlohnt sich nicht der Mühe. Die brauchbarste ist immer
noch die erste von H. C. v. D. GABELENTZ, Éléments de la grammaire
mandchoue, Altenburg, 1832, aus der die Nachfolger mehr oder weniger
richtig und flüchtig abgeschrieben haben. Eine Ausnahme macht die russische Grammatik von J. ZACHAROV (St. Petersburg, 1879), welche die voll-
ständigste, aber auch die weitschweifigste ist. Dringend zu warnen ist vor
C. DE HARLEZ, Manuel de la langue mandchoue, Paris, 1884, worin es von
den gröbsten Irrtümern, Flüchtigkeiten und Druckfehlern auf jeder Seite
nur so wimmelt. Hier nur ein Beispiel von der ganz unglaublichen Confusion dieses »Gelehrten«. Auf pp. 158—159 ist das Vorwort zur Geschichte
der Yüan transkribirt abgedruckt, d. h. aus KLAPROTH's Chrestomathie,
pp. 122—126 blindlings abgeschrieben mit unzähligen Fehlern und unter
vollständiger Auslassung zweier Zeilen (Z. 5 und 9, p. 125 bei KLAPROTH),
der Genitiv *bithe-i* wird in drei auf einander folgenden Zeilen in drei verschiedenen Transkriptionen *bithe i*, *bithe-i* und *bithei* gedruckt. Auf p. 227
wird nun dieses Stück unter dem Titel *Histoire des trois royaumes* übersetzt, während eine Fussnote zu diesem Titel auf die Geschichte der Yüan
hinweist! Und wiederum im Inhaltsverzeichnis (p. 230, letzte Zeile) wird
auf diese Übersetzung mit *Histoire du royaume mongol* verwiesen, während an keiner Stelle gesagt wird, dass es sich um das *Vorwort* zum
*Yüan-shih* handelt; ausserdem ist die Übersetzung unter aller Kritik, kaum
ein Satz richtig, selbst die Eigennamen sind anders als in dem transkri-

**Schrift.** — Die Manju haben zwei Perioden der Schrift-
entwicklung durchgemacht: in der ersten ahmten sie einfach
die Schrift der Mongolen nach, in der zweiten passten sie in

---

birten Text geschrieben (z. B. Text: *Udaxi*, Übersetzung: *Utari*; beides
falsch statt *Udari*). — Ein recht brauchbares Buch ist A. WYLIE, Translation of the Ts'ing Wan K'e Mung, a Chinese Grammar of the Manchu
Tartar Language; with introductory Notes on Manchu Literature, Shanghai, 1855. Dies ist die Übersetzung eines praktischen Handbuchs des
Manju von einem Chinesen WU-KO SHOU-PING aus der Manchurei (1730)
in vier Büchern; ich weiss nicht, ob man sich nicht aus diesem Werk
des alten chinesischen Schulmeisters mehr lebendiges und förderliches
Sprachmaterial aneignet als aus allen europäischen Grammatiken zu-
sammengenommen. Wer einiges Sprachtalent hat, kann sich ohne alle
Hülfsmittel eine so einfach und durchsichtig gebaute Sprache, wie das
Manju, leicht aus den Texten selbst abstrahiren. Lehrreich ist auch die
aus dem *San ho pien lan* von H. C. v. D. GABELENTZ übersetzte Manju-
chinesische Grammatik (Zeitschrift für die Kunde des Morgenlandes,
Band III, Göttingen, 1840, S. 88—104). Desselben Autors Mandschu-
Deutsches Wörterbuch (Leipzig, 1864) ist recht zuverlässig und reicht,
obwohl Specialwörterbuch zu Se-shu, Shu-king und Shi-king, im allgemeinen auch für andere Schriften aus. Ziemlich vollständig ist das Wörterbuch von J. ZACHAROV (Полный Маньчжурско-русскій словарь, St. Petersburg, 1875, XXX, 64 und 1129 p.), doch kann man es kaum als wissenschaftliche Leistung rühmen, da weder die Bedeutungsentwicklung der
Wörter behandelt, noch die Quellen citirt und kritisch verwertet, noch
genügend chinesische Äquivalente gegeben werden. Das wirkliche Wörterbuch des Manju ist noch der Zukunft vorbehalten. Bis dahin fährt man
am besten, sich an die einheimischen Werke der Lexikographie zu halten.
Der Vollständigkeit wegen seien noch die folgenden Grammatiken genannt: Linguae Mandshuricae institutiones quas conscripsit, indicibus or-
navit, chrestomathia et vocabulario auxit FRANCISCUS KAULEN, Ratisbonae,
sumptus fecit G. Josephus Manz, Lipsiae ex officina Niesiana, 1856 (Chrestomathie in Originaltypen, pp. 97—139, fünf Stücke aus der Bibel, eins
aus dem Chung-yung, eins aus *Yüan-shih* nach KLAPROTH, zwei aus Bibl.
Reg. Berol. Libr. Sin. Nr. 32); LUCIEN ADAM, Grammaire de la langue
mandchoue, Paris, 1873, 137 p. (mit zwei analysirten Texten, eine Erzählung nach ST. JULIEN und Vorwort zum Gedicht auf Mukden); P. G. v.
MÖLLENDORFF, A Manchu Grammar with analysed Texts, Shanghai, 1892,
53 p., 4° (mit Text, Übersetzung und Analyse der «Hundert Lektionen»). —
Chrestomathien gibt es von J. KLAPROTH (Chrestomathie mandchoue, Paris,
1828, 273 p.; nicht sehr zu empfehlen, auf die Übersetzungen ist kein
Verlass); von VASILJEV (St. Petersburg, 1863, 228 p.); von A. O. IVANOVSKI
(zwei Hefte, St. Petersburg, 1893, 1895; nicht gesehen); von A. POZDNÉJEV

1*

mehr selbständiger Weise die mongolische Schrift ihrer Sprache
an. Die Einführung des mongolischen Alphabets ist mit dem
Namen des ersten Manju-Herrschers T'ai-tsu[1]) verknüpft, der
im Jahre 1599 dem Erdeni-Baksi, einem in der mongolischen
Literatur bewanderten, zum gelben Banner gehörenden Gelehrten,
und seinem Minister Gagai Carduci den Befehl gab, eine Schrift
für die Manju zu ersinnen und zu diesem Zwecke die mongolische zu Grunde zu legen. Alle Bewunderung verdient hier der
praktische Scharfblick des T'ai-tsu, der bei dieser Gelegenheit
unumwunden aussprach, dass Manju dem Mongolischen verwandt sei, dass die Manju-Wörter aus denselben Silben beständen
wie die mongolischen (also eine Ahnung phonetischer Verwandt-
schaft). Mit Recht schloss er daher, dass auch Manju mit mongolischen Buchstaben geschrieben werden könne. Diese Einsicht
erwies sich in der Tat als von grosser Tragweite, besonders

---

(Опытъ собранія образцовъ маньчжурской литературы). Von letzterem
Werke sind bisher einzelne Druckbogen in den Nachrichten des Orien-
talischen Instituts von Vladivostok erschienen; mir liegen pp. 1—272
vor; Titel und Inhaltsverzeichnis sind meines Wissens noch nicht heraus,
und das Urteil über die Arbeit muss bis nach ihrem Abschluss aufgespart
werden; ich habe aber schon auf dieselbe in meiner Skizze Rücksicht genommen. Th. T. MEADOWS, Translations from the Manchu, with the original Texts, prefaced by an Essay on the Language, Canton, 1849, habe
ich nie gesehen. In dem blossen Vereinigen von Manju-Texten (ohne die
chinesischen Versionen) zu Chrestomathien kann kein wesentlicher Nutzen
erblickt werden. Manju-chinesische Bücher sind in Peking so leicht und
zu so billigen Preisen (abgesehen von einigen Seltenheiten und umfang-
reichen Werken) zu erlangen (jedenfalls viel leichter als die drei russischen Chrestomathien), dass jedem das Manju Studirenden nur dringend
zu raten ist, nur nach chinesischen Ausgaben zu arbeiten; dieselben haben
den Vorteil, dass sie schön, elegant und korrekt gedruckt sind, dass man
sich den Ärger und Zeitverlust über die Druckfehler der europäischen
Nachdrucke spart, dass man Manju und chinesisch, übersichtlich in Pa-
rallelreihen geordnet, bequem nebeneinander lesen kann, dass man ferner
einen ästhetischen Genuss an den schönen Formen der Schrift und der
Grazie der unübertroffenen chinesischen Buchtechnik hat, während sich
die europäischen Manju-Drucke durch plumpe Typen, schlechten Satz und
miserables Papier auszeichnen.

[1]) Manjurisch *Nurhaci*, lebte 1559 bis 1626. — Über die Beziehungen der Manju zu den Mongolen vergl. E. H. PARKER, Manchu Relations
with Mongolia, China Review, Vol. XV. 1887, pp. 319—328.

wenn wir an das auf die chinesischen Schriftzeichen basirte
schwerfällige System der Niŭchi zurückdenken. Nachdem Gagai
für ein Staatsverbrechen die Todesstrafe erlitten hatte, beendete
Erdeni allein das Werk, und das von ihm für das Manju frucht-
bar gemachte Alphabet der Mongolen kam bald in allgemeinen
Gebrauch.[1]) Damit war der Triumphzug der syrischen Schrift
vollendet, die so von den Gestaden des Mittelmeers auf dem
Wege über die Uiguren und Mongolen die Ufer des Amur und
des Stillen Oceans erreicht hatte: noch jetzt bewahrt an der
Westküste Sachalins der Ainu-Häuptling eines kleinen Dorfes
eine manjurisch geschriebene Urkunde.[2])

Ganz wie bei den Mongolen, war auch für die Manju die
Notwendigkeit, mit ihren Nachbarn, zunächst mit den Mongolen
und Korea, in schriftlichen Verkehr zu treten, die Triebfeder
für die Einführung der Schrift. Der nächste Schritt führte denn
zur Einsetzung eines sogenannten «gesetzgebenden Amts» *(kooli
selgere yamun)*, dessen Aufgabe es war, alle die innere Verwal-

---

[1]) Vergl. A. WYLIE, A Discussion of the Origin of the Manchus, and
their written Character (Chinese Researches, Part IV, Shanghai, 1897,
bes. pp. 261—271; dies ist Wiederabdruck des ersten Teils der Einleitung
seines Buches Translation of the Ts'ing Wan K'e Mung, Shanghai, 1855,
pp. I—XXXVI), ZACHAROV, Полный маньчжурско-русскій словарь, St. Pet.,
1875, pp. XI—XIII. — ERDENI hat keine neuen Schriftzeichen erfunden,
sondern einfach das mongolische Alphabet, so wie es war, übernommen;
seine Arbeit bestand in der richtigen Verteilung der mongolischen Zeichen
auf die entsprechenden manjurischen Laute.

[2]) FR. SCHMIDT (Beiträge zur Kenntnis des russischen Reiches,
Band XXV, St. Petersburg, 1868, S. 93) berichtet aus dem Jahre 1860,
dass der Ainu-Älteste des Dorfes Naiero ein in manjurischer Sprache ab-
gefasstes Schreiben bei sich aufbewahre, das sein Vater früher auf einer
Tributreise nach Sansin am Sungari von der manjurischen Obrigkeit erhalten hatte, und in welchem er zu einem Ältesten der Ainu bestellt war;
selbst war er nicht mehr in das Land der Manju gekommen und konnte
sich auch nicht erinnern, Manju bei sich gesehen zu haben. Als ich mich
im Winter 1898 auf einer Hundeschlittenfahrt an der Ostküste von Sa-
chalin befand, hatte ich den Plan, vom Posten Manue aus nach Naiero
hinüberzufahren, um eine Abschrift von jenem Dokument zu nehmen,
wurde aber leider an der Ausführung meines Vorhabens durch einen hef-
tigen Schneesturm verhindert, der den Weg verweht und unpassirbar gemacht hatte.

tung, wie die auswärtigen Beziehungen betreffenden Schriftstücke zu redigiren. Die Seele dieses Instituts war der unermüdliche Dahai. der nur zwanzig Jahre alt, die Stellung des kaiserlichen Sekretärs bekleidete. In beiden Sprachen wohl bewandert, verfasste er eigenhändig alle wichtigeren Staatspapiere und fand noch Musse für die Bearbeitung der ersten Übersetzungen aus dem Chinesischen ins Manju.[1]

Unter T‘ai-tsu wurde die Schrift noch ohne diakritische Zeichen gebraucht. Literarische Dokumente aus jener Zeit in diesem Alphabet haben wir nicht, doch nach einer interessanten Handschrift aus dem Jahre 1741 können wir ungefähr eine Vorstellung von seinem Aussehen gewinnen. Dies ist ein Unikum der Pariser Nationalbibliothek, betitelt *Tongki fuka aku hergen-i bithe* «Buch der Wörter, die ohne Punkte und Kreise geschrieben werden», in drei Bänden.[2] Aus dem Vorwort zu diesem Manuskript geht deutlich hervor, dass damals (1741) zur Zeit K‘ien-lung’s in den kaiserlichen Archiven Aktenstücke in manjurischer Sprache aufbewahrt wurden, die ohne alle diakritische Zeichen geschrieben waren. und dass die Ansicht der damaligen Gelehrten dahin ging, dass sich die spätere punktirte Schrift aus dieser unpunktirten entwickelt habe. Schon damals

---

[1] ZACHAROV, Vorwort zum Wörterbuch, p. XIII.

[2] Zuerst erwähnt von L. LANGLÈS (Alphabet mantchou, 3. Aufl., Paris, 1807, p. 59, Note), der eine Seite der Handschrift reproducirt (teilweise wiedergegeben von A. WYLIE, l. c., p. 263). Gründlich erörtert ist sie von A. POZDNÉJEV, Разысканія въ области вопроса о происхожденіи и развитіи маньчжурскаго алфавита (Извѣстія Восточнаго Института, Vol. II, Vladivostok, 1901, pp. 118—140). Die Hälfte dieser Arbeit ist mit einer Widerlegung der Irrtümer von AMIOT (1770) und LANGLÈS inbezug auf die Geschichte der Manju-Schrift angefüllt; man sollte doch den Jahrhunderte alten Toten endlich im Grabe Ruhe gönnen. Der gelehrte russische Professor hat aber in seinem Disputationseifer ganz und gar die Arbeit von A. WYLIE übersehen, in welcher der Sachverhalt schon richtig dargestellt war, so dass der Sauerkohl des achtzehnten Jahrhunderts nicht mehr hätte aufgewärmt zu werden brauchen. Der zweite Teil von POZDNÉJEV’s Arbeit ist indessen wertvoll: er teilt das Vorwort der oben erwähnten Handschrift im Manju-Text und Übersetzung mit und die sich daraus ergebenden Schlussfolgerungen. Leider gibt er keine Proben oder Faksimiles der Handschrift selbst, die eine Publikation zu verdienen scheint.

hatten selbst die Akademiker Schwierigkeiten, diese ursprüng-
lichen Schriftformen zu lesen, und der Wunsch, dieselben der
Nachwelt aufzubewahren, resultirte in jener auf kaiserlichen
Befehl veranstalteten Sammlung von Wörtern ohne graphische
Unterscheidungszeichen, wie man sie in den alten Urkunden
vorfand. Aller Wahrscheinlichkeit stammten diese noch aus der
Zeit des T'ai-tsu selbst.[1]

---

[1] Wirkliche zeitgenössische Proben dieser Schriftgattung sind auf
den Münzen der Perioden T'ien-ming (1616—1626) und T'ien-ts'ung (1627—
1643) vorhanden. Proben der ersteren tragen auf dem Avers die chinesische Legende *T'ien-ming t'ung-pao* «Cirkulirende Münze der Periode T'ien-
ming». auf dem Revers in Manju *Abka-i fulingga Han jiha* «Kaiserliche
Münze der Periode Abkai fulingga» (d. i. Bestimmung des Himmels =
T'ien-ming). Eine Abbildung dieser Münze findet man bei A. WYLIE, Coins
of the Ta-Ts'ing, or Present Dynasty of China (Journal of the Shanghai
Literary and Scientific Society, No. I, 1858, reprinted, Shanghai, 1887,
p. 45). Das Beispiel ist sehr instruktiv, da hier in der Schreibung von
*Han* der das *h* später von *k* unterscheidende Haken, in *fulingga* die Punkte
hinter *u* und *g*, in *jiha* wiederum der Haken bei *h* fehlen. Münzen dieses
Typus sollen bereits im Jahre 1616 geprägt worden sein. Vom Typus der
Periode T'ien-ts'ung sind zwei Formen bekannt geworden, die sich nach
der gegenwärtigen Ausdrucksweise als «grosser» und «kleiner Cash» be-
zeichnen lassen. Der «grosse» ist publicirt von S. W. BUSHELL (A Rare
Manchu Coin, China Review, Vol. VI, 1877, pp. 143—144; wiederholt in
Journal of the China Branch of the Royal Asiatic Society, New Series,
Vol. XV, Shanghai, 1880, pp. 196, 217). Die Legende ist ausschliesslich
Manju; Avers: *Sure* (nicht *Sura*, wie BUSHELL durchgängig schreibt) *Han
ni jiha* «Münze des Sure Han» (Äquivalent von T'ien-ts'ung, d. i. der
Verständige); Revers: *juwan emu yan* «Elf Unzen (oder Taels)», d. i. Ge-
wicht und Wert der Münze. Die kleinere Form (gleichfalls illustrirt und
beschrieben von BUSHELL, Additional Coins of the Present Dynasty, Ibid.,
Vol. XXXIII, 1899—1900, p. 30) hat dieselbe Manju-Lesung in derselben
Schriftform (nur kleiner) auf dem Avers, doch die Rückseite ist unbe-
schrieben. Die Schreibung *Sure* auf dieser Münze, auf welcher das *u* durch
mongolisches *ü* (d. h. Kreis mit angesetztem Querstrich) dargestellt ist,
liefert einen untrüglichen Beweis dafür, dass die Schrift der Manju von
1599 bis 1632 nichts anderes als eine Copie der mongolischen war. Dagegen sind die Münzen der folgenden Periode Shun-chih (1644—1661),
soweit sie Manju-Legenden tragen (und das ist nur bei wenigen Ausgaben
der Fall) mit diakritischen Buchstaben geschrieben. — Über die Numis-
matik der gegenwärtigen Dynastie vergl. ausser den erwähnten grundlegenden Arbeiten von WYLIE und BUSHELL einige Artikel von H. F. SCHE-
PENS (China Review, Vol. XXII, pp. 556—557, 598—605) und E. W. TUWING

Doch alle aus dem siebzehnten Jahrhundert stammenden
Bücher, die wir als die Inkunabeln der Manju-Literatur bezeichnen können, das älteste manjurische epigraphische Denkmal in
Korea vom Jahre 1639, die Nephrittafeln und Münzen derselben
Periode gehören der zweiten verbesserten Entwicklungsphase der
Schrift an, die Dahai inaugurirt hatte. Im Jahre 1627 war T'ai-
tsung seinem Vater T'ai-tsu gefolgt; die literarischen Bestrebungen der Manju hatten damals schon wesentliche Fortschritte
gemacht, und die Unvollkommenheiten der mongolischen Schrift
machten sich ihren Literaten empfindlich bemerkbar. Der Kaiser
selbst soll den Anstoss zum Gebrauch unterscheidender Merk-
male für die doppeldeutigen Buchstaben gegeben haben, doch
es ist wahrscheinlicher, dass diese Idee dem talentvollen Dahai
selbst entsprungen ist, als eine sich notwendig ergebende Über-
zeugung aus seiner lang fortgesetzten Beschäftigung mit der
einheimischen Schrift. Dahai formte denn im Jahre 1632 diakritische Punkte (*tongki*) und Kreise (*fuka*), die der Unsicherheit
und Verwirrung der Laute ein Ziel setzten, und auch besondere zusammengesetzte Zeichen zur Wiedergabe von chinesischen
und Sanskrit-Lauten. Wörter, wie *aga* «Regen» und *aha* «Sklave»,
*toro* «Pfirsich» und *doro* «Weg», die früher dasselbe Aussehen
im Schriftbilde boten, waren von nun an deutlich geschieden.
Dahai's Erfindung wurde in demselben Jahre durch kaiserliches
Dekret sanktionirt und blieb von da ab ohne wesentliche Ver-
änderungen in Kraft.[1] Zu bemerken ist noch, dass die Manju

(Ibid., Vol. XXIII, pp. 105 114); A. WEYL, *Dreisprachige chinesische
Münzen*, welche in und für Ost-Turkestan geschlagen sind (Abdruck aus
Berliner Münz-Blätter, Berlin, 1882, 10 p.; ganz nach BUSHELL gearbeitet,
Manju-Umschreibungen sehr ungenau); H. N. STUART, *Catalogus der munten
en amuletten van China, Japan, Corea en Annam* (Batavia, 1904, pp. 90
et seq.; leider ohne Abbildungen). Trotz dieser Arbeiten scheint der Gegenstand noch nicht erschöpft zu sein; es gibt noch viele unpublicirteVarianten
teilweise sind die Prägestätten noch festzustellen, die Manju-Legenden
genauer zu transkribiren und die verschiedenen Schriftformen derselben
zu studiren.

[1] Wie HARLEZ (Manuel de la langue mandchoue, p. 6) dazu kommt,
die Vervollkommnung des Alphabets einem Befehl und den Angaben des
Kaisers K'ien-lung zuzuschreiben, ist unerfindlich, aber bei all der Ober-
flächlichkeit und Gedankenlosigkeit dieses Autors nicht im geringsten
wunderzunehmen.

selbst ihre Schrift nicht in einzelnen Buchstaben analysiren,
sondern sie in Form eines in zwölf Klassen eingeteilten Sylla-
bars besitzen, dessen Kenntnis für uns nur ein historisches Interesse hat; de facto ist ihre Schrift eine echte Lautschrift.
Für den schriftlichen Verkehr gibt es, wie im Mongolischen,
eine vom Druck etwas abweichende Currentschrift.

Als der Kaiser K'ien-lung im Jahre 1748 sein Gedicht auf
Mukden chinesisch und Manju publicirt hatte, liess er einen
Ausschuss gelehrter Männer zusammentreten und betraute sie
mit der Aufgabe, in Anlehnung an archaische chinesische
Schriftarten altertümliche Manju-Alphabete zu erfinden. Das
Komitee brütete drei Jahre über dieser Arbeit, an welcher der
Kaiser selbst tätigen Anteil nahm. Die Hauptvorlage bildete
eine In schrift auf dem Siegel des Kaisers T'ai-tsung, die vermutlich in mongolischer Quadratschrift geschrieben war, denn
unter den neu construirten Schriften befindet sich eine auf
letzterer basirte manjurische Quadratschrift. Im ganzen wurden
so 32 pseudo-antike Alphabete für das Manju construirt, und
des Kaisers Gedicht wurde darauf in diesen 32 Manju-Schriften
und in 32 alten chinesischen Schriften gedruckt, und es wurde
befohlen, dass die manjurische Quadratschrift von da ab auf
den Siegeln des Kaisers, der Ministerien, Tribunale und aller
Beamten über der sechsten Klasse, sowie auf den Siegeln der
Patente, die hohe Ernennungen und erblichen Rang verleihen,
angewendet werden sollte.[1]

ZACHAROV hat die Bildung dieser manjurischen Quadratschrift etwas spöttisch behandelt, nicht ganz mit Recht. Denn
einerseits blieb sie keineswegs, wie er annimmt, eine bloss persönliche Spielerei des Kaisers, sondern fand eine, wenn auch
beschränkte, praktische Anwendung auf staatlichen und amtlichen Siegeln.[2] Diese Siegel haben ein gewisses historisches und
antiquarisches Interesse, und wenn sie, wie es denn zu geschehen

---

[1] Vergl. ZACHAROV, Vorwort zum Wörterbuch, pp. 63—64. Nach
MÖLLENDORFF, Nr. 238, soll diese Ausgabe aus 64 Bänden bestehen. Ein
vollständiges Exemplar scheint in keiner europäischen Bibliothek vorhanden zu sein (vielleicht in Paris?); ich habe nur vier Bände des Werkes erlangen können.

[2] Beispiele bei A. WYLIE, l. c., pp. 268—271.

pflegt, sich in unsere Sammlungen verirren, ist es berechtigt
und wünschenswert, die Lesung der Schrift festzustellen. An-
dererseits ist der kaiserliche Hang nach antiquirten Manju-
Skripten im Zusammenhang mit der chinesischen Ästhetik, mit
der Vorliebe für das Ornamentale und Altertümliche in der
Schrift, für ihre malerische und grosszügige Wirkung zu verstehen und zu würdigen. Die neuen Schriften hatten lediglich
einen künstlerischen Zweck, waren auf ästhetischen Genuss berechnet und nur eine Begleiterscheinung all jener auf die Renaissance der Kunst und der Antike gerichteten Bestrebungen,
welche die grosse Epoche K'ien-lung's begleiten und verschönern.

Im Druck haben sich die Manju an das chinesische Verfahren des Holztafeldrucks angeschlossen; mit beweglichen
Typen ist Manju nur in Korea gedruckt worden. In Technik
und Stil stimmen die Manju-Bücher genau mit den chinesischen
überein, nur mit dem einen Unterschiede, dass sie, entsprechend
den von links nach rechts laufenden vertikalen Columnen der
Schrift, in den Seiten, ganz wie unsere Bücher, von links nach
rechts laufen (nicht wie die chinesischen von rechts nach links);
und das ist auch in den Manju-chinesischen Paralleldrucken der
Fall, in denen die chinesischen Worte mit den entsprechenden
manjurischen möglichst zusammen gedruckt sind. Alle Manju-
Bücher sind in Peking hergestellt worden; weitaus die meisten
sind kaiserlicher Initiative entsprungen und mit grosser Eleganz
gedruckt, andere sind auch von Pekinger Buchhandlungen veröffentlicht worden. Die ältesten erhaltenen Manju-Drucke stammen aus dem Jahre 1647.[1]

---

[1] Das Unikum eines Manju-Druckes besitzt *scheinbar* die Universitätsbibliothek von Cambridge in England. In dem von H. A. GILES bearbeiteten Catalogue of the Wade Collection of Chinese and Manchu Books
in the Library of the University of Cambridge (Cambridge, 1898, p. **143**,
Nr. G 215) liest man einigermassen überrascht von einem Manju-Werke
*Hsieh wên ts'ing kung yao yü* mit chinesischem interlinearen Text, welches
das Datum 1564 tragen soll. Das sich daraus ergebende Phänomen, dass
ein Manju-Werk 35 Jahre vor der Einführung der Schrift bei den Manju
(1599) bestanden habe, wird indessen mit keinem Worte erörtert. Das
Rätsel löst sich jedoch leicht, wenn wir in MÖLLENDORFF's Essay vom Jahre
1889, der doch dem Verfasser jenes Katalogs bekannt gewesen sein muss,

**Bedeutung und Probleme der Manju-Studien.** — Die Frage nach dem praktischen Nutzen des Manju für den Sinologen ist wiederholt für und wider erörtert worden.[1] Ganz abgesehen davon, dass weder das Leben noch die Wissenschaft nach rein utilitarischen Gesichtspunkten zugeschnitten sein müssen, und die Wissenschaft auch solche gar nicht verfolgt, liegen für jeden Einsichtigen die Vorzüge einer Kenntnis des Manju für das Studium der chinesischen Sprache und Literatur auf der Hand, obwohl, wie wir bald sehen werden, dies in keiner Weise das wesentliche Problem ist, welches uns die manjurische Literatur aufgibt. Die Ansicht der Missionare des achtzehnten Jahrhunderts, die da glaubten, dass die Manju-Literatur die chinesische ersetzen, und die chinesische ausschliesslich auf Grund jener studirt werden könne, ging natürlich zu weit und beruht auf einer Verkennung des wahren Sachverhalts, denn es war nie und nimmer die Bestimmung der Manju-Übersetzungen, die chinesischen Vorlagen zu verdrängen oder überflüssig zu machen, sondern ihr Zweck war ausschliesslich, das Verständnis des Originals für die Manju zu erleichtern. Das gilt in erster Linie für die Verdolmetschungen der klassischen Literatur; niemals sind die Manju-Texte derselben in China separat, sondern stets als Interlinearversionen zusammen mit dem chinesischen Text gedruckt worden. Das zeigt deutlich an, dass beide zusammen gelesen werden sollten; natürlich haben die Manju die klassischen Schriften im Original studirt und sich der danebenstehenden Übersetzung einfach als eines bequemen technischen Hülfsmittels zum rascheren und deutlicheren Verständnis bedient. Das ist denn auch der Grund, weshalb diese Übersetzun-

---

unter Nr. 77 dasselbe Werk citirt finden, mit dem Datum 1714 für die Manju-chinesische Ausgabe, während sich das Datum 1564 ausschliesslich auf das zweite und dritte chinesische Vorwort einer früheren, nur chinesischen Ausgabe beziehen. Die Richtigkeit des Datums von MÖLLENDORFF ergibt sich ferner aus dem Katalog des Asiatischen Departementes in St. Petersburg (1844), wo auf p. 64, Nr. 430, derselbe chinesische und manjurische Titel mit dem chinesischen Datum «53. Jahr der Periode K'anghsi», d. i. 1714, angeführt wird.

[1] Dabei wurde implicite der logische Fehler begangen, als wenn das Manju nur für die Sinologen geschaffen worden wäre.

gen ganz wörtliche Paraphrasen des Originals sind, und man
braucht sich nicht über die »sklavische» Übersetzungstechnik
zu ereifern, die in der Natur der gestellten Aufgabe lag, ein
schulmässiges Lehrmittel zu schaffen. Daraus folgt aber auch
ganz klar, dass auch wir, wenn dieses Mittel den Manju för-
derlich war, denselben Nutzen daraus ziehen können. Man bilde
sich natürlich nicht ein, dass wir bereits alles besser machen
könnten als die Manju, und dass ein Europäer, auch wenn er
Jahrzehnte lang in China herumgesessen hat, tiefer in den Geist
des Chinesentums eingedrungen wäre als die Manju, die ihn
mit der Muttermilch einsaugen. Wenn die Manju eine Stelle *so*,
und nicht anders, übersetzt haben, dann werden sie auch trif-
tige Gründe dazu gehabt haben. Nur selbstgefälliger Dünkel und
die Arroganz der Ignoranz können zu dem engherzigen Dogma
gelangen, dass das Manju eine quantité négligeable sei. Ganz
objektiv lässt sich ferner sagen, dass es ein Grundprincip ge-
sunder philologischer Kritik und Methode ist, dass zur Bearbeitung jedes literarischen Denkmals sämtliche darauf bezüglichen
vorhandenen Quellen, welchen Charakters sie auch sein mögen,
zu Rate gezogen und ausgenützt werden müssen. Diese Regel
ist auch auf unseren Fall anzuwenden: wenn ein chinesisches
Werk, zu dem eine Manju-Übersetzung vorhanden ist, eine Bearbeitung erfahren soll, dann ist es eben eine unabweisbare
Forderung philologischer Methode, die Manju-Übersetzung zu
studiren, eben weil sie vorhanden ist, als ein Dokument, das
sich auf die zu behandelnde Quelle bezieht und ihr Verständnis
wesentlich zu erleichtern und aufzuklären vermag. Das ist freilich nur ein Nebengewinn, den das Manju abwirft, und nicht
der Zweck des Manju-Studiums. Die ganze Frage, was und wieviel
es für die chinesische Literatur nützt, ist ziemlich müssig und
nichtssagend im Vergleich zu dem Problem, das uns das Vorhandensein der manjurischen Literatur stellt, und dieses Problem sind nicht die Chinesen, sondern die Manju selbst. Wie
über alle anderen Völker der Erde, so wünschen wir auch über
ihre Sprache, Geschichte, Kultur, und die Stellung ihrer Kultur
zu anderen Stämmen und Gruppen Klarheit zu erhalten. Ein
Volk, das, ursprünglich ein unbedeutender Stamm des östlichen
Sibiriens, China erobert, diesem Reiche zwei seiner grössten

Herrscher geschenkt und noch jetzt die politischen Geschicke
Chinas leitet, verdient unsere Anteilnahme im höchsten Masse
und hat gerechten Anspruch auf eine gründliche und gewissen-
hafte Erforschung aller von ihm hinterlassenen Denkmäler.
Mit Ausnahme der Khitan und Niüchi, von deren Sprache
wir eine nur sehr fragmentarische Kenntnis haben, ist seine
Sprache die einzige schriftlich fixirte und literarische der grossen
tungusischen Gruppe und allein schon um ihrer selbst willen
von grösstem Interesse. Und das Hauptproblem ist hier noch
zu lösen, das in der Frage gipfelt, wie das Manju vor dem Con-
takt mit der chinesischen Kultur ausgesehen hat. Jeder, der die
Wörterbücher von v. d. Gabelentz und Zacharov nur flüchtig
überblickt, wird von der grossen Zahl der Bedeutungen betroffen
sein, die den einzelnen Wörtern, besonders den Verben, zugeschrieben werden, und in diesem Punkte liegt bekanntlich die
einzige Schwierigkeit der Sprache; in den Übersetzungen aus
dem Chinesischen mussten naturgemäss unzählige neue Gedanken
ausgedrückt werden, und der natürliche Fluss der Sprache wurde
in ein neues ungewohntes Bett gelenkt. Zur Wiederherstellung
des ursprünglichen Sprachschatzes sind drei Aufgaben erforderlich, einmal die Feststellung und Ausscheidung der aus dem
Chinesischen empfangenen Lehnwörter,[1] sodann die Festsetzung
der ursprünglichen Bedeutungen und der später unter chinesischem Einfluss adoptirten oder forcirten Bedeutungen, und drit-
tens Vergleichung des Wortschatzes mit den verwandten tungusischen Sprachen, woraus sich das vorchinesische Gewand
am besten herausarbeiten liesse.[2] Da das Manju jetzt als tote
Sprache angesehen werden muss, und das gesamte Sprachmaterial in der einheimischen Lexikographie völlig abgeschlossen
vorliegt, empfiehlt es sich vorzüglich für Studien dieser Art.

---

[1] Gute Vorarbeiten von E. v. Zach, Über Wortzusammensetzungen
im Mandchu, WZKM, Band XI, 1897, S. 242—248; Manchurica, China
Review, Vol. XXIV, 1899, pp. 47—48, 196--198, 268—269.

[2] Eine gute Vorarbeit liegt vor in dem Werk von Wilhelm Grube,
Goldisch-deutsches Wörterverzeichnis mit vergleichender Berücksichtigung
der übrigen tungusischen Dialekte, Anhang zum III. Bande (2. Lieferung)
von L. v. Schrenck's Reisen und Forschungen im Amur-Lande, St. Petersburg, 1900.

Kenntnis des Chinesischen ist nun eigentlich wieder zum vollen
Verständnis der manjurischen Literatur erforderlich, aber es
wäre durchaus ungerecht und einseitig, sie darum für ein Sonder-
pachtgebiet der Sinologen abstecken zu wollen. Das Interesse
für die Sprachwissenschaft und die altaischen Sprachen ist gegenwärtig so gross, dass es durchaus berechtigt ist, wenn Forscher, die z. B. vom Türkischen oder Mongolischen kommen
oder die tungusischen Sprachformen studiren wollen oder irgendwelche andere Zwecke im Auge haben, sich eine Kenntnis des
Manju, ohne Vermittlung des Chinesischen, zu verschaffen be-
strebt sind. Dies ist meiner Ansicht nach auch sehr wohl möglich, und solche Bestrebungen sollten mit allen Kräften gefördert werden. Glücklicher Weise ist für solche Fälle auch ein
Textmaterial in Bereitschaft, das mit grossem Nutzen ganz ohne
chinesische Kenntnisse dem Studium des Manju als sichere
Grundlage dienen kann. Und hier möchte ich mir im Interesse
solcher Sprachforscher wie des Fortschritts der Manju-Studien
überhaupt ein Wort pro domo erlauben. Nach dem, was oben
über die Technik der Übersetzungen aus der klassischen Literatur bemerkt wurde, dürfen wir keine grossen Erwartungen
hegen, in diesen viel ureigenes reines Sprachmaterial vorzufinden,
und der Sprachforscher sollte sich meines Erachtens davon fern-
halten, da hier chinesischer, und kein manjurischer Stil vorliegt;
auch wird sich der Anfänger aus dieser Lektüre eher eine Aversion gegen die Sprache holen als sein Interesse an derselben
allmählich gesteigert finden.

Die Gedanken der chinesischen Philosophen sind gewiss
sehr edel und erhebend, und ihr wohl durchdachtes System
der Ethik ist aller Hochachtung wert. Aber selbst der gedul-
digste Mensch der Gegenwart kann so viel Tugend in so starken
Dosen auf die Dauer nicht vertragen, die Tugend wird schliesslich langweilig,[1] und das sogenannte Laster, weil es eben all-

---

[1] Ich selbst muss bekennen, dass ich die grösste Bewunderung für
die chinesische Philosophie, wie für das chinesische Volk überhaupt hege;
aber ich möchte den Europäer sehen, den beim Lesen der confucianischen Schriften noch nie das Gefühl der Langeweile beschlichen hat. Bei
der Lektüre buddhistischer Literatur (recht lang ausgesponnene Jātaka
ausgenommen habe ich dieses Gefühl durchaus nicht.

gemeiner und menschlicher ist, hat nun einmal die Anziehungs-
kraft des Interessanten. Den Manju und Chinesen geht es darin
gerade so wie uns: die moralischen Klassiker werden auf der
Schulbank eingepaukt, weil sie im Examen verlangt werden und
den Weg zu Rang und Amt öffnen (ganz wie bei uns), während
*Kin P'ing Mei, Liao chai chih i* und andere inhaltlich fesselnde
Bücher die häusliche Lektüre bilden. Wer kann es ihnen
schliesslich verdenken? Die Manju-Übersetzungen dieser Roman-
literatur nehmen sich daher ganz anders aus als die der klassischen Literatur und sind in einem fliessenden, allgemein ver-
ständlichen Stil geschrieben, welcher der wirklichen Umgangssprache der Manju sehr nahe kommen muss. Hier ist die Sprache
keine blosse Paraphrase, keine tote Aneinanderreihung übersetz-
ter Wörter, sondern ein lesbares, lebendiges Gebilde. Die Manju-
Version des *Kin P'ing Mei* ist ohne den chinesischen Text edirt
worden, nur Personen- und Ortsnamen und einige seltenere,
dem Gedankenkreis der Manju fremde Wörter sind darin chinesisch glossirt: d. h. die Manju-Ausgabe bildet für sich ein selb-
ständiges, geniessbares Ganze und kann ohne das Original verstanden werden. Ich bin der Meinung, dass diese Übersetzungen der Erzählungsliteratur den zukünftigen Forschungen über
die Manju-Sprache zu Grunde gelegt, und dass sich unsere Anschauungen von derselben wesentlich nach diesen, nicht aber
so sehr nach der Kunstsprache der klassischen Literatur, bilden
sollten. Man vergleiche eine mongolische Erzählung mit einer
aus dem Chinesischen oder Tibetischen übersetzten mongolischen Schrift und frage sich, wie uns das Mongolische erschiene,
wenn wir jene nicht, und nur diese, besässen; dieselbe Anwendung dürfen wir, cum grano salis, auf das Manju machen. *Kin
P'ing Mei* und *Liao chai chih i* müssen daher als Quelle der
echten Sprache (so unverfälscht darin wenigstens, als wir sie
überhaupt haben, und sie den Umständen nach sein kann) zum
Studium empfohlen werden, und es ist kein Grund vorhanden,
warum das Studium nicht von vornherein mit diesen beginnen
sollte, aus denen der Leser, einer anziehenden Lektüre gewiss,
zugleich ein höchst lehrreiches Bild des chinesischen Kulturlebens gewinnen kann.

Dem Manju kommt ferner eine nicht zu unterschätzende

Bedeutung für das Studium der Geschichte Chinas während der
letzten drei Jahrhunderte zu. Die Geschichte der gegenwärtigen
Dynastie kann, auch was ihre chinesischen Quellen betrifft, ohne
Kenntnis des Manju nicht gründlich verstanden und gewürdigt
werden. Ebenso wie ein erfolgreiches Verständnis der Geschichte
der Yüan-Dynastie Bekanntschaft mit der mongolischen Sprache
voraussetzt, so enthalten alle Dokumente der Ts'ing-Dynastie
Manju-Namen, Ausdrücke und Anspielungen auf specifisch manjurische Dinge und Einrichtungen, mit denen man sich nur
durch eine Kenntnis der manjurischen Sprache selbst völlige
Vertrautheit erwerben kann; es genügt nicht, die Manju-Namen
und Termini in ihrer chinesischen Transkription zu copiren,
sondern sie müssen in ihrer richtigen Lesung restituirt werden.
Zahlreiche Dokumente der Pekinger Ministerien, in erster Reihe
natürlich solche, die sich auf Angelegenheiten der Acht Banner,
auf die Verwaltung der Manchurei und Mongolei beziehen, sind
auch ausschliesslich in manjurischer Sprache aufgesetzt worden;
solcher Aktenstücke sind schon viele ans Licht gekommen,[1]) in
Zukunft werden sie jedenfalls zu einer weit grösseren Bestimmung berufen sein und interessantes Material zur Geschichte
der Dynastie und ihres militärischen Verwaltungssystems bringen.
Ich brauche kaum zu erwähnen, dass auch die Siegel und Münzen
dieser Periode ausser chinesischen auch manjurische Legenden
tragen, und dass der Numismatiker wie der Sammler historischer und archäologischer Objekte der Kenntnis der Manju-
Sprache nicht entraten können.[2]) Für die Bearbeitung des epi-

---

[1]) Beispiele bei G. SCHLEGEL und E. v. ZACH, Zwei mandschu-
chinesische kaiserliche Diplome, T'oung Pao, Vol. VIII. 1897, pp. 261—
308; M. F. A. FRASER, A Manchu Ukase, Journal of the China Branch of
the Royal Asiatic Society, Vol. XXX, 1895—1896, pp. 161—176; F. W. K.
MÜLLER, Zeitschrift für Ethnologie, Band XXXIV, 1902, S. (252)—(255).

[2]) Der Museumsbeamte, der mit den Äusserungen und Erzeugnissen
des realen Kulturlebens Chinas zu thun hat, wird natürlich am ersten
in die Lage versetzt sein, auch vom Manju praktischen Gebrauch zu
machen; bald verlieren sich zu ihm einige verstreute Bände eines zu be-
stimmenden Werkes, bald Stickereien, Bilder, Amulete, Medaillen mit
Manju-Aufschriften, und das ihn täglich bestürmende Heer der China-
sammler ist ja während der letzten sieben Jahre zu einer wahren Hoch-
flut angeschwollen.

graphischen Materials, das unter der gegenwärtigen Dynastie eine gewaltige Ausdehnung erreicht hat,[1] ist die Kenntnis des Manju ganz selbstverständlich und conditio sine qua non.

Aber auch damit ist der Inhalt der Manju-Studien noch nicht erschöpft. Wenn schon die historischen Dokumente den Einfluss der manjurischen Sprache zeigen, so hat sie noch viel tiefer der Umgangssprache des nördlichen Chinas ihren Stempel aufgedrückt. Wir wissen jetzt, dass die Sprache der Hauptstadt Peking von manjurischen Worten und Redensarten wimmelt, und dass viele sogenannte Pekingismen eben nichts anderes als Manjuismen sind.[2] Eine wissenschaftliche Darstellung der nordchinesischen Volkssprache wird daher auch auf das Manju Rücksicht zu nehmen haben, und auch der in Peking tätige Dolmetscher wird stets gut daran tun, mit einer gewissen Kenntnis des Manju ausgerüstet zu sein, wenn auch mehr zu idealen als zu praktischen Zwecken; die Fähigkeit, einem hohen Manju-Würdenträger ein par Sätze in der Sprache seiner Ahnen zu sagen, ihm gelegentlich zu zeigen, dass wir das verlorene Erbe angetreten haben und bewahren, gehört zu den kleinen Würzen des Daseins, die man sich selbst gern gönnen sollte. Kein Reisender im nördlichen China endlich, der nicht zu dem umnachteten Heer der Globetrotter gehört, sollte sich um das Vergnügen bringen, die zahlreichen Manju-Aufschriften auf Toren, Tempeln, Grabdenkmälern u. s. w. lesen zu können, ebenso, wie es uns Freude macht, bei einer Reise in Europa lateinische Inschriften auf Gebäuden und Monumenten zu verstehen.

Das Problem der manjurischen Kultur ist weit schwieriger zu lösen als das der Sprache. Gebricht es doch leider an allen Denkmälern, die den unverfälschten Geist der Manju atmen. Ihre Literatur ist Übersetzungsliteratur, ihre Gedanken sind den Chinesen erborgt, und in den wenigen scheinbar originalen Werken bezieht sich die Originalität wesentlich auf die Composition, aber nicht auf den Inhalt, der wiederum ein Nachhall

---

[1] Vergl. die Skizze der mongolischen Literatur unter «Neuere Inschriften».

[2] E. v. ZACH, Lexicographische Beiträge, Band I, Peking, 1902, S. 51—54.

chinesischer Ideen ist. So achtunggebietend der plötzliche und
rasche Aufschwung der Manju zu den Höhen der chinesischen
Civilisation auch ist, so beklagenswert ist vom Standpunkt der
Wissenschaft, dass sie es von vornherein nicht für der Mühe wert
erachtet haben, die alten Erzeugnisse ihrer Volksliteratur der
Nachwelt zu überliefern. Bedingungslos haben sie sich dem pro-
saischen Confucius in die Arme geworfen und über der neuen
Liebe das alte Volkslied am heimischen Herde treulos sitzen
lassen. Keine Volksweise, keinen Heldensang. keine alte Mär.
kein Gebet, das der Schamane nächtlich beim flackernden Feuer
der Jurte zur Zaubertrommel sang, haben sie in ihrer Sprache aufgezeichnet. Diese Tatsache ist überaus merkwürdig und ein in
der Geschichte der uralaltaischen Völkerfamilie vielleicht einzig
dastehender Fall. Die Finnen, die Magyaren, die Türken und
selbst die bigotten Mongolen haben uns kostbare Vermächtnisse
aus ihrer heidnischen Vorzeit hinterlassen: keine Spur von
alledem bei den Manju, und es ist wahrlich schwer, eine ad-
äquate Erklärung für diese Erscheinung zu finden. Dass sie
vor der Berührung mit China nichts, rein nichts besessen haben
sollen, ist schlechterdings undenkbar, denn ihre nächsten Verwandten, die Golden am mittleren und unteren Amur. besitzen
noch jetzt einen reichen Schatz von Mythen und Sagen, besonders Heldenmärchen. Freilich gibt es gegenwärtig eine ganze Anzahl russificirter Tungusenstämme in Sibirien, die ihre alten
Sagen fast oder ganz vergessen haben und auf die Aufforderung
des Reisenden, Geschichten zu erzählen, ihre Kenntnisse der
Bibel auskramen. Hier haben wir vielleicht eine Analogieerschei-
nung zu den Manju, von denen sich wohl nur denken lässt,
dass die neue Welt von Ideen, die auf sie einstürmte, sie so
mächtig ergriff, dass die alte darunter zusammenbrach, und dass
die absorbirenden Geschäfte nach der Besetzung von China, die
sie fortwährend auf dem qui vive hielten, ihnen keine Musse
liessen, sich auf ihr Altertum und ihr Volksleben zu besinnen.
Manjurische Eigenart ist dem Chinesentum erlegen, die Manju
sind zu Chinesen geworden. Bei dieser betrübenden Sachlage
muss es als eine wissenschaftliche Grosstat bezeichnet werden,
dass Prof. WILHELM GRUBE in seinem grundlegenden Werke
«Zur Pekinger Volkskunde» alles sorgfältig gesammelt hat, was

noch gegenwärtig von manjurischen Sitten und Bräuchen be-
steht, und so noch die letzten Reste des Geistes eines in seiner
Selbständigkeit ganz dahinschwindenden grossen Volkes für die
Forschung gerettet hat.[1])

Hier mögen denn auch einige Worte über die Geschicke
der Manju-Sprache Platz finden. Für das siebzehnte Jahrhundert
dürfen wir wohl im allgemeinen annehmen, dass das Manju noch
ein von der ganzen Nation lebendig empfundenes und allgemein
gesprochenes Idiom war. Doch bereits in der ersten Hälfte des
folgenden Säkulums sind Spuren eines allmählichen Rückganges,
eines Gebietsverlustes der Sprache wahrnehmbar. Als im Jahre
1708 auf kaiserlichen Befehl «der Spiegel der Manju-Sprache»
(*Manju gisun-i buleku bithe*) von einem gelehrten Ausschuss bearbeitet wurde, dienten als Grundlage die damals vorhandenen
Bücher, verschiedene alte Handschriften und bei *alten Leuten*
angestellte Umfragen.[2]) Dass man gerade zu solchen als den Be-
wahrern der Sprache seine Zuflucht nehmen musste, ist bemerkenswert, und die Situation wird noch weit klarer ausgesprochen
von dem Manju-Gelehrten DAIGU im Vorwort zu seinem vortreff-

---

[1]) Chinesische Nachrichten über die Sitten der Manju findet man
in dem historischen Werke *K'in ting Man-chou yüan liu k'ao* («Auf kaiserlichen Befehl verfasste Untersuchungen zur Geschichte der Manju»,
8 Vols., 1777) in Buch 16—19; doch sind die meisten hier gegebenen
Nachrichten Auszüge aus älteren Historikern, die sich auf andere tungusische Stämme beziehen. Eine bessere Quelle ist das *Man-chou se li tsi*
(«Sammlung der vier Rituale der Manju», 4 Vols., 1801, mit Vorwort von
1796), verfasst von SO NING-NGAN, worin die Festtage, die Geister und ihr
Kultus, Opferdarbringung, Bräuche bei Begräbnis und Trauer beschrieben
werden. WYLIE (Translation of the Ts'ing Wan K'e Mung, Shanghai, 1855,
p. XLII) erwähnt ohne nähere Angaben ein Manju-Buch *Daicing gurun-i
uheri kooli* «Bräuche der Dynastie Ta Ts'ing», das mir nicht bekannt ist,
aus dem aber POZDNÉJEV in seiner Sammlung von Manju-Texten einige
Abschnitte mitteilt.

[2]) J. B. DUHALDE (A Description of the Empire of China, Vol. II,
London, 1741, p. 265) erzählt gleichfalls, dass die alten Leute der Acht
Banner befragt wurden, und setzt hinzu, dass, wenn unbefriedigt, man
diejenigen consultirte, die gerade von den äussersten Teilen ihres Landes
gekommen waren; Belohnungen wurden für die Entdeckung alter Worte
oder Phrasen ausgesetzt, und man trug Sorge, sie denen einzuprägen, die
sie vergessen oder vielmehr nie gelernt hatten.

2*

lichen Wörterbuch vom Jahre 1722: er bekennt darin, dass ihn zu seinem Werke die Erwägung angetrieben habe, dass die Manju schon zu seiner Zeit ihre heimische Sprache zu vergessen begannen, dass er selbst, obwohl von einem Geschlechte aus der Manchurei und von einer langen Linie manjurischer Beamter abstammend, in seiner Jugend diese Sprache nicht gekannt, sondern sie erst später erlernt habe, da dies eine zum Eintritt in den Staatsdienst unerlässliche Forderung war.[1]) In der Verwaltung, der Strafgesetzgebung, den Gerichten war die manjurische Sprache noch geläufig, sie war Beamtensprache und der Weg zur Erlangung eines höheren Amtes. Wie sich damals das Volk zu seiner Muttersprache gestellt hat, ist schwer zu sagen; auf der einen Seite wird man annehmen dürfen, dass der am meisten mit Chinesen in Berührung kommende Teil, vor allem die hauptstädtische Bevölkerung, am meisten der Gefahr des Vergessens ausgesetzt war, auf der anderen Seite, dass die im nördlichen Chihli, in der Manchurei und an isolirten Punkten ansässige Landbevölkerung vielleicht am längsten der lebendige Träger manjurischer Rede geblieben ist.

Aus der 1747 datirten Vorrede des Kaisers K'ien-lung zum «Ritual der Manju» erfahren wir, dass die Schamanen der früheren Zeiten, die alle aus der Mandschurei stammten, die Sprache von Kindheit auf erlernt hätten; später, als dies aufhörte, beschränkte sich ihre Kenntnis derselben auf das mechanische Auswendiglernen von Gebeten, die auf diese Weise immer unrichtiger und verständnisloser überliefert wurden.

In der zweiten Hälfte des achtzehnten Jahrhunderts nehmen die Klagen über das Vergessen des Manju immer mehr zu, das nur noch von Beamten wie eine fremde Sprache erlernt wurde, um die gebührende Qualification für Amt und Würde zu erlangen. Die Regierung liess es an Edikten und Ermahnungen nicht fehlen und setzte Belohnungen in Form hoher literarischer Grade auf manjurische Schriften aus. Der natürliche Umschwung der Dinge konnte aber damit nicht geändert werden. Die Manju-Sprache war dem Untergang ge-

---

[1]) ZACHAROV, Vorwort zum Wörterbuch, p. XVI.

weiht, die all-absorbirende Kraft des Chinesentums hatte sie verdrängt.[1])

In Ili und anderen fernen Stationen, wo die Manju Truppen in Garnison gelegt haben, soll auch jetzt noch Manju gesprochen werden, und W. RADLOFF hat dort Aufzeichnungen aus dem Munde von Manju gemacht. Ebenso mag es entlegene Winkel in der Manchurei geben, wo hie und da noch ein der Sprache mächtiges Individuum lebt. In Peking und in der Provinz Chihli dagegen ist sie unwiderruflich tot, und es dürfte selbst zu bezweifeln sein, ob sie noch unter den Mitgliedern der kaiserlichen Familie ein künstliches Dasein fristet.

Nach J. EDKINS[2]) hat noch jetzt jedes der Acht Banner seine Schule in Peking, in welcher die Knaben Unterricht im Chinesischen und Manju erhalten; doch nur eine geringe Zahl der Schüler bemüht sich um die letztere Sprache, und wenn, als eine Vorstufe zur Beförderung. In der Regel hat ein Lehrer zwanzig Knaben unter sich. In einer grossen Schule von etwa 160 Knaben sind acht Lehrer, von denen nur einer Manju lehrt so dass die Zahl der Manju Studirenden auf etwa ein Achtel der Schüler geschätzt werden darf.

Wie die mongolische, so ist auch die manjurische Sprache in Korea studirt und gepflegt worden. MAURICE COURANT hat darüber interessante Mitteilungen gemacht und eine sorgfältige Liste der in Korea gebrauchten oder hergestellten Manju-Bücher redigirt.[3]) Aus seinen Untersuchungen geht hervor, dass die Ko-

---

[1]) Es war der Process eines langsamen natürlichen Todes, wie man ihn z. B. hier in Amerika fast alltäglich beobachten kann. In vielen deutschen Familien sprechen die Eltern deutsch, die Kinder nur englisch. Lange kämpfen die Eltern dagegen an, bis sie resignirt den Kampf aufgeben. In anderen Familien behaupten die Kinder noch deutsch zu verstehen, aber nicht sprechen zu können, oder sie vermögen es zu radebrechen, in der nächsten Generation wird natürlich nur englisch gelispelt. Die Erscheinung ist betrübend, aber der Grund liegt auf der Hand; im Schul-, wie im Geschäftsleben ist das Englische erforderlich, und den meisten Eltern fehlt es an der Zeit, der Energie und dem Geist, zu Hause die Muttersprache zu pflegen.

[2]) Description of Peking, Shanghai, 1898, p. 34.

[3]) Bibliographie coréenne, Vol. I, Paris, 1894, pp. 79—92. Supplément à la bibliographie coréenne, Paris, 1901, p. 4.

reaner schon früh die Sprache der Niüchi betrieben, dass spä-
testens 1469 das Studium derselben im officiellen Dolmetscher-
amt von Söul vertreten war,[1] dass es damals Konversations-
bücher und Übersetzungen chinesischer Bücher in dieser Sprache
gab, die in den Examina gebraucht wurden. Im siebzehnten
Jahrhundert wurde das Niüchi vom Manju verdrängt, und seit
1636, also schon acht Jahre vor der Einnahme Chinas durch
die Manju, bediente man sich ihrer Sprache in officiellen Briefen,
wie in der Unterhaltung. Bereits in jenem Jahre wurde ein
Werk zur Erlernung des Manju von Koreanern verfasst, die
während des Krieges in der Manchurei Gefangene gewesen waren.
Die Fehler dieser ersten Arbeit wurden 1760 von einem koreanischen Dolmetscher mit Hülfe eines Manju-Sekretärs von Nin-
guta verbessert, und das neue Werk, welches Manju-Gespräche
mit koreanischer Transkription und Übersetzung enthält, wurde
1765 gedruckt. Das Vorwort desselben erinnert daran, dass das
Studium der Manju-Sprache für die Koreaner unerlässlich sei,
während es in der koreanischen Vorrede einer Manju-Übersetzung von Auszügen aus dem *San kuo chi* heisst, dass das
Manju die wichtigste aller (d. h. fremden) Sprachen für die
Koreaner sei. Im Jahre 1764 belief sich die Zahl der ihnen zur
Verfügung stehenden Lehrbücher auf zwanzig Bände. Interessant
ist vor allem die Tatsache, dass die Koreaner auch Manju-Bücher
mit beweglichen Typen gedruckt haben, und zwar drei Werke
im Jahre 1703, während man in China inbezug auf das Manju
nie auf diese Idee verfallen zu sein scheint; bekanntlich haben
die Koreaner seit 1403 den Typendruck mit Vorliebe und Eifer
gepflegt.

   **Bestand und Inhalt der manjurischen Literatur.** —
Bibliographien der Manju-Literatur sind wiederholt versucht
worden, so von C. DE HARLEZ,[2] zuletzt von P. G. v. MÖLLEN-
DORFF,[3] letztere Arbeit im allgemeinen nützlich und annähernd

----

[1] L. c., Vol. I, pp. CLXXV, 88.

[2] Notes de bibliographie tartare-mandchoue, Mémoires de la Société
des études japonais, Vol. V, 1886, pp. 61—63, 132—135, 208—209; Vol.
VI, 1887, pp. 54—56, 126.

[3] Essay on Manchu Literature, Journal of the China Branch of
the Royal Asiatic Society, Vol. XXIV, 1889, pp. 1–45.

vollständig.[1]) Dagegen ist die Statistik, die er von dem Bestande der Literatur entwirft, aus der er 249 Werke herausrechnet, auf einer etwas trügerischen Grundlage aufgebaut. Nicht nur, dass verschiedene (und oft identische) Editionen desselben Werkes oder die blosse Abschrift eines Druckes unter besondere Nummern gebracht sind, figuriren als besondere Werke der Manju-Literatur auch geschriebene Examensaufsätze (Nr. 105—107), Diplome (Nr. 173, 185—187) und die einzelnen in VASILJEV's Chrestomathie abgedruckten Verträge mit Russland (Nr. 177, 179, 188, 196, 213, 217, 218). Scheidet man dieses und anderes nicht hierher Gehörige aus, so gelangt man zu einer reinen Zählung von 168 verschiedenen Werken; rechnet man dazu etwa ein Dutzend solcher, die MÖLLENDORFF entgangen oder unbekannt gewesen sind, so gibt die Zahl von 180 einen wohl annähernd korrekten Schätzungswert von dem materiellen Bestand der Manju-Editionen. Man sieht, dass es sich nicht um eine Literatur handelt, die durch Quantität imponirt, obwohl einzelne Werke ganz respektable Dimensionen annehmen.[2]) Man erkennt daraus ferner, dass das Gebiet dieser Literatur ein begrenztes sein muss, und dass sie nicht im entferntesten ein getreues Spiegelbild der ganzen immensen chinesischen Literatur in ihrem weitesten Umfang darstellt und darstellen kann. In der Tat, sie ist nur ein kleiner Ausschnitt aus diesem grossen Gebiete, ein Florilegium, eine Auswahl des Besten, d. h. dessen, was die Manju-Kaiser im Interesse ihrer Bannerleute für das Beste hielten. Gerade nun in der Wahl, die sie getroffen, zeigt sich in charakteristischer Weise der praktische Blick der Manju. Abgesehen von den sprachlichen Hülfsmitteln, die sie ins Leben gerufen, teils um ihre eigene Sprache zu erhalten, teils um

---

[1]) Doch schlecht angeordnet, nicht frei von Ungenauigkeiten in Einzelheiten und mit einer etwas starken Portion von Druckfehlern.

[2]) So z. B. «Die heiligen Belehrungen» (enduringge tacihiyan) des Kaisers K'ien-lung, die nach MÖLLENDORFF, Nr. 207, 150 Bände umfassen soll. Diese Angabe stützt sich aber nur auf den Umstand, dass er selbst nur die Bände 145—150 dieses Corpus besass. Vollständige Ausgaben sind wohl nirgends vorhanden. Ich habe in Peking 18 Bände erlangt, nämlich Band 1—6, 199—204 und 271—276. Wie aus dem Vorwort und Inhaltsverzeichnis hervorgeht, umfasst das Werk dreihundert Bände.

durch Vermittlung derselben die chinesische zu lehren, besteht die von ihnen gepflegte Literatur fundamental aus den Schriften der praktischen chinesischen Moralphilosophie, den Sätzen der confucianischen Schule, die dem Volk der Manju einzuimpfen keine Mittel gescheut wurden, und denen sich ihr Geist als in hohem Grade anpassungsfähig erwies. Daher die Energie, mit der die Übersetzung der sogenannten Klassiker betrieben wurde, daher die »heiligen Edikte« und immer mächtiger anschwellenden »heiligen Belehrungen«, daher die vielen Spruchsammlungen mit ihren schönen Sentenzen von Tugend, Moral und praktischer Lebensweisheit, deren sich das Volk befleissigen muss, wenn es dem Herrscher und den Beamten gut gehen soll. Obwohl sich die grossen Manju-Kaiser K'ang-hsi und K'ienlung für alle Religionen interessirten und sich der Dienste der Jesuiten wie der Lamas bedienten, waren sie in Wahrheit freie Menschen und keiner Sorge um die religiöse Erziehung ihres Volkes hingegeben. Zu einer religiösen Literatur hat es daher die Sprache der Manju nicht gebracht, die par vorhandenen buddhistischen und taoistischen Traktätchen sind nicht der Rede wert. Die Rücksicht auf den allgemeinen gesunden Menschenverstand (und teilweise auf den beschränkten Untertanenverstand vielleicht), auf die praktischen Bedürfnisse des Hauses und Herdes, war der leitende Gesichtspunkt für ihr eklektisches Verfahren: an den tiefsinnigen Denkern der Taoisten gingen sie achtlos vorüber, den Lehren des Çâkyasohnes gewannen sie keinen Geschmack ab; die lebenslustigen, weinfröhlichen Dichter der T'ang und Sung hätten die Reinheit der Moral des Volkes trüben können. Aber das Volk gab sich mit der unter kaiserlicher Censur verabreichten trockenen Kost nicht immer zufrieden; man sehnte sich nach Unterhaltung und guten Geschichten, und angesteckt von der chinesischen Lebensfreude, entdeckte man solche mit Entzücken in den das pochende Leben schildernden Novellen und Romanen der Nachbarn, die dann heimlich übersetzt und mit einem vom Überdruss der Schulethik doppelt gesteigerten Reizhunger verschlungen wurden. Man muss beachten, dass diese Gattung der Literatur das Unternehmen Privater war und nicht wie die vorher erwähnte unter allerhöchstem Protektorat erschien; im Gegenteil, die Kaiser wider-

setzten sich diesen Bestrebungen und fanden in ihrem Interesse,
wie sie glaubten, es für gut, die belletristischen Erzeugnisse als
«unmoralisch» zu brandmarken. Auch in der Auswahl der historischen und geographischen Literatur waren die Kaiser wesentlich vom nationalen oder dynastischen Interesse bestimmt; es
ist interessant, dass sie sich aus der Reihe der officiellen Reichs-
annalen gerade drei, die der Khitan oder Liao, die der Niüchi
oder Kin und die der Mongolen oder Yüan — für die sie das
Gefühl der historischen Zusammengehörigkeit hatten — zur
Bearbeitung in ihrer Sprache aussuchten, und dass sie sich für
den Rest der Geschichte Chinas mit den zusammenfassenden
Darstellungen des Yüan Huang und des Se-ma Kuang begnüg-
ten. Was sonst auf diesem Gebiete vorhanden ist, betrifft das
Zeitalter der Dynastie selbst. Rechnen wir dazu die Übersetzungen einiger Schriften über Kriegskunst, des Strafgesetzbuchs,
einiger anderer Codifikationen und der Sammlungen der Aktenstücke der Ministerien, so ist der geistige Horizont, innerhalb
dessen sich die Manju-Literatur bewegt, ausreichend gekenn-
zeichnet.

Was die Qualität der Übersetzungen selbst betrifft, so hat
G. v. d. Gabelentz[1] geurteilt: «Die meisten dieser Übertra-
gungen können als authentische, manche von ihnen als meister-
hafte gelten». Und Zacharov:[2] «Ihrem Fleiss und ihrer Arbeit-
samkeit im Studium der chinesischen Literatur kann man die
Achtung nicht versagen, und in dieser Hinsicht haben sie ihrem
Volke einen grossen und wichtigen Dienst erwiesen. Sie haben
die klassischen chinesischen Bücher genau und mit Verständnis
übersetzt und dadurch vor den vielbändigen Kommentaren bewahrt, über denen Lernende und Lehrende in China fast ihr
ganzes Leben verbringen; sie haben ferner die Geschichte Chinas,
wenn auch nicht vollständig, übersetzt, dann die Gesetze, die
man mit Kommentaren lesen muss, nicht nur zur Erklärung von
Ereignissen, sondern auch von Wörtern. Alle diese Übersetzungen sind klar und korrekt, so dass man jedem die chinesische
Sprache und Literatur Studirenden behufs gründlicher Kenntnis

---

[1] Chinesische Grammatik, Leipzig, 1881, S. 18.
[2] Vorwort zum Wörterbuch, p. XX.

raten kann und muss, die chinesischen Bücher mit den manjurischen Übersetzungen zu studiren.»

**Entwicklung der manjurischen Literatur.** - Ein kurzer
historischer Abriss der Übersetzungstätigkeit der Manju mag
die Entwicklung ihrer geistigen Bestrebungen veranschaulichen.
Als ihr erster Schriftsteller wird gewöhnlich der Bildner ihrer
Schrift, Dahai, in Anspruch genommen. Das stimmt nicht genau,
denn Nurhaci selbst hat vor ihm in seiner Sprache geschrieben.[1]
Dies beweist die vom Kaiser K'ang-hsi veranstaltete Sammlung
seiner «heiligen Belehrungen» oder «weisen Ermahnungen».[2]
kleine Skizzen, die der Kaiser mit genauen Daten versehen im
Laufe der Jahre hingeworfen hat, die ältesten aus dem Jahre
1616. Dies sind somit die frühesten Denkmäler in manjurischer
Sprache, und nicht nur von diesem Gesichtspunkt aus, sondern
auch inhaltlich sind sie von grossem historischen Interesse. Sie
beweisen, dass Nurhaci, der zielbewusste Realpolitiker, sehr früh,
wenigstens der officiellen Erscheinung nach, den Manju abge-
streift und den Chinesen angezogen, dass er sich bereits das
ganze chinesische Denken in seiner praktisch-philosophischen
Richtung zu eigen gemacht und seinen Untertanen in Wort und
Schrift predigte. Lange bevor sie den Boden Chinas betraten,
muss daher der Umwandlungs- und Assimilationsprocess der
Manju vor sich gegangen sein, wenigstens in grossen Zügen
eingesetzt haben. In Nurhaci's Aufzeichnungen liegt ein grosser
Teil der chinesischen Terminologie bereits fertig vor, so dass
man der eminenten Bildungs- und Anpassungsfähigkeit der Manju
nur die höchste Bewunderung zollen kann, und in dem kurzen
Zeitmass, dem geschwinden Tempo, in dem sie sich den Geist
des Chinesentums eroberten, liegt ein sehr bemerkenswertes

---

[1] Auch der Kaiser K'ANG-HSI sagt von ihm im Vorwort zum *Manju
gisun-i buleki bithe:* «Voll erhabener Menschlichkeit, in Harmonie mit
Himmel und Erde, verfasste er als erster manjurische Bücher; kraft
seiner grossen Bildung leuchtete er wie Sonne und Mond.» *(Ten-i gosin,
abka na-de acanaha, Manju bithe-be deribume banjibufi, amba šu, šun biyai
gese eldeke.)*

[2] *Tai-tsu Dergi Hōwangdi-i* (= Kao Huang-ti) *enduringge tacihiyan.*
4 Vols., mit Vorwort von K'ang-hsi, datirt 1686, und darauf folgendem
Vorwort von K'ien-lung, datirt 1739. Ohne chinesischen Text.

kulturgeschichtliches Faktum. Aus den Skizzen des Kaisers spricht ein durchaus edler, geläuterter Sinn, eine milde Humanität, die sich selbst auf die Kriegsführung erstreckte. Ein interessantes Dokument in dieser Hinsicht, das gleichzeitig als Literaturprobe aus seinem Buche dienen mag, ist der folgende Tagesbefehl, den er im Jahre 1618 vor seinem Feldzug gegen die Stadt Fu-shan (östlich von Mukden) an seine Armee erliess.

»Im dritten Jahre der Periode Abkai Fulingga (= T'ienming, 1618), einem gelben Pferd-Jahr, im vierten Monat, an einem Tage des schwarzen Tigers.

Der Kaiser stellte ein Heer auf, und im Begriff, gegen die Ming-Dynastie zu Felde zu ziehen, erliess er folgenden Befehl an die *Beise* und Offiziere: Nicht mit Freude beginne ich diesen Krieg. Doch sieben grosse schwere Ungerechtigkeiten bleiben zu sühnen.[1] Kleine Ungerechtigkeiten lassen sich leicht beseitigen, doch bei grosser Unbill beginnt der Krieg. Wenn im Kriege Gefangene gemacht werden, beraubt sie nicht der Kleider, die sie anhaben! Weiber sollt ihr nicht wegschleppen! Ehegatten sollt ihr nicht von einander trennen! Die sich auflehnen, sollen gewiss den Tod erleiden, doch die sich nicht auflehnenden Menschen sollt ihr sicherlich nicht in leichtsinniger Weise niedermachen!»

Noch zu Lebzeiten des Nurhaci war Dahai eifrig mit Übersetzungsarbeiten beschäftigt; er erhielt den kaiserlichen Auftrag, das *Ming hui tien*, «die Statuten der Ming-Dynastie» und mehrere Schriften über Kriegswissenschaft zu übersetzen, und vollendete dieses Werk im Jahre 1631. Seine Übersetzung des *Ming hui tien* scheint verlorengegangen zu sein; wenigstens ist kein Druck derselben bisher ans Licht gekommen. Bei seinem Tode (1632) fand man in seinem Nachlass eine Reihe angefangener

---

[1] Aus anderen Quellen ist bekannt, dass Nurhaci zur Rechtfertigung seines Feldzugs ein Schriftstück aufsetzen liess, in dem er sieben grosse Ungerechtigkeiten aufzählte, die er von der Hand der Ming erlitten hatte. Dies liess er dem Heere vorlesen und dann verbrennen, um es zum Himmel aufsteigen zu lassen und den Schutz des Himmels für sein Unternehmen zu gewinnen. Vergl. z. B. J. MACGOWAN, A History of China. Shanghai, 1897, p. 506. — Die *Beise* sind die Fürsten kaiserlichen Geschlechts zweiten Ranges.

Übersetzungen vor, das *Tung kien kang mu*, *Leu t'ao*, *Mong-tse*,
die Geschichte der Drei Reiche (*San kuo chi*) und ein Sūtra
des Mahāyāna. Diese Aufzeichnungen haben späteren Autoren
als Grundlage gedient, die Übersetzungen sind, mit Ausnahme
des *Leu t'ao* und des Sūtra, vollendet und edirt worden. Literarische Prüfungen im Manju wurden im Jahre 1634 einge-
richtet unter einer Kommission von sechzehn in der Manju,
mongolischen und chinesischen Sprache und Literatur bewan-
derter Gelehrter.[1]) Im Jahre 1644 wurde der Minister HIFE be-
auftragt, die dynastischen Historien der *Liao*, *Kin* und *Yüan*,
sowie das *San kuo chi* zu übersetzen, die 1647, im dritten Jahre
der Einnahme Pekings durch die Manju, publicirt wurden;[2])
von den Liao- und Kin-Annalen wurden je dreihundert, von
der Geschichte der Yüan sechshundert Exemplare gedruckt.[3])

---

[1]) A. WYLIE, Translation etc., p. XXXVII.

[2]) ZACHAROV, l. c., scheint sich im Irrtum zu befinden, wenn er die
Zeit dieser Übersetzungen in die Tage des Dahai verlegt. Die obigen Zeit-
angaben stützen sich auf das manjurische Vorwort zum *Yüan-shih*, wonach
MÖLLENDORFF's Ansetzungen (Nr. 142—144) zu berichtigen sind. — Das
*Liao shih* wurde von H. C. V. D. GABELENTZ (Geschichte der grossen Liao,
aus dem Mandschu übersetzt, St. Petersburg, 1877, mit musterhaftem Index), das *Kin shih* von C. DE HARLEZ (Histoire de l'Empire d'or, Aisin
gurun-i suderi bithe, Paris, 1886) übersetzt. Fragmente aus dem *Yüan
shih* abgedruckt in KLAPROTH's Chrestomathie mandchoue, pp. 121—192.
Es kann nicht genug betont werden, dass das hier in Rede stehende *San
kuo chi* die wirkliche historische Geschichte der drei Reiche ist, und nicht
der Roman, wie MÖLLENDORFF, Nr. 233, glaubt; auch die von DAHAI unvollendet hinterlassene Übersetzung ist die Geschichte, und nicht der
Roman. Ich kann diese Versicherung um so bestimmter geben, als ich
selbst in Peking nach langer Bemühung aus dem Privatbesitz eines Banner-
mannes ein Exemplar des manjurisch-chinesischen *San kuo chi* vom Jahre
1647 erlangt habe. Auszüge daraus sind auch in Korea gedruckt worden
(s. M. COURANT, Bibliographie coréenne, Vol. I, pp. 87—89). Dagegen sind
meine vielfachen Erkundigungen nach der Manju-Übersetzung des Romans
völlig erfolglos geblieben. Dass eine solche vorhanden ist, entnehme ich
dem Buche von TÉODORE PAVIE, Histoire des Trois Royaumes, roman
historique traduit sur les textes chinois et mandchou de la Bibliothèque
Royale, 2 Vols., Paris, 1845, 1846.

[3]) C. DE HARLEZ (Manuel etc., p. 228) ist das Missgeschick passirt,
die gar nicht misszuverstehende Stelle im Vorwort des *Yüan-shih* zu übersetzen, dass das *Liao* und *Kin shih* in 300 Kapiteln, das *Yüan-shih* in

Im Jahre 1647 folgte die erste Manju-Ausgabe des Strafgesetzbuchs (*Ta Ts'ing lü li*, Neuausgabe, 1766).[1]) Zu der nämlichen Zeit wurden auch acht Schulen zur Erziehung der manjurischen Jugend gegründet, je eine für die acht Banner, in welche die Manju zerfallen.

Im Jahre 1654 erschien das erste Werk der klassischen Literatur, das «Liederbuch» (*Shih king*), Manju und chinesisch. Es ist diese ältere Version, die H. C. v. D. GABELENTZ[2]) transkribirt herausgegeben hat, die, wie auch die Ausgabe der «Vier Bücher» von 1691, noch mit einer Menge chinesischer Wörter überladen ist. Später machten sich immer mehr puristische Bestrebungen geltend, und in den Neuausgaben der Klassiker unter K'ien-lung sind die Lehnwörter meist durch einheimisches Sprachgut ersetzt worden.[3])

In den Jahren 1655 und 1656 veröffentlichte der Kaiser Shun-chih einige ethische Abhandlungen zum Besten seiner Beamten.[4]) Ein grosser Aufschwung der Literatur begann in der

---

600 Kapiteln abgefasst worden sei. Und dabei gibt WYLIE (Notes on Chinese Literature, p. 16) eine Tafel der 24 Geschichtswerke mit der Angabe der Kapitelzahl für jedes, und dabei hat HARLEZ selbst das aus 135 Kapiteln bestehende *Kin shih* übersetzt.

[1]) Einige Fragmente daraus abgedruckt bei C. DE HARLEZ, Manuel etc., pp. 130—133, übersetzt von W. BANG, Le Muséon, Vol. IX, 1890, pp. 475—477 (ohne den chinesischen Text und die vorhandenen Übersetzungen desselben zu Rate zu ziehen).

[2]) Sse-schu, Schu-king, Schi-king in mandschuischer Übersetzung mit einem mandschu-deutschen Wörterbuch. Leipzig, 1864 (Abhandlungen der Deutschen Morgenländischen Gesellschaft, Band III, Nr. 1—2).

[3]) Für das *Ta Hsio* hat v. D. GABELENTZ als instruktiven Fall zu der Neuübersetzung die Varianten der ersten Ausgabe hinzugefügt.

[4]) Ich erlangte in Peking einen von MÖLLENDORFF nicht erwähnten Band von hervorragend schönem Druck, in Gross-Folio, in gelbe Seide gebunden, betitelt *Yü ch'i k'üan shan yao yen (Han-i araha sain-be huwekiyebure oyonggo gisun* «Vom Kaiser [Shun-chih] verfasste wichtige Worte zur Ermahnung zum Guten»), datirt 1655, chinesisch und Manju, mit einem Vorwort von FU I-TSIEN, Gross-Sekretär (*Ta hsio shih*) im Geheimarchiv der *Han lin* (besser als «Literaturamt» denn als «Akademie» zu bezeichnen) mit dem Titel *Féng chéng ta fu*. Diese Ausgabe ist vermutlich identisch mit dem von CHAVANNES (Bulletin de l'Ecole française d'Extrême-Orient, Vol. III, 1903, p. 549) nach der Peking Zeitung erwähnten, vom gegenwärtigen Kaiser veranstalteten Neudruck. POZDNÉJEV hat in seinem

Ära des Kaisers K'ang-hsi, der die von Dahai begonnenen Arbeiten wieder aufnahm und zu Ende führen liess. Drei Jahre
nach seinem Regierungsantritt (1665) kam die Übersetzung des
*T'ung kien tsi lan*, einer Geschichte Chinas bis zum Ende der
Yüan-Dynastie (verfasst von Yüan Huang am Ende des sechzehnten Jahrhunderts) heraus, chinesisch und Manju,[1] einer
der vorzüglichsten Manju-Drucke, die je hergestellt worden sind;
1691 erschien das *T'ung kien kang mu*, ausschliesslich in man-
jurischem Text (48 Bände). Dies ist gleichfalls eine allgemeine
Geschichte Chinas vom vierten Jahrhundert v. Chr. an, ursprünglich verfasst von Se-ma Kuang (1066) und später von dem
Philosophen Chu Hsi (1130--1200) umgearbeitet und condensirt.
Der Kaiser K'ang-hsi liess nach einander zwei Übersetzungen
von diesem Werke veranstalten: da ihm die erste nicht vollständig und treu genug erschien, liess er die Arbeit noch einmal von vorne beginnen. In der neuen Version fügte man Noten
hinzu, um dunkle Stellen zu erklären, ohne sich jedoch Zusätze
im Text zu erlauben.[2] DE MAILLA bediente sich ebenso dieser
Übersetzung als des Originals bei der Abfassung seines grossen
Werkes über die Geschichte Chinas.[3]

Im Jahre 1676 erschien das *Chu tse tsieh yao* in vier
Bänden, Auszüge aus den philosophischen Schriften des Chou,
Chang und der beiden Ch'êng mit Chu Hsi's Kommentar, in

---

«Versuch einer Sammlung von Proben der manjurischen Literatur», pp.
1—28, den Manju- (ohne den chinesischen) Text, mit Ausnahme des Vor-
worts, abgedruckt (ohne nähere Angaben über das Werk).

[1] Die Behauptung von MÖLLENDORFF (l. c., p. 6), dass «unter den
drei ersten Kaisern, als Manju noch gesprochen wurde, nur der Manju-
Text publicirt wurde, und dass solche Ausgaben unter K'ien-lung auf-
hören, um bilinguen Drucken Platz zu machen», ist durchaus unzutreffend
Das 1647 publicirte *San kuo chi* ist mit dem chinesischen Text abgedruckt;
ebenso sind unter K'ien-lung Manju-Texte ohne chinesische Version erschienen. Der ganze Gesichtspunkt MÖLLENDORFF's beruht einfach in seiner
Phantasie. Die Frage, warum bilinguer oder einsprachiger Text, hat mit
dem Gesprochen oder Nicht-gesprochen-werden des Manju nichts zu tun,
sondern hängt einzig und allein davon ab, ob die Manju-Übersetzung nur
mit der chinesischen Version oder ohne dieselbe verständlich ist.

[2] L. LANGLÈS, Alphabet mantchou, 3. Aufl., Paris, 1807, pp. 70. 71.
[3] Histoire générale de la Chine, 13 Vols., Paris, 1777—1785.

Manju (übersetzt von CHU CHI-PI) und chinesisch;[1] ein Jahr
später die erste Ausgabe der Vier Bücher (*Se shu*) in zwölf
Bänden, gleichfalls in beiden Sprachen, die von allen Manju-
Ausgaben die meisten Auflagen erlebt hat: vier Jahre später
(1681) folgte das *Shu king* (15 Bände).

Das Jahr 1682 ist wegen der Edition des ersten Manju-
chinesischen Wörterbuchs denkwürdig. Dies ist das *Ta Ts'ing
ts'üan shu (Daicing gurun-i yooni bithe)*, das nach ZACHAROV'S[2]
sachverständigem Urteil in alphabetischer Ordnung nach einem
trefflichen Plane angelegt, aber, begreiflicher Weise, nicht vollständig ist. Weniger klar ist seine Bemerkung, dass spätere
Bearbeiter besonders solche darin befindliche Wörter unterdrückt hätten, von denen eine «alte Aussprache» gegeben ist.
Bemerkenswert ist, dass in diesem Werke zum ersten Male kurze
grammatische Bemerkungen über das Manju niedergelegt sind,
die später weiter benutzt und ausgebildet wurden. Im Jahre
1683 wurde die klassische Literatur durch Herausgabe des
*I king* vervollständigt: 1686 wurde eine Manju-Chrestomathie
(64 Vols.), enthaltend ausgewählte historische Aufsätze aus den
Annalen bis herab auf die Sung-Dynastie, übersetzt von MING-
KUNG, veröffentlicht (*Ts'ing wên ku wên yüan kien*); 1694 das
*K'in ting Ta Ts'ing hui tien*, gleichzeitig auf chinesisch und
Manju in besonderen Ausgaben, ein auf kaiserlichen Befehl ver-
fasstes Handbuch, das die staatlichen Einrichtungen der gegenwärtigen Dynastie beschreibt und das wichtigste Quellenwerk für
ihr Verwaltungsrecht und ihre Verwaltungspolitik bildet, später
wiederholt revidirt, vermehrt und aufs neue herausgegeben.

So hatte im siebzehnten Jahrhundert die Manju-Literatur
einen guten Anlauf genommen, wenn sie sich auch noch in
bescheidenen Grenzen hielt; ihre Blüte erlebte sie aber im
achtzehnten Jahrhundert, als die Flut der Manju-Bücher bis
zum Ende der langen Regierung des Kaisers K'ien-lung immer
mächtiger anschwoll, um im neunzehnten langsam zu verebben
und zu ersterben.

---

[1] Übersetzt von C. DE HARLEZ (Journal Asiatique, 1887, Janv.-Févr.,
pp. 37—71).

[2] Vorwort zum Wörterbuch, p. XV.

Der Anfang des neuen Jahrhunderts ist vor allem durch die Pflege der schönen Literatur durch die Manju ausgezeichnet. Im Jahre 1708 erschien die Übersetzung des Sittenromans *Kin P'ing Mei* (48 Vols.), die man vielleicht als die Glanzleistung der manjurischen Sprache und Literatur bezeichnen kann, und deren Studium niemand, der die Sprache aus echter Quelle schöpfen will, ausser Acht lassen sollte. Die Verfasserschaft der Manju-Übersetzung wird auf WYLIE's[1]) Autorität hin gewöhnlich einem Bruder des Kaisers K'ang-hsi[2]) zugeschrieben, was aber noch des Nachweises oder der Bestätigung aus einer einheimischen Quelle bedarf, da das Vorwort der Manju-Übersetzung über diesen Punkt schweigt. Die Bannflüche, welche die moralisch entrüsteten Kaiser K'ang-hsi und K'ien-lung gegen diesen Roman schleuderten, obwohl sie selbst weise genug waren, das zu tun, was darin so offen und rückhaltlos geschildert wird, haben seine grosse Popularität nicht zu brechen vermögen; im Gegenteil, auch in China schmecken die verbotenen Früchte und die Eva's Äpfel am besten. Als künstlerische Leistung gehört dieses Erzeugnis, wie WYLIE bekennt, zu den höchsten seiner Gattung, und zu einem ähnlichen Urteil gelangt auch WILHELM GRUBE.[3]) Dass der Roman unmoralisch ist, muss direkt in Abrede gestellt werden; er ist ebenso wenig unmoralisch als ein Werk von Zola oder Ibsen und eben wie diese das Kunstwerk von der Hand eines Meisters und grossen Menschenforschers, der die Menschen mit ihren Leidenschaften schildert, wie sie sind, nicht, wie sie nach der Ansicht der Frömmler und Heuchler sein sollten. Und dass der Verfasser am Ende im höchsten Sinne ethische Tendenzen verfolgt hat, wird ganz unzweideutig aus der manjurischen Vorrede offenbar, wo zweimal ausdrücklich gesagt wird, dass die hundert Kapitel dieses Buches ebenso viele Warnungen darstellen sollen.[4])

---

[1]) Translation etc., p. XLII: Notes on Chinese Literature, 2. Ausg., pp. 202—203.

[2]) Nicht K'ien-lung, wie MÖLLENDORFF, Nr. 235, in seiner Confusion behauptet, trotzdem er das Datum, wenn auch fälschlich, auf 1709 angibt.

[3]) Geschichte der chinesischen Literatur, Leipzig, 1902, S. 430—431.

[4]) Die Literatur über *Kin P'ing Mei* s. bei CORDIER, Bibliotheca Sinica, 2. Aufl., Vol. III, Paris, 1906, Spalte 1772, wo hinzuzufügen ist

Dem *Kin P'ing Mei* folgte im Jahre 1711 eine Manju-Übertragung der auch als Drama bearbeiteten Novelle *Hsi hsiang ki*,[1]) und zwei Jahre später eine Sammlung von 36 Gedichten des Kaisers K'ang-hsi auf den Sommerpalast in Jehol, in zwei Bänden, von denen der eine die chinesische, der andere die manjurische Version enthält. Das Werk ist mit 36 Holzschnitten illustrirt, nach Malerien des Shēn Yü von Lēng Mei aus Kiao-chou in Shantung gestochen.[2]) O. FRANKE[3]) bemerkt, dass K'ien-lung im Jahre 1741 zu Ehren seines Grossvaters K'ang-hsi ein zweibändiges Prachtwerk herausgegeben habe, das künstlerische Illustrationen der 36 schönen Punkte des Sommerpalastes von Jehol nebst poetischen Beschreibungen derselben enthalte. Es mag ja sein, dass zu dieser Zeit eine Neuausgabe des Werkes veranstaltet worden ist, aber es kann

---

G. v. D. GABELENTZ, ZDMG, Band XVI, 1862, S. 543—546, der einige Proben übersetzt hat. Beim Verkauf der Sammlung KLAPROTH soll die chinesische Ausgabe 511 Franken, die manjurische 720 gebracht haben. In Peking kostet erstere im Durchschnitt 10, letztere 30 mexikanische Dollars; doch hängen naturgemäss die Preise alter Ausgaben ganz von Glück und Zufall ab. Beide von mir in Peking erworbenen Ausgaben sind jetzt in New-York; die manjurische (48 Vols.) ist brillant gedruckt, weit besser als die chinesische (23 Vols.), enthält aber nicht die interessanten Holzschnitte, mit denen die zwei ersten Bände der letzteren angefüllt sind, und die allein ein sittengeschichtliches Repertorium bilden. Nach GRUBE (l. c.) hat H. C. V. D. GABELENTZ nach der Manju-Version eine vollständige deutsche Übersetzung geliefert, «die aus naheliegenden Gründen bisher leider unveröffentlicht geblieben ist». Die Publikation dieser Arbeit wäre dringend wünschenswert, einmal im Interesse der manjurischen Studien, sodann im Interesse der Sexualpsychologie der Chinesen und des Fortschritts dieses wichtigen Gebietes der Wissenschaft überhaupt, das in den letzten Jahren einen so grossen Aufschwung genommen hat. Die stupide Feigenblattprüderie sollten wir getrost den Herren Engländern und Amerikanern überlassen: die Kulturgeschichte wird doch nicht für Betschwestern und höhere Töchterschulen geschrieben. Das K'in P'ing Mei ist ein einzigartiges und ohne Zweifel das wertvollste sexualpsychologische Dokument des chinesischen Geistes, über dessen Vorhandensein wir alle Veranlassung haben uns zu freuen.

[1]) Handschriftliche Ausgabe (chinesisch und Manju) in vier Bänden in New-York.

[2]) F. HIRTH, Fremde Einflüsse in der chinesischen Kunst, S. 56.

[3]) Beschreibung des Jehol-Gebietes, Leipzig, 1902, S. 61, Note 1.

kein Zweifel sein, dass die Verse von K'ang-hsi, nicht von K'ien-
lung, herrühren, und dass auch die Holzschnitte aus jener Zeit
stammen. In der mir vorliegenden manjurischen Originalausgabe,
in welcher der Name K'ien-lung's überhaupt nicht erscheint, ist
das Vorwort K'ang-hsi's vom Jahre 1711 datirt, das Buch aber
wohl nicht vor 1713 gedruckt, da einige Nachschriften 1712
datirt sind; ebenso werden darin die Gedichte dem Kaiser
K'ang-hsi zugeschrieben.[1])

Im Jahre 1710 erschienen Übersetzungen zweier militär-
wissenschaftlicher Werke, der Kriegskunst des SUN-TSE (sechstes
Jahrhundert v. Chr.) und der des WU-TSE (viertes Jahrhundert
v. Chr.), zusammen in vier Bänden.[2]) Als Übersetzer nennt
sich am Schlusse beider Schriften ein Manju namens SANGGA
von Šanggiyan Alin («Weisses Gebirge», chin. Ch'ang pai shan).[3])

---

[1]) Eine photolithographische Reproduktion des chinesischen Bandes,
Shanghai, 1895. Die Holzschnitte sind hier auf die Hälfte verkleinert und
verlieren noch mehr von ihrem ursprünglichen Reiz, da sie auf schlechtes
unharmonisches Papier gedruckt sind.

[2]) Beide Abhandlungen sind unter Benutzung der Manju-Übersetzung
von M. AMIOT übertragen worden (Mémoires concernant les Chinois, Vol.
VII. Paris, 1782, pp. 45—224).

[3]) So entnehme ich einer von mir in Peking erlangten handschrift-
lichen bilinguen Ausgabe. Ebenda wird das Datum in der Form *Daicing
gurun Elhe Taifin šanggiyan tasha aniya* (d. i. im Weissen Tiger Jahr der
Periode K'ang-hsi der Dynastie Ta Ts'ing, entsprechend dem Jahre 1710)
gegeben. Es scheint zweifelhaft, ob die Arbeit des Sangga je gedruckt
worden ist. AMIOT (l. c., pp. 8, 9) erwähnt zweimal ausdrücklich, dass er
nach einem Manuskript gearbeitet habe. Aus dem Katalog von St. Petersburg (p. 67, Nr. 444) geht nicht hervor, ob das dort vorhandene Exemplar
Druck oder Handschrift ist; dagegen geht aus dem hinzugesetzten Manju-
Datum (49. Jahr der Periode K'ang-hsi) hervor, dass es sich um dieselbe
Übersetzung aus dem Jahre 1710 handelt. Eine neue Übersetzung von
Sun-tse's Kriegskunst soll nach MÖLLENDORFF, Nr. 226, ein Manju KIYING
im Jahre 1846 herausgegeben haben. — AMIOT (l. c., p. 225) erwähnt auch
eine Manju-Übersetzung des *Se-ma fa* («Kriegskunst des Se-ma»: WYLIE,
Notes on Chinese Literature, 2. Ausg., p. 90) ebenfalls aus dem Jahre
1710; ich habe dieselbe nicht gesehen, sie sollte indessen in Paris vor-
handen sein, wo sie, von AMIOT hingesandt, 1769 angelangt ist (ibid., p. V).
Ein anderes Werk über Taktik, das *Leu t'ao* (WYLIE, l. c., p. 89) soll von
DAHAI übersetzt worden sein; es scheint aber unvollendet in seinem Nachlass geblieben und nie beendigt worden zu sein. AMIOT, der einen Auszug

Es kann nicht genug bedauert werden, dass die Manju nicht ihr eigenes System des Kriegswesens in ihrer Sprache dargestellt haben; aber sie mochten das Gefühl gehabt haben, dass die Chinesen, während sie diesen in der Praxis der Kriegsführung überlegen waren, wie auf anderen Gebieten, so auch hier wenigstens in der Theorie die grössere Meisterschaft besassen.

Die Philosophie der Sung-Zeit scheint die Manju besonders gefesselt zu haben. Die bekannte Sammlung *Hsing li tsing i*, ein Corpus philosophischer Schriften, wurde 1717, von einer Vorrede des Kaisers K'ang-hsï begleitet, ausschliesslich in Manju, in vier Bänden[1]) gedruckt. Im Inhalt und in der Anordnung des Stoffes stimmt diese Ausgabe genau mit der von WYLIE[2]) gegebenen Analyse des *Hsing li ta ts'üan shu* überein, der älteren Vorlage vom Jahre 1415. Das im Jahre 1732 gedruckte *Ho pi hsing li* enthält nur die vier ersten Werke der vorigen Sammlung, die darin das erste und die Hälfte des zweiten Buches ausfüllen.[3])

Von dem philosophischen Werke *Hsing li chên tsüan*, das HsÜN TÊH-CHAO 1753 publicirte, hat H. C. v. D. GABELENTZ[4]) das erste Buch übersetzt, das von der Seele des Menschen handelt.

Die klassische Literatur wurde im achtzehnten Jahrhundert weiter bearbeitet, die älteren Übersetzungen erneuert und neu

---

aus diesem Werke mitgeteilt hat (l. c., pp. 303—315), macht in diesem Falle keine Anspielung auf eine Manju-Übersetzung.

[1]) Nicht in zwölf Bänden, wie MÖLLENDORFF, Nr. 82, angibt; das Werk umfasst zwölf Bücher oder Kapitel.

[2]) Notes on Chinese Literature, 2. Ausg., p. 85.

[3]) Danach G. v. D. GABELENTZ, Thai-kih-thu, des Tscheu-Tsï Tafel des Urprinzipes mit Tschu-Hï's Commentar, Dresden, 1876; die Übersetzung auch bei W. GRUBE, Geschichte der chinesischen Literatur, Leipzig, 1902, S. 334—336; W. GRUBE, Ein Beitrag zur Kenntnis der chinesischen Philosophie, T'ûng-Šu des Čeu-Tsï, mit Čü-Hï's Commentare, Leipzig, 1882. Aus dem *Hsing li tsing i*: W. GRUBE, Zur Naturphilosophie der Chinesen. Lï Khï, Vernunft und Materie (Mélanges asiatiques de l'Acad. de St.-Pét., Vol. VIII, 1881, pp. 667—689). Sämtliche drei Schriften geben den chinesischen, wie den Manju-Text. S. ferner H. CORDIER, Bibliotheca Sinica, 2. Aufl., Vol. II, Paris, 1906, Spalte 1424.

[4]) Zeitschrift für die Kunde des Morgenlandes, Band III, Göttingen, 1840, S. 250—279.

**3\***

herausgegeben. Im Jahre 1727 erschienen das «Buch der Kindesliebe» (*Hsiao king*), Manju und chinesisch,[1] und Chu Hsi's
*Hsiao hsio;* 1737 eine Auswahl aus dem *Shi king* und *Shu king;*
1756 kam die neue Übersetzung der «Vier Bücher» (*Se shu*)
auf Befehl des Kaisers K'ien-lung heraus (in Chinesisch und
Manju, 6 Vols.); 1760 eine neue Ausgabe des *Shu king* (gleichfalls als Bilingue, 6 Vols.); 1766 das *I king*[2] (ebenso, 4 Vols.);
1769 eine neue Ausgabe des *Shi king* (ebenso, 4 Vols.); 1784
das *Li ki* (ebenso, 12 Vols.); 1785 *Ch'un Ts'iu* (ebenso, 48 Vols.).

Das *San tzŭ king*, dessen mongolischer Bearbeitung bereits
gedacht wurde, soll nach Möllendorff (Nr. 78) bereits im siebzehnten Jahrhundert übersetzt worden sein. Ich vermag leider
nichts Bestimmteres darüber zu sagen, mir ist nur die mit
Kommentar versehene Parallelausgabe vom Jahre 1795 (in zwei
Bänden) bekannt, deren Manju-Übersetzung von T'ao Ko-king
herrührt.[3]

Eine Sammlung von Maximen, die nach Materien geordnet
sind, ist das *Ko yen tsi yao (Koolingga gisun-i oyonggo-be sosoho
bithe)* in zwölf Bänden, das ohne Datum und Übersetzername
publicirt ist (wenigstens in der mir zur Verfügung stehenden
Ausgabe).

Die unter dem Namen des heiligen Edikts bekannten ethi-
schen Belehrungen des Kaisers K'ang-hsi wurden zum ersten
Male mit manjurischer Übersetzung von seinem Sohne und
Nachfolger Yung-chêng 1724 herausgegeben. Ein vorzüglicher
Neudruck derselben, mit Hinzufügung der mongolischen Version,
wurde 1873 in vier Bänden veranstaltet.[4]

---

[1] Neue Ausgabe 1851, 2 Vols. (mit dem *Chung king*), Vorwort von
Ting-hsien Mêng-pao.

[2] Über die Manju-Übersetzung des *I king* vergl. C. de Harlez.
Journal Asiatique, Juillet-Août. 1896, pp. 177—178.

[3] F. Turettini hat in der Sammlung *Ban-zai-sau* den chinesischen
Text mit der japanischen, manjurischen und mongolischen Version (Genf,
1876), den Text des manjurischen Kommentars ebenda (1892—1894) herausgegeben. Vergl. E. Teza, Del commento alle «Tre Parole» secondo la ver-
sione mangese (Rendiconti dell' Accademia dei Lincei, Vol. III, 1894,
pp. 447—466).

[4] Dieser Druck ist noch leicht in Peking erhältlich und allen europäischen Ausgaben weit vorzuziehen. Betreffs Übersetzungen s. H. Cor-

An weisen väterlichen Ermahnungen haben es alle Kaiser der Ts'ing-Dynastie nicht fehlen lassen und sie in breitester Schreibseligkeit ihren Landeskindern hinterlassen. Die des Kaisers T'ai-tsu, der sich noch in massvollem Umfange hielt, sind bereits erwähnt worden: die des Kaisers K'ang-hsi erschienen 1741,[1]) die K'ien-lung's, der sich nicht weniger als dreihundert Bände geleistet hat, im Jahre 1807. Auch von seinen Nachfolgern Kia-k'ing, Tao-kuang, Hsien-fêng und T'ung-chih sind solche Collectaneen publicirt worden. Um eine Vorstellung von den Themata zu geben, die in diesen Schriften behandelt werden, lasse ich eine kurze Analyse von K'ien-lung's Sammelwerk auf Grund des Inhaltsverzeichnisses im ersten Bande folgen: die Tugenden des Weisen (Band 1 6); Kindesliebe des Weisen (Band 7—12); Wissen des Weisen (Band 13—14); amtliche Tätigkeit des Weisen (Band 15—26); Verehrung des Himmels (Band 27—29); sich ein Vorbild an den Vorfahren nehmen (Band 30—32); literarische Belehrung (Band 33—40); militärische Tätigkeit (Band 41—51); militärische Ausrüstung (Band 52—54); Verträglichkeit mit Verwandten (Band 55—57); Umgang mit Menschen (Band 58—69); Liebe zum Volke (Band 70—85); Hingebung in Amtsgeschäften (Band 86 87); Bittgesuche (Band 88 89); Untersuchungen der Beamten (Band 90—103); Verwaltung des Reichtums (Band 104—110); Respekt vor der Strafe (Band 111—121); Untersuchungen von Grundstücken (Band 122 125); Wasserregulirung (Band 126—136); dasselbe mit dem Zusatz: Ausnutzung der Wasservorteile (Band 137); Nachlass von Naturalabgaben und Verteilung der Steuern (Band 138—153); dasselbe mit dem Zusatz: Darlehen, um die

DIER, Bibliotheca Sinica, 2. Aufl., Vol. II, Paris, 1906, Spalte 1426. Als Datum der Editio princeps gibt auch WYLIE (Notes on Chinese Literature, 2. Ausg., p. 88) das Jahr 1724. Ich besitze jedoch eine reich illustrirte Ausgabe in zwanzig Bänden (Shêng hsün hsiang tsieh «Das heilige Edikt, durch Bilder erläutert») mit einem Vorwort von K'ang-hsi aus dem Jahre 1681 und einem zweiten Vorwort des Kreisvorstehers von Fan-ch'ang hsien (in T'ai-p'ing fu, Provinz Anhui) aus demselben Jahre.

[1]) Übersetzt in Mémoires concernant les Chinois, Vol. IX, Paris, 1783, pp. 65—281, mit einem Vorwort des Kaisers Yung-chêng vom Jahre 1730.

Abgaben hinauszuschieben (Band 154—161); Verwahrung von
Gütern (Band 162- 167); dasselbe mit Zusatz: Verkauf zu rich-
tigem Preise (Band 168—169); Gnadenbezeigungen (Band 170 -
171); Güte gegen das Heer (Band 172- 179); Viehwirtschaft
(Band 180—182); Belehrung der Beamten (Band 183—192);
strenge Handhabung der Gesetze (Band 193- 208); Hochschät-
zung des Landbaues und Seidenbaues (Band 209—214); För-
derung der Ceremonien und Musik (Band 215—216); Ausdehnung des Ackerlandes (Band 217- 222); richtige Beobachtung der
Formen der Sitte (Band 223- 236); Achtung vor dem Alten und
Überkommenen (Band 237—242); Lob der Gerechtigkeit (Band
243—244); Beobachtung der Opferceremonien (Band 245—246);
ehrenvolle Behandlung der Greise (Band 247—250); Vernich-
tung des Falschen und Bösen (Band 251- 260); Achtung vor
allem Wissen (Band 261—264); Belehrung der Eunuchen (Band
265- 266); wie man die auswärtigen Provinzen in Ruhe hält
(Band 267—276); Verwaltung der Grenzen (Band 277—288);
Liebe zu den Beamten (Band 289—294); wie der Feldherr sein
Heer anspornt (Band 295  300). Man sieht, dass diese Essays
sich nicht nur auf das ethische Gebiet erstrecken, sondern auch
Politik, Verwaltung, Ökonomie, Militärwesen und Kultus be-
rühren.

Auch die historischen Werke, selbst diejenigen, welche Zustände der Manju behandeln, sind aus dem Chinesischen übersetzt, so die grosse Geschichte der Acht Banner in 250 Büchern,
die 1727 begonnen und 1739 abgeschlossen wurde.[1] Man gab
davon eine besondere chinesische und besondere manjurische
Version heraus. Es gibt auch eine Genealogie der Manju-Fami-
lien vom Jahre 1744 in 45 Bänden.[2]

Erwähnung verdient die Sammlung der Edikte des Kaisers Yung-chêng (1723—1735), da einige Fragmente daraus
übersetzt worden sind.[3]

---

[1] A. WYLIE. Notes on Chinese Literature, 2. Ausg., p. 71.

[2] MÖLLENDORFF, Nr. 158.

[3] Von C. DE HARLEZ in Le Muséon, 1884, pp. 619—626, Travaux
de la sixième session du Congrès des Orientalistes à Leide, Vol. II, 1883.
pp. 143—149, und ZDMG, Band XLI, 1887, S. 311—318 (sechs Edikte).
Einige Texte im Manuel de la langue mandchoue, pp. 119—129. Wann

Ein ganz interessantes Werkchen ist das *Yi hsio san kuan T's'ing wên kien*, was man frei wiedergeben könnte «Drei Fliegen mit einem Schlag», indem man durch sein Studium drei Dinge auf einmal erlernen soll, chinesisch, Manju und noch allerlei nützliche Sachen. Die uns erhaltene Ausgabe stammt aus dem Jahre 1746 (4 Vols.); da sie aber als Neudruck bezeichnet ist, muss das Buch schon früher vorhanden gewesen sein. In der Einleitung wird das Syllabar mitgeteilt; der Stoff ist encyklopädisch nach dem bekannten chinesischen Einteilungsschema behandelt: Himmel, Jahreszeiten, Erde, Fürsten, Beamte, Verwaltung, Ritual u. s. w., und man erhält in der Tat eine in leichtem Stil geschriebene knappe Übersicht des chinesischen Kulturlebens.

Im Jahre 1747 liess der Kaiser K'ien-lung das «Opferritual der Manju» (*Hesei toktobuha Manjusai wecere metere kooli bithe*) nur in manjurischer Sprache publiciren. Dieses Werk ist zuerst von LANGLÈS[1]) bekannt gemacht, später von C. DE HARLEZ vollständig übersetzt worden.[2]) MÖLLENDORFF, Nr. 132, gibt als Datum dieses Werkes das Jahr 1765 an, auf welche Autorität gestützt, ist mir nicht bekannt; das einzige Datum, das sich in dem Buche selbst finden lässt, ist das der Vorrede des Kai-

---

und wie seine Vorlage publicirt worden ist, hat HARLEZ immer verschwiegen.

[1]) Rituel des Tatars-Mantchoux, rédigé par l'ordre de l'empereur Kien-Long, et précédé d'un discours préliminaire composé par ce souverain; avec les dessins des principaux ustensiles et instrumens du culte chamanique: ouvrage traduit par des extraits du tatar-mantchou, et accompagné des textes en caractères originaux. Paris, de l'imprimerie de la République: An XII. = (1804, v. s.), 4°, 74 p., 10 Tafeln. Auf p. V: Fait partie du tome VII-e, première partie, page 241—308, des Notices et Extraits des Manuscrits de la Bibliothèque nationale.

[2]) La religion nationale des Tartares (sic!) orientaux Mandchous et Mongols, comparée à la religion des anciens Chinois, d'après les textes indigènes, avec le rituel tartare (!) de l'empereur K'ien-long, traduit pour la première fois. Bruxelles, 1887. Die schönen Holzschnitte des Originals sind nur zum geringen Teil und ganz elend reproducirt und noch schlechter beschrieben. Der Abschnitt über die Religion der Mongolen ist durchaus mangelhaft und oberflächlich: von einer Benutzung von BANZAROV's grundlegenden Arbeiten und der übrigen russischen Literatur gar keine Rede.

sers, die 1747 datirt ist. HARLEZ hat in seiner Übersetzung dieses Datum einfach unterdrückt und die Zeit der Abfassung überhaupt nicht erörtert. Im allgemeinen hat er das Werk recht missverstanden, indem er darin ein Originalwerk der Manju-Literatur und den authentischen Codex der Manju-Religion im sechzehnten und siebzehnten Jahrhundert sieht. Wie aber schon ein Blick auf den sechsten, die Illustrationen enthaltenden Band lehrt, sind alle Tempel, Opfergeräte und Utensilien, ausgenommen vielleicht zwei oder drei Instrumente schamanischen Ursprungs, echt chinesisch, und so ist auch der Inhalt des Werkes selbst, dessen Grundlage die altchinesische Staatsreligion und ihr Kultus, von buddhistischem Einfluss berührt, bildet. Dass hier nur wenige Spuren der ursprünglich schamanischen Religion der Manju vorhanden sind, wird jedem klar, der den Schamanismus der Golden und der übrigen sibirischen Völker kennt, und es ist auch nicht einzusehen, wie dieses Werk ein getreuer Spiegel manjurischen Schamanismus hätte werden können, da derselbe zur Zeit seiner Abfassung längst erloschen, und die Schamanen nach dem Geständnis K'ien-lung's in der Vorrede die alten Traditionen verständnislos hersagten oder schon vergessen hatten. Eine neue kritische Übersetzung dieses Werkes wäre sehr wünschenswert, da HARLEZ' Übersetzungen aus dem Manju unzuverlässig sind.

Im Jahre 1748 schenkte der Kaiser K'ien-lung der Welt sein Gedicht zur Verherrlichung von Mukden,[1] das zuerst von AMIOT übersetzt worden ist.[2] Derselbe hat auch ein anderes

---

[1] Nicht in einem Bande, wie MÖLLENDORFF, Nr. 237, bemerkt, sondern in zwei Bänden, von denen der eine den chinesischen, der andere der Manju Text enthält.

[2] Éloge de la ville de Moukden et de ses environs; poème composé par Kien-Long, empereur de la Chine et de la Tartarie, actuellement régnant. Accompagné de notes curieuses sur la géographie, sur l'histoire naturelle de la Tartarie orientale, et sur les anciens usages des Chinois, composées par les éditeurs chinois et tartares. On y a joint une pièce de vers sur le Thé, composé par le même Empereur. Traduit en françois par le P. AMIOT; et publié par M. DEGUIGNES. Paris, 1770. XXIV, XXXVIII und 381 p. — Manju Text in KLAPROTH's Chrestomathie mandchoue, pp. 63—99; Übersetzung, pp. 235—273. Text auch in POZDNÉJEV's Sammlung, pp. 143–162.

Gedicht desselben Herrschers auf die Unterwerfung des Kin-ch'uan Gebietes im Westen Chinas übertragen.[1]) Von Poesie ist hier freilich wenig die Rede, ausgenommen, dass sie sich im Stil ankündigt; das Gedicht auf Mukden ist eine Art versificirter Landeskunde dieses Gebietes. Es ist bekannt, dass sich der Kaiser dieserhalb eine gute Censur von Voltaire zuzog (in der Epistel mit dem Anfang: Reçois mes compliments, charmant roi de la Chine). Uebrigens sind diese beiden Stücke nur Beispiele aus der zahllosen Masse chinesisch-manjurischer und chinesischer Gedichte, die das fruchtbare Talent des Kaisers während seines langen Lebens hervorgebracht hat.

Unter den Auspicien K'ien-lung's erschien 1766 in acht Bänden ein interessantes geographisches Wörterbuch (*K'in-ting Hsi yü t'ung wên chi*), das in sechs Sprachen (chinesisch, Manju, mongolisch, tibetisch, kalmükisch und Djagatai) die Ortsnamen Central- und Westasiens, im ganzen 3111 Artikel, aufzählt.[2])

Das achtzehnte Jahrhundert ist vor allem durch den grossen Aufschwung der Lexikographie ausgezeichnet. Nachdem LIU-SHUN und SANG-KO 1703 in vier Bänden ein kurzes chinesisch-manjurisches Vokabular und SANGGE 1706 ein manjurisch-chinesisches in acht Bänden herausgegeben hatten,[3]) wurde 1708 das erste grosse Wörterbuch (*Manju gisun-i buleku bithe*) unter kaiserlichen Auspicien publicirt; es ist in 36 Kapitel eingeteilt

---

[1]) Hymne **tartare-mantchou** chanté à l'occasion (auf dem Titel occaission) de la conquête du Kin-tchouen. Traduit en français et accompagné de notes, par M. AMIOT et publié par L. LANGLÈS. Paris, 1792.

[2]) Danach JULES THONNELIER, Dictionnaire géographique de l'Asie centrale offrant par ordre alphabétique, les transcriptions, en caractères mandchoux et chinois, des noms géographiques. Prolégomènes, pays en dehors de la domination chinoise. Paris (Maisonneuve), 1869. Lithographirt, 51 p., 4°. — E. v. ZACH (Lexicographische Beiträge, Band I, Peking, 1902, S. 83—98, und Band III, ib., 1905, S. 108—135) hat danach die tibetischen Orts-, Berg- und Flussnamen bearbeitet. — Von anderen geographischen Schriften in Manju verdient der Reisebericht des Gesandten T'u LI-SHEN zu den Kalmüken (*I yü lu*) Erwähnung, der wiederholt übersetzt worden ist (s. die Literatur bei H. CORDIER, Bibliotheca Sinica, 2. Aufl., Vol. I, Paris, 1904, Spalte 637).

[3]) MÖLLENDORFF, Nr. 31, 33.

mit 280 Unterabteilungen, mit chinesischer Übersetzung der
Manju-Wörter und einer Erklärung in Manju. Im folgenden
Jahre (1709) erschien in sechzehn Bänden eine andere Ausgabe
gänzlich in Manju. Dabei ruhte auch die Arbeit privater Gelehrter nicht. Die nützlichste und erfolgreichste war das Werk
des DAIGU SHIH-SE von Ch'ang Pai, das *Manju gisun-i yongki-
yame toktobuha bithe* vom Jahre 1722, in zwölf Bänden. Hier
ist der Wortschatz alphabetisch angeordnet. Er gibt ferner grammatische Beobachtungen, eine anscheinend sehr vollständige Liste der zahlreichen, für das Manju so charak-
teristischen Interjektionalwörter (*muruseme alhôdara gisun*),
Synonyme und die Fachausdrücke im Sprachgebrauch der
sechs Ministerien, dann ein Manju-chinesisches Wörterbuch.

Chronologisch das nächste Werk ist das *Ts'ing wên hui
shu (Manju isabuha bithe)*, verfasst von LI YEN-KI, 1751, in
zwölf Bänden, verlegt von der Firma *San huai t'ang* in der
Gasse des *Lung fu*-Tempels (*Lung fu se hu-t'ung*) zu Peking.
Sein Wörterbuch ist in der Reihenfolge des Manju-Alphabets,
mit chinesischen Erläuterungen, angelegt.

Im Jahre 1771 producirte ein von K'ien-lung eingesetzter
Gelehrtenausschuss ein grosses Wörterbuch in 48 Bänden (*Han-i
araha nonggime toktobuha Manju gisun-i buleku bithe*), das, wie
der Titel besagt, mit «Zusätzen» versehen ist und eine um 5000
Wörter vermehrte Auflage des kaiserlichen Wörterspiegels von
1708 darstellt. Im zweiten Bande ist daher K'ang-hsi's Vorwort
aus jenem Jahre wiederholt, während der erste das des Kaisers
K'ien-lung (datirt 1771) und der dritte das Syllabar enthält.
Der Rest zerfällt in drei Abteilungen: die erste gibt das
Wörterbuch K'ang-hsi's (Band 4—35) nach Kategorien geordnet, jedes Manju-Wort in chinesischer Transkription und
Übersetzung, letztere wieder in Manju-Umschrift, dann eine
Begriffsdefinition des Wortes in Manju; die zweite Sektion
(Band 36—43) enthält einen alphabetisch geordneten Index
der Manju-Wörter, mit Hinweisen auf die Kapitel des vorher-
gehenden Teils; die dritte (Band 44—48) die Liste der neu
geschaffenen Wörter, mit Index. Unter diesen befinden sich
viele künstliche Ableitungen aus dem Chinesischen, die mehr

philologischer Zeitvertreib sind und nie wirkliches Sprachgut wurden.[1]

Auf der Basis der beiden Wörterbücher von 1751 und 1771 verfasste I-HIN im Jahre 1786 ein alphabetisches Ergänzungswörterbuch (*Ts'ing wên pu hui, Manju gisun-be niyeceme isabuha bithe*) in acht Bänden. Dieses Werk verdient Beachtung, denn es sind darin über 7900 neue Wörter registrirt, d. h. die 5000 neuen des K'ien-lung Thesaurus und 2900 aus Büchern und alten Akten ganz neu aufgestöberte und durch modernere ersetzte veraltete Ausdrücke. Das Graviren der Druckplatten begann 1799 und verzögerte sich so lange, dass das Werk erst 1802 ans Licht kam.[2]

Die mehrsprachigen Wörterbücher, in denen ausser Manju noch mongolisch, tibetisch und türkisch aufgenommen sind, wurden bei Gelegenheit der mongolischen Literatur erörtert: sie gehören auch nicht hierher, da sie für das Manju nichts Neues ergeben, sondern sich alle an den Plan des grundlegenden Werkes von 1771 anlehnen.

Von sämtlichen, im Vorhergehenden aufgezählten Wörterbüchern sind vollständige Exemplare in New-York vorhanden. Ausserdem erwarb ich in Peking ein, wie es scheint, nie gedrucktes Manuskript von hervorragender Schönheit in vier starken Bänden, unter dem Titel *Hsin tsuan Ts'ing yü (Ije banjibuha ubaliyambure Manju gisun,* «Neu verfasste Manju-Sprache»). Dieses Wörterbuch ist nach Kategorien geordnet, deren Titel durch Rotschrift hervorgehoben sind; abweichend von allen anderen Werken nimmt das Chinesische die obere, das Manju die

---

[1] K'ien-lung's Vorwort übersetzt von L. LANGLÈS, Alphabet mantchou, 3. Aufl., Paris, 1807, pp. 76—81 (Text, pp. 81—89); K'ang-hsi's von C. DE HARLEZ, Manuel de la langue mandchoue, Paris, 1884, pp. 223—225 (Text, pp. 107—110; am Schluss übersetzt HARLEZ das Datum trotz seines richtigen Textes 40. statt 47. Jahr K'ang-hsi und hält es für überflüssig, das europäische Datum 1708 hinzuzufügen). Derselbe hat ZDMG, Band XXXVIII, 1884, S. 634—641 einige Beispiele aus diesem Wörterbuch mit den Definitionen übersetzt.

[2] Vergl. ZACHAROV, Vorwort zum Wörterbuch, p. XVIII. MÖLLENDORFF, Nr. 40, stellt die Sache irrtümlich so dar, als sei das Buch 1786 erschienen und 1802 in neuer Auflage. Des Verfassers Vorwort ist 1786 und eine zweite Vorrede an die Leser 1802 datirt.

untere Hälfte der Seite ein. In Cambridge ist ein zweibändiges handschriftliches Wörterbuch vorhanden, das alle chinesischen Wörter und Redensarten des grossen Thesaurus des Kaisers K'ang-hsi enthält und mit demselben ein vollständiges chinesisch-manjurisches Lexikon bildet.[1]

Im neunzehnten Jahrhundert ist die Manju-Literatur im Rückgang und Verfall begriffen, obwohl es nicht an gutem Willen und Rettungsversuchen der dem Untergang geweihten Sprache fehlt. Das moralische Element, Schul- und Lehrmittel überwiegen. Im Jahre 1816 verfasste WANG TING einen ethischen Traktat unter dem Titel «Ermahnungen in sechs Angelegenheiten», in dem er Lebensregeln darüber gibt, wie man für sich selbst und sein Haus sorgen, wie man seine amtlichen Pflichten erfüllen, wie man auf dem Lande leben, seine Geschäfte verrichten und den Charakter der Menschen erforschen soll. MÊNG PAO übersetzte dieses Handbuch des praktischen Lebens ins Manju, und eine neue Ausgabe desselben war 1851 erforderlich.[2]

In demselben Jahre (1816) verlegte die Buchhandlung *Wên shêng t'ang* in Peking eine Sammlung der in den sechs Ministerien von Peking gebräuchlichen Fachausdrücke,[3] eine Arbeit, die schon im Wörterbuch des DAIGU von 1722 vorlag.

Ein nützliches, auf der Grundlage des *P'ei wên yün fu* aufgebautes Werk erschien im Jahre 1821 unter dem Titel «Meer der Manju und chinesischen Sprache» (*Ts'ing Han wên hai*) in zwanzig Bänden, verfasst von Kua-êrh-kia pa-ni-hun aus Girin und herausgegeben von seinem Sohn P'u-kung in der Manju-Garnison von Nanking. Dieses bisher kaum beachtete Wörterbuch, das in keiner europäischen Bibliothek vorhanden

---

[1] GILES. A Catalogue of the Wade Collection, Cambridge, 1898, p. 140, Nr. G 110–111.

[2] Nach MÖLLENDORFF, No. 99.

[3] *Leu pu ch'êng yü (Ninggun jurgan toktoho gisun-i bithe),* 6 Vols., ohne Vorwort und Angabe des Verfassers. Je ein Band ist einem der sechs Ministerien gewidmet. MÖLLENDORFF, Nr. 14, erwähnt ein Werk dieses Titels in vier Bänden (wohl versehentlich) vom Jahre 1843, das wohl nur neue Auflage ist.

zu sein scheint,[1]) dürfte beim Gebrauch des *P'ei wên yün fu*
von grossem Nutzen sein.

Die künstliche Aufrechthaltung der Sprache im Beamtenstande führte hin und wieder noch zu schriftstellerischen Ver-
suchen auf dem Gebiete des Administrationswesens; so legte
ein Manju T'UNG JUI aus Shên yang (Manju: *Simiyan*, in Girin)
die wichtigsten Grundsätze der Civilverwaltung dar (*Li chih tsi
yao*, Manju *Hafan-i dasan-i oyonggo-be isabuha bithe*).[2])

Auf Befehl des Kaisers Tao-kuang wurden im Jahre 1824
von einem Gelehrten-Ausschuss drei wichtige Werke publicirt,
die es sich zur Aufgabe setzten, die in den Annalen der Liao-,
Kin- und Yüan-Dynastie vorkommenden fremdsprachlichen Namen von Personen, Ämtern, Lokalitäten und andere Wörter festzustellen.[3]) Dieselben sind gut geordnet, in den chinesischen
Transkriptionen der Originale und dann in einer (nach der
Meinung der chinesischen Gelehrten) der wirklichen Aussprache
möglichst nahe kommenden Manju-Umschrift gegeben. Es kann
wohl kein Zweifel sein, dass mit Hülfe dieser Glossare die Lektüre jener drei historischen Werke bedeutend erleichtert und
gefördert wird, und dass sie wertvolle Materialien für das Studium der Sprachen der Khitan und der Niüchi[4]) enthalten.

---

[1]) Exemplar in New-York. Die Universitätsbibliothek in Cambridge
besitzt nur sieben Bände (GILES, Catalogue of the Wade Collection, p. 145).

[2]) Gedruckt in zwei Bänden, 1822, nicht 1845, wie MÖLLENDORFF,
Nr. 210, angibt. Nach Nr. 211 ist 1857 eine neue Auflage mit Zufügung
einer mongolischen Übersetzung erschienen, die ich nicht gesehen habe.

[3]) *K'in ting Liao shih yü tsieh* («Auf kaiserlichen Befehl verfasste
Erklärung der Wörter in der Geschichte der Liao»), 4 Vols., 10 Kapitel.
*K'in ting Kin shih yü tsieh*, 4 Vols., 12 Kapitel. *K'in ting Yüan shih yü
tsieh*, 8 Vols., 24 Kapitel, alle drei 1824. MÖLLENDORFF, Nr. 16, nennt davon nur das zweite, mit dem Zusatz: no year; er kann demnach das
Buch gar nicht gesehen haben, da das volle Datum auf dem Rande eines
jeden Blattes aufgedruckt ist. — A. O. IVANOVSKI hat aus Nr. 2 und 3 die
solonischen Wörter zusammengestellt (Mandjurica, I, St. Petersburg, 1894.
pp. X, XI).

[4]) Die oft wiederholte Behauptung, dass die Manju die Nachkommen der Niüchi seien, ist ganz unbegründet und unerwiesen. Alles, was
sich nach unserer geringen Kenntniss des Niüchi sagen lässt, ist, dass die
beiden Sprachen eng verwandt sind und wahrscheinlich eine Verwandt-
schaft derselben Linie repräsentiren, nicht aber dass die eine aus der an-

Doch sind natürlich die Rekonstruktionen immer kritisch nach-
zuprüfen.

Im Jahre 1836 erschien, mit einem Edikt des Kaisers
K'ien-lung vom Jahre 1772 bevorwortet, das vollständige Syllabar des Manju mit chinesischer Transkription (1 Band),[1] das
für uns dadurch einen gewissen Wert hat, dass am Schlusse
ein Verzeichnis geographischer Namen der Manchurei, der Mongolei, des Kukunōr-Gebiets, Kansu und Turkistans angehängt ist.

Auch die Übersetzung eines Teiles der Erzählungen aus
der Sammlung *Liao chai chih i* durch Wu-fei ist ein erfreu-
liches Zeichen des literarischen Interesses der neueren Manju.
Das Werk wurde chinesisch und Manju im Jahre 1848 in 24
Bänden gedruckt und enthält 129 ausgewählte Geschichten von
den 300 der Originalausgabe von 1740.

Das historische Interesse scheint jedoch unter den modernen Manju ganz erloschen zu sein; irgendwelche geschicht-
lichen Werke haben sie im neunzehnten Jahrhundert nicht mehr
übersetzt. Ein Manju, Hindge, schrieb 1849 eine Abhandlung
darüber, «wie man Geschichte studiren soll».[2] Als Geschichts-
werk ist kaum eine Sammlung zu bezeichnen, die eine Liste
der Namen der mongolischen und türkischen Fürsten enthält
und eher einem Handbuch wie dem Gothaer Hofalmanach entspricht. Sie wurde, wie aus dem Vorwort der manjurischen Ausgabe zu schliessen ist, 1839 auf Befehl des Kaisers Tao-kuang
gedruckt und kam 1849 an die Öffentlichkeit. Das von mir erlangte Exemplar besteht aus 24 Bänden in Gross-Folio Format,
von denen zwölf den chinesischen, zwölf den manjurischen Text
enthalten.[3]

---

dern historisch entwickelt ist. Vergl. C. de Harlez, Niu-tchis et Mandchous, rapports d'origine et de langage (Extrait du Journal Asiatique,
Paris, 1888, p. 31).

[1] *K'in ting Ts'ing Han tui yin tse shih* (Möllendorff, Nr. 6; Exemplar in New-York).

[2] So Möllendorff, Nr. 167. Leider besitze ich das Buch selbst
nicht. Dagegen ist das Vorwort und ein Teil des Manju Textes in Poz-
dnējev's Versuch einer Sammlung von Proben der manjurischen Literatur, pp. 28 et seq., abgedruckt, woraus hervorgeht, dass die Schrift des
Hingde nur der Neudruck einer Arbeit vom Jahre 1730 ist.

[3] Möllendorff, Nr. 165, citirt das Werk nur nach dem Katalog

Doch selbst in neuerer Zeit ist der Sinn für die Manju-Literatur nicht ganz erstorben. Dies beweist vor allem eine mit ausnehmender Sorgfalt und Schönheit in Paralleltexten gedruckte Anthologie aus der historischen Literatur (*Fan i ku wên*), die ein Manju namens Ting-hsien Mêng-pao[1]) im ersten Jahre der Periode Hsien-fêng (1851) in sechzehn Bänden veranstaltet hat, nach dem Vorbild der vom Kaiser K'ang-hsi 1686 edirten Sammlung historischer Musterschriften. Die ersten drei Bücher enthalten Kapitel aus dem *Tso chuan;* dann folgen Abschnitte aus der Geschichte der *Han* (Buch 4), aus *Se-ma Ts'ien* und dem *San kuo chi* (Buch 5), aus der Geschichte der *Tsin* (Buch 6) und Biographien aus den Annalen der *T'ang* (Buch 7—12; Buch 8—11: Biographie des *Han Yü*), der *Sung* (Buch 13—15) und der *Ming* (Buch 16).

Auch im letzten Drittel des Jahrhunderts sind ab und zu noch Manju-Werke publicirt worden. Das Jahr 1873 sah eine Neuausgabe des Heiligen Edikts in Manju, mongolisch und chinesisch, das Jahr 1876 sogar das Originalwerk eines Manju, des oben erwähnten Mêng Pao (in chinesisch und Manju). Dies ist ein moralisches Erbauungsbuch unter dem Titel «Wichtiges zur Belehrung der Welt» (*Ubaliyambuha jalan-de ulhibure oyonggo gisun-i bithe*) in vier Bänden, mit Vorwort vom Jahre 1867, in welchem der Verfasser wohlgemeinte Ratschläge in Versen an Knaben und Mädchen richtet und «ein Lied von der Ermahnung zur Tugend» und «ein Lied von guten Menschen» singt. Es ist wie der Schwanensang jener hausbackenen Moralphilosophie, welche die Manju sich mit besonderer Vorliebe aus der chinesischen Kultur erkoren und bis zur Bewusstlosigkeit und über das Ende ihrer Sprache und ihres Volkstums hinaus immer

---

von St. Petersburg, hat sich aber versehen, indem er es danach auf 60 Vols. angibt. In jenem Katalog heisst es aber: *ninju debtelin ninggun dobton*, d. h. sechzig Kapitel (oder Bücher) in sechs Bänden. Das kann indessen auch nicht stimmen, da das manjurische Vorwort von 72 Kapiteln spricht.

[1]) *Mêng-pao* ist nicht, wie Möllendorff, Nr. 72, irrtümlich angibt, der Übersetzer, sondern nur der Herausgeber der auf Grund früherer Übersetzungen veranstalteten Sammlung; die Namen der betreffenden Übersetzer gibt er zu jedem Abschnitt besonders an.

und immer wieder gepredigt haben. So konnte 1879 auch noch eine neue Auflage der Spruchsammlung *Ming hsien tsi*, gleichfalls als Trilingue, herauskommen. Ein recht nützliches Werkchen publicirte im Jahre 1885 ein Manju, GIORJE HSI-TʻAI, in zwei Bänden. Es führt den Titel «Kompass der Partikeln der Manju-Sprache» (*Tsʻing wên hsü tse chi nan pien*) und behandelt, kurz gesagt, alle nominalen und verbalen Affixe, deren Gebrauch durch gute Beispiele illustrirt wird (mit chinesischer Übersetzung).[1] Die Schrift zeigt jedenfalls, dass ein gewisses Bedürfnis nach solchen Lehrfäden vorhanden ist, und dass die Sprache auch jetzt noch theoretisch studirt wird. Die letzte mir bekannte Erscheinung auf dem Gebiete der manjurischen Literatur ist der Neudruck des grossen mongolisch-chinesisch-manjurischen Wörterbuchs vom Jahre 1891, wovon schon in der Skizze der mongolischen Literatur die Rede war.

MÖLLENDORFF (Nr. 134) lässt auch den ganzen Kanjur und Tanjur in den Reihen der Manju-Literatur mitaufmarschiren und behauptet ohne Angabe irgendeiner Quelle, dass «diese grosse Sammlung auf Befehl des Kaisers Kʻien-lung in vier Sprachen, chinesisch, Manju, tibetisch und mongolisch gedruckt worden sei, in jeder Sprache in 348 Bänden folio». Die einzige mir denkbare Autorität, auf die sich diese kühne Aufstellung stützen könnte, ist eine Notiz von A. WYLIE,[2] der sich allerdings weit vorsichtiger ausdrückt, indem er sagt: «Nach einer Angabe aus der Feder des Vaters Hyakinth erscheint es, dass alle heiligen Bücher der Tibeter im vorigen Jahrhundert unter Aufsicht einer kaiserlichen Kommission ins Manju übersetzt worden sind.» Von dem Vorhandensein eines manjurischen Kanjur sind wiederholt Gerüchte im Umlauf gewesen; wenn er vorhanden ist, handelt es sich jedenfalls um ein handschriftliches Exemplar. Dass aber der Tanjur je ins Manju übersetzt worden sein sollte (cui bono?),

---

[1] Von MÖLLENDORFF nicht erwähnt; doch citirt er unter Nr. 4 ein «Gedicht auf die Manju Partikeln», ohne Jahr und Autor, nicht gedruckt, das vielleicht dem obigen Werke nachgebildet ist.

[2] Translation etc., p. XLVII, der selbst nach MEADOWS' Essay on the Manchu Language, p. 11, citirt, eine Schrift, die mir nicht zugänglich ist.

ist stark zu bezweifeln;[1] die ganze Frage muss vorläufig auf sich beruhen bleiben, bis handgreifliche Beweise vorliegen. Die Manju haben sich nie für den Buddhismus und noch weniger für den Lamaismus sehr erwärmt; die nüchternen Maximen des Confucius sagten ihrem rationalistischen Wesen mehr zu. Diesen verstandesmässigen Zug teilen sie mit der ganzen Gruppe der tungusischen Völker, und im Verkehr mit den ihnen stammverwandten Golden am Amur ist er mir stets besonders auffallend gewesen.[2] Mit der buddhistischen Literatur der Manju lässt sich nun auch kein grosser Staat machen; sie beschränkt sich auf einige wenige kurze Sūtra, während die Übersetzung (resp. teilweise Transkription) der Namen der tausend Buddha mehr eine philologische Spielerei war denn einem realen Bedürfnis entsprach. DAHAI soll im Jahre 1633 ein Mahāyānasūtra in seine Muttersprache übersetzt haben, welches, wird nicht gesagt. Ein rein graphisches Curiosum ist der von A. IVANOVSKI publicirte Text des tibetischen *Prajñāpāramitāhṛdaya* in manjurischer Transkription.[3]

Auch die taoistische Literatur ist von den Manju wenig gepflegt worden. Schon Nurhaci hat seine Beamten vor den Lehren der Tao-se gewarnt.[4] Die bekannte volkstümliche Schrift

---

[1] Ein Index des Kanjur und Tanjur in Manju scheint dagegen vorhanden zu sein (P. CORDIER in Bulletin de l'École française d'Extrême-Orient, Vol. III, 1903, p. 605).

[2] American Anthropologist, 1900, p. 319.

[3] Тибетскій текстъ въ маньчжурской транскрипціи. in Восточныя Замѣтки, St. Pet., 1895, pp. 261—267 (mit drei phototypischen Tafeln). — C. DE HARLEZ hat den Manju Text des Vajrachhedikāsūtra veröffentlicht (WZKM, Band XI, 1897, S. 209—230, 331—356). Ich selbst besitze einen chinesisch-manjurischen Paralleltext (gedruckt in vier Bänden, ohne Datum und Vorrede, wahrscheinlich neueren Ursprungs) des *Ti-tsang p'u-sa pén yen king ( Na-i niyamangga fusa-i de forobun-i nomun)*. Die Titel von zwölf buddhistischen Manju-Schriften werden in Mélanges asiatiques de l'Académie de St.-Pét., Vol. I, pp. 419—421, mitgeteilt.

[4] *Tai-tsu Dergi hówangdi-i enduringge tacihiyan*, Vol. I, pp. 1b, 2a: «Wenn die *Beise* (die kaiserlichen Prinzen zweiten Ranges) und Beamten den Kaiser als ihren Vater ansehen, werden sie nicht die verdorbene Gesinnung der Tao-se ( *Doose* ) lieben». Und weiter: «Wenn sie ein trügerisches und schlechtes Herz nähren, wenn sie die verdorbene Lehre der

*T'ai shang kan ying pien*[1]) hat im Jahre 1759 eine Übersetzung
durch FUYANTAI erfahren; ausserdem zählt MÖLLENDORFF noch
sieben kleine unbedeutende taoistische Traktate auf. Eine Über-
raschung war es daher, als E. v. ZACH[2]) den Text einer Manju-
Übersetzung des unter Lao-tse's Namen gehenden *Tao-têh king*
veröffentlichte, von deren Existenz früher nichts bekannt gewesen
ist. Leider macht der Herausgeber keine bibliographischen Angaben über seine Vorlage; vermutlich handelt es sich um ein
handschriftliches Unikum, in dem solche Details fehlen.

**Katholische Literatur.** — Die Jesuitenmissionare, die Ende
des sechzehnten Jahrhunderts nach China kamen, erkannten
frühzeitig die Bedeutung des Manju; ein grosser Teil von ihnen
machte sich die Sprache zu eigen und schrieb in derselben. In
diesen literarischen Unternehmungen wurden sie vom Kaiser
K'ang-hsi ermuntert und begünstigt. Er beauftragte sie mit der
Abfassung von Abhandlungen über Musik und Wissenschaften
in chinesischer und Manju-Sprache. Die Väter GERBILLON und
BOUVET, die nach sieben- oder achtmonatlichem Studium ziem-
liche Fertigkeit im Manju erlangt hatten, machten sich zunächst
an eine Übersetzung der Elemente des Euklid in diese Sprache,
«die leichter und weit hübscher ist als die chinesische», wie
BOUVET sagt. Den Elementen fügten sie alle Sätze bei, die sie
zur Belehrung des Monarchen für notwendig hielten. Nachdem
er in die Elemente der Geometrie eingedrungen war, befahl er
ihnen, ein vollständiges Handbuch der praktischen Geometrie
nebst der ganzen Theorie auf Manju zu verfassen. Er wollte sich

---

Tao-se befolgen, wird der Kaiser sie zur Rechenschaft ziehen und ihnen
Amt und Rang nehmen».
    [1]) Wiederholt übersetzt, von ST. JULIEN, Le livre des récompenses
et des peines, Paris, 1835; von J. LEGGE, T'ai Shang, Tractate of Actions
and their Retributions (Sacred Books of the East, Vol. XL, pp. 233—
246); von T. SUZUKI und P. CARUS, Treatise of the Exalted One on Re-
sponse and Retribution, Chicago und London, 1906. S. auch H. CORDIER,
Bibliotheca Sinica, 2. Aufl., Vol. I, Paris, 1904, Spalte 726. — Eine Manju
Übersetzung existirte vielleicht schon 1673 (s. KLAPROTH, Chrestomathie
mandchoue, p. 221).
    [2]) Manchurian Translation of Lao-tzu's Tao-tê-ching, Romanized
Text. China Review, Vol. XXV, 1901, pp. 157—162, 228—234.

auch der scholastischen Philosophie widmen, so dass die beiden
Missionare die alte und neue Philosophie von Duhamel für ihn
ins Manju übersetzten. Sie verfassten auch in derselben Sprache
an zwanzig kleine Abhandlungen über Medicin und Anatomie,
die in hohem Grade die Aufmerksamkeit des Kaisers erregten.[1]
Im Jahre 1723 übersandte P. Dominique Parennin (1665—1741)
der Akademie der Wissenschaften in Paris Exemplare von Manju-
Übersetzungen einer Abhandlung über Anatomie,[2] Cursus der
Medicin und ein Werk über Physik, die aus der Mission der
Jesuiten stammten. In seinem Begleitschreiben bemerkt er, dass
sie auf besonderen Befehl und unter Aufsicht des Kaisers übersetzt worden seien, und dass er sich bis dahin zehn Jahre lang
mit Übersetzungen aus dem Manju in europäische Sprachen
und mit Übersetzungen aus dem Französischen, Lateinischen,
Portugiesischen und Italienischen ins Manju beschäftigt habe.[3]

Eines der Hauptwerke von MATTEO RICCI (1552 -1610)
«Wahre Lehre von Gott», zuerst 1601 auf Chinesisch publicirt,
wurde 1758 in Manju-Übersetzung herausgegeben.[4] Ebenso ist
das Buch von GIULIO ALENI (1582—1649) über den «Wahren
Ursprung aller Dinge» (chinesisch, 1628) ins Manju übersetzt
worden:[5] ebenso die Schrift von FERDINAND VERBIEST (1623—
1688) «Auseinandersetzung der wichtigsten Lehren der Kirche»

---

[1] L. LANGLÈS, Alphabet mantchou, Paris, 1807, pp. 71—72.

[2] Es handelte sich um eine Übersetzung der Anatomie des PIERRE
DIONIS mit Nachbildungen der anatomischen Tafeln des THOMAS BARTHOLIN
(Mémoires concernant les Chinois, Vol. VII, Paris, 1782, p. 12, Note).
Diese Übersetzung scheint nie gedruckt worden zu sein; CORDIER (L'imprimerie sino-européenne en Chine, Paris, 1901, p. 36) notirt sie nicht
unter PARENNIN's Namen. Die Mitteilungen zur Geschichte der Medizin
und Naturwissenschaften (Band VI, Nr. 4, 1907, S. 448) brachten kürzlich
nach der Kölnischen Zeitung die Nachricht, dass PARENNIN's chinesisches
Manuskript der Anatomie im Besitz eines Kaufmanns in Kopenhagen
aufgefunden worden sei, und dass der Carlsbergfonds eine grössere Summe
zur Drucklegung desselben bewilligt habe.

[3] A. WYLIE, Translation of the Ts'ing Wan K'e Mung, Shanghai,
1855, p. XIV.

[4] H. CORDIER, L'imprimerie sino-européenne en Chine, Paris, 1901,
p. 39, Nr. 225.

[5] H. CORDIER, l. c., p. 3, Nr. 13.

4*

(chinesisch, 1677). deren Manju-Version in einem kaiserlichen
Edikt vom Jahre 1805 denuncirt und verboten wurde.[1] VERBIEST
gehörte sicher zu den besten Kennern des Manju unter den
Jesuiten. Im zweiten Bande der Relations de divers voyages.
herausgegeben 1696 von THEVENOT, ist eine Manju-Grammatik
unter dem Titel Elementa linguæ Tartaricæ abgedruckt mit dem
Namen GERBILLON als dem des Verfassers. Es wird jedoch von
anderer Seite behauptet. dass sie von VERBIEST verfasst sei. was
WYLIE als möglich zugibt, da in dem in China herausgegebenen
Katalog von Jesuiten verfasster Werke unter den Arbeiten von
VERBIEST eine Grammatica Tartarica erwähnt sei. Er soll auch
ein Wörterbuch des Manju verfasst haben.[2] Endlich verdient
noch die Schrift von DE MAILLA (1669—1748) über «das Heilige
Jahr» Erwähnung. gleichfalls aus dem Chinesischen (1738) ins
Manju übersetzt und 1805 verboten.[3]

Auch Teile der Bibel sind von Jesuiten in Manju bearbeitet worden, doch scheinen keine bibliographischen Nachrichten
darüber aufbewahrt zu sein. Doch hat J. EDKINS[4] einen Teil
der Genesis und den ganzen Exodus in Manju aufgefunden und
kurz analysirt, woraus hervorgeht, dass es sich um eine katholische Arbeit handelt, leider aber sind keine Angaben darin betreffs Übersetzer und Zeit enthalten.[5]

**Bibelübersetzungen.** — Die Russische Bibelgesellschaft
erörterte die Frage einer Manju-Übertragung der Bibel bereits
im Jahre 1816, doch infolge Überbürdung mit anderer Arbeit
geschah damals nichts. Im Jahre 1821 erhielt Dr. Pinkerton von
der British and Foreign Bible Society den Auftrag, diese Arbeit
ins Werk zu setzen, und engagirte zu diesem Zwecke Stepan
Lipovzov. den Manju-Dolmetscher der Russen. der vierzehnjäh-
rige Studien in dieser Sprache gemacht hatte. Eine Übersetzung

---

[1] H. CORDIER, l. c.. p. 59, Nr. 349; A. WYLIE, Notes on Chinese
Literature, 2. Aufl., p. 177.

[2] A. WYLIE, Translation etc., pp. LI, LII.

[3] H. CORDIER, l. c., p. 33. Nr. 164; A. WYLIE, Notes etc., p. 178.

[4] Manchu Christian Literature, China Review, Vol. XXIV, 1899,
pp. 72. 73.

[5] Vielleicht handelt es sich um die Bibelübersetzung von POIROT
(s. weiter unten).

des Evangeliums Matthäi wurde 1823 veröffentlicht, und Ende
1825 lag das ganze Neue Testament beendigt vor, doch die russische Regierung entschloss sich nicht zum Druck. W. Swan von
der London Missionary Society begab sich 1832 nach St. Petersburg, mit dem Auftrag, ein im Besitz der Heiligen Synode be-
findliches Manuskript zu copiren, das Teile des Alten Testaments,
Matthäus und die Apostelgeschichte enthielt und Ende des 18.
Jahrhunderts von C. P. Louis de Poirot in Peking mit einer
chinesischen Interlinearversion angefertigt worden war. Im Jahre
1833 wurde George Borrow zwecks Manju-Studien nach St. Petersburg geschickt, wo er und Swan die Abschrift eines Teils des
Alten Testaments fertig stellten.[1] E. Stallybrass, der sie in China
verwenden wollte, machte eine zweite Abschrift davon, die 1850
beendet war. 1834 erlangte George Borrow als Beauftragter der
Britischen Bibelgesellschaft Erlaubnis von der russischen Regierung, das ganze Neue Testament unter Lipovzov's Censoraufs-
sicht zu drucken. Die Arbeit wurde in zehn Monaten beendet,
und eine Auflage von tausend Exemplaren wurde 1836 nach
London versandt. Unter der Aufsicht von ALEXANDER WYLIE
wurden Matthäus und Markus in Lipovzov's Version 1859 in
Shanghai publicirt. Fernerhin haben keine Gründe vorgelegen,
weitere Teile der Schrift in Manju zu veröffentlichen.[2]

---

[1] Diese Handschrift ist nie gedruckt worden; sie befindet sich
noch im Besitz der Bibelgesellschaft in London.

[2] Vergl. ABEL-RÉMUSAT, Journal Asiatique, Vol. I, 1822, p. 256;
Vol. II, 1823, p. 250; A. WYLIE, Translation of the Ts'ing Wan K'e Mung,
Shanghai, 1855, pp. XLV—XLVII. [A. WYLIE], Memorials of Protestant
Missionaries to the Chinese, Shanghai, 1867, p. 174. M. BROOMHALL, The
Chinese Empire, A General and Missionary Survey, London 1907,
pp. 417—418.

# 060

印度的约拿传说

# THE MONIST

A QUARTERLY MAGAZINE

DEVOTED TO THE PHILOSOPHY OF SCIENCE

VOLUME XVIII.

CHICAGO
THE OPEN COURT PUBLISHING COMPANY
1908

*Reprinted with the permission of the original publishers*

KRAUS REPRINT CORPORATION
New York
1966

*Printed in U.S.A.*

# THE JONAH LEGEND IN INDIA.

STUDENTS of the Old Testament may be interested in learning about a migration of the Jonah Legend eastward into India.

Two versions of it occur in the Tibetan work *bKa-babs bdun-ldan,* a History of Buddhism in India from the eleventh century A. D. to the reign of Akbar, written by the Lama Tāranātha in 1600.[1] This author is well known to students of Buddhism by another work on the history of Buddhism in India compiled in 1608, which has become easily accessible by the Russian translation of W. Wassiljef and a German version of Anton Schiefner. The former book represents the continuation of the latter.

The legends refer to the late period of Buddhism characterized by a group of eighty-four saints or rather sorcerers known under the designation Mahāsiddha. Their activity seems to embrace the time roughly from the eleventh to the thirteenth century. They play an important rôle in the mythology of Lamaism.[2] One version of the legend is connected with the name of Naropa who, in all likelihood, died in A. D. 1035.[3] The story is very brief. Naropa, says our text (p. 37, 7), had consecrated a Man-

---

[1] The Tibetan text has been edited by the Pandit Sarat Chandra Das at Darjeeling (printed at the Bengal Secretariat Press), 1895, 76 pp. The book has not yet been translated.

[2] A. Grünwedel, *Mythologie des Buddhismus in Tibet und der Mongolei,* p. 40.

[3] This date has been computed by E. Schlagintweit, *Abhandlungen der Bayer. Akademie,* 1896, p. 602.

dala of the God Hevajra[4] and studied certain methods of meditation, and while he once was in a state of contemplation, he was carried away by a stream and swallowed by a fish; but as he perceived in the belly of the fish the Mandala of Heruka,[5] he did not suffer any harm and was cast out again.

The other story (p. 58, 1) is somewhat more detailed. "Mînapa (i. e., the fisher), a pupil of the Saint Kakkutipa, was a fisherman in Kāmarūpa in the east of India. While he used to meditate a little after the 'wind' method,[6] as practised by the fishermen, he once threw his fishing-hook at a fish, and pulling the line, he was snatched and swallowed by the fish. By virtue of his deeds and meditations, however, he did not die, but drifted on the river Rohita towards Kāmarūpa. There, on the little hill Umagiri, Maheçvara[7] preached to the goddess Umā[8] instructions on the 'wind' Yoga. As the fish came into that river, the fisherman in the belly of the fish listened to the sermon, meditated, and obtained many accomplishments (*guna*). It once happened that this fish was caught by fishermen and killed, when a man turned out. The former king had then already died, and meanwhile, since the birth of his (Mînapa's) son, thirteen years had elapsed: thus it was found that he had spent twelve years in the belly of the fish. Thereupon, father and son betook themselves to the master Carpatipa, requested a sermon from him, meditated, and obtained both the siddhi. The father is known as the Siddha Mînapa, the son is called the Siddha Ma-ts'in-dra-

[4] Grünwedel, *loc. cit.*, pp. 104, 105.

[5] Identical with Hevajra. See Schiefner, *Tāranātha*, p. 233.

[6] A term of mysticism denoting the drawing in and holding one's breath to prepare for meditation and finally the power of holding back one's breath for a great length of time, by which faculty miracles and many extraordinary things may be performed, not only those of religious significance, but also of good practical purposes, as in the above case the method of the fishermen is doubtless suggestive of a good catch of fish. See also R. Garbe, *Sāmkhya und Yoga*, pp. 44 et seq.

[7] Shiva.          [8] Shiva's consort.

pa." The latter word is apparently identical with the
Sanskrit *matsyendra*, "the lord of the fish."

As traditions of men being swallowed by a fish seem not
to be found in ancient India, and as the above two stories
relate to a period when Mohammedan power was established, we shall probably not err in supposing that it may
have been the Arabs who spread the story in India. Indeed, the legend of Jonah is narrated in the Koran (Sūrah,
XXXVII, 139-148) and quoted in four other passages
of it.[9]

                              Dr. Berthold Laufer.
New York.

[9] See Hughes, *Dictionary of Islam*, p. 249.

# 061

中国鸽哨

# SCIENTIFIC AMERICAN

[Entered at the Post Office of New York, N. Y., as Second Class Matter. Copyright, 1908, by Munn & Co.]

Vol. XCVIII.—No. 22.
Established 1845.

NEW YORK, MAY 30, 1908.

10 CENTS A COPY
$3.00 A YEAR.

This Aeroplane is the First Dynamic Flying Machine to Traverse a Considerable Distance and Carry Two Men at a Speed of 45 Miles an Hour.  A Similar Aeroplane to be Built for the United States Government by the Last of August Must Remain in the Air for an Hour and Cover 40 Miles in That Time.

THE WRIGHT 30 HORSE POWER AEROPLANE IN FLIGHT ABOVE THE NORTH CAROLINA COAST.—DRAWING PREPARED FROM DESCRIPTIONS BY OBSERVERS OF THE EXPERIMENTS.—[See page 395.]

## CHINESE PIGEON WHISTLES.

BY DR. BERTHOLD LAUFER, COLUMBIA UNIVERSITY.

Traditions imbibed in school, through education and by reading, are apt to sway our lives and our thoughts, and to influence strongly our judgment of other peoples. An almost fixed standard of attributes involuntarily arises in our minds when the names of French, Spaniards, Negroes, Indians, strike our ears, and it is often hard to see why such and such an adjective, expressing such and such a quality, became inevitably linked in our thoughts with the names of certain nations. Thus, we are wont to speak of the Chinese as a sober, practical, and prosaic people, and to view them throughout in that light. Immensely rational they are (this cannot be gainsaid), secular, and worldly-minded, bestowing all their efforts on useful temporal affairs; but nevertheless these people are by no means lacking in purely emotional matters of great attractiveness. It is needless to turn to their poetry and art, in which they are at their best, regarded from this viewpoint; even in affairs of minor importance their soul reveals to us traits of poetical quality of no small degree.

As early as the eleventh century one of their greatest poets sang:

"Upon the bridge the livelong day
I stand, and watch the goldfish play."

The domestication of the goldfish, the first species of which reached England only in 1691, and of the wonderful paradise-fish as well, is justly ascribed to the Chinese; and it is remarkable to notice that their attempts in this direction and the amazing results achieved

so that when the birds fly the wind blowing through the whistles sets them vibrating, and thus produces an open-air concert, for the instruments in one and the same flock are all tuned differently. On a serene day in Peking, where these instruments are manufactured with great cleverness and ingenuity, it is possible to enjoy this aerial music while sitting in one's room.

There are two distinct types of whistles—those consisting of bamboo tubes placed side by side, and a type based on the principle of tubes attached to a gourd body or wind-chest. They are lacquered in yellow, brown, red, and black, to protect the material from the destructive influences of the atmosphere. The tube whistles have either two, three, or five tubes. In some specimens the five tubes are made of ox-horn instead of bamboo. The gourd-whistles are furnished with a mouthpiece and small apertures to the number of two, three, six, ten, and even thirteen. Certain among them have, besides, a number of bamboo tubes, some on the principal mouthpiece, some arranged around it. These varieties are distinguished by different names. Thus, a whistle with one mouthpiece and ten tubes is called "the eleven-eyed one."

As to the materials and implements used in the manufacture of pigeon-whistles, there are small gourds that serve for the bodies; halves of large gourds (a particular species imported from Shantung to Peking for this industry), from which stoppers are made that fit into them; and four kinds of bamboo—cylindrical pieces of a large species that grows in the south, for making the mouthpieces of the large tubes; thin sticks for making those of the small ones; hard bamboo for

### A New Italian Method of Preserving Eggs.

Consul D. I. Murphy, of Bordeaux, forwards the following synopsis from a French journal on a new method of preserving eggs, which, he says, appears to have the double merit of cheapness and simplicity. The article was based upon the experiments of Doctor Campanini, as reported by him in a bulletin issued by the Italian minister of agriculture.

Dr. Campanini, after reviewing the various known means of preserving eggs—by salt water, lime water, silicate of potash, vaseline, and cold storage—described his experiments, which showed better results than all others.

His theory is that to preserve eggs some system must be adopted that will absolutely prevent the exchange between the air outside and that inside the egg—for it is this continual exchange that causes putrefaction.

Dr. Campanini selected perfectly fresh eggs and covered them with lard, so as to effectually stop up all the pores. The shells were thus rendered impermeable, the exchange of air was prevented and, the obstruction of the pores not permitting the evaporation of the water, there was no loss of weight. The whites and yellows of the eggs retained their color perfectly and the taste was not modified in the slightest degree. When properly coated with lard—not too thickly—the eggs are put in baskets or boxes upon a bed of tow or fine odorless shavings and so arranged that there will be no point of contact between them—otherwise a mold will develop and putrefaction result. The packing room should be perfectly dry, the question of temperature not being important. By his process Dr. Campanini kept a quantity of eggs for a whole year

A Pigeon With a Whistle Wired to Its Tail. The Whistles Look Clumsy But Are Very Light.

Pigeon Whistles, With the Gourds and Bamboos From Which They Are Made; and the Tools Used in Making Them.

### CHINESE PIGEON WHISTLES.

were not prompted by any utilitarian views they had in mind, as neither fish is of any practical advantage. On the contrary, their skilful breeding, so eagerly pursued, is due solely and exclusively to the æsthetic tendency of the Chinese in their art of living, and to their highly cultivated sense of beauty, which delights in the bright coloration of the skin of these fishes, the graceful form of their bodies, and the restless motions of their long, flowing fins. This is the more worthy of note, as the only fish among us which has been placed within the range of domestication, the carp, is granted this privilege merely from its proxy connection with the kitchen.

While the almost Darwinian experiments to which Chinese breeders have subjected the goldfish, and their unbounded admiration of this little creature in its hundred and one forms and variations, illustrate well the intimate relation of the people to the element water, their friendly associations with the world of birds are no less close and sympathetic. The lover of birds does not permanently confine his pet in its prison-cage, but he takes it out with him on his walks, carrying it on a stick, to which one of its feet is fastened by means of a thread long enough to allow it ample freedom of motion. Where the shade of some stately tree bids him welcome, he makes a halt, and permits the bird to perch and swing on a supple twig, watching it for hours.

One of the most curious expressions of emotional life is the application of whistles to a flock of pigeons. These whistles, very light, weighing hardly a few grammes, are attached to the tails of young pigeons soon after their birth, by means of fine copper wire,

the large tubes themselves, and a soft kind for the smaller ones. The separate pieces are fastened together by means of fish-glue, which is applied with an iron nail. A razor-like knife is used for splitting the bamboo sticks, and a chisel to break the harder pieces. For the general work a dozen spatulas are required, and awls are used for drilling the small mouthpieces. There is also a whetstone for grinding the implements, the same as is employed in other industries and by professional knife-grinders, and a saw with a slightly curved blade for cutting the gourds. The smallest whistles are of course most difficult to produce. One workman is said to be able to turn out about three specimens a day, which shows that the work requires some time and skill.

The explanation of the practice of this quaint custom which the Chinese offer is not very satisfactory. According to them, these whistles are intended to keep the flock together, and to protect the pigeons from attacks of birds of prey. There seems, however, little reason to believe that a hungry hawk could be induced by this innocent music to keep aloof from satisfying his appetite; and this doubtless savors of an after-thought which came up long after the introduction of this usage, through the attempt to give a rational and practical interpretation of something that has no rational origin whatever; for it is not the pigeon that profits from this practice, but merely the human ear, which feasts on the wind-blown tunes, and derives æsthetic pleasure from this music. And here, again, it seems to be a purely artistic and emotional tendency that has given rise to a unique industry and custom applied to nature-life.

—through a very hot summer and a very cold winter—and they were perfectly preserved. He says that 4 cents' worth of lard suffices to coat 100 eggs, and that anyone could easily prepare that number of eggs in one hour's time.

### Wasting Water Assets.

California has learned, says the California Cultivator, not only that the ground may become waterlogged by over-irrigation, but that ill-considered drainage and the inconsiderate use of water from wells may so lower the underground water-plane as to threaten the reversion of large areas to unproductiveness.

The Geological Survey has demonstrated that all the subterranean waters of 775 square miles in southern California are connected, and that every well taps a common supply; and on this the water-plane, which was twenty-three feet below the surface of the soil in 1898, is now fifty feet below.

People can live beyond their means in respect to water, as well as timber, oil, natural gas, fish, and game.

The output of copper in the Ural district in 1906 amounted to 4174 tons, as compared with 3,610 tons in the preceding year, thus showing an increase of 564 tons. These figures cannot be considered as satisfactory from the Russian point of view, taking into consideration the high prices which prevailed last year for copper. Moreover, on comparing the total output of copper in 1906 with that of the years 1904, 1903 and 1902, a decrease in the production is noticeable.

汉代中国陶器
附：书评四则

# CHINESE POTTERY OF THE HAN DYNASTY.

PUBLICATION

of the

East Asiatic Committee of the American Museum
of Natural History — The Jacob H. Schiff
Chinese Expedition.

———————

# CHINESE POTTERY OF THE HAN DYNASTY.

BY

## BERTHOLD LAUFER.

LEIDEN,
E. J. BRILL Ltd.,
PRINTERS AND PUBLISHERS,
1909.

# ACKNOWLEDGMENTS.

In 1901 Mr. Jacob H. Schiff donated to the American Museum of Natural History funds for making investigations and collections in China. The administration of these funds was intrusted by Mr. Schiff to a committee organized under the auspices of the American Museum of Natural History. The following gentlemen were members of the Committee: Morris K. Jesup, Chairman; Edward D. Adams; Hermon C. Bumpus; Nicholas Murray Butler; Clarence Cary; C. C. Cuyler; William E. Dodge; John Foord; E. H. Harriman; James J. Hill; Arthur Curtiss James; Seth Low; Clarence H. Mackay; Howard Mansfield; James R. Morse; William Barclay Parsons; Henry Clay Pierce; George A. Plimpton; Jacob H. Schiff; John H. Winser, Treasurer; Franz Boas, Secretary.

The work endowed by Mr. Schiff was intrusted to Dr. Berthold Laufer, who spent nearly three years, from 1901 to 1904, in China, making collections of specimens, and carrying on investigations.

Printed by E. J. BRILL. — LEIDEN (Holland).

# CONTENTS.

V

# LIST OF ILLUSTRATIONS.

―――

## TEXT FIGURES.

VI

PLATES.

# PREFATORY.

THE main bulk of the pottery forming the basis of the present paper was collected by me in Hsi an fu, Shensi Province, in August, 1903, while on a mission in China for the American Museum of Natural History. The material is supplemented by a number of other pieces received on loan from private collectors. First of all, I am under great obligations to Mr. Thomas B. Clarke, who very kindly loaned to the Museum 61 pieces of this pottery, 54 of which are here illustrated and described. I have further to tender my heartiest thanks to Mr. Marsden J. Perry of Providence, R.I., to Mr. R. H. Williams and Mrs. E. C. Bodman of this city, all of whom generously placed their ceramic treasures at my disposal. For one very interesting specimen I am indebted to Messrs. Yamanaka & Co.

All together, 111 pieces of Han pottery are here figured and discussed, 41 of which belong to the Museum's collections, and 70 of which are loaned objects. There are here, besides, 6 pieces of Chou pottery, 13 Han tiles, 20 pieces of Sung pottery, 3 porcelains, 9 pieces of modern common pottery, and 5 implements illustrating the manufacture of pottery, making a total of 167 pieces of Chinese pottery. Of reproductions of bronzes from actual specimens, there are 17, 14 of which are in the collection of bronzes of the Museum, and 3 were loaned by Mr. Ōkazaki. Figures of other bronzes and of various other objects are derived from en-

XV

gravings in Chinese books. Those from the "Hsi ch'ing ku chien," the descriptive catalogue of the Ch'ien lung collection (published in 1749), are made from photographs of the quarto edition (43 cm by 27 cm), printed in Japan in 1888, which is an exact facsimile of the original, while the modern Shanghai photo-lithographic reprint measures only 25.5 cm by 15.5 cm. On the whole, this paper contains 249 illustrated objects.

The illustrations to this paper have been prepared by Mr. Rudolf Weber. My gratitude, which I herewith take pleasure in expressing publicly, is due to Miss H. A. Andrews for her assistance in the preparation of the manuscript, and to Miss M. L. Taylor for seeing it through the press.

The last half of the book was sent to press during the author's absence in Asia, which made it impossible for him to read the proofs himself. The manuscript of this work was completed in April, 1906.

NEW YORK CITY, January, 1908.

BERTHOLD LAUFER.

# I. — INTRODUCTORY.

THE first mention of Han pottery in European literature is made by S. W. BUSHELL, in his book "Oriental Ceramic Art" (New York, 1897). On p. 10 he says, "Bricks and tiles are among the most useful of ceramic products. They may even rank as historical monuments when inscribed. The Chinese antiquary collects them in chronological series to show the changes in the style of the written character, or puts one upon his writing-table for daily use, excavated into the shape of an ink-pallet. They were first moulded, with the date inscribed on one side, during the Han dynasty.

"Some of the pottery of the period is also inscribed. There is, for instance, a bottle-shaped vase of dark-reddish stoneware in the Dana collection in New York, moulded in the shape of a bronze ritual vessel of the time, enamelled with a deep-green iridescent glaze [p. 11] much exfoliated, which is engraved on the surface with a date corresponding to 133 B. C., the second year of the period *Yüan-kuang.* A similar vase in the British Museum, although it has no inscription upon it, evidently dates from about the same time, and specimens of this kind are not uncommon in Chinese collections. The vase illustrated in Fig. 49 is a good example of this class, an ancient stoneware of brownish-red paste, invested with a thin but lustrous glaze of camellia-leaf green, which came from the collection of *Chang Yin-huan,* formerly Chinese minister at Washington, as a relic of the Han dynasty."

C. H. READ, in "Relics from Chinese Tombs" (*Man,* 1901,

I

I

Vol. I, p. 17), discussed some objects of an early Chinese tomb in Shensi Province, consisting of two pottery bowls, a bottle or vase, and a mirror. These were correctly attributed by S. W. BUSHELL (*l.c.*, p. 67) to the Han time. "With regard to the pottery, there is no reason," he remarks, "as far as I know, why it should not be attributed to the same early period (Han dynasty). The vase, with its stippled brownish-black glaze shot with invisible green, stopping short in an irregularly curved line before it quite reaches the foot, would certainly be referred by a Chinese collector to the Han dynasty. The material generally used in the production of the color being an impure [p. 68] native cobaltiferous ore of manganese containing iron, the iron gives a brownish tinge to the black body, and changes the cobalt to green.

"The small red glazed bowls are of a much rarer type, and I have never seen their like in any Chinese collection. Of finished technique, they exhibit a smooth glaze of remarkably uniform color, due doubtless to iron peroxide, one of the earliest pigments used in Chinese ceramics. Are they not, by the way, wine-cups, buried with the owner's wine-vessel? The wine-cup of the Han dynasty was usually fashioned of glazed earthenware, replacing the bronze, jade, and horn cups of earlier times; under the *T'ang*, wine-cups were made of gold, chiselled silver, carved rock-crystal and other hard stones, glass, and porcelain; and under the *Sung* (A.D. 960–1279), self-colored porcelain came into general vogue, such color being selected as would enhance the natural tints of the wine or tea for which they were intended to be used.

"The prevailing color of the pottery of the Han dynasty was a bright-green monochrome tint, produced by the addition of copper oxide to a siliceous flux. A dull black comes next, being that of the lac-black circular dish described in the *T'ao Shuo*, the well-known Chinese book on pottery, as having been discovered in the tomb of the Empress *Tao Hou*, a consort of the celebrated *Wu Ti* (140-87 B.C.) of the former Han dynasty. From the evidence of this recent find, it seems that we may venture to add a pale vermilion to the brief list of self-colored glazes of this early period."

A number of pieces of Han pottery were exhibited at the St. Louis Fair. In the "Catalogue of Chinese Exhibits at the Louisiana Purchase Exposition" (St. Louis, 1904), three pottery vases of the Han dynasty, with the designation "silver gilded," are noted on p. 50; and two "green earthenware vases," three "earthenware incense-burners," and one "earthenware pot" (Nos. 744-749 of the Hupeh Provincial Exhibit), all ascribed to the Han dynasty (206 B.C. to A.D. 25), on p. 132.

Seventeen pieces of Han pottery are noted in the "Illustrated Catalogue of Antique and Modern Chinese and Japanese Objects of Art," etc., of the YAMANAKA sale (New York, 1905, p. 149, Nos. 837-853). Eleven of these are reproduced and described in this paper.

In the "Catalogue *de luxe* of the Art Treasures collected by THOMAS E. WAGGAMAN," revised and edited by TH. E. KIRBY (New York, 1905), three Han vases (Nos. 601-603) are enumerated, one of which is reproduced in that catalogue.

A few single pieces of Han ceramics seem to have come into European collections during recent years without always being correctly recognized as such.

In the catalogue, "Objets d'art et peintures de la Chine et du Japon réunis par T. HAYAIISI" (Part II, Paris, 1903, p. 106, No. 511), a piece of Han pottery of the type of the hill-jar (but without cover) is described and illustrated, but classified among porcelains, and assigned to the twelfth century. From the illustration and description it is quite evident that it is a Han piece: "Brasero cylindrique, porté par trois pieds de monstre et contourné d'une frise en relief composée de tigres héraldiques, alternant avec un motif de vagues. Email vert irisé, partiellement rongé par son séjour dans la terre. Haut. 0.15; larg. 0.19." A few other specimens briefly described on this page may belong to the same period; but, since they are not figured, this question cannot be decided.

Of pieces of Han pottery in other museums, the British Museum possesses the vase mentioned above in Bushell's book, a reproduction of which is published in this paper, and the three specimens described in the above article in "Man." In the Metropolitan Museum of Art, New York, there are two specimens of Han pottery exhibited in Hall 31. The one is a globular vase, with narrow base, and long-extended, straight, cylindrical neck (No. 178); the other one, a granary urn (No. 175 A).

The pottery described on the following pages (excepting in the prefatory chapter, which deals with pottery dating prior to the Han period) is the mortuary pottery of the time of the Han dynasty. The localities where the finds have thus far been made are limited to the present province of Shensi, the centre of the Chinese culture of that period. Specimens of this pottery first came to light, towards the end of the seventies, in *Hsi an fu*, according to the statements of the dealers in antiquities living there, and consequently no mention of it is made in the archæological or ceramic literature of the Chinese.[1] In the beginning, nobody cared for these pieces,

---

[1] The only allusion in Chinese literature to a find of pottery, presumably of the Han dynasty, which I came across, is in the following passage from the *K'uei hsin tsa shih* 癸辛雜識 (former part of the fourteenth century, A. WYLIE, Notes on Chinese

and they were indifferently thrown away for a few *cash*, until, after some years, larger cargoes of them reached Peking, where they brought enormous prices among Chinese lovers of antiquity. Nowadays it is the unanimous opinion of all Chinese judges, that this pottery represents genuine and most precious relics of the Han time.

The actual proofs for this fact could naturally be gathered only from an investigation of an extensive number of specimens, and these should now be summarized as follows : —

1. Inscriptions occurring on this pottery agree in the *ductus* of the characters, as well as in their style and composition, with those on the bronzes of the Han time, as will be seen in the chapter on this subject.

2. The dates furnished by some of these inscriptions inva-

---

Literature, 2d ed., p. 198; Professor Hirth, however, informs me that CHOU MI, the author of this work, lived under the Southern Sung dynasty in the thirteenth century, according to the Imperial Catalogue, Chap. 141, p. 34), quoted in *T'u shu chi ch'êng*, Vol. 1126, *k'ao kung tien*, Book 251 古玩部雜錄, p. 3 b: 伯機云。長安中有耕者得陶器於古墓中。形如臥繭。口與足出繭。腹之上下其色黝黑勻細。若石光潤如玉。呼為繭瓶。大者容數斗。小者僅容數合。養花成實。或云三代秦以前物。若漢物則苟簡不足觀也。 "*Po Chi* narrates: 'In *Ch'ang an* there lived a farmer who had found pottery vessels in an ancient grave. They were shaped like reclining silkworm-cocoons. On the mouth and the feet of the vessels, cocoons were brought out. The upper and lower parts of the body were intensely black in color, and equally thin; they were like stone, and glossy like nephrite. They were called "silkworm-cocoon jars." The larger ones contained as much as a peck; the smaller ones, hardly one *ko* (a tenth of a pint). They had a perfect oxidation. Several said that they were things originating in the times of the Three Dynasties [*Hsia, Shang,* and *Chou*] before the *Ts'in.* They are like objects of the Han, but there are indeed no written documents sufficient to allow of an investigation.'" The "reclining silkworm" is the name for an ornament of the shape ◎ or ◎ or ◎ (see *Ku yü t'u p'u*, Book 6, pp. 1-4, where it appears on two jade tablets of the *Chou* time; ibid., Book 5, pp. 9, 11; Book 59, pp. 7, 9, 11; Book 60, p. 11; *K'ao ku yü t'u*, Book 下, p. 11).

riably refer to the Han time. With regard to some types, the hill-censer and hill-jar, historical considerations lead to the same period.

3. The subjects and style of art displayed on the reliefs of vases agree with those of the stone bas-reliefs of the Han period.

4. Many shapes of pottery vases are repetitions of others in bronze, and of just such bronze forms as flourished particularly, or even came up, during the Han time.

5. In a chapter of the "Annals of the Later Han Dynasty"[1] the various objects made of pottery and placed in the graves of the dead are enumerated, and a great many of these we find among this pottery excavated from the graves. This is an historical confirmation of its having been mortuary pottery, and in particular the burial pottery of this period. The fact that this was indeed the case is corroborated by the objects themselves, many of which are clay imitations or models, on a small scale, of objects of daily life, which certainly could have served no other purpose.

6. The unanimous statement of all Chinese concerned is, that this pottery was dug up out of graves of the Han time in Shensi Province, which, considering all points previously enumerated, deserves the fullest confidence.

7. The great antiquity of this pottery is plainly shown by the very character of the pieces, the hardened incrustations of loess adhering to them, the decomposition or oxidation of the glazes into gold and silver iridescence[2] both testifying to a long and continuous burial under ground.

---

[1] This account has been given in text and translation by DE GROOT (The Religious System of China, Vol. II, pp. 401–405).

[2] The oxidation of metals and pottery is well known to the Chinese, as may be ascertained from the example quoted (p. 5, note). That my translation of the term

The high value which must be attached to these objects is thus obvious. They reveal —

(1) An hitherto unknown field of Chinese ceramics.

(2) The ideas of the ancient Chinese connected with the burial of the dead and the fittings of the grave.

(3) The material culture of the Han period as to agriculture, mechanics, architecture, domestic life, domestic animals, etc.

(4) The art of the Han period in the ceramic forms and decorative motives, which in a measure reach far beyond the special interests of Chinese archæology by pointing to an early influence of ancient Siberian or Turkish art on that of China.

The most important result of a study of the forms of this pottery seems to be that the Han period was extremely rich in fine ceramic shapes, and that many forms made in porcelain at later times were already in existence then. The study of the development of porcelain forms, which still remains a task of the future, must therefore take its starting-point from the Han pottery. We now see actual evidence for

---

養花 ("to produce flowers") by "oxidation" is correct, becomes evident from the following passage in *T'u shu chi ch'êng*, Vol. 1126, *k'ao kung tien*, Book 251 古玩部彙考, p. 2 b: 古銅瓶鉢養花果。古銅器入土年久受土氣染以之養花。花色鮮明如枝頭開速而謝遲。或謝則就瓶結實若水銹。傳世古則爾。陶器入土千年亦然。 "On the production of flowers and fruits (i.e., oxidation) on old bronze vases and basins. Old bronze vessels placed in the earth for a long series of years receive the vapors of the earth, which color them by producing flowers. The color of these flowers is pure and clear, like blossoms which open speedily and die slowly. Some die off, and then the jar forms fruits like rust arising from water (i.e., a further process of oxidation). Thus they come down to posterity in an antiquated condition. The same is the case also with *pottery vessels* placed in the earth for a thousand years."

the use of the potter's wheel in that age, for all the vessels appear to be turned on the wheel.[1] In the art of glazing, the workmen of those days were very proficient, the most frequent color being green, which occurs in a great variety of shades,[2] then brown, yellow, white, and bluish gray. The process of craquelé was also in use during the Han times, and the crackled glazes even prevail.

As regards the chronology of this pottery, the dates furnished by two pieces are of primary importance, — the one, 133 B.C., found by Bushell on a vase of the Dana collection (see p. 1); and the other, 52 B.C., on a jug obtained by me (see chapter on Inscriptions, No. 5). There is another vase bearing the year-period *Shên chüeh*, i.e., 61–57 B.C.; but the reading of this inscription is still obscure (see Inscriptions, No. 7). On the basis of these data, we might fairly presume that this pottery originated in the second and first centuries before our era, although it may well be that some pieces belong to the first century A.D., which may be considered as the terminus *ad quem*. From internal evidence

---

[1] The potter's wheel 鈞, 均, or 鋆 has doubtless been known to the Chinese since remote times. The expression occurs repeatedly in SSŬ-MA CH'IEN, the *Ch'ien Han shu*, and in the philosophers HUAI NAN TZŬ and CHUANG TZŬ (see the quotations of E. CHAVANNES, Les mémoires historiques de Se-ma Ts'ien, Paris, 1905, Vol. V, p. 26, note, and COUVREUR, Dictionnaire classique de la langue chinoise, p. 952 c), particularly used of the evolution of heaven creating the beings in the same way as the potter turns the objects of clay on his wheel. The "potter's-wheel heaven" is the central one in one of the nine regions of heaven. According to S. W. BUSHELL (Oriental Ceramic Art, p. 9), the *K'ao kung chi* in the *Chou li* gives names and measurements of several kinds of cooking-vessels, sacrificial vases and dishes, in the fabrication of which the different processes of fashioning upon the wheel and of moulding are clearly distinguished.

[2] A continuation of the green-glazed pottery of the Han dynasty is perhaps given in the Japanese collections in the great storehouse at Nara, which, according to EDWARD S. MORSE (Catalogue of the Morse Collection of Japanese Pottery, Cambridge, 1901, p. 31), "reveal the existence of soft green-glazed pottery one thousand years ago, though Ninagawa was inclined to believe that this pottery was brought from China, and I am inclined to the same opinion. Fragments of a hard green-glazed pottery dug up in Ōmi are figured by Ninagawa and accorded an age of nine hundred years."

it is possible to fix the date of the type of the hill-censer (*po shan lu*) in the first part of the first century B.C.

In the following study it has been the aim of the author to furnish contributions, not so much to Chinese ceramics as to the archæology or culture of the period in question, and to illustrate ancient Chinese culture by means of these finds. An attempt has been made to determine their historical position; to interpret them from the ideas of contemporaries; to connect these, when evidence is given, with those of older periods; and to follow them up, as far as possible, through subsequent ages. What is offered in the following pages is very little: the field is new, and it can hardly be expected that everything is taken at the first leap or always correctly understood. Many problems are here only slightly alluded to, and their working-up must be reserved for a more favorable opportunity.

In a prefatory chapter we shall deal with some pieces of pottery dating back prior to the Han dynasty. As regards the Han material, it has been divided into two large groups, — (1) models in pottery in imitation of actual objects,[1] and (2) vessels. Dealing as we do, in the first part, with mills, roofed urns, jars representing draw-wells, and stoves, we shall obtain an insight into the agriculture, architecture, mechanics, and domestic life respectively of the time. In the examination of the various forms of vessels, we continue in our inroads over the last-mentioned domain, finally ascending, in the vases and their relief-work, to the higher expressions of art.

---

[1] Regarding the tendency of making substitutes, chiefly of clay, for real objects under the Han dynasty, see DE GROOT, The Religious System of China, Vol. II, p. 708.

# II. — POTTERY PRIOR TO THE HAN DYNASTY.

THE tripod-like vessel seen in Fig. 2, Plate 1 (height, 15.9 cm; diameter of mouth, 13.4 cm), is the only one of its type known to me. It is made of a gray clay of a very hard consistency, which on the surface has assumed a black color. This clay is not found in the pottery of the Han time. It is unglazed, and the whole technique of this piece also differs widely from that observed in the Han pottery. Chinese antiquarians of *Hsi an fu* suggest its origin as in the period of the *Chou* dynasty (1122–255 B.C.). Incrustations of hardened loess outside and inside bear witness to long burial under ground.

The entire surface is covered with incised diagonal lines intersecting each other, and thus forming a number of small rhomboids in relief. The vessel has a flaring rim. It rests on three short feet of triangular form. It seems to have been freely moulded by hand; at least, no traces of its having been shaped on the wheel can be discerned. This piece of pottery is apparently in imitation of the bronze kettle *li* 鬲 of the *Chou* time. A few of these *li* are assigned to the *Shang* dynasty, one of which is depicted in the "*Hsi ch'ing ku chien*" (Book 31). There also are in the same book fifteen *li* of the *Chou* dynasty. In the "*Po ku t'u*"[1] (Book 19), sixteen *li* are illustrated.

Fig 3, Plate 1, represents a bronze *li* of the *Chou* dynasty,

---

[1] The edition used by me is that printed in 1752.

10

PLATE I.

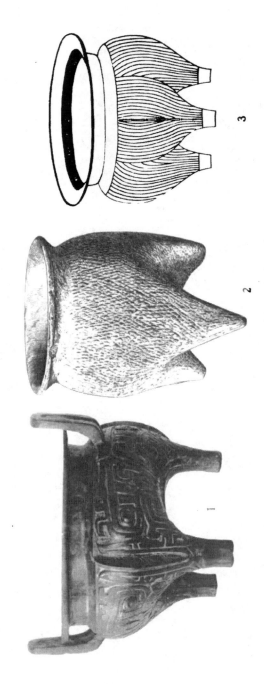

Figs. 1, 3, Bronze Tripod Vessel; Fig. 2. Chou Pottery Tripod Vessel.

after an engraving in "*Liang lei hsien i ch'i t'u shih*" 兩罍軒彝器圖釋 (Book 7, p. 18 a). According to the description of the author, the vessel is 4 *ts'un* 8 *fên* (17.6 cm)[1] high, 3 *ts'un* 7 *fên* (12.9 cm) deep. Its periphery is 6 *ts'un* 2 *fên* (22.4 cm), and the diameter of the mouth 6 *ts'un* (21.6 cm). Among the different varieties of *li*, this piece is the type which comes nearest to the above specimen of pottery. The rim flares in the same way, and the feet have the same characteristic shape. The lines appearing black in the woodcut, which seem to be engraved in the bronze, correspond to the network of lines in the clay.

Fig 1 on Plate 1 represents a bronze *li* from the Ōkazaki collection, and is attributed by Ōkazaki to the later *T'ang* time (about A.D. 900). It is an exact reproduction of a *Chou t'ao t'ieh li* 周饕餮鬲; i.e., a *li* with the design of a glutton or *t'ao t'ieh*. While the loop-handles in the vessels sketched in the "*Hsi ch'ing ku chien*" (Book 31) rise vertically from the rim, those in this tripod are attached at right angles to the body of the vessel. The piece is 11.9 cm high up to the rim, and 14.4 cm up to the handles. The diameter of the mouth measures 13.4 cm, which agrees with that of our pottery *li*. Tradition has it that "about nine hundred years ago, when the Taianiji temple was built in Nara, a priest who came from China brought this vessel with him, and used it for cooking during his visit. When he returned, he left it in the temple" (ŌKAZAKI). The entire surface is covered with a green patina, and the make-up of the piece is somewhat coarse and primitive.

---

[1] I reckon as follows: 1 ch'ih 尺 = 35.8 cm; 1 ts'un 寸 = 3.6 cm  1 fên 分 = 0.4 cm.

These vessels called *li*[1] have various decorations and forms. There are pieces with three and four feet, and with and without ears on the rim, or handles on the sides. The chief characteristic of this type lies in the peculiar shaping of the feet, which are not attached to the bottom, as in the *ting* 鼎, but feet and body form a unit, — a feature peculiarly adapted to the shape of this type. One might almost say that the whole vessel is composed of three triangular pieces, globular in the upper part, and gradually tapering downward. Looking at Fig. 2, Plate I, from below, the points of the three feet form a triangle, and seem to diverge one from another. There are slight depressions between the feet on the outer side (see Figs. 1 and 2, Plate II, showing the under side of Figs. 2 and 1, Plate I), and the three triangles slope in concave outlines to the narrow centre of the bottom. Inside of the vessel are three deep cavities running down into the feet. In this description our pottery *li* (Fig. 2, Plate I) agrees exactly with the bronze kettle (Fig. 3 of the same plate). The fact that the triangular shape of the feet was an essential characteristic in the type of the kettle *li* right from the beginning, is obvious from the earliest writing of the symbol for *li* on a bronze *li* of the *Shang* dynasty, where it appears in the form ⟨symbol⟩.[2]

A very lucky find made by Mr. Frank H. Chalfant of the American Presbyterian Mission in *Wei hsien*, Shantung Province, and presented to the American Museum of Natural History with another collection of mortuary *Sung* pottery, which is described in an appendix to this paper, enables me

---

[1] According to COUVREUR, "marmite dont les pieds allaient en s'écartant les uns des autres." GILES: "(Read *li* 4), a large iron caldron; (read *ko* 4), an earthen pot."

[2] Reproduced from *Chi ku chai chung t'ing i ch'i kuan shih*, Book II, p. 19 a.

PLATE II.

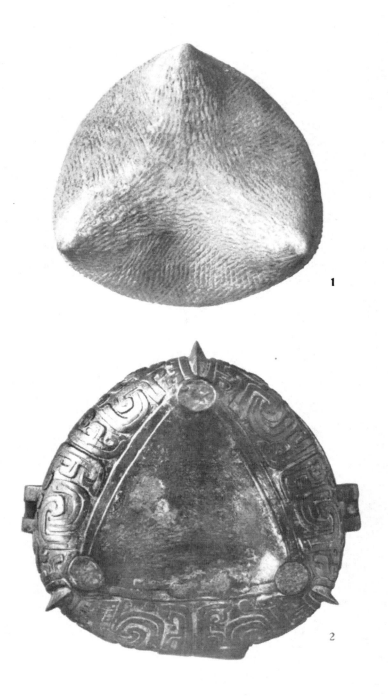

1

2

Under Sides of Tripod Vessels shown in Figs. 2 and 1, Plate I.

to publish here four pieces of pottery and one brick (Plate III), which may be assumed with fair reason to have come down from a time prior to the Han dynasty, either from the latter part of the *Chou* or early *Ts'in* dynasty. This lot was excavated on the soil of the ancient city of *Lin tzŭ* 臨淄 in *Ch'ing chou fu*, Shantung, which was the capital of the feudal kingdom *Ch'i* 齊 until 221 B.C., when the city was destroyed by *Ts'in shih Huang ti* 秦始皇帝.[1] The locality of this find; the ancient aspect and primitive qualities of this pottery; the character *Ch'i* stamped on the brick, and the potter's mark stamped on the *tou* vase, both in the style of writing of the *Chou* period, — leave no doubt that these antiquities are venerable relics of the time when the state of *Ch'i* was still flourishing.

The brick represented in Fig. 3, Plate III, is probably from the ruins of some public structure. It is 25.5 cm long, 12.2 cm wide, and 6 cm thick, and is of a light-grayish color, almost the same as that of the pottery from the same finds. The ancient character stamped on its surface corresponds to modern 齊 *Ch'i*. The greatest length of the character is 10.5 cm; its greatest width, 6.5 cm; its depth, 0.3 cm. The character denotes the name of the ancient feudal state *Ch'i*, and it must be presumed that a brick bearing this seal could have been made only at the time of the existence of this state. This brick may probably claim the honor of being recognized as the oldest extant brick of Chinese antiquity, in that the inscribed bricks illustrated in Chinese archæological literature are not older than the Han time.

Fig. 1, Plate III, represents a globular water-jar 24.5 cm high,

---

[1] See ÉDOUARD CHAVANNES, Les mémoires historiques de Se-ma Ts'ien (Paris, 1905), Vol. V, p. 224.

13 cm in diameter at the mouth, and 8 cm in diameter at the bottom. It has a short neck and a rim that is turned downward. Pieces of the rim were chipped out when found. The rounded body gradually tapers towards the bottom, the outline of which is somewhat rounded. There is a slight round depression in the bottom, and the bottom itself is notable for its small size. In Chinese pottery I have met with no other bottom of this description. The lower part of the surface and the bottom are decorated with irregular horizontal and oblique lines, sometimes with cross-hatchings, in a similar way as in Fig. 2, Plate I; so that I am almost inclined to see in this mode and method of ornamentation a special characteristic of the pottery of the *Chou* dynasty.[1]

Fig. 2, Plate III, shows an oval-shaped water-jar made from the same kind of gray clay. Its height is 25.2 cm; diameter of the mouth, 10 cm; width of rim, 1.7 cm; diameter of bottom, 14 cm. The rim is straight, flat, somewhat slanting, and projects over the constricted neck. It has no decoration except a few short rows of small indentations on one side.

In the next illustration (Fig. 4, Plate III) is seen an earthenware bowl for ordinary domestic use. It is 10.2 cm high, with a diameter at the mouth of 13.5 cm, and at the bottom of 10 cm. The rim at the top is thickened. The lower portions around the base have been flattened out by means of a flat stick. The surface and interior walls are covered with fine parallel lines. A peculiar greenish color, like that of moss, shows up in many places, both outside and inside.

Fig. 5, Plate III, represents a shallow dish connected by a tall standard, and belongs to a type of vessels called *tou* 豆 (see Index, under *tou*).

---

[1] Similar modes of decoration may be seen in old Japanese and Corean pieces in the Morse collection of Japanese pottery in the Fine Arts Museum of Boston.

PLATE III.

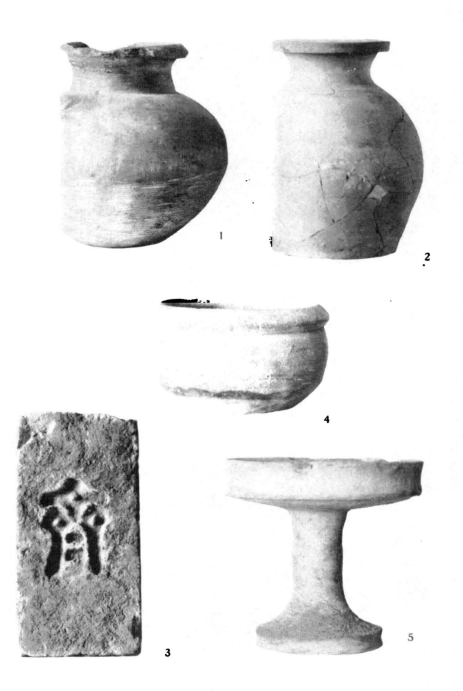

Excavated Pottery of the Chou Period.

Fig. 1. Han Pottery Mill; Fig. 2. Modern Stone Hand-Mill.

# III. — DESCRIPTION OF HAN POTTERY, IMITATIVE FORMS.

## MILLS AND OTHER AGRICULTURAL OBJECTS.

FIG. 1, Plate IV, represents a clay imitation of a grist-mill. The specimen is 7 cm high, its diameter being 17.5 cm. It consists of two parts, representing the lower and upper millstone. The under part, or receiver, is shaped somewhat like a dish, for catching the ground grain as it falls down. There is a perforation (not visible in the illustration) along the side of the dish, presumably for the purpose of letting the grain fall through into a basket below. In the centre of the specimen is a conical knob (representing the iron spindle of the real mill), which fits into a cavity on the under side of the upper part, or runner. There are five concentric rows of shallow oval cavities on both the lower part and the under side of the upper part. These are apparently a substitute for the furrows on real millstones to make the faces of the two stones rough. On the upper side of the runner we see the same motive brought out; but, as it serves no real purpose here, it must be considered to be purely decorative. The runner is furnished with a short handle having a small indentation, as if for inserting the bar by means of which the stone is revolved; and in the two semicircular depressions on the top of this part are two rectangular perforations through which to pour the grist. The nether part, or bedstone, is turned on the wheel, as is evident from the wheel-marks in the hollow, seen from below.

15

The object is coated with an olive-green glaze. The two grinding surfaces are unglazed.

The object seen in Plate v, which is 9.5 cm high, represents a miniature mill like the preceding. It also consists of two parts, — the bedstone, resting on a receiving-bowl (the middle figure), and over it the runner. The upper surface of the runner, here without handle, is glazed, and has two round holes in two oval depressions; while Fig. 1 in the text shows the unglazed under side of the runner, on which the turning-motion is symbolically indicated by a network of thirty-five incised lines radiating from the periphery of a central circular band. In the middle is a round shallow cavity, as was observed also in the previous specimen.

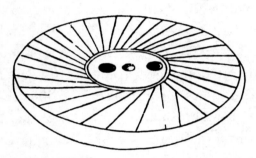

Fig. 1. Under Side of Mill-Runner shown in Plate v.

The tenon on the bedstone, however, is lacking in the present case, and, besides, this whole surface is entirely smooth. The lowest figure in Plate v shows the bottom of the mill, seen from below. The central hollow is 5.8 cm deep; the diameter of the opening, 7 cm. The color of the glaze is light green. There are portions of the bottom now without glaze, which naturally appear light in the illustration. The runner is covered with many brown and light-yellow spots, seemingly due to an accident during the baking process.

The two preceding miniature clay mills of the Han period are exact counterparts of the modern stone hand-mill which is in use all over northern China. The specimen in the Museum, shown on Plate IV, Fig. 2, was obtained in the village *Pei hsin ngan*, thirty *li* west from Peking. The

PLATE V.

Han Pottery Mill (*a*, Runner; *b*, Bedstone; *c*, Bedstone Seen from Below).

total height of it is 24 cm; the greatest width, 60.5 cm. The diameter of the upper stone varies from 28.7 cm to 29.2 cm. It is 10.2 cm high. The bedstone and receiver are cut out of one and the same piece of stone. The spindle is a round wooden peg 2.5 cm high, surmounted by an iron ring. The socket on the lower side of the runner is 3.5 cm deep, with a diameter of 3 cm, and is surrounded likewise with iron. The surfaces of both millstones are jagged. Each is divided into eight triangles, in which deep-cut grooves run alternately in a horizontal, diagonal, and vertical direction. The grist-hole in the runner is about 4 cm distant from the periphery, and has a diameter of 3.5 cm. The wooden handle is placed in a vertical position, and is 17.7 cm long. We must realize that the handle of the Han mills must have been of the same sort. The only apparent point of difference between the modern and the Han type is the spout of the receiver, from which the flour easily falls into a basket standing below it.

A rice-mill from Shasi, of the same type but of somewhat different shape, has been described and illustrated by ENRICO H. GIGLIOLI in the "Archivio per l'antropologia e la etnologia" (Vol. XXVIII, 1898, pp. 376, 377). This mill rests on a wooden vat in which three cudgels are laid crosswise as a support to the stone mill. The vat is for receiving the ground rice. The lower disk has a diameter of 34 cm, and is 6.8 cm high. Its surface is regularly grooved with radiating furrows. In a central cavity is a wooden pivot that fits into a socket in the upper disk. The runner is narrower and higher than that in our specimen, having a diameter of from 32 cm to 33.5 cm, and being 15.5 cm high. Another point in which this mill varies from those mentioned before is that the wooden handle by which the

2

runner is turned is fastened into a piece of wood driven into a cavity in the side of the stone.

These stone mills are as a rule the property of the village community, and are almost invariably operated by women.

It is a most striking fact that the Chinese treatises on husbandry,[1] in those sections which treat of agricultural implements, — among others also of mills, — entirely neglect to describe this hand-mill, which, on my journeyings, I found in general use in the provinces of Chihli, Shantung, Honan, and Shensi. A stone mill of the same construction, only of much larger size, and worked by animal power, is employed in the same regions.[2] According to the purpose intended, there are two principal types of mills[3] in China, which have not always been clearly distinguished by foreign writers. One, called *lung* 礱, is for husking grain; the other, called *mo* 磨, for pounding hulled grain into flour.[4] Only the latter consists of two millstones, while in the former,

---

[1] The best synopsis of these will be found in E. BRETSCHNEIDER, Botanicon Sinicum (Journal of the North China Branch of the Royal Asiatic Society, 1881, N.S., Vol. XVI), Part 1, pp. 75–86.

[2] An illustration of such a mill, after a Chinese sketch (source not given and not known to me), is published in A. FORKE, Von Peking nach Ch'ang-an und Lo-yang (Mitteilungen des Seminars für orientalische Sprachen, Vol. I, Sect. 1, p. 95).

[3] Besides grist-mills, the Chinese have oil and sugar mills. The oil-mill is of approximately the same construction as the grist-mill. The roasted sesamum-seeds are poured into a conical basket placed in the centre of the upper stone, and, shaken from a hole in the side of the basket by the motion of the mill, they pass through an aperture in the middle of the upper stone to the surface of the lower one, where, being ground, they give out the oil, which flows over the lower stone into a reservoir below it. Compare CLARKE ABEL, Narrative of a Journey in the Interior of China (London, 1819), p. 138; DE GUIGNES, Voyages à Peking (Paris, 1808), Vol. I, p. 290, Vol. II, p. 117; Catalogue of the Collection of Chinese Exhibits at the Louisiana Purchase Exposition, St. Louis, 1904, p. 9. For a description of the sugar-mills, see CLARKE ABEL, l.c., p. 200; Port Catalogues of the Chinese Customs' Collection at the Austro-Hungarian Universal Exhibition, Vienna, 1873, p. 370, Shanghai, 1873 (this catalogue is a rich mine for the ethnography of China).

[4] The dictionaries of GILES, COUVREUR, and PALLADIUS, do not state this essential difference between the terms *lung* and *mo*, but translate them indiscriminately by "to grind, to pound," "mill."

clay or wood is substituted for the stone. The type called
*nien* 碾 (cylinder or roller mill) is a variation of the *mo*,
but suited only for hulling grain. As to the mode of motion,
the *lung* may be driven by human and animal labor; the
*mo* and *nien*, by the human hand, by animal and water
power.[1] Following is a brief survey of the subject, based
on the principal sources of Chinese literature.

---

[1] Windmills, which have sprung up in Europe only since the middle ages (the first
mention being in an Anglo-Saxon document of 833), have been unknown in China.
JOHAN NEUHOF, however (Die Gesandtschaft der ostindischen Gesellschaft in den
vereinigten Niederländern an den Tartarischen Cham, Amsterdam, 1669, p. 121), makes
mention of windmills with mat sails for the irrigation of rice-fields. They are illustrated
on p. 122; but neither this sketch nor his description ("Alhie befinden sich auch viele
Windmühlen/ gar klüglich auff besondere Manier angelegt; welche mit Matten-segeln/
auff einer Spille/ umbgetrieben werden/ das Wasser mit mahlen hinweg zu leiten") is
clear enough to draw any conclusive evidence. At all events, these windmills differ
widely in principle and mechanism from ours, and this is the only instance I found
regarding this subject in European literature on China. The fact that windmills were
actually unknown to the Chinese is proved by the treatise on European machinery,
奇器圖說, by the famous Jesuit mathematician JOHN TERENCE (see WYLIE,
Notes on Chinese Literature, p. 144; HAVRET, La Stèle chrétienne de Hsi an fou, Vol. II,
pp. 154, 328; H. CORDIER, L'imprimerie sino-européenne en Chine, p. 50: Havret,
and Cordier following him, make Terence a Swiss on the ground that he was born
in Constance; but Constance belongs to the grand duchy of Baden, and never
belonged to Switzerland), published in 1627, and reprinted in Vols. 66 and 67 of the
Cyclopædia 守山閣叢書. In Book 3, Plate 4, on p. 21 b, he introduced to
the Chinese for the first time a picture and explanation of the foreign windmill. It
is true that WANG CH'ÊNG 王徵, who put the Jesuit's work into literary form,
and then himself wrote a short treatise on mechanics (reprinted in Vol. 67 of the
same cyclopædia), gives a drawing of a Chinese windmill, which is called *fêng wei*
風磑, and which consists of four square pieces of rush matting arranged around a
wooden pole driven through the runner of a stone mill. WYLIE (l.c., p. 145) says
that his book treats of native machinery; but from a perusal of it I am under the
strong impression that it contains nothing more than mechanical chimeras of the author's
imagination, which have no foundation at all, and were undoubtedly inspired through
his close occupation with the work of Terence. Neither is he able to state concerning
either his windmill or the fanciful objects in his other sketches — among which he anticipated the automobile (自行車) — when and where they were ever used. In a
modern Chinese school-book *Tzŭ k'o t'u shuo* 字課圖說, printed in Shanghai
in 1901, it is said, under the character *mo*, "In China, mills are turned by human
strength or by ox and horse power, while in Europe they avail themselves of the power
of *wind*, water, and steam, to move these machines. By this means greater results are

*Mo*, the mill proper (Fig. 2), is thus described in the
"*Nung chêng ch'üan shu*"[1] and "*Shou shih t'ung k'ao*:"[2]
"The dictionary *T'ang yün*[3] makes *mo* equivalent to *wei*[4]
(or explains the character *mo* 'mill' by *wei* 'to grind').  The
dictionary *Shuo wên*[5] says *mo* means 'grinding between
stones' (or a stone mill).   The *Shih pên*[6] says *Kung Shu-pan*[7]

---

obtained through the saving of human strength." (中國以人力或牛
馬力運之。泰西則借風水火力以動其機故用
人力少而成功多。) GILES and COUVREUR translate the term 風車 by
"windmill."  This is a mistake.  風車 or 扇車 or 風扇車 is the well-known
fanning or winnowing mill (also called "van") which is illustrated and described in
the Chinese agricultural books (also in the *T'ien kung k'ai wu*, Book 4), and is still
used in modern China as well as in Corea, in Japan, in the Philippines, and in the
United States.   The proper term for "windmill" is *fêng wei* (see above) or *fêng mo* (see
COUVREUR, Dictionnaire français-chinois, 1884, p. 623; A. DEBESSE, Petit Dictionnaire
Chinois français, Shanghai, 1900, p. 53; L. AUBAZAC, Dictionnaire français cantonnais,
Hongkong, 1902, p. 207).   The term *fêng ch'ê* means, as Giles correctly puts it,
"whirligig," a toy much played with by children in Peking during New-Year's time.

[1] 農政全書, Complete Treatise on Husbandry, by HSÜ KUANG-CH'I 徐光
啟 (1562–1634: see GILES, Biographical Dictionary, p. 306), in 60 books, and published in 1640.  See WYLIE, Notes on Chinese Literature, 2d ed., p. 94; and, for an
analysis of the contents, E. BRETSCHNEIDER, Botanicon Sinicum, Part I, p. 82.

[2] 授時通考, the most comprehensive work on agriculture, in 78 books,
published in 1748 (see WYLIE, Notes on Chinese Literature, 2d ed., p. 95; and
BRETSCHNEIDER, l.c., p. 84).  In the 52d book the woodcuts of the *Kêng chih t'u*
are repeated.   Most of the other illustrations are taken from the *Nung chêng ch'üan shu*.

[3] Published in A.D. 750 by SUN MIEN (see China Review, Vol. IV, p. 340).

[4] GILES's Dictionary gives 磑 only the meaning "to break to pieces;" but COUVREUR
explains it correctly by "*moulin, moudre; mortier; piler, broyer;*" as did also PALLA-
DIUS (Vol. I, p. 77).   E. H. PARKER, who speculates on this character to some extent
in the China Review (Vol. XXV, p. 191 b), might have ascertained from Palladius
that 水磑 is nothing but a water-mill.

[5] An ancient dictionary by HSÜ SHÊN, written at the end of the first century,
and completed in A.D. 100.

[6] Ancient historical records, now lost, from the Emperor *Huang ti* down to *Nan
Wang*, probably written in 206–202 B.C.  *Ssü-ma Ch'ien* made use of this work in the
*Shih chi* (see E. BRETSCHNEIDER, Botanicon Sinicum, Part I, p. 187; ÉD. CHAVANNES,
Les mémoires historiques de Se-ma Ts'ien, Vol. I, p. CXLI).

[7] The above quotation is thus given from the *Shih pên* in the *T'u shu chi ch'êng*,
Vol. 1125, *k'ao kung tien*, Book 245, *mo wei pu hui k'ao*, p. 1: 公輸般作

Fig. 2. Grist-Mill (*mo*) (from *Nung chêng ch'üan shu*).

made the mill (*wei*). In local dialects it is sometimes called *ch'i*. The common term is *t'ien mo* (i.e., 'the mill making

磨礱之始。編竹附泥。破穀出米。曰礱。鑿石上下合。研米麥爲粉。曰磨。二物皆始於周。 "*Kung Shu-pan* was the first to make mills. He made a plaiting of bamboo, which he filled in with mud. He crushed grain and produced it unhulled. This mill was called *wei*. He chiselled out stones, which he placed together one on the other. He ground hulled rice

a crashing sound') or also *t'ung mo* (i.e., 'the smooth, polished mill'). The framework [on which the mill rests] is called *t'i*.[1] Nowadays there are, further, the following words: the term *chu* (literally, 'the master, the principal part'), used in reference to the mill, meaning 'the navel' (i.e., the centre); the term *chu* ('to pour in,' 'to put in'), as regards the mill, meaning the aperture [in the runner]; the term *chuan* ('to turn,' 'the object by which the mill is turned'), in reference to the mill, meaning the wooden pole; the term *ch'êng* ('to receive,' 'the receiver'), as regards the mill, meaning the tray (i.e., the outer open rim of the bedstone into which the flour falls); the term *tsai* ('the support,' it might be translated 'frame-holder'), with reference to the mill, meaning the framework [on which it rests]. Frequent use is made of animal power for dragging. In some cases the people resort to a water-wheel. In other cases they drive a framework into the ground and set up an axle [?] below.[2] If they [the mills] are turned by animal power, they

and wheat into flour. This mill is called *mo*. These two objects had their origin under the *Chou* dynasty." *Kung Shu-pan*, commonly known under the sobriquet "*Lu-pan*," was a famous mechanic of the state of *Lu*, and a contemporary of Confucius (see GILES, Biographical Dictionary, p. 548; PELLIOT in Bulletin de l'École française d'Extrême-Orient, Vol. II, 1902, p. 143). The invention of the two kinds of mill ascribed to him is apparently a mere popular tradition, although it is strange to find it recorded in such an early work as the *Shih pên*. He might be credited with having brought about some improvements in the mills then existing; provided, of course, that he was an historical personage. Millstones are mentioned as early as in the *Yü kung* of the *Shu king* and in the *Shih king* (see further CHAVANNES, l.c., Vol. I, p. 123), and, to judge merely from the whole aspect of earliest Chinese culture, there can be no doubt that mills should be traced as far back as to prehistoric times. It is worthy of note that in the above passage *wei* 磑 stands for a hulling-mill, now called *lung*.

[1] This meaning of the character *t'i*, pronounced also *chai* and *tsê*, is not given in the dictionaries, which explain it as "to pick, take with the hand, take off, take away." I presume that 摘 stands here for 商 *ti* ("stem, root, foot, base").

[2] *Tsuan* or *tsun* 鐏. GILES: "the butt-end of a spear." COUVREUR: "garniture de cuivre qui termine la poignée dune lance." PALLADIUS: "end of a knife or spear to hold with the hand." K'ANG HSI: 銳底曰鐏取其鐏也。"What is

are called 'water-mills of the dry land.' Compared with the ordinary [i.e., hand] mills, they are in particular a saving of strength. On the surface of all mills they use a *lou-tou* [literally, a 'peck,' through which the grain runs; i.e., the hopper]. It contains the wheat which falls down through the aperture. Then it is whirled around between the sharp teeth [of the millstones]. The wheat-husks are made into bran. Afterwards the ground wheat is winnowed through a sieve, thus yielding a perfect flour. And this was the beginning of flour-cakes in the world."[1]

The "earth mill" (*t'u lung*), reproduced in Fig. 3, is described as follows in the "*Shou shih t'ung k'ao*" (Book 40, p. 4 a): "The hulling-mill (called *lung*)[2] is an implement

---

[1] beneath the point [of a weapon or knife] is called *tsuan*, so that *tsuan* is the part by which it is held (Commentary on *Li Ki*)." I do not understand the meaning of this word in the above connection.

[1] 礦唐韻作磨礎也。說文云。礦石礎也。世本曰。公輸班作礎。方言或謂之磑。通俗之曰塡磨。曰硐磨。牀曰摘。今又謂主磨曰臍。注磨曰眼。轉磨曰榦。承磨曰榮。載磨曰牀。

多用畜力輓行。或借水輪。或掘地架木。下置鐏軸。亦轉以畜力。謂之旱水磨。比之常磨特爲省力。

凡磨上皆用漏斗。盛麥下之眼中。則利齒旋轉。破麥作麩。然後收之篩羅。乃得成麵。世間餅餌。自此始矣。

[2] The fact that the character *lung* 礱 ("hulling-mill"), formed with the radical 石 ("stone"), is also written 䃃 (pronounced in the same way) with the radical 瓦 ("clay"), may suggest the probability that such mills were also made of clay, as is actually demonstrated by the earth mill described above, and observed by Syrski in southern China. With regard to the change of the material used in the construction of mills, there is an interesting note in the *Nung chêng ch'üan shu* (Book 23, p. 50 b): 礱礎 字從石。初本用石。今竹木代者。亦便。 "The characters

for pounding grain to remove the husk.[1]    [For this purpose]
enclosures are made of plaited bamboo.   Inside of them
mud and earth are piled up.   In shape the whole resembles a small grist-mill (called *mo*).   Teeth made of bamboo
and wood are then arranged in close proximity.   In breaking
the [unhulled] grain, be careful not to injure the hulled
fruit.[2]   To turn the mill round, a piece of wood inserted in
a hole in the upper part of the mill is used.   The shaft
is moved to and fro by means of a cord attached to the
cross-beams of the roof.   A man[3] keeps it revolving by
bending his elbows.   In the course of a day more than
forty piculs[4] of grain can be crushed in this way."[5]

---

*lung* ('a husking-mill') and *lei* ('to pound') are formed with the radical 石 ('stone').
Originally stone was utilized [for the manufacture of mills], which is nowadays replaced
by bamboo and wood, these being more handy." This remark must, of course, be
taken for what it is worth, — the reflection of a modern author on a supposed ancient
evolution as mirrored in his own mind.  Moreover, it is not quite correct; for considering the fact that the stone mills serve the purpose of pounding grain into flour,
and the earth and wood mills merely that of hulling, it cannot be concluded without
further historical proof that the latter, since they differ in their objects, were evolved
from the former.

[1] *Nung chêng ch'üan shu* (Book 23, p. 50 b) and *T'u shu chi ch'êng*, Vol. 668,
*i shu tien*, Book 9, *nung pu hui k'ao*, both giving the same quotation, insert here
the sentence 淮 人 謂 之 礱。江 浙 之 間 謂 之 礱。 "The people liv-
ing on the borders of the Huai River (in Honan and Anhui) call it *lung*, and in the
territory of the Yangtse and Chê Rivers it is likewise so called."  From this passage
it might be inferred that this pounding-mill was (and still is?) a local implement of
the eastern parts of central China.

[2] A mill for husking rice is also mentioned by STAUNTON (Macartney's Embassy,
Vol. II, p. 395).  The grain, he says, is passed between two flat stones of a circular
form, the upper of which turns round upon the other, but at such a distance from it
as not to break the intermediate grain.  The latter wording corresponds exactly with
the above expression of the Chinese author.

[3] Literally, "a man's strength" (人力).  I take 人 in a singular sense, to
judge from the illustration (Fig. 3) accompanying the text, in which only one man is
engaged in pulling the shaft; but *Nung chêng ch'üan shu*, in the passage quoted above,
reads here 眾 力, as if a plurality of men, or several men in succession, were required.

[4] *Nung chêng ch'üan shu*, as quoted above, has the reading 斛 instead of 石,
thus giving an equal measure of capacity for the corresponding weight.

[5] 礱 礳 穀 器。所 以 去 殼 也。編 竹 作 圍。內 貯 泥

土礱

Fig. 3. Earth Mill (*t'u lung*) (from *Shou shih t'ung k'ao*).

It is difficult to gain a clear understanding of this mill

土。狀如小磨。仍以竹木排爲密齒。破穀不致
損米。就用拐木簽貫礱上。掉軸以繩懸檁上。
人力運肘以轉之。日可破穀四十餘石。

from the explanation of the text;[1] but from an examination
of the original engraving in the "*Shou shih t'ung k'ao*"[2]

---

[1] Foreign critics are certainly prone to upbraid Chinese writers for the deficiencies
in their records and for their uncritical attitude. That is all very well; but is our
own (proudly styled "scientific") literature any better? To speak on the subject
under consideration, I have hunted through nearly everything that European authors
have written on Chinese agriculture, and have not found a single description of even
one agricultural implement sufficiently full for really scientific purposes. By mention-
ing a few characteristics or some of the effects of its utilization, these writers imagine
they have described an object. In G. H. MASON, The Costume of China (London,
1800), illustrated by 60 engravings, on Plate XLV a miller turning a hand-mill is
represented. The plate is worse than any Chinese engraving, and quite unintelligible.
The under millstone is colored violet, the upper one lilac! JUSTUS DOOLITTLE, in his
much-quoted book, Social Life of the Chinese (London, 1868), p. 32, gives the following
description of a mill somewhat similar in construction to the one described above: "The
hull is removed from rice by a kind of mill, turned by hand, consisting of two parts.
The upper part, which is not very heavy, is made to move slowly around upon the
lower by a man pushing and pulling upon the handle. One end of the handle is
suspended by a cord attached to something in the room. By simply pulling and
pushing this handle in a certain way, the upper part revolves. The rice, unhulled,
is put upon the upper part, and passes through a hole down to the surfaces, which
touch and rub against each other. The rice comes out from the side, and falls into
a basket. What is not perfectly hulled by this process is then pounded in a large
stone mortar. This operation always removes the last of the hulls from the rice."
Of what material the "two parts" and the whole apparatus consist, which is
certainly essential to know, this author does not state. The Chinese author
informs us that the cord is attached to the cross-beams of the roof, which to the
English writer simply dwindle into a "something." According to him, the farmer
pushes the handle "in a certain way," while the Chinese account teaches that he turns
the mill by bending his elbows. Even less satisfactory is Sir JOHN DAVIS, who, in
his work China (Vol. II, London, 1857, p. 257), gives a sketch of a rice-mill not unlike
that of Doolittle, without adding a word of descriptive matter. WELLS WILLIAMS
(The Middle Kingdom) and DYER BALL (Things Chinese), in their essays on Chinese
agriculture, observe strict silence concerning mills, and set the minds of their readers
at ease by the remark that "their agricultural utensils are few and simple, and are
probably now made similar to those centuries ago" (WILLIAMS); and "the agricultural
implements of the Chinese are few in number, as well as of simple construction"
(DYER BALL). In studying the same subject in Chinese and in European literature, I
find more and more that Chinese literature is not as bad as it is made out to be,
even by sinologues, and that European literature is not as good as we like to boast it
is, in comparison with the Chinese and other nations. In investigating things Chinese,
it is safe to consult Chinese accounts in each and every case, which, in spite of all their
drawbacks (these I readily admit), are simple, plain, matter-of-fact, and to the point,
while the corresponding European notes prove in many cases superficial, misleading, or
entirely erroneous.

[2] I was fortunate enough to secure in Peking a volume containing 171 beautifully
executed plates, which seem to be the original illustrations to the *Shou shih t'ung k'ao*.
In the modern photo-lithographic edition of this work, the plates, which betray nothing
of the beauty and artistic execution of the originals, are so much reduced and so

(reproduced in Fig. 3), I am able to add a few more suggestions. There are two disks of earth surrounded by a bamboo plaiting, the nether one of which is flat and low, and rests on a frame over the ground, on which a mat is spread. The upper part is high, cylindrical, and of less diameter than the nether one, so that an uncovered space is left, on which the hulled grain falls. The hopper, in the form of a bamboo basket, is attached to what appears to be a flat board resting on the runner. To the end of this board the handle is attached, and the moving mechanism is quite clear. There is unmistakably an aperture in the runner, through which the grain poured into the hopper passes to the surface of the stationary part. The bamboo and wooden teeth mentioned in the above text are stuck into the earth or clay on the inner faces of both millstones, as may be seen also from the brief description of this mill given by S. SYRSKI in K. v. SCHERZER's "Fachmännische Berichte der österreichisch-ungarischen Expedition nach Siam, China und Japan" (Stuttgart, 1872, Appendix, p. 74). He says literally: "The hand-mill consists of a lower thicker and an upper thinner cylinder made of plaited work and clay. Their two surfaces, which are contiguous, are thickly set with pieces of hard wood. This hand-mill is more easily operated than that employed in Europe by the peasantry, since with the Chinese mill the power acts on a longer radius than is the case with the European mill. A similar mill, but of larger dimensions, is moved by a buffalo harnessed to a side-pole." (See also DOOLITTLE's account in footnote 1, p. 26.)

The "*Nung chêng ch'üan shu*" gives an engraving called

---

indistinct, that no investigation can be based upon them. An approximate equivalent for the originals is offered in the agricultural sections of the *Tu shu chi ch'ing*. The illustration above alluded to will be found in Vol. **1125**, *K'ao kung tien*, Book **245**, *mo wei pu hui k'ao*, p. **14 a**.

*lung* (Fig. 4), similar to the *t'u lung* and *mu lung* of the "*Shou shih t'ung k'ao*," but with the same description as the

Fig. 4.   Hand-Mill (*lung*) (from *Nung chêng ch'üan shu*).

latter; and a mill of the same type may be found on Plate XXI of the "*Kêng chih t'u*."[1]

---

[1] 御製耕織圖。 Drawings of Tillage and Weaving by Imperial Command

The "wooden" mill (Fig. 5) is described thus in the "*Shou shih t'ung k'ao*" (Book 40, p. 4 a), in accordance with the "*Nung chêng ch'üan shu:*" "The wooden hulling-mills (*mu lung*) are in frequent use. They are made of the wood of the pine-tree. In shape they resemble a large grist-mill (*ta mo*). Teeth are inserted in holes bored in both disks of the mill. A spindle is set up straight in the lower disk, while a socket [to receive it] is hollowed out in the upper one. On the ground a framework is set up, to which the pulling-contrivance is attached. It is pulled by several hands [relieving one another successively], and kept in revolution. The hulled grain results from the removal of the husks. The mill emits the sound '*yin yin,*' like the rolling of thunder.

(published 1696, under the Emperor *K'ang hsi*). Regarding this work, see F. HIRTH, Über fremde Einflüsse in der chinesischen Kunst, p. 57. We have here three editions of this book: (1) the original edition, with 45 engravings (the first engraving of later editions is wanting), printed on strong yellow cartoon, size 31.5 cm by 24.5 cm, in the form of a folding album, with manuscript preface and three seals attached to each engraving (the technique of execution is masterly); (2) a later undated edition on thin paper with very rough cuts, preface printed in red, same size as the original; (3) the reduced photo-lithographic edition of Shanghai (1887). Professor Hirth (l.c.) sees in these engravings a tendency towards correct observation of perspective, which he attributes to the influence of European art transmitted by Jesuit painters at the Imperial Court. It must be considered, on the other hand, that the work undertaken by *K'ang hsi* is only the reproduction of a much earlier set of engravings made by *Lou Shou* in 1210 (see WYLIE, Notes on Chinese Literature, 2d ed., p. 93, who says erroneously that it was recut during the *K'ien lung* period). The original edition of these illustrations has probably been lost, but the descriptive text is reprinted in the cyclopædia 知不足齋叢書, from which it is evident that the number of engravings in the work of *Lou shou* was 45. This number agrees with the original *K'ang hsi* edition (see above). In collating the two books, I found also that the title and letterpress description to each plate of the *K'ang hsi* work was simply literally copied from the older book of the *Sung* dynasty; so that I am led to conclude that its engravings also may have been kept intact rather than subjected to radical changes. At all events, to settle the question of a possible Jesuit influence in the *K'ang hsi* drawings, as proposed by Hirth (see now HIRTH, Scraps from a Collector's Note-Book, *T'oung Pao*, 1905, p. 398), it would be necessary to submit the edition of 1210 to a minute comparison with the former. Undoubtedly also the work of *Lou shou* had its predecessors, and drew from earlier models. The Emperor *Shih Tsung* of the *Posterior Chou* dynasty (A.D. 954—959) had a pavilion, on the walls of which paintings of the various labors of husbandry were displayed (see C. PÉTILLON, Allusions littéraires Vol. I, p. 319, Variétés sinologiques, No. 13, Shanghai, 1898).

Fig. 5. Wooden Mill (*mu lung*) (from *Shou shih t'ung k'ao*).

The farmers work with united energy; and when they sing
in chorus, one giving the first note, and the others joining
in with him, their work becomes a happy pleasure."[1]

[1] 木䃀。多用松木爲之。形如大磨。兩扇皆鑿
齒。下合植筍。穿貫上合。場中植架懸掉軸。以

龍
磨

Fig. 6.   Hulling-Mill (*lung mo*) (from *Shou shih t'ung k'ao*).

Another modification of the hulling-mill (Fig. 6) is explained under the name *lung mo* in the "*Shou shi t'ung k'ao*."
"A worn-out grist-mill, the surface of which is covered with

眾力曳轉。去穀出米。殷殷如雷聲。田家通力
合作。雜以倡和之聲。慶成事也。

rushes, can take the place of a hulling-mill (*ku lung*). It does not injure the rice. Either men or animals turn it round. It is called *lung mo* (i.e., a combination of a *lung* and a *mo*). If animal power is employed for pulling, the animals work the axle-tree of the big wooden wheel. A leather cord or a heavy rope passes over and under the wheel, and is connected again with the upper part of the mill. The revolution of the wheel causes the rope to revolve, and this revolution of the rope accordingly turns the mill. Calculations show that one rotation of the wheel corresponds to over fifteen rotations of the mill, whereas in the employment of human labor its speed is greatly reduced."[1]

Another apparatus used in hulling grain consists of a round base two feet high, called *ts'ao* 槽, made of coarse stone and brick work, over which a stone roller is passed. It is known under the name *nien* 輾 or 碾 ("roller-mill"). The roller is in the shape of either a cylinder or a wheel. According to a note of Hsü Kuang-ch'i, author of the "*Nung chêng ch'üan shu*," the base and the roller are made of wood in Kiangsi Province, while stone rollers are used in Shansi.[2] The cylindrical stone rollers are about three (Chinese) feet in diameter, and about five feet long. The roller is enclosed in a wooden frame attached to a long pole that revolves round a spindle in the centre of

---

[1] 有廢磨上級已薄。可代穀礱。亦不損米。或人或畜轉之。謂之礱磨。復有畜力輾行。大木輪軸以皮弦或大緪。繞輪兩周。復交於礱之上級。輪轉則緪轉。緪轉則礱亦隨轉。計輪轉一周。則礱轉十五餘周。比用人工既速且省。

[2] A good photograph of this Shansi roller-mill will be found in a book of F. H. Nichols, Through Hidden Shensi (New York, 1902), on the plate facing p. 92.

the base. The pole is dragged by a team of two animals. This contrivance is designated as *hai ch'ing nien* 海青輾. Another engraving in the "*Nung chêng ch'üan shu,*" and No. 78 in the above-mentioned single volume of plates, but missing in the modern editions of the "*Shou shih t'ung k'ao,*" is called simply *nien*, and shows two wheels, through each of which is a pole that slides along the outer side of the base, and each is drawn by an animal. The invention of this contrivance, according to a quotation from the "*Hou Wei shu*" 後魏書, is ascribed to a certain *Ts'ui Liang* 崔亮. In the "*Shou shih t'ung k'ao*" only (not in the "*Nung chêng ch'üan shu*"), there is, further, a hand-roller (小碾) worked alternately by two women over a stone block (*shih tun* 石墩). It is used only for the four kinds of millet, not for paddy and wheat.

The "Annals of the Three Kingdoms" 三國志 name *Tu yü*[1] 杜預 as the inventor of mills driven by water. The text says that *Tu yü* made double water-mills, in consequence of which the price of grain fell extremely low in the district around *Lo yang.*[2] In the "*Wei shu*" (chap. 66), the annals of the *Wei* dynasty (A.D. 386–532), and in the "*Pei shih*" (chap. 44), the aforementioned *Ts'ui liang* 崔亮 is credited with the construction of water-mills based on the idea conceived of by *Tu yü*. In the biography of *Ts'ui liang* it is said, "When *Liang* was in *Yung chou*, he read the biography of *Tu yü*, from which he saw that this man had made eight mills, and had thereby largely benefited his con-

---

[1] He lived A.D. 222—284 (GILES, Biographical Dictionary, p. 785).

[2] 杜預作連機水碓。由此洛下穀價豐賤。*Tu shu chi ch'êng*, Vol. 1125, *k'ao kung tien*, Book 245, *mo wei pu chi shih*, p. 1 b. For an engraving of this type of water-mill, see ibid., *mo wei pu hui k'ao*, p. 12 a, and in the other books on agriculture.

3

temporaries. Thereupon he taught the people to make husking stone-rollers. In his capacity as major-domo (*p'u yeh*), he made in *Chang ch'iao, Fang ch'iao,*[1] *Tung yen,* and *Ku shui* water stone-rollers and mills. In ten places this invention proved a tenfold advantage to the needs of the country."[2]

Water-mills appear in Europe in the beginnings of the *imperium Romanum* (first century A.D.).[3] The first mention of them is made in STRABO with regard to King Mithridates, who is said to have had a water-mill near his residence.[4] This fact hints perhaps at the derivation of this invention from the Hither Orient; but, so far as I can see, no special investigation of this subject has been made. It would lead too far, and beyond the scope of the present paper, to enter here into the construction of Chinese and other water-mills, and to trace their historical distribution. A few notes may suffice. It is interesting to note that they sprang up in China about the same time they came into general vogue in Europe, and that the contrivance of the water-mills of the ancients appears to be based on the same principle as the Chinese ones. Although there is nothing in the Chinese accounts to indicate that they were introduced from

---

[1] *Chang ch'iao* is a post-town near *Sui ning*, Kiangsu (PLAYFAIR, The Cities and Towns of China, No. 221); *Fang ch'iao*, a market-town near *Yü yao*, Cheh-kiang (ibid., No. 1698). But *Chang fang ch'iao* may be the name of *one* village not further known.

[2] 魏書崔亮傳。亮在雍州讀杜預傳。見爲八磨嘉其有濟時用。遂教民爲碾。及爲僕射奏于張方橋東堰穀水造水碾磨。數十區其利十倍國用便之。— *T'u shu chi ch'êng*, l.c., p. 2 a.

[3] O. SCHRADER, Lexikon der indogermanischen Altertumskunde, p. 512.

[4] H. BLÜMNER, Technologie und Terminologie der Gewerbe und Künste bei Griechen und Römern, Vol. I, pp. 46 et seq.

[5] Water-mills are referred to by STAUNTON (Macartney's Embassy, Vol. II, p. 395), R. FORTUNE (Two Visits to the Tea Countries of China, London, 1853, Vol. II, p. 37), and A. HOSIE (Three Years in Western China, London, 1897, p. 88).

abroad, but are attributed to *Tu yü* and *Ts'ui liang* as individual inventions, yet I am inclined to think that the contrivance must have been made before in a region intermediary between the Roman Empire and China, and that it likewise spread thence to the east and west. In Japanese literature, mills are first mentioned under the date A.D. 610, in connection with the name of a Buddhist priest, *Tam ch'i*, sent by the King of Corea, who knew the five Chinese classics, was, moreover, skilled in preparing painter's colors, paper, and ink. He also made mills; and this was, according to the Japanese Annals (*Nihongi*, written in 720), apparently the first time that mills were made.[1] In the same history it is mentioned that in A.D. 670 water-mills were made, and therewith iron smelted.[2] In the "*Kōjiki*" preceding the "*Nihongi*," and dating from the year 712, no mention is made of any kind of mills, but we hear there only of the pestle and mortar.[3] These facts seem to furnish sufficient evidence for supposing that mills and water-mills arrived in Japan from China by way of Corea. In about A.D. 635 water-mills were introduced into Tibet from China, owing to the marriage of the Tibetan king *Srong btsan sgam po* to a Chinese princess, who brought many inventions of her native country along to her new home.[4] These data evidently show that water-mills must have been in general use in China during the seventh century A.D.

---

[1] W. G. ASTON, Nihongi, Chronicles of Japan, Vol. II, p. 140. Aston makes the annotation, "It is not quite clear what sort of mills is intended, probably hand-mills."

[2] W. G. ASTON, l.c., Vol. II, p. 290, with the note, "The word for 'mill' does not mean merely a water-wheel, but something for pounding or grinding; perhaps the trituration of the ore is intended."

B. H. CHAMBERLAIN, Ko-ji-ki or Records of Ancient Matters (Supplement to Vol. X of the Transactions of the Asiatic Society of Japan, Yokohama, 1883, p. XXV; p. 201, footnote 6; p. 251).

W. W. ROCKHILL, Notes on the Ethnology of Tibet (Report of the U. S. National Museum for 1893, Washington, 1895, p. 672).

Fig. 7 shows a piece of pottery which may be said to represent scenes of agricultural work of the Han period. It is a square clay plaque, 2 cm thick, the sides of which measure 20.4 cm and 20.8 cm. It has a sort of bevelled

Fig. 7. Pottery Plaque representing Scenes of Agricultural Work (from specimen in Clarke collection).

edge, which encloses four smaller squares. The one in the lower right-hand corner of the illustration represents, I believe, a threshing-floor, which is usually made of earth-pisé.[1]

---

[1] Also called *chunam* in the East (YULE, Hobson-Jobson, 1903, p. 218). STAUNTON (Macartney's Embassy, Vol. II, p. 107) describes the threshing-floor as a platform of hard earth and sand, prepared in the open air, and mentions, besides the process of flailing and treading out the grain, the moving of a roller over it. According to S. WELLS WILLIAMS (The Middle Kingdom, New York, 1901, Vol. II, p. 9), "the threshing-floors about Canton are made of a mixture of sand and lime, well pounded upon an inclined surface enclosed by a curb; a little cement added in the last coat makes it impervious to the rain; with proper care, it lasts many years, and is used by all the villagers for thrashing rice, peas, mustard, turnips, and other seeds, either with unshod oxen or flails. Where frost and snow come, the ground requires to be repaired every season; and each farmer usually has his own." The threshing-floor is called 場院，禾場 or 穀場. Threshing with flails on the area is depicted in the 17th engraving of the *Kêng chih t'u* and in the agricultural books above quoted. Flails are seldom employed nowadays, and are entirely unknown in the villages around Peking, and, as it would seem, in almost all provinces of northern China. The most common method of winnowing practised there is to throw up the grain, intermixed with the chaff, into the air, whereupon it falls down on a mat on the ground, while a current of wind carries away the chaff. A shallow basket, called *po* 簸, is used for this purpose; and the whole process is named *po yang* 簸揚 or 颺. No. 20 of

In two diagonally opposite corners of this area are two small concentric circles in relief. Back of this square the bedstone of a grist-mill is moulded, the runner having been lost. Comparison with the mill pictured in Fig. 1, Plate IV, identifies this object beyond doubt. We see here the same knob in the centre, four concentric rows of small oval depressions, and a deep channel or reservoir around the millstone for receiving the flour. On the left of the illustration we observe a grain-crusher for hulling cereals, occupying two squares. In the front square is the mortar, indicated by a cavity 2 cm in depth and 2.7 cm in diameter. The pestle, if there ever was one, is not represented in the specimen. It may have been lost; no traces of any injury to the object, however, can be discovered. It is rather to be presumed that the potter shirked the difficult task of representing the pestle, since the significance of this object was easily intelligible without it. In the back square four truncated pyramids, 2.8 cm high, are arranged in the form of a rectangle. These are connected with one another by low walls of clay, except on the front side, which is left open for the reception of the lever. The two front pyramids represent the fulcrum on which the lever rests. On the

the engravings in the *K'ing chih t'u* affords a good illustration of it. The same process is followed in Corea, where, according to I. BIRD BISHOP (Korea and her Neighbors, p. 162), all grain is winnowed by being thrown up in the wind. The same was done in ancient Egypt, Greece, and Rome. "Women winnow the corn by throwing it up quickly by means of two small bent boards. The grain falls straight down, while the chaff is blown forwards" (see A. ERMAN, Life in Ancient Egypt, London, 1894, p. 432). By means of a wooden shovel or a plaited van (Lat. *vannus*) the threshed-out grain was thrown from the ground into the air when the wind was blowing somewhat strong. The wind then carried the light chaff beyond the threshing-floor, or, if the latter was more extensive, to a place on it intended for the chaff; while the heavier grains fell down to the ground or into a basket made ready for the purpose (see H. BLÜMNER, Technologie und Terminologie der Gewerbe und Künste bei Griechen und Römern, Vol. I, p. 9; O. SCHRADER, Lexikon der indogermanischen Altertumskunde, p. 966). This method seems to have originated everywhere from natural observation.

inner sides of the two front pyramids are two oblong
incisions that serve as sockets for the pivot on which the
lever moves. In the top surface of the two rear pyramids
are two small shallow depressions. In reconstructing from
this model the actual implement as it was then applied in
agriculture, we must presume that these four pyramids were
wooden logs, and that the two rear ones served as supports
for a wooden framework, the cross-beam of which furnished
a hold for the hands of the laborer, who operated the
treadle with his feet. The shallow depressions would seem
to indicate that the two vertical poles forming the framework
were inserted in them. The treadle is a flat board, being
the continuation of the lever. An engraving from the
"*Shou shih t'ung k'ao*" (Fig. 8) represents a contrivance
called *kang tui* 坰碓 (i.e., "the pestle of the earthen vat").
According to the Chinese description of it, a pit more than
two feet deep is dug in the ground, and a large earthen
vat placed in it, over which a bamboo basket-strainer is put.
The pestle is worked by a wooden lever moving up and
down on a pivot between two vertical bars. The illustration
shows a man working the treadle. He alternately ascends
and descends with it, his left foot resting on the middle
part of the treadle, thus insuring a certain steadiness and
uniformity of motion. The pounding is done in the basket,
in which the husks remain, while the hulled grain sifts
through into the vat. Three piculs of grain can be pounded
at a time.

In the work on agriculture above quoted, other contriv-
ances for such mortars are also pictured, among them one
driven by a water-wheel.[1] An excellent sketch of a rice-

[1] A brief notice of such a rice-crusher, driven by water, is given by SVEN HEDIN
(Durch Asiens Wüsten, Vol. 1, p. 295).

Fig. 8. Tread-Mill (*kang tui*) (from *Shou shih t'ung k'ao*).

crusher, answering in some degree to our Chinese engraving

and description, is given in GEORGE STAUNTON'S "Macartney's Embassy" (London, 1797, Vol. II, p. 397). His text reads as follows: "To remove the skin or husk of rice, a large strong earthen vessel, or hollow stone, in form somewhat like that which is used elsewhere for filtering water, is fixed firmly in the ground; and the grain, placed in it, is struck with a conical stone fixed to the extremity of a lever, and cleared, sometimes indeed imperfectly, from the husk. The stone is worked frequently by a person upon the end of the lever." It is not mentioned of what material the two bars holding the lever are made. From the engraving it would almost seem that they are made of earth. They are low, flat, and rounded over the upper surface. The man working the treadle steadies himself by seizing hold of a horizontal bamboo bar laid across two vertical bamboo poles stuck into the ground, in exactly the same way as we were led to reconstruct the grain-crusher of the Han time from the clay model. This confirms the correctness of our reconstruction, and proves that the type of the modern grain-crusher existed as early as the Han period.

The piece of pottery shown in Plate vi was kindly loaned to us by Messrs. Yamanaka & Co. of New York City.[1] It represents a farmer's shed with a rice-pounder inside of it. The piece is 17.2 cm high, and the length of the ridge-pole 22.8 cm. The long sides of the shed measure 18.5 – 19.5 cm, the open one being 11.1 cm high, and the opposite one 9.5 cm high. The end walls are 15.5 – 16.1 cm wide by 11.2 – 11.3 cm high, and the thickness of the walls is about 1 cm. The one long side of the shed is left open for technical reasons, and in order that a free

---

[1] This was five months after the description of the previous specimen was written, so that it may serve as a confirmation of the above interpretation of the latter.

PLATE VI.

Pottery Farm-Shed with Rice-Pounder, Interior and Exterior.

view into the interior might be had. In reality, of course, this side corresponds to the back wall. It cannot be supposed that the entrance was from this side, since in that case the door and windows on the opposite side would be meaningless. On the front side we note the rectangular opening for the door, with a low threshold, and two window-holes not far below the eaves, their top sides being on the same level as the top of the door. In each of the two narrow end walls a window-opening is likewise cut out at the same height as the two front windows; and above each of these, extending from the roof down, is a triangular open space which might be regarded as a gable window, allowing the air to circulate freely through the building. All these arrangements agree in a striking manner with the modern farm-sheds of northern China. The roof consists of two sloping sides separated by a thick ridge-pole that merges at either end into a mound-like knob. Each slope of the roof is divided into five sections by four ribs of *imbrices*, the sections between being covered with flat tiles (*tegulæ*). Each rib shows distinctly, by means of grooves, four of these half-cylindrical tiles. Along the edge of the gable ends of the roof are represented two heavy massive rafters which project over the walls, and merge at either end, like the central ridge-pole, into a mound-like knob or thickened elevation. These beams should be conceived of as wood-work. The framework on which the roof rests is not indicated.

The mechanism of the object within the shed, representing a rice-pounder, is clear from the description previously given. The features which distinguish this pounder from that shown in Fig. 8 are, that here the mortar stands on the floor-level, and the lever, which rests in a horizontal position (i.e., it is not in use) on the fulcrum (an upright flat clay slab), is represented. The lower end of the pestle fills the mortar, and the wart in front of the pestle indicates the end of the lever passing through it. The lever gradually broadens towards the fulcrum, and is of rectangular section. It must be assumed that this lever was also worked by means of the feet. It is not shown in the clay how the lever was attached to the fulcrum, if at all: possibly it rested loosely on it without being connected to it, as here represented.

The whole outside of the piece is glazed a light green, the greater part of which has dissolved into a silver and gold iridescence interspersed with light and dark blue specks. Inside, the bottom is partially glazed, as are also the mortar and pestle (but not the fulcrum) and the inner wall along which the rice-pounder is placed. The piece seems to have been in the furnace several times, probably three, since spur-marks (see Index) are visible on three of the walls, and particularly on the roundish knobs on the gable ends, which are the most prominent parts on each side. No spur-marks are to be found on the bottom.

In connection with this shed, I may here insert the pottery model of a house (Plate VII) to further illustrate the architectural features of that period. The piece in question is owned by Mr. Marsden J. Perry of Providence, R. I. The main part of the house is 25 cm long; the roof, 32.5 cm. It is posed on four low feet forming bear-heads (see p. 57). The door in the middle is ajar, the one leaf slightly pushed inward, the other wide open toward the outside, thus allowing an insight into the interior. Along the back wall a bunk (*k'ang* 炕) is built, such as is still found in a large number of homes of northern China. The door is surrounded by a framework of beams into which big-headed nails seem to have been driven. The latter appear also along the edges

PLATE VII.

Pottery House.

of the front walls. The roof is much simpler than the previous one. It lacks the gable ends, and two side-rafters slant down obliquely from the ridge-pole. There is no indication of tile-work, and it might therefore be supposed that it is meant to represent a thatched roof. The glaze has entirely dissolved into a silver iridescence.

The modern Japanese rice-pounder comes very near that of the Han time, except that the large cylindrical mortar is of wood, and rests on the ground. It is sketched in Fig. 9,

Fig. 9. Modern Japanese Rice-Pounder, after photograph (Museum Catalogue No. 2875) obtained in Japan.

made from a colored photograph obtained in Japan. The pestle is of wood; and, to increase its striking-power, a stone weight is attached to the upper part of the lever by means of ropes. Here we also see the four wooden logs which we made out in rebuilding that of the Han period. A narrow board is laid over the two rear logs, on which the laborer takes a firm stand with his left foot, while his hands grasp the cross-bar at the top of the high wooden frame just in front of him. His right foot operates the treadle. Compare also a humorous Japanese sketch of a man working a rice-pounder, who, by his energy, has jerked the fulcrum from its socket, reproduced in CHR. DRESSER, "Japan, its Architecture, Art, etc." (London, 1882), p. 463.

Glancing back at the three agricultural objects combined in the piece of pottery shown in Fig. 7, we notice that they illustrate the three principal consecutive stages in the work of the husbandman after the harvest. The representation of the threshing-floor, the crusher, and the mill, suggest the three kinds of labor performed by the farmer, — threshing, husking, and grinding his crops.

Fig. 10 shows one of the most curious pieces of Han pottery. It symbolizes, so to speak, a peaceful rural scene of that period. It appears to represent a row of five sheep in a sheepfold squatting down in front of a mill. The heads of the animals, with their horns turned in, are exceedingly well expressed. The pen, in all likelihood an imitation of a real one of that time, is shaped somewhat like a horseshoe, with a roof over that part in which the sheep are. There is a rectangular opening in front, probably the entrance to the fold. The object in the foreground of the pen is a mill moulded in one piece, and not in two pieces, as in the mills represented in Plates IV and V. It consequently shows only the runner; and that it must be thus interpreted, is shown by comparing it with Fig. 1, Plate IV, and Plate V. It has a handle like that of the former and of the modern stone mill, and the two semicircular depressions agree exactly with those there seen. The greatest height of the pen is 10 cm; that of the low front wall at the opening, 5.4 cm; the greatest width is 18 cm; the roof is 17.3 cm long and 6.5 cm wide.

In connection with this unique piece of pottery, it may not be out of place to discuss a character of the Chinese language which may throw some light on the subject represented by this specimen. This character is *yang* 垟, composed of the two parts *t'u* 土 ("earth") and *yang* 羊 ("ram,"

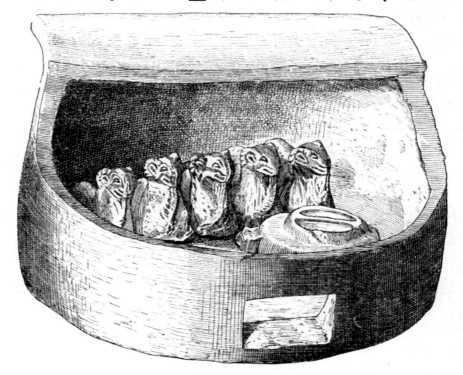

Fig. 10. Sheepfold containing Five Sheep in Front of a Mill (from specimen in Clarke collection).

"sheep"). In hunting through the ordinary dictionaries, we find the meaning of this term thus explained. WILLIAMS (p. 1072 b) says, "The elf or sprite that guards a spot; Confucius was asked the meaning of a sheep found in digging a well, when he said it was a *fên yang* 羵垟 or local brownie." As regards *fên*, he remarks on p. 130 c, "羵 is regarded as another form of 墳 in its meaning of a sprite shaped like a half-formed ram." GILES translates 羊羊 by "an elf, a sprite," and *fên yang* by "the 'genius' of a river or mountain." COUVREUR has *yang* 羊羊 (1) "brebis d'argile qui fut trouvée dans un puits," and quotes 史記孔子世家;[1] (2) 玉羊羊 "brebis de jade qu'on trouvait sur le mont *Hua shan* au Shensi (according to 易說)." The above term *fên* is interpreted by COUVREUR "Brebis à grosse tête. Démon qui prend la forme d'un bouc." PALLADIUS, in his "Chinese-Russian Dictionary," explains *fên* as a sprite of the earth in the shape of a ram, and quotes as synonymes 羵羊, 井羵, and 蘷羵 *k'uei fên*.

Turning now to Chinese sources, we meet in K'ANG HSI's Dictionary, under 羊羊, with the quotation from SSŬ-MA CH'IEN mentioned by Couvreur, as follows: "*Chi Huan-tzŭ* bored a well, and found there an earthenware vessel. Inside of the latter there was a sheep. He questioned Confucius [about it], who said, 'I have heard that the supernatural beings of the earth are the *fên yang*.'"[2] It follows from this, that, according to the idea of Confucius, *fên yang* was understood to be a spirit of the earth, and not, as Giles says, of a river or mountain; and we further see that also Couvreur is wrong in his remark, as the clay sheep in the above quotation has no causal connection with a well, but was accidentally discovered in the ground when a well was being dug. The word *fên* 羵 is explained by K'ANG HSI: "(1) After the *Chêng yün* 正韻: 土中怪羊 'a supernatural

---

[1] The Genealogy of Confucius in the Annals of Ssŭ-ma-Ch'ien, chap. 47.

[2] 季桓子穿井獲一土缶。其中有羊。問于仲尼曰。以丘所聞土之怪則羵羊也。(Gloss: 或作羊。)

sheep in the earth;' (2) equivalent to *fĕn* 憤 ('zeal, energy')."
The "*P'ei wĕn yün fu*" (Chap. 22 b, under 羊) gives the
following quotation from the "*Lu yü*" 魯語, which contains
a fuller version of the above saying of Confucius: "*Chi
Huan-tzŭ* bored a well and obtained an object like an earthenware vessel. Inside of the latter there was a sheep. He
made inquiry of Confucius, who replied, 'I have heard thus:
the supernatural beings of trees and stones are called *k'uei*
(demons in the form of a dragon with a single paw) and
*wang liang;* the supernatural beings of the water are called
dragons and *wang hsiang* (literally, "shapeless"); the supernatural beings of the earth are called *fĕn yang.*'"[1]  In the
latter case we note the different writing of the characters
from that in the passage in *Ssŭ-ma Ch'ien*, 墳羊, which
means literally "sheep of the mound or grave."[2]  The writing
羵垟 is, of course, a secondary stage, made up of "sheep
and mound, earth and sheep." The character 垟 could have
two significations: (1) "a sheep of the earth," (2) "a sheep
made of earth." I agree with Couvreur, that the latter was
the original meaning, from which I think the former, with the
sense of a sheep-formed demon, was subsequently derived,
when the real understanding of such clay sheep was lost.
What *Chi Huan-tzŭ* found in his unearthed jar could certainly
have been nothing but an unpretentious clay sheep. In our
language, he had luckily hit upon an archæological find of

---

[1] 季桓子穿井獲如土缶．其中有羊焉．使問
之仲尼對曰．丘聞之．木石之怪曰夔蝄蜽 (pro
魍魎)．水之怪曰龍罔象．土之怪曰墳羊．

[2] Also *K'ang hsi* remarks expressly that 羊 can be written in lieu of 垟, and
that 墳 is a substitute for 羵.

a probable prehistoric specimen of pottery. He submitted it to the judgment of the sage, who naturally was a poor diagnostician in such matters, and, much embarrassed, improvised an explanation suited to the occasion, expressing a popular belief in pronouncing the clay sheep a local sprite. It is generally known that prehistoric objects are connected with similar superstitious ideas in all countries, and there are abundant proofs of this fact in Chinese literature. The original meaning of the term *fên yang*, therefore, is "a clay sheep buried in a mound with the dead." This was undoubtedly an ancient custom in China, which is not surprising, in view of the fact that the sheep was the most esteemed of the domestic animals of the ancient Chinese, who expressed everything good and beautiful with the character *mei* 美, composed of the symbols for "sheep" and "great." I also believe that we are justified in linking the representation of sheep in the pottery of the Han period with the finds of clay sheep of the old Chinese records.

In the "*Lung chou t'u ching*" 龍州圖經 it is said, "In the district of *Ch'ien yüan hsien* in *Lung chou* (in *Fêng hsiang fu*, Shensi Province), there is the temple of the spirit of the earth sheep. In days of old, when the Emperor *Ts'in shih* made a tour of inspection, he saw two white sheep fighting. He despatched an envoy to drive them away. When he arrived on the spot, they had transformed themselves into heaps of earth. The envoy, taken aback, returned. The Emperor *Shih* then saw by a lucky chance two men bowing at a corner of the road, and questioned them. They responded, 'We are not men, but the spirits of the earth sheep; and we have come here to pay a visit to the ruler.' No sooner was their speech finished than they disappeared. The Emperor *Shih* accordingly ordered a temple

to be erected, in which sacrifices have been and are offered
to them up to the present day without interruption."[1]

Sheep were moulded in bronze as early as in the time
of the *Chou* dynasty. In the "*Po ku t'u lu*" (Book 27,
p. 18 b) we find the engraving of a cover to a *lien* 匳 from
that period, on which three reclining sheep are formed, their
heads turned towards the knob in the centre of the lid.[2]
In the "Illustrated Book of Ancient Jades"[3] (*Ku yü t'u p'u*),
Book 72, p. 3 b, is illustrated a single squatting sheep carved
of nephrite, the head carried high (羊首昂起), and with
horns turned back and downward. It is of almost the same
description as the sheep in the above piece of Han pottery.
The author, who wrote in the second half of the twelfth
century under the *Southern Sung*, attributes this jade sheep
to the time of the *Tsin* 晉 (A.D. 265-419) or *T'ang*
(A.D. 618-905) dynasty. It was used as a *shui chu*
水注; i.e., as a receptacle for holding the water poured

---

[1] 隴州汧源縣有土羊神廟。昔秦始皇開御道
見二白羊鬪。遣使逐之。至此化為土堆。使者
驚而回。始皇乃幸其所見二人拜於路隅。始
皇問之答曰。臣非人。乃土羊之神也。以君至
此故來相謁。言訖而滅。始皇遂令立廟。至今
祭祀不絕。 *T'u shu chi ch'êng*, Vol. 593, *po wu hui pien, ch'in ch'ung tien,*
Book 114, 羊部外編, p. 1 b.

[2] 蓋起三羊回頸相向頂。 The same illustration is reproduced in
*T'u shu chi ch'êng*, Vol. 1123, *K'ao kung tien*, Book 228.

[3] The full title of the book runs thus: 宋淳熙敕編古玉圖譜
"Illustrated Book of Ancient Jades, compiled at the imperial command in the period
*Shun hsi* (A.D. 1174-1189) of the *Southern Sung* dynasty." Original preface dated
1176. 100 books in 16 volumes. Composed by LUNG TA-YÜAN 龍大淵 and others.
Edited under *Ch'ien lung* in 1779. The copy of the Bibliothèque nationale of Paris is
in 7 volumes, and apparently incomplete (see M. COURANT, Catalogue des livres chinois,
etc., Paris, 1900, Vol. I, p. 67 a). See also BUSHELL, Chinese Art, Vol. I, p. 139.

4

over the stone slab on which the ink for writing is rubbed. The water drips from the sheep's mouth.[1]    Others of these jade *shui chu* were made in the shape of geese, striped toads, tortoises, lotus-flowers, melons, etc.    Also single stone and bronze figures of sheep were occasionally discovered, as shown by two passages in the "*Ssŭ ch'uan t'ung chi*" 四川 甬志 (Book 60, pp. 34 a and 42 a).    In the period *Wan li* (A.D. 1573–1619), 25 *li* to the southwest of the district city *P'i*, a stone sheep was excavated when the temple *Lu yüan* ("Temple of the Deer Park") was repaired: the name of the temple was therefore altered into *Shih yang ssŭ;* i. e., "Temple of the Stone Sheep."[2]    In the other passage, mention is made of a bronze sheep in the *Ch'ing yang kung* ("Palace of the Dark Sheep") outside of the suburbs to the south of *Ch'êng tu*, the capital of Ssŭch'uan Province.    It is described as about two feet high, and as an elegant and very antique piece of work.[3]    The following five characters are engraved on the base: 藏梅閣珍玩 *ts'ang mei ko chên wan.*[4]    The sheep has only one horn, and resembles a unicorn (角端).    Its interior is hollow, and there are holes in its sides.    If it is rubbed between the hands, it becomes brilliant,[5] which is considered to be evidence of its antiquity.    In the same paragraph a book,

---

[1] 羊口出水以供硯滴。
[2] Quotation from 郫縣志。在縣西南二十五里鹿苑庵 萬歷年間修寺掘石羊一枚。因更名石羊寺。
[3] 製極古雅。
[4] The inscription probably means "much-prized object of amusement of the hall *Ts'ang mei.*"
[5] 摩挲有光。

"*Yün chan chi ch'êng,*"[1] is quoted in the statement that in a Taoistic temple there was of old a stone sheep. *Wang Yü-yang*[2] saw it when he went there for the first time, but on his second visit it had been lost.[3]

## GRANARY URNS.

Public granaries have been instituted in China since the oldest times.[4] There were round (*ch'üan* 囷) and square ones (*ts'ang* 倉 or *ching* 京).[5] Fig. 11 is reproduced from the engraving of a granary (*ch'üan*) in the original edition of the "*San ts'ai t'u hui*" (published in 1607 by WANG CH'I),

---

[1] 雲棧紀程。

[2] Or *Wang Shih-chêng* 王士禎, 1634–1711 (see GILES, Biographical Dictionary, p. 838).

[3] 觀中舊有石羊一。王漁洋初至曾見之。再至則失之矣。

[4] The great antiquity of granaries in China is evident from the fact that the names for the square and the round granary, for the public (*wei* 胃 ) and the imperial granaries (*lin* 廩 ), appear also as ancient names for asterisms (see G. SCHLEGEL, Uranographie chinoise, pp. 337–346). According to MÊNG TZŬ and SSŬ-MA CH'IEN, the Emperor *Yao* made *Shun* construct a storehouse and a granary. See CHAVANNES, Les mémoires historiques de Se-ma Ts'ien, Vol. I, p. 74; also Vol. III, p. 544. Compare also J. H. PLATH, Nahrung, Kleidung und Wohnung der alten Chinesen (München, 1868), p. 83; PLATH, Die Beschäftigungen der alten Chinesen (München, 1869), p. 34; PAUTHIER and BAZIN, Chine moderne, p. 594. The organization of a granary system in the district cities is said to have originated in a memorial submitted to the government by *Chu Hsi* in A.D. 1181 (see HIRTH, T'oung Pao, Vol. VII, p. 306). In this connection, mention might be made of the refuge-towers (*tiao lou*) of Yünnan, erected by the Chinese to protect themselves from the frequent raids of the warlike aborigines. In times of peace they serve the purpose of granaries. According to the description of S. POLLARD (Refuge Towers of North Yünnan, East of Asia, Shanghai, 1903, Vol. II, No. 2, p. 175), they may be built of dressed or undressed stone, burnt or sun-dried bricks, and adobe. The roofs are covered with tiles or thatch. The towers vary in height from forty-five to seventy-five feet; the base being about twenty-two feet square, and the walls three feet in thickness. Sometimes the base is larger, measuring even as much as twenty-two by fifty feet. Occasionally the tower and adjoining houses are surrounded by a strongly built wall with a moat outside. Their walls are often battlemented, like the ordinary city walls.

[5] See *Li ki* (edited by COUVREUR), Vol. I, p. 381.

section on architecture 宮室, Book II, p. 22. A descrip·
tion of it is not given in the text, but from the drawing the
roof would appear to be represented as thatched.[1]

These grain-towers suggested the production of a certain
type of Han ceramics. We
meet here, first of all, with the
actual representation of the
model of a grain-tower (Fig.
12), and then with a variety
of urn whose principal shape
is derived from the idea of the
same structure.[2]

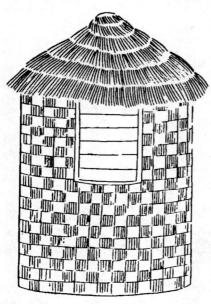

Fig. 11.  Grain-Tower (from an engraving
in *San ts'ai t'u hui*).

This tower is placed in a
bowl-shaped vessel, and over
to one side of it. The form
of the bowl is not exactly
circular, as the diameter varies
from 22.2 cm to 23.4 cm.
Its height is 8.4 cm. It has
a slanting, somewhat flaring
rim 1.4 cm wide, the outer edge of which curves downward.
The tower, which extends from the bottom of the bowl up,
reaches to a height of 15.5 cm, and has a circumference of
23 cm. A stairway 9.8 cm long, on the outside wall of the
bowl (indicated by an oblong slab of clay with eight incisions
denoting steps), leads up to the rectangular entrance to the
tower, through which a glance at the inside may be had. A
little beyond the threshold we see the bottom somewhat deeper

---

[1] Compare G. SCHLEGEL, Uranographie chinoise, p. 240.
[2] The colloquial term for these urns, among the antiquarians of *Hsi an fu* is *wu
ku kuan* 五穀罐; i.e., "jar for the five grains," which are rice, wheat, *Panicum
italicum*, *Panicum miliaceum*, and the Soya bean.

down, the entire lower part of the tower being filled up with clay. The portion under the roof is hollow. The roof itself is rectangular (the two long sides being 10 cm and 11.3 cm long respectively, and the short sides, 3 cm). It has a flat rectangular ridge-pole, from which four triangular ridges

Fig. 12. Model of a Grain-Tower (from specimen in Clarke collection).

slope downward, each decorated with five parallel ribs indicating tile-work. The ribs symbolize apparently the rows of imbrices; and the intervals between them, the tegulæ.[1] The ridges project over the tower-wall a distance of 1.7 cm. Below, on the inner side of the tower, is a small opening, in front of which stands an animal, apparently a pig, to judge from the general shape of its head and body, its short tail, and long drooping ears. Its head is bent down, as if feeding on some grain supposed to be dropping out

---

[1] To obtain a correct idea of the arrangement of Chinese tiling, see MORSE, On the Older Forms of Terra-cotta Roofing-Tiles (Boston, 1892), particularly Figs. 9, 11, 17.

of the opening of the tower.[1]   Its feet are extremely broad and clumsy.   It is a freehand moulding, and was stuck on the bottom of the bowl while the clay was still wet.   The place on which the pig is standing is marked by cross-hatchings.   The stairway and the tower were also formed separately, the joining between the latter and the inner edge of the bowl being still visible.   The body of the tower was made wholly by hand, as its rough and irregular shape amply prove.   The roof was also made separately, as shown by the irregular masses of clay smeared between its under side and the wall of the tower.   The bowl itself was turned on the wheel.

From a first view of this piece in its present condition, one gains the impression that it is unglazed; but on closer inspection it will be seen that over the outside of the bowl, the tower, and the pig, there was originally a light-greenish glaze, thin traces of which are now left here and there. The clay is of a very light reddish hue.

Plate VIII illustrates three granary urns viewed from above, thus showing only the roofs.   Fig. 1 represents the simplest kind of roof, which does not project over the wall of the jar.   It is divided into five parts by five ribs denoting the imbrices, but without further ornamentation, so that this piece might be explained as the result of a conventionalization developed from the type with genuine projecting roof.   The jar to which this top belongs is one of the largest and most remarkable, being 32.7 cm high, with a diameter at the mouth of 9.4 cm, and at the bottom of 16.1 cm.

In Fig. 2 of this plate we see on the roof four flat

---

[1] This subject calls to mind the character *huan* 豢 , which originally meant "to feed pigs with grain."   Thence it came to have the signification of feeding or fattening any animal for killing.   It is even used as in the *Tso-chuan*, of feasting a man with a view of bringing him to destruction, and of the rearing of female domestic slaves.   See T. WATTERS, Essays on the Chinese Language (Shanghai, 1889), p. 248.

PLATE VIII.

Granary Urns Viewed from Above.

rectangular pieces over four of the ribs. These probably mark the ridges. Around the roof, running between the ribs, are two faint parallel lines, in most places hardly visible, to indicate three rows of flat tiles. This jar is 24.3 cm high; the diameter of the mouth, 4.7 cm.

Fig. 3, Plate VIII, shows a very peculiar roof. The round opening is surrounded by a square, from the angles of which four oblique ribs emanate. Each of the four spaces between these oblique ribs is filled with twelve parallel ribs running at right angles to the sides of the square. The urn which this roof covers measures 21.8 cm in height, the diameter of the opening being 3.2 cm. Its three feet have been broken off.

The fact that this type of roof comes nearest to that on the actual grain-tower in Fig. 12 furnishes clear evidence that it was derived from the roof of the real grain-tower. The mode of tiling, as brought out on these vessels, reveals the interesting fact that the method of tiling the roof was exactly the same at the time of the Han dynasty as in modern China. Moreover, it is clear that the roofs were then not turned up or curved at the eaves, as is seen now in temples, palaces, and public buildings.[1]

---

[1] "Chinese architecture, then, had nothing to do at first with the imitation of tent forms. The suggestion has been made that the concave shape of the eaves of Chinese buildings shows that the people love to remember the nomad life which they once led when they occupied tents of a conical form or shaped like a house-roof. No such concave curve is seen in any old roofs in sculptures hitherto brought to light. . . . The love of fantastic curves in the lower part of roofs came into vogue later, and must be sought rather in Buddhism." — J. EDKINS, Chinese Architecture (Journal of the China Branch of the Royal Asiatic Society, N.S., Vol. XXIV, p. 259). See also v. FRIES, The Tent Theory of Chinese Architecture (ibid., pp. 303–306), who justly opposes the theory of tent-origin. — The turned-up corners of a roof are called *fei yen* 飛簷 ("flying eaves"). In a paper on Chinese Architecture by I. LAMPREY, in the Journal of the Royal Institute of British Architects (1867, p. 164), it is suggested that these curves "had some connection with that graceful curve we notice in the branches of the trees of the fir species."(!)

Plate IX gives the side views of three granary urns. That shown in Fig. 1 is 31.4 cm high, and has a diameter at the mouth of 4.6 cm, and at the bottom of 14.9 cm. The roof is slanting, and projects over the wall of the jar 2.5 cm. It is divided into sections by twenty-four ribs, each section between a pair of ribs including three flat tiles, indicated by slightly raised semicircles. The glaze is a lustrous olive-green, and finely crackled. On the under side of the bottom is a large isosceles triangle, the sides of which are each formed by a pair of parallel incised lines. The portion covering this triangle is glazed, while the parts outside of it remain unglazed. Perhaps it served as a potter's mark.

The roof of the next granary urn (Fig. 2, Plate IX) corresponds to the much conventionalized type seen in Fig. 1 of the preceding plate, except that it has only four ribs. The urn is coated with a yellowish-brown crackled glaze, showing in places a greenish tint. It is 28.5 cm high, the diameter of the mouth being 5.6 cm, and that of the bottom 14.5 cm.

The last of the three figures on this plate (Fig. 3) represents a granary urn 33.15 cm high, with a diameter at the mouth of 5.2 cm, and at the bottom of 17.9 cm. The roof projects over the wall of the vessel 3 cm. This is one of the largest and most elaborate specimens of this group. The roof has four ridges (each about 4 cm wide and 5 cm long) and twenty-six ribs, between which three tile-marks in the form of raised curved lines have been distinctly moulded. The body of the urn is divided off by three bands of grooves, the upper band consisting of four circles, the middle and lower ones of three circles each. These grooves were, and are still partially, filled with a black glaze. The glaze of the roof is a light leaf-green;

PLATE IX.

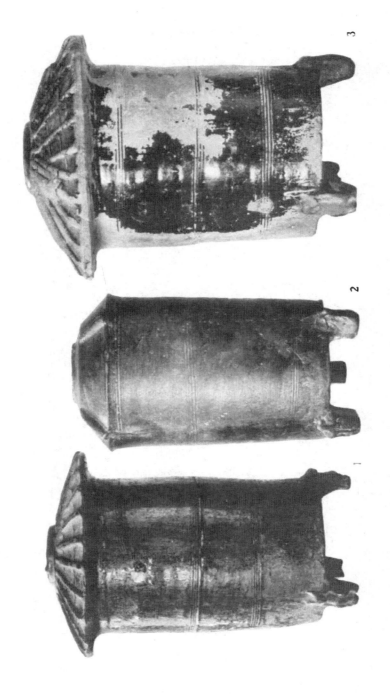

Granary Urns.

PLATE X.

Granary Urn.

that of the body, dull green, with a network of white crackles. In the large portions which appear white in the illustration the glaze has disappeared.

The specimen illustrated on Plate x is in the private collection of Mr. R. H. Williams of New York, and is remarkable for the extraordinary distinctness with which the figure of the bear[1] over its three feet is moulded, which occurs in all these jars. The outline of the head, the pointed muzzle, the upright ears, and the paws, are clearly brought out. Three bands of double grooves adorn the body. The roof is divided off by six ribs, each 5.3 cm

---

[1] The explanation of this figure as being a bear doubtless goes back to an ancient tradition. In the *Po ku t'u*, Book 20, p. 7 b, a *chiao tou* 銚斗 resting on bear-feet 熊足 is illustrated and thus explained: 足飾以熊者。秦漢以來奩具印鈕每作此狀。蓋熊男子之祥取其有所堪能故也。 "The feet are decorated with bears. From the time of the *Ts'in* and *Han*, the vessels called *lien* and the knobs of seals were made into such figures. The bear was a good omen for securing male offspring, owing to its symbolical significance of strength and endurance." The same idea is expressed in the description of a *lien* of the *Chou* time posed on three such bear-feet, in *Po ku t'u*, Book 27, p. 19. From the close agreement of this figure in the bronzes with the corresponding figures in the pottery, I am strongly inclined to apply the same explanation to the latter. Fig. 13 is reproduced from the *Hsi ch'ing ku chien* (Book 38, p. 28), and represents the full figure of a bear (called "flying bear") in a bronze of the *T'ang* period, explained as the stand for a gnomon, *piao tso* 表座 (see ibid., p. 30 b, where the *Po ku t'u* is referred to, and quoted very inexactly. The passage alluded to is found in this work in Book 27, p. 37 b; see also p. 38 b, from which it follows that sun-dials were in use at the time of the *Chou* dynasty). Regarding a symbolical meaning of the bear-ornament in con-

Fig. 13. Figure of a "Flying Bear" on a Bronze of the *T'ang* (from *Hsi ch'ing ku chien*).

唐飛熊表座

nection with the eagle, see S. W. BUSHELL, Chinese Art (London, 1904), Vol. I, p. 149.

long, but no tile-marks are indicated. The vessel is 32.4 cm high, and its feet 2.2 cm, while the bears are 4.5 cm high. The diameter of the mouth is 7.3 cm. The glaze is dark green, mottled brown spots being scattered here and there.

The majority of these urns show no decoration on their bodies. Only two ornamented pieces have thus far become known to me.

One is the urn shown in Fig. 1, Plate XI, which is 24 cm high, with a depth of 19.9 cm, and a mouth 5.5 cm in diameter. The roof projects 2 cm over the wall of the jar, and is divided into four parts by four ridge-poles. Two curved parallel lines are plainly visible between the single ribs, thus clearly expressing three layers of tiles. Around the body of the vessel, spiral ornaments, partially made up of undulating lines, are brought out in low relief. This field is enclosed above and below by double grooves. The urn rests on three feet in the form of squatting bears. It is covered with a dark-green crackled glaze.

The other urn is pictured in Fig. 2 of the same plate. Its total height is 27 cm; depth, 22.9 cm; diameter of the mouth, 5.2 cm, and of the bottom, 16 cm. The roof projects 2.2 cm over the jar, and is divided off by twenty ribs. Slight traces of the lines mentioned before as marking the tiles are still visible. Otherwise this specimen corresponds exactly to the previous one, also as regards the motive of the decoration on the body. Here, however, two rows or sections, separated by a group of three grooves, and filled in with compound spirals, are represented.

Plate XII illustrates three vessels more or less resembling the granary urn in shape, but all lacking in its essential characteristic, — the roof. Fig. 1 on this plate is a cylindrical jar 27.9 cm high, with a rim 0.9 cm wide and measur-

PLATE XI.

Decorated Granary Urns.

PLATE XII.

Pottery Vessels Related to the Type of Granary Urn.

ing around it 36.9 cm. The diameter of the opening is 9.6 cm; that of the bottom, 12.9 cm. The shoulders measure 52.9 cm around; and the base, 41.8 cm. These measurements show that the greatest circumference of the jar is around the uppermost groove, and that the vessel gradually narrows slightly from there towards the base. Another groove runs around the middle of the body. The neck of the jar is formed by a slanting portion (3.1 cm long), from which rises a narrow upright rim (1 cm wide). The original color of the glaze can hardly be defined now, as it has been much injured by underground chemical action, which has changed the original dark green partly into a light sea-green, partly into a yellowish-white. The clay is of a deep dark-red.

Another of these jars, measuring 30.4 cm in height, is shown in Fig. 3 of this plate. The diameter of its mouth is 10.4 cm; that of the bottom, 14.6 cm; and it has a rim 3 cm wide. The cylindrical part of the jar reaches a height of 19.3 cm, and is decorated with four grooves. It is continued upward into a constricted neck with concave outlines, surmounted by a flat projecting rim. On two opposite sides of this rim are two round perforations, probably for the passage of a cord by which to carry the vessel. The glaze, a bright green, is well preserved, and covers a space of 4.6 cm on the interior downward from the rim.

Fig. 2, Plate XII, represents a jar 21.1 cm high, with a diameter at the mouth of 6.2 cm, and at the bottom 10.7 cm. The periphery around the upper groove measures 44.9 cm; around the base, 36.8 cm. These measurements show that the shape is not exactly cylindrical, but that it narrows slightly towards the base. The upper part is dome-shaped. Around the body there are two belts or bands consisting of four deep parallel grooved lines. The jar, of reddish-gray clay, is covered with a light-brown or fawn-colored glaze intermixed with whitish, red, and black spots.

Plate XIII represents presumably also a vessel derived from the type of the granary urn, with which it has in common the straight cylindrical shape and the symmetrical division of the body by means of grooves. The neck and rim are of a form similar to that of the vessel represented in Fig. 1 of the preceding plate. The peculiar feature of this specimen is its square base. The four corners have been chamfered, so that its form, strictly speaking, is octagonal. The sides are 11.5 cm long. The bevelled corners or sides vary in length, two being 1.7 cm long, while the other two measure 1.2 cm and 1.6 cm respectively. Only one of the long sides forms a straight line, while the others run in a somewhat concave curve. This base is not of uniform thickness, the maximum being 2.2 cm in the middle, while towards the corners it is only from 1.8 cm to 2 cm thick. There are three extraordinarily deep grooves around this vessel, — two encircling the middle of the body, and one just below the neck. The sloping shoulder is 2.8 cm long; the straight neck rising from it, 1.4 cm high. The total height of the vessel is 32.1 cm. It is made of a light-reddish clay coated with a dull-green glaze, the original color of which seems to be entirely preserved, while its brilliancy has disappeared. The occurrence of this type at the time of the Han is noteworthy, because at later periods this form was imitated in porcelain.

A certain amount of similarity between this type of ancient Chinese granary urn and the so-called "hut-urns" found in middle Italy and northern Germany, and attributed to the

PLATE XIII.

Pottery Vessel Related to the Type of Granary Urn.

later bronze period, is obvious, though we must admit
fundamental differences between the two. The Chinese
idea of these vessels is autochthonous and a natural
outgrowth of their ancestral worship and the funeral observances connected with that belief. The deceased was in
need of food to continue his existence beyond the grave,
and this food was to be well preserved for a long time to
come. Solidly erected towers furnished the living generation
the means of storing up grain for times of famine or drought.
What, then, could have better served the purposes of the
grave than an adaptation of this durable structure to the
clay? The vessel for holding "the five kinds of grain"
thus naturally assumed the shape of the granary: it became
the storehouse on which the departed soul could rely for
its subsistence, imperishable and inexhaustible. Its firmness
and durability were not least expressed by the representation
of the tiled roof over the jar, which is eminently a Chinese
idea, for such roofs are never found on the European hut-
urn. All of the latter show, as MORSE justly remarks,[1] not
only a sloping roof, but one that is thatched as well, with
the characteristic cross-pieces on the ridge. This feature
forms the most striking difference in the two types. Another
point of variance has reference to the end they are to serve,
the European urns being cinerary vessels, the habitations of
the dead. Further, the Chinese jars represent exclusively
grain-towers, and not houses; while in Europe the form of
the house was imitated in round and square, mostly irregular
and crude shapes. The Chinese potter of the Han period,
however, was a full-fledged artist, to whom beauty of form
was a prevailing law. The conception of the tower, roughly
built of earth and brick, assumed under his clever hands a

---

[1] On the Older Forms of Terra-cotta Roofing-Tiles (Boston, 1892), p. 4.

well-proportioned, smooth, cylindrical form, systematically divided off by grooves at certain distances apart, modestly ornamented, set on three feet, and crowned with an adequate roof partially covering the opening of the jar, — in fact, everything, even the details, arranged, measured, and proportioned intentionally to form a work of art. The underlying ideas which the Chinese and European types, accordingly, share in common, are that of producing a vessel, and that of so shaping it that it shall at the same time represent a structure, — cognate conceptions originating entirely independently, but resulting, in spite of some outward coincidences, in different expressions of form.

### DRAW-WELL JARS.

Under the term "draw-well jar" I include three vessels which in shape remind one of a basket with a handle above. These become of intrinsic archæological value when we look upon them as representations in clay of a mechanical device for raising water from a well.

The most elaborate one, in which the idea of a well is strikingly obvious, is seen in Plate XIV, which represents a piece of pottery glazed a leaf-green. It is 40 cm high, the diameter of the mouth being 15 cm, and that of the bottom varying from 12.5 cm to 13.1 cm. The jar proper is 13.6 cm high, and has somewhat the form of an hour-glass, with a flaring rim 2.5 cm in width. On the front edge of the rim is a small water-bucket (5.6 cm in height) resembling in shape the one used nowadays in China. It is separately moulded, and stuck to the jar with its bottom resting on the rim, and its side against the slab back of it. The high structure over the jar, fastened to two opposite

PLATE XIV.

**Pottery Draw-Well Jar with Pulley and Water-Bucket.**

sides of the rim, represents the well-frame with the contrivance for drawing water. In the upper portion, enclosed in a frame, we notice the pulley, a small wheel with a deep groove in the circumference, over which the rope passes, from the ends of which the water-buckets are suspended. Above and below the pulley are two holes bored through the clay, indicating merely the spaces left between it and the frame. The pulley is protected by a tiled rectangular roof (7.5 cm by 5 cm) with gabled ends, having a central ridge-pole and two sloping ridges. The tile-work is formed in the same way as that on granary urns (see p. 53). The sides of the well-frame are represented in the pottery by two long slabs, each 19 cm long, 1.3-1.5 cm thick, 4.3 cm wide at the base, and 2.2 cm wide at the top. Two fantastic animal heads project over the sides at the upper end: they are perhaps dragon-heads,[1] and are somewhat in

[1] The "earliest known representation of the Chinese dragon" is seen by Dr. BUSHELL (Chinese Art, London, 1904, Vol. I, p. 37) in a bas-relief of the *Hsiao t'ang shan*, which he makes the first century before Christ, while it can be said with safety only that these bas-reliefs are somewhat earlier than A.D. 129. Besides, this conventional design of the dragon is hardly either the oldest or the oldest known. There are numerous bronze vases and bells of the *Chou* dynasty on which these monsters occur, and are at least explained as dragons in Chinese archæological books. Compare, also, the dragons carved into jade tablets of the *Chou* dynasty in *Ku yü t'u p'u*, Book 7, pp. 3, 5. Whatever their Chinese names or the interpretation of Chinese scholars may be, is not a matter of great consequence for a typological research; but first of all the iconography of all these various monsters must be minutely studied, and their types well defined, before anything final can be said regarding this motive. Particularly the question whether it is a Chinese invention, or derived from a foreign art, cannot be answered until this condition is fulfilled. On a relief of Gandhāra, GRÜNWEDEL (Alterthümer aus der Malakand- und Swat-Gegend, Sitzungsberichte der Preussischen Akademie der Wissenschaften zu Berlin, 1901, IX, p. 215; p. 14 of the separate print) found a type of dragon with long body, four distinct feet, wings with jagged edges, long neck, and horns on its forehead, exactly corresponding to the Chinese dragon, and entirely deviating from the usual forms of dragon-like creatures in the art of Gandhāra, all of which have fish-tails. The two front dragons (there are four all together) hold a cut jewel in their jaws, — a motive much employed in eastern Asia, where this jewel is known as the "dragon's saliva." The relief in question, Grünwedel thinks, might have gotten the dragon type from a Persian source, for also the reliefs on the tomb of *Wu liang*, which likewise show representations of

the style of gargoyles.[1]   They have evidently been shaped with the hand and stuck on, traces of the joining being still visible.

To make this specimen more intelligible, and to prove that it corresponds to a contrivance actually applied in Eastern culture, I here insert two illustrations from EDWARD S. MORSE's excellent book "Japanese Homes and their Surroundings" (New York, 1904), with the kind permission of the author and of the publishers, Messrs. Harper & Brothers.

---

dragons, doubtless display, according to his opinion, Persian forms of style, in spite of the contradictory view of the sinologues. "That the creature of the Chinese dragon, so fully developed in style, is said to be a production of the unimaginative Chinese," concludes Grünwedel, "was always a mystery to me. However, with the remarks here outlined I do not wish to have said anything positive, but only to have referred to this curious, extraordinary representation." To solve this question, it would be necessary to investigate whether the jewel-holding dragon is pre- or post-Buddhistic in China. It may very well be that certain dragon motives may have arrived in China with Buddhism; but that does not prove that all dragon forms of China, and particularly those of pre-Buddhistic times, have been adopted from India or Persia. We see that the Chinese represent lions (see Index), elephants, and monkeys before any contact with India; and, when Buddhism brought to them other designs of the same animals, they readily imitated them in the Indian style. One may compare the remarkable difference which exists between the figures of elephants supporting bronze vessels of the *Chou* time, or the elephant on the bas-reliefs of the *Hsiao t'ang shan*, and the later Chinese-Indian elephants. The representation of dragons might have undergone a similar course of development. Recently the same problem was touched upon by A. FOUCHER, in his book L'art gréco-bouddhique du Gandhāra (Paris, 1905, Vol. I, p. 214), who likewise sees in the Gandhāra dragons with the jewel "the probable origin of the eternal subject of Chinese art." For the rest, he holds the theory that the Chinese dragon has been borrowed from Assyria, which, in this form, I cannot approve of, unless it shall be proved to us by which way this motive may have spread from Assyria to China. Which is the historical and geographical route leading from Assyria to China? — that remains the question to be answered. And I am convinced that there never was such a direct way, since there is not the slightest evidence for it, and that alleged and real coincidences between Chinese and Assyrian art must be explained through the channels of intermediary provinces, particularly through the domain of ancient Siberian or Turkish art and that of Turkistan. On the dragon with the pearl, see particularly E. CHAVANNES, De l'expression des vœux dans l'art populaire chinois (Journal Asiatique, 1901, p. 193).

[1] In a similar manner two dragon-heads form the two ends of a chariot on the sculptures of the *Hsiao t'ang shan* (CHAVANNES, La sculpture sur pierre en Chine, p. 75 and Plate XXXVII, first panel); and from one side of a roof of a pavilion represented on a bas-relief of *Wu liang*, a dragon-head, apparently a wood-carving in reality, projects (ibid., Plate X, upper panel, to the right).

Fig. 14 shows a Japanese wooden well-frame erected over a well, to support a pulley in which runs a rope with a bucket attached to each end. Fig. 15 represents one of the old wells still seen in the Kaga Yashiki, in Tokio. These two wells are strikingly analogous to the piece of pottery figured above. We see here a pulley (in Fig. 14 in a frame, in Fig. 15 free) and a small roof over it, both of the same shape as there.[1] From the Japanese well in Fig. 14 we might draw some further conclusions with regard to the ideas which the potter meant to bring out in our specimen. "Japanese wells are made," Professor Morse remarks (p. 298), "in the shape of barrels, of stout staves five or six feet in height; wells of great depth are often sunk in this way; the well made in this manner has the appearance, as it projects above the ground, of an ordinary barrel or hogshead partially buried." I suppose that the jar in the above specimen

Fig. 14. Japanese Wooden Well-House (from Morse's "Japanese Homes and their Surroundings").

Fig. 15. Old Well seen in Kaga Yashiki, in Tokio (from Morse's "Japanese Homes and their Surroundings").

[1] See also the illustrations of wells in the excellent book of F. BALTZER, Das japanische Haus, eine bautechnische Studie (Berlin, 1903), p. 41.

5

is to symbolize the enclosure above ground around the
mouth of the well, which in China is frequently made of
stone, in the shape of a circular trough.[1]   From our specimen we can but presume that the well-frame in ancient
China was erected over the edge of the well-curb, while in
Japan it is built over the ground.   Another difference is,
that in Japan the pulley-frame is suspended from a horizontal
beam, while, on the strength of this and the two following
specimens, we are led to infer that in the Han time the
pulley was above this beam.   On the other hand, in an attempt to reconstruct the outward appearance of the framework of the Han period, which was no doubt also of wood,
we must certainly not rely too much on the pottery representation, as the potter, from his point of view, was compelled to modify the original appearance in some degree
to adapt it to the artistic idea of a handle, as is obvious
in the two following specimens.   Nevertheless we may safely
presume that the real framework was either of a similar
shape (as shown by the above sketches from Japan), or that
the vertical side-pillars were slanting, and united above by
an arch, as illustrated in the Han pottery (Plate XIV).

The question now arises, if the "pulley-well," as we shall
call it for convenience, was made use of in ancient China,
and is still used in Japan, can we trace its occurrence in
modern China or in Chinese literature?

There are three methods of raising water from a well
in the north of China nowadays.   The simplest and most
frequent way is by means of an oval wicker basket at the
end of a long rope.   The water is poured out of the basket
into wooden buckets, two of which are carried suspended

---

[1] A half-tone picture of such a stone well-curb is given in ARTHUR H. SMITH,
Village Life in China, on the plate facing p. 16.

from the ends of a pole.[1]  The second device is the well-sweep.[2]  A beam is suspended by a rope at its centre, in the manner of a scale-balance, from the horizontal branch of a tree.  One end is weighted with a stone, and to the other end the rope holding the water-bucket is fastened. When the sweep is at rest, the stone weight is on the ground and the bucket in the air; when the sweep is jerked upwards, the bucket is lowered into the well, and hoisted by winding up the rope.  This method is feasible, of course, only in shallow wells.

The book "*T'ien kung k'ai wu*"[3] contains a good sketch of a well-sweep made entirely of bamboo poles.  Even the rope is replaced here by a bamboo pole.

The oldest mention and description of the well-sweep — "a contrivance made of wood, heavy behind and light in front; it draws up water as you do with your hands, but in a constantly overflowing stream; you pull it, and down it comes; you release it, and up it goes" — is found in two passages of the philosopher CHUANG TZŬ, who lived during the last half of the fourth and the first half of the third century before Christ.[4]  The oldest representations of

---

[1] *Tan chang* 擔杖 or *pien t'iao* 扁挑, a carrying-stick with a chain and hook at each end.

[2] Called *chieh kao* 桔槹.  The description of it as given above is based on the illustration and text in the *San ts'ai t'u hui*, section on implements 器用, Book 10, p. 25.  The same engraving is copied in the *Nung chêng ch'üan shu*, *Shou shih t'ung k'ao*, and *T'u shu chi ch'êng*.  A sketch of a Japanese well-sweep will be found in E. S. MORSE, Japanese Homes and their Surroundings (New York, 1904), p. 73.

[3] 天工開物, a treatise on technology, by SUNG YING-HSING 宋應星, in 18 books.  Preface dated 1637.  The edition used by me was printed in Japan in 1771.  The synopsis of this work given by ST. JULIEN (Histoire et fabrication de la porcelaine chinoise, Paris, 1856, pp. viii, lxxi) is incomplete.

[4] See GILES, Chuang Tzŭ, Mystic, Moralist, and Social Reformer (London, 1889), pp. 147, 181.  Both passages are cited in the *T'u shu chi ch'êng*, in the chapter on well-sweeps.

well-sweeps are found on the bas-reliefs of the Han period, in the province of Shantung.[1]   There are three in the stone chamber of the *Wu* family in the district of *Chia hsiang* (Plates IV and XIV, 4th section on the right; and Plate XXIII in the book of CHAVANNES), and another one on the bas-reliefs of the *Hsiao t'ang shan* in the district of *Fei ch'êng*[2] (Plate XXXIX, last section but one on the left). The first three agree with one another.   The accompanying illustration, reproducing one of them (Fig. 16), is made from

Fig. 16.   Representation of a Well-Sweep, on a Han Bas-Relief of the *Hsiao t'ang shan* (from a rubbing).

a rubbing of the bas-relief, which corresponds to Plate XIV in the work of CHAVANNES.   The cross-beam of the well-sweep

---

[1] See ÉD. CHAVANNES, La sculpture sur pierre en Chine au temps des deux dynasties Han (Paris, 1893), pp. 20, 46, 58.

[2] The bas-reliefs on the hill called *Hsiao t'ang shan* 孝堂山, situated about sixty *li* northwest of the city *Fei ch'êng* in western Shantung, consist of eight stone slabs representing remnants of a mortuary chamber.   They are neither dated, nor accompanied with contemporary inscriptions; but the place was frequently visited by early pilgrims, who left dedicatory and laudatory inscriptions ( 善題字 ) incised in the stones.   The oldest of these is dated A.D. 129; so that these monuments must

moves in an oblong perforation in the upper part of the
vertical beam.   The counterbalance is a stone laid on the
end of the beam.   The rope with the bucket is plainly

---

have existed prior to this time which is the only safe argument that can be produced.
The authors of the *Kin shih so*, without apparent reason, refer them to the first century
B.C.; and both PALÉOLOGUE (L'art chinois, p. 132) and Dr. BUSHELL (Chinese Art,
London, 1904, Vol. I, p. 36) are inclined to accept this theory.   WANG CH'ANG,
however, the author of the *Kin shih tsui pien* (published in 1805), dates them from
the commencement of the first century A.D., depending chiefly on the close resemblance which these sculptures bear, in composition and style, to those of the tombs
of the *Wu* family.   CHAVANNES (La sculpture sur pierre en Chine, Paris, 1893, p. xxii)
joins him in this opinion, and Professor HIRTH (Fremde Einflüsse in der chinesischen
Kunst, München, 1896, p. 70) follows Chavannes.   At all events, the bas-reliefs of
the *Hsiao t'ang shan* are somewhat earlier than those at the foot of the *Wu tsė shan*,
which may be roughly dated about A.D. 150.   There is no reason to suppose, with
K. WOERMANN (Geschichte der Kunst aller Zeiten und Völker, Leipzig, 1900, Vol. I,
p. 524), that the difference in style between the two groups of monuments requires
their separation by a longer space of time; as such differences of style, if they really
exist, are quite compatible with contemporaneous productions, especially in such a
vast country as China, where such manifold manifestations of art, and such various
styles and influences acting from different quarters, have sprung up since earliest times.
Moreover, the difference in technique — the *Wu* bas-reliefs are flat low reliefs, raising
the figures two millimetres above the level of the stone, while the slabs of the *Hsiao
t'ang shan* are incised carvings — also caused a variation of style. — The stone chamber
石室 (colloquially 石頭房子) in which the *Hsiao t'ang shan* bas-reliefs
are arranged is said to be in a temple, *Yü huang tien* 玉皇店, on the summit
of the mountain, as rubbing-men in *T'ai an fu*, who had visited the place for business,
told me there.   There is a village at the foot of the mountain, called *Hsiao li pu*. —
The mountain is not indicated on Richthofen's map of Shantung, nor on the map of
the Prussian Geographical Survey, but it will be found on the Chinese maps of *T'ai
an fu* in the *Shan tung t'ung chi* and *T'ai an fu chi*. — In the village of *Liu*
劉村 (locally called *Liu chã' ts'un*), in *Chia hsiang hsien*, about two *li* northwest
from the village of *Hu t'ou*, there are three stone slabs, according to CHAVANNES (La
sculpture sur pierre en Chine, Paris, 1893, p. xxiii), one of which is reproduced in
his Plate XLIII.   Another is reproduced in the *Kin shih so*, section *Shih so*, Vol. 4.
At the time when I visited this place (Jan. 10, 1903), only one of these stone slabs
was left there, in the temple *Hung fu yüan* 洪福院, situated on a low height
at the northern end of the village.   As the representation on this stone is different
from the two, — that described by Chavannes and that in the *Kin shih so* respectively, — I presume it must be the third, described by neither.   The relief on the
stone in question is divided into three panels.   In the upper one there is, on the left,
a man in profile, holding a snake with both uplifted hands; on the right, a man with
sword in his right hand, and hammer-shaped axe in his left.   The middle panel represents three men squatting on the ground, raising their hands to salute.   Another man,
standing, with two attendants behind, is bowing to this group.   The lower panel

visible. We see, further, a well-curb, presumably of stone, of the hourglass shape.[1] A man is holding, with his left hand, a cylindrical jug on the edge of the curb, and is about to pour into it the water which a woman has just drawn up in the well-bucket.[2]

depicts a variant of the story about the search for a tripod vessel (see p. 75). This stone slab lies horizontally on the ground of the temple courtyard, and is exposed to the destructive influences of the air. On my inquiry for the other two slabs, I was shown another stone, likewise thrown on the ground, but with no traces of pictorial representations, exhibiting only a few effaced characters, of which some, like 天下 and 太平, were readable. Certainly it was not a relic of the Han time. As regards the third stone, there was a hazy recollection among some people that there had formerly been one, but it had been sold by the *Ho-shang* of the temple while in straitened circumstances. — The slabs of *Liu ts'un* agree in their technique with those of the *Hsiao t'ang shan*, and, like those, bear incised carvings.

[1] This or a similar form seems to have suggested to our potter the peculiar shape of the jar in Plate XIV; and the coincidence in motive — the jug on the edge of the well-curb — both on the Han bas-relief and in the Han pottery is also noteworthy.

[2] The rubbings taken from the bas-reliefs are gradually deteriorating, as seen from a detailed comparison of two sets of rubbings, obtained by me in China in 1901 and 1904 respectively, with those acquired by Professor Chavannes in 1891, which latter are in many particulars more distinct than mine. Again, collating Chavannes' plates with the woodcuts in the *Kin shih so* (published in 1821), the remarkable superiority of the latter, in completeness and clearness of detail with regard to the decorative elements, may be recognized. The reason is obvious. Previous to 1821, when the brothers *Fêng Yün-p'êng* and *Fêng Yün-Yüan* had access to these bas-reliefs, they were in a much better state of preservation. Carelessness on the part of the Chinese, together with the ravages of time, has considerably damaged them, and even caused partial destruction; and the action of the rain and humidity, penetrating the stone building through the soil below, has obliterated the under portions of the stone slabs just above the floor. At the time of my visit there (January, 1903), many slabs had fallen down, and lay scattered around on the floor. A second destructive element is the work of the professional rubbing-men 打碑的, who, after taking impressions of the stones, are too easy-going to clean off the ink, which naturally soaks through the paper. At last, from constant repetition of this process to fill the demand of the Chinese market, the stone reliefs gradually become covered with a thick layer of ink and dirt. Many of the finer lines, therefore, are now hidden under this mass. On this account, modern rubbings do not give exact and faithful facsimiles of the original reliefs: hence I think, that, for a minute study of the subject, we ought not to neglect to consult for our investigations the reproductions in the first edition of the *Kin shih so*, — although these are somewhat reduced in size and the drawings are deficient as to their relative proportions, and also in some cases arbitrary changes in details seem to have been made, — and avail ourselves of the modern rubbings chiefly for verifying the original measurements of objects. The blackness of the reproduction of the reliefs is unfortunate. This black appearance (although of course merely incidental, resulting

On Plate XXIII of CHAVANNES' work a well-sweep is shown which is incomplete, owing to extensive damage done to the stone in this part. The big blank appearing here on Chavannes' plate is indeed on all modern rubbings; but, fortunately, this whole scene is preserved complete in the *Kin shih so*, section *Shih so*, Vol. IV. We see there a well-sweep constructed in the same style as that in Fig. 16, and a well-curb of similar shape. Its upper part is ornamented with three consecutive circles. The well-sweep of the *Hsiao t'ang shan* is somewhat differently constructed. The vertical beam is surmounted by a fork, in which the balance moves; and the stone weight is stuck on to the end of the pole. The well is merely outlined as a rhomboid.

As to the location of wells, we recognize from the bas-reliefs that they were situated in close proximity to the kitchen. The preparation of a festive meal is represented on them several times. An animal, probably a pig, standing erect, is fastened to the vertical beam of the well-sweep, and a man is about to slaughter it. Ravens are fluttering around,

---

from the ink on the rubbing or print) may give rise to curious misconceptions, leading one rather to understand these reliefs as silhouettes or shadow-pictures. Nowadays the reliefs are in fact blackened all over, the result of the inking in taking the rubbings; but that should not deceive us as to their maiden appearance in bygone days. The original color of the stone of which they were sculptured, and which is still quarried in this region, was gray; and the impression gained by a scrutiny of this work on the spot is that of a real plastic relief fully expressive of life (and by no means of shadowy silhouettes), which is an essential feature entirely lacking in the Chinese rubbings. Hence it is inadequate to reproduce these reliefs in black, as this gives merely a reproduction of the rubbing, but not of the relief as such. It would be preferable to leave the black of the rubbings blank, and to sketch only the engraved lines, i.e., those which appear white in the rubbing; the figure thus becoming the primary design, after which the relief was subsequently executed; for it was undoubtedly first drawn by the artist of the Han time, which design was then pasted over the stone slab, the outlines of the sketch being chiselled in the stone. An attempt to reflect the plastic character of the reliefs in gray has been made by HELMOLT, in his Weltgeschichte, Vol. II, plate opposite p. 67.

as if eagerly waiting for their share in the spoils. From the other side a man is swiftly running towards the well with arms outstretched, to receive, for transportation to the kitchen-room, the water which a woman has raised. In the courtyard in front of the house, people are busy bringing in dressed fowl, and washing them in a tub; while in the kitchen a woman is squatting down in front of the range, as may be seen in Fig. 22, p. 87.

The third contrivance, used in deep wells, is the winch or windlass.[1] According to the illustration and description given in the "*San ts'ai t'u hui*" (section on implements 器用, Book 10, p. 16), both of which have been merely copied in the later books on agriculture, the axis rests in the angles formed by the intersection of the wooden bars in each of the two pairs in the framework. It is weighted down with a stone at one end; and on the opposite end is the barrel or sheave[2] wound around with the rope, one end of which is tied to the handle of the water-bucket. The barrel is rotated by means of a wooden crank-handle[3] inserted from above, and manipulated with both hands of a man. Two ropes also are used at the same time; so that, of the two buckets suspended from them, the empty one descends and the full one is hoisted. This type of well I have frequently observed in the province of Shensi. There, at least in the cases which came under my notice, the axis or winding-

---

[1] Called *lu lu* 轆轤. Most of our dictionaries confound the terms "pulley" and "windlass," and attribute both meanings to *lu lu*. The pulley, however, is called *hua ch'ê* 滑車, as may be seen in the treatise on mechanics by TERENCE (quoted on p. 19), who always makes, and correctly, a strict distinction between the two (see also A. WYLIE, Terms used in Mechanics, in J. DOOLITTLE, A Vocabulary and Handbook of the Chinese Language, Foochow, 1872, Vol. II, p. 177).

[2] Called *ch'ang ku* 長轂, a long wheel-nave.

[3] Called *ch'ü mu* 曲木.

roller is not supported by a wooden frame, nor has it any support from below, but it is driven horizontally into a heap of earth or stones or into a wall erected at its side. The stone weight is not suspended from the axis by means of a rope, but is flat, round, or square, and has a perforation through which the axis passes. Sometimes, in addition, the axis is supported by a bowlder laid under it on the ground. Occasionally I also saw winches without any stone weights or supports. There are two other slight points of variation from the windlass of the "*San ts'ai t'u hui.*" A short iron chain consisting of narrow oblong links is wound around the sheave, and continued into a long rope. The crank is attached to the end of the sheave, much shorter, and turned slightly upwards, which makes handling more convenient and quicker.[1] I never noticed the use of the pulley-well in China,[2] however, which fact of course is not sufficient to warrant the supposition that it does not exist. Neither in Chinese nor in European literature can I find any reference to this subject; but there is no doubt that the pulley itself is well known in China. As Sir JOHN FRANCIS DAVIS, in his now scarce book "China: A General Description of that Empire and its Inhabitants" (London, 1857, Vol. II, pp. 255, 256), has given a good sketch and description of it, I here reproduce both (Fig. 17).

"The pulley is applied on board their vessels, but always with a single sheave, and apparently more for the purpose

---

[1] A. FORKE (Von Peking nach Ch'ang an und Lo yang, Mitteilungen des Seminars für orientalische Sprachen, Vol. I, Sect. 1, p. 102) illustrates a field-well consisting of four winches, the axles of which are arranged around a wooden stake erected over the opening of the well.

[2] SARAT CHANDRA DAS, in his Journey to Lhasa and Central Tibet (ed. by W. W. Rockhill, London, 1904), p. 118, mentions wells in Tibet, where water-carriers were drawing water in hide buckets attached to a rope about 150 feet long, passing over a pulley.

of giving a particular lead to the ropes than with a view to the mechanical advantage to be gained by it. . . . It is remarkable that they should seem always to have possessed that particular application of the principle of the wheel and axle by which the greatest power is attained within the least space, and at the same time with the greatest simplicity, as well as strength of machinery. The cylinder consists of two parts of unequal diameter, with a rope coiled round both

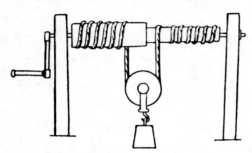

parts in the same direction, the weight to be moved being suspended by a pulley in the middle. Every turn of the cylinder raises a portion of the rope equal to the circumference of the thicker part, but at the same time lets down a portion equal to

Fig. 17.   Sketch of a Pulley (from Sir John Francis Davis's "China").

the circumference of the thinner; and, as the weight is suspended by a pulley, it rises at each turn through a space equal to only half the difference between the span of the thicker and thinner parts of the cylinder. The action of the machine, therefore, is very slow; but the mechanical advantage is great in proportion, or, in other words, 'power is gained at the expense of velocity,' according to an invariable law of mechanics."

Another modern application of the pulley occurs in fishing-nets suspended from a wooden frame, with a wheel and axle to lower and raise the net.[1]

Notwithstanding the lack of historical data regarding the pulley-well, we can nevertheless read in the piece of pottery shown in Plate XIV, as from an ancient document, that, as

---

[1] JAMES LEGGE, The Chinese Classics, Vol. IV, Prolegomena, p. 149.

early as the Han period, the mechanical contrivance of the pulley was known to the Chinese, and employed in the raising of water from the well,[1] and that even then China took a high place in the appliance of mechanical arts.[2] This conclusion represents the most important fact gained from an investigation of this piece of pottery. Aside from its high scientific value, it is a most exquisite work of art. From the presence of the two gargoyles it must be concluded that wells of that time were artistically decorated with such figures. This fact throws an entirely new light on the much misjudged art of the Han period. As this ceramic object must be considered as having been interred with the dead, we may presume that it was buried there with the idea of symbolizing a well to furnish water for the departed.

An interesting application of the pulley in connection with another contrivance in the Han time is illustrated on one of the bas-reliefs of the *Hsiao t'ang shan*. A tripod-vessel has been drawn from the depths of a river, and has to be hauled ashore out of the boat in which it was first placed. The banks of the stream are rugged, and walled up with row upon row of stones.[3] On either side of the river are four men pulling a long rope which runs through two pulleys, each on top of a pole.[4] In the scene executed on the

---

[1] On this ground we are also compelled to admit that the Japanese must have adopted the pulley and its appliances from China.

[2] "In the science of mechanics and machinery, the Chinese, without possessing any theoretical rules, practically apply all the mechanical powers, except the screw, with considerable effect." — Sir JOHN F. DAVIS, China: A General Description of that Empire and its Inhabitants (London, 1857), Vol. II, p. 255.

[3] 一 河 兩 岸 壘 石 爲 礎, as the authors of the *Kin shih so* (*Shih so*, Vol. I) say in their explanation of this bas-relief. For further details regarding this subject of the search for the tripod, see ÉD. CHAVANNES, l.c., pp. 58, 78; and Les mémoires historiques de Se-ma Ts'ien, Vol. I, p. 296; Vol. II, pp. 94, 154; Vol. III, pp. 460, 482, et seq.

[4] The authors of the *Kin shih so* describe this affair somewhat stiffly by saying:

relief, the tripod-vessel has almost reached the level of the pulleys, when, just at the moment when the workmen are naturally making their supreme effort, the rope yields, and one ear of the vessel breaks off. It threatens to fall, and a boatman in the river, in the twinkling of an eye, pushes his pole against it to support it. This scene is here reproduced (Plate xv) from the *Kin shih so*. On Plate XL in the book of Chavannes, which reproduces the same bas-relief, the pulley does not come out clearly. In fact, the artist has represented a pulley only on the left beam, and has neglected to draw any on the right one. However incorrect the drawing may be, there can be no doubt that the object over the left beam, through which the rope passes, is meant to represent a pulley, and, naturally, this whole operation could have been effected by means of no other contrivance. This bas-relief representation and the moulding of the pulley in the Han pottery corroborate each other, and confirm our view that this piece of mechanics existed at that early date.

This fact is not surprising, considering the extensive work of engineering in river-channels and wells for the irrigation of fields, undertaken by the Han Emperor *Wu* (140–85 B.C.), as we read in Ssŭ-ma Ch'ien's history.[1] At that time there lived professional hydraulic engineers, and then for the first time a canal was built with wells bored at intervals and intercommunicating below. The deepest of these wells were sunk to a depth of about four hundred feet.

---

兩邊各四人曳之。其繩貫于植木之上。穿孔而
出。 "On both sides four men pull it; their rope runs through the upper part of a stake perforated by a hole, out of which the rope comes."

[1] See CHAVANNES, Les mémoires historiques de Se-ma Ts'ien, Vol. III, pp 525–537.

PLATE XV.

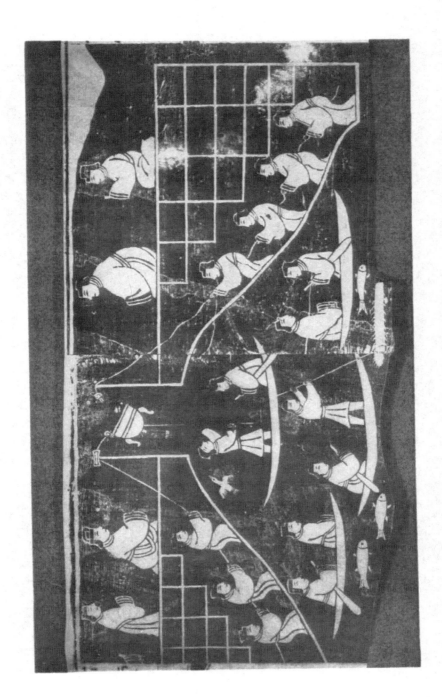

The Search for the Tripod Vessel (on a Han Bas-Relief), to show Application of Pulley.

PLATE XVI.

Pottery Draw-Well Jars.

The vessel shown in Fig. 1, Plate xvi, is of cylindrical shape, and has a flat projecting rim 2.6 cm in width. Its total height to top of the handle is 31.3 cm; height to the rim only, 15.9 cm; and the mouth measures 14.5 cm in diameter. The surface is decorated with three grooves, the lowest one just along the bottom edge, the uppermost being a double groove. From the rim rises a semicircular handle (round in cross-section), which is surmounted by a structure (6 cm high by 5.8 cm wide) that appears to be an imitation of the roofed pulley. The pulley is not as distinctly moulded as in the foregoing specimen. Its two sides have been torn asunder, so to speak, so that they resemble warts; and two round holes are bored just where we should expect to see the groove of the wheel. The frame enclosing the sheave, together with the roof, has more the character of a building (the well-house) than that in Plate xiv, in that the vertical framework on either side seems to be so formed as to look like a pair of wooden pillars surmounted by a horizontal cross-beam. While the roof in the previous figure over-shadows the pulley-frame with its flat and straight under surface, this roof closely abuts on the house below. It also lacks the gable ends; and the two ridges, which do not project so far, form eaves, indicated by the row of tile-work beneath them. The whole vessel, except its interior, is glazed a dark green, and covered with a network of fine crackles. A *Su hua* (colloquial) name for this vessel, which I obtained at *Hsi an fu*, is *t'i lung* 提籠 ("basket to lift"). The handle is called *t'i liang* 提梁, as is the case also in archæological literature.

The piece of pottery represented in Fig. 3, Plate xvi, is of the same type as the preceding. Its total height to top of the handle is 25 cm; height to rim, 13.6 cm; diameter of mouth, 13 cm; width of rim, 2.7 cm. The handle is not so high, the curve being more that of an ellipse. The pulley-frame has a trapezoidal shape, and the pulley itself is much like that in Plate XIV. Here, however, there are no perforations, which is the principal difference between this piece and the other two, and makes this type an instructive example of advanced conventionalization. The roof is without gables, and the ridges project only slightly. The glaze, originally a very light leaf-green, has undergone a change, being now a magnificent iridescent gold and silver that completely covers its exterior. That it was long buried under the ground is indicated by the thick layers of extremely hard loess incrustations on the inner walls and bottom.

The ceramic object seen in Fig. 2, Plate XVI, is 14.5 cm high to the top of the handle, and 8.9 cm to the rim; the diameter of the mouth being 7 cm, and of the bottom 10.9 cm. Whether it should be classed with the three preceding vessels may at first sight seem questionable, as it lacks their essential characteristic, the roofed pulley. The arched handle attached to the flat projecting rim makes it approach in some degree those represented in Figs. 1 and 3 on the same plate. The form of the vessel, and the handle, might suggest its derivation from a basket. The lower portion only of the handle is round, the top and sides having been flattened out, probably by means of a bamboo stick. The flat area on top is rectangular in form, and is left unglazed; and it would almost seem as if there had once been a top-piece here with a pulley and roof, as in the preceding objects, but which is now broken off. Otherwise the glaze covers the entire handle. It is curious that the inside of the vessel should be also glazed on the bottom only, but not on the sides. The glaze, originally dark green, as still visible on a few spots, has changed into a silver-white gloss with iridescent flecks scattered here and there. To the bottom of this jar inside, a small jug lying on its side has been stuck, apparently representing a bucket at the bottom of the well. To show this peculiar feature, a sectional drawing of this piece is here reproduced (Fig. 18).

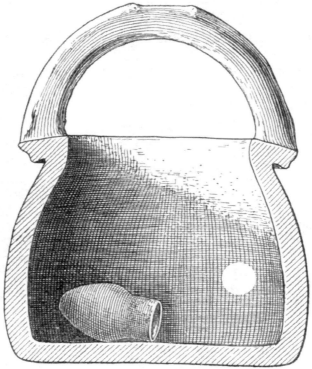

Fig. 18. Section of a Pottery Draw-Well Jar, showing Bucket at Bottom (from specimen in Clarke collection).

### COOKING—RANGES.

The burial of clay cooking-stoves (*wa tsao* 瓦竈) in the imperial graves of the Han period is expressly mentioned in the "Annals of the Later Han Dynasty."[1] Two were used for the Emperor, but there can be no doubt that they were

---

[1] De Groot, The Religious System of China, Vol. II, p. 403.

then a favorite mortuary object also for all classes of people. As we see from the description of the six following specimens, they are models of large kitchen ranges, much reduced in size, and have never been intended for any practical use. This is definite proof that they could be nothing but mortuary outfits, which in their technique and ornamentation bear the unquestionable stamp of the Han time, which is unequivocally confirmed by nearly coeval historical records.

Fig. 19. Miniature Pottery Kitchen Stove (from specimen in Clarke collection).

There are two principal forms of such ranges, — quadrangular ones, and those shaped like an elongated horseshoe.

Fig. 19 shows a miniature kitchen stove of rectangular shape. It is 12.5 cm high to base of chimney, and 21.7 cm to top of chimney. The sides measure 29.1 cm and 22.5 cm in length. The clay of which it is made is 0.6–1.2 cm thick. It is covered with an olive-green glaze, with earth incrustations partially adhering to it. The interior is not glazed, and shows a dark-reddish clay. On the top surface there is a large hole (diameter, 13.3 cm), over which a clay cooking-vessel with curved handle was found, the only example of this kind hitherto known. Since it exactly fits the opening, it undoubtedly belongs to this stove. For a description of this piece see p. 104. There is a chimney at one end, which juts slightly over the front edge. It is hollow, the smoke-hole having a diameter of 1.4 cm. On the side opposite the

Fig. 20. Pottery Stove, seen from the Top.

chimney is an opening cut out very unevenly and of rectangular form; the long sides being 6.3 cm, and the short sides 4.6 cm, in length. This evidently symbolizes the door through which the fuel is put.

Fig. 20 shows a rectangular stove, seen from the top, the largest of its type. It is 45.6–46 cm long, 35.7–36.1 cm wide, and 16 cm high. There are three cooking-holes, —

6

the single one having a diameter of 11 cm; the other
two, of 8.1 cm and 8.5 cm respectively, — each covered
underneath with a bowl.  The entrance to the fire-chamber
is horseshoe-shaped, being 9.1 cm wide at the base, and
7.4 cm high.  Above this entrance, on the edge of the
stove, a vertical rectangular wall (10.6 cm long, 3.2 cm
high, and 1.1 cm thick) rises.  On the opposite side, back
of the largest cooking-hole, there is a round perforation,
1 cm in diameter, probably to denote the place of the
smoke-hole.  The glaze is light green, crackled, and has
dissolved in many places into a silvery white.  The piece
seems to have been in the furnace at least three times, as
there are spur-marks on three sides, — three on one of the
long sides, eight on the front side, and two on the side
opposite the latter.

On Plate XVII, three stoves of the elongated horseshoe
form are illustrated.  They have no bottoms, and are hollow.
Each has two cooking-holes with sides slanting up from
the surface of the stove.  On the under side, bowl-shaped
bottoms have been stuck on, as in the previous specimen, to
fit over the holes from below, the whole thus forming two
receptacles.  The three specimens show kindred forms of
ornamentation.  Fig. 2 represents a stove, unglazed, and
made of grayish clay.  It is 8.6 cm high, 19.8 cm long,
and has a maximum width of 15.6 cm.  The opening to
the fire-chamber (7.3 cm by 5 cm) is cut out in rectangular
form down to the bottom edge of the front end.  The diameters of the two receptacles measure over the opening 4.2
cm and 3.2 cm respectively.  There is a small knob at the
rear, presumably to denote the place of the chimney.  The
ornaments are all in low relief.  On either side of the door
to the fire-room (not visible in the illustration) is a trotting

PLATE XVII.

3

2

1

Pottery Stoves.

tiger, the two running in the same direction.   On the surface, on the right side, are several animals, a quadruped,
and some birds, but much effaced and indistinct.   Between
the two cooking-holes is an implement resembling a brush
for cleaning pots, such as is still used in China.   On the
other side we see a ring with a dot in the centre; an object
consisting of a raised rectangular piece with short handle,
which I think is a large kitchen knife, represented also on
the Han bas-reliefs, and a most common type of knife everywhere in the China of nowadays; also two concentric circles
filled with radii; an implement, probably a poker; an object
which I believe is the representation of an inverted spoon
(the backs of spoons of the Han time have exactly the same
outlines and appearance as this ornament); and a rectangle
containing two fishes[1] side by side, and eight domino-like
dots arranged in two rows of four each.   The rectangle might
represent a tray on which the fishes were placed.

The next stove (Fig. 1, Plate XVII) is 9 cm high, 18.9 cm
long, and 16.2 cm wide.   The rectangular opening on the
front (not visible in the illustration) is 5.9 cm by 4.2 cm.
The recesses are 4.8 cm deep, with respective diameters of
3.2 and 3 cm.   The place of the chimney is indicated by a
round knob at the rear end.   Around the top surface is a

---

[1] The fish is indubitably one of the most ancient ornamental motives in Chinese
art.   I have here inserted (Fig. 21) a Han bronze vase after the *Hsi ch'ing ku chien*
(Book 21, p. 19), called "vase with wild ducks and fishes," showing ducks holding
eels in their bills, and others with fishes in front of them, besides rows of swimming
fishes (probably carp) with tortoises interspersed.   This will also illustrate the prototype and presumable origin of the fish-holding cock in the ornamentation of the Amur
tribes (B. LAUFER, The Decorative Art of the Amur Tribes: see, for example, Plate VI,
Fig. 2; Plate XXVIII, Fig. 1 a; Plate XXX, Figs. 1 a, 2 a, etc.).

The motive of the double or twin fishes 雙魚, which plays a prominent part
on the bronze *hsi* 洗 of the Han, is so well known that I need not refer particularly to it.   Compare also the passage in *Hsi ch'ing ku chien*, Book 21, p. 1 b, where
the fish on a Han vase is stated to be different from those on the fish-cauldrons

row of geometric ornaments in low relief, principally based on the figure of rhomboids, one within another.   Of the ob-

漢
麀
魚
壺

Fig. 21.  Bronze Vase of Han Time (from *Hsi ch'ing ku chien*).

魚 鼎 of the *Chou* time.  On the Han bas-reliefs a great variety of fishes is well outlined; and the cleverness of Chinese artists in drawing fish already appears there ni such full display that almost the single species can easily be discriminated.

PLATE XVIII.

Pottery Stove with Bronze Kettle.

jects in relief, there are, on the left side, three pieces resembling inverted spoons; on the other side, three concentric circles painted under the glaze and an instrument not unlike a mallet. The glaze is a deep lustrous green over a light-red clay, with many earth and loess incrustations.

The third stove on this plate (Fig. 3) is 9.3 cm high and 20.4 cm long, with a maximum width of 16.6 cm. It is very similar to that seen in Fig. 1. The knob is higher and pointed, and closer to the cooking-hole. On the right side are three inverted spoons and a ring in which two parallel vertical lines are crossed by three horizontal lines; on the left side, an object resembling a poker, an empty ring, a spoon, and a flat knob encircled by a ring. The glaze is a light green, which has partly dissolved into a silvery iridescence, particularly along the sides.

The stove in the next illustration (Plate XVIII) measures 13 cm in height, including the feet, and 8.8 cm without them. It is 30.2 cm long and 14.7 cm wide. It differs from the three preceding stoves in having a bottom raised on four feet in the form of squatting bears. This bottom extends out at the base into a hearth 8 cm in length, from which a square opening leads into the fire-chamber. This entrance has a slanting arm at each side, of the same form as those seen nowadays (only made of earth) in front of the doors of Chinese houses. A rhomboid pattern runs above the door, the three middle ones with their diagonals represented. Along either side of the door are two loop-shaped ornaments.

The front recess is 6.4 cm deep and 4.2 cm in diameter over the opening. With it was found a small bronze kettle fitting it, which has a projecting rim around the middle of the body, like the bronze type known as *fu* 鍑, and is covered with spots of light-green and dark-blue patina.

The rear receptacle is 5.6 cm deep, and has a diameter of 3.5 cm. Behind it there is a chimney 3 cm in height. The decorations on the surface are clearer or better preserved than those on the stoves in Plate XVII. A band of rhomboids runs along the front edge, the three in the centre having raised knobs inside. On each side of the first cooking-hole we notice a fish. Between the two cooking-holes is a poker. On one side of this is an inverted spoon; on the other, an empty ring. Back of the ring is an object much resembling a brush for cleaning pots, like that alluded to on p. 83. Further, there are on the opposite side a raised ring with radii and a mallet-like object.

This stove is made of a light-grayish clay, and is unglazed. Feet, bottom, sides, and interior are full of hardened loess incrustations, which bear full evidence of its long burial under ground.

The general signification of the ornaments on these stoves was suggested to me as follows. On the bas-reliefs of the *Wu tsê shan* we find the representation of the interior view of a kitchen of the Han time, which is here reproduced (Fig. 22) after the "*Kin shih so,*" section *shih so*, Vol. 4. There a large range of quadrangular shape is seen resting on a base with slanting side. This range is also provided with two receptacles with sides rising above the surface in a bowl-like shape, exactly as in our specimens of pottery. Over the front receptacle there is a low pan with an object protruding from it, which is nothing else but the curved handle of a *spoon*. Over the other one is a large covered kettle. A woman holding a *poker* in her left hand is squatting in front of the stove. A *fish* is suspended from the capital of the pillar, and two dressed fowl and one pig's head are hanging from the ceiling of the room. Another object, at the left of the pig's head, I cannot define. Generally I am inclined to presume that the objects moulded on the four stoves are paraphernalia or appurtenances of the kitchen; that the animals, birds, and fishes are there simply to

Fig. 22. Interior View of a Kitchen of the Han Time, as represented on a Han Bas-Relief of the *Wu tsê shan* (from *Kin shih so*).

represent articles of food for cooking; that the long implement is a poker for stirring up the fire, as we see, from the bas-relief, was then the practice; and that the rings with radii may symbolize covers for the kettles, or pieces of matting on which to place vessels.

Certainly the kitchen ranges in actual use during the Han
time were not ornamented at all, but were plain, like the

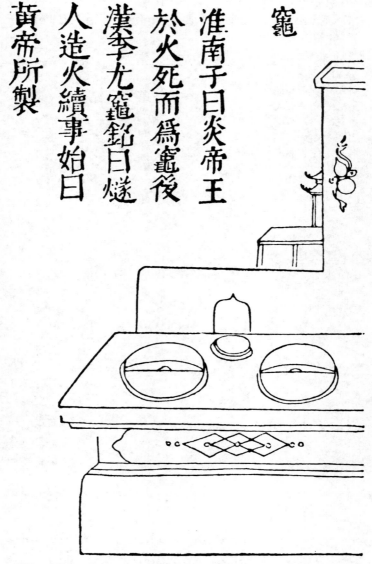

Fig. 23.  Engraving of Kitchen Range (from *San ts'ai t'u hui*).

stoves in Figs. 19 and 20.  The decoration was an addition
to the mortuary models only, with the intention of producing

a more realistic picture of the kitchen furniture on the range, and thus of increasing the number of utensils useful to the dead.

Fig. 23 affords a view of the form of a modern kitchen range, taken from the original edition (1607) of the cyclopædia "*San ts'ai t'u hui*," section on architecture 宮室, Book 1, which pictures a kitchen (p. 25 a) and also a kitchen range (p. 25 b). The text over the engraving reads as follows: "*Huai Nan tzŭ* (philosopher of the second century before Christ) says, '*Yen ti* (i.e., the Emperor *Shên nung*, 2838–2698 B.C.) ruled by virtue of the fire, and when dying made kitchen ranges.' The *Tsao ming* of *Li yu* of the Later Han says, '*Sui jên* created the fire, but the first to continue his work [by making kitchen ranges] was *Huang ti* (2698–2598 B.C.)'"[1]

---

[1] *K'o chih ching yüan* (Book 19, p. 16 a) gives a quotation from the *Li yu tsao ming* thus: "*Sui jên* made fire and kitchen ranges;" and from *Huai Nan-tzŭ*, "*Huang ti* made kitchen ranges, and became after his death god of the kitchen ranges." On *Yen ti* see CHAVANNES, Les mémoires historiques de Se-ma Ts'ien, Vol. I, pp. 7, 13; on *Sui-jên*, ibid., p. 5. The latter name means "man of the fire-drill," and personifies the invention of fire-making by friction. — An illustration of a modern cooking-stove built of brick may be seen in A. H. SMITH, Village Life in China, p. 262. See also H. E. M. JAMES, The Long White Mountain (London, 1888), p. 236.

# IV. — DESCRIPTION OF HAN POTTERY. — VESSELS.

## BRAZIERS.

As to means of heating and cooking, models of kitchen stoves have already been discussed. Here three specimens remain to be mentioned, two of which served the purpose of warming only, and one very likely also that of cooking. These, however, are no artificial models of larger stoves, but represent the real objects in actual use. It therefore seemed best to include them in this section.

The pottery object shown in Fig. 24 evidently represents

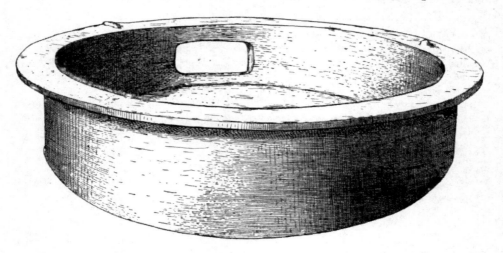

Fig. 24. Pottery Fire-Pan (from specimen in Clarke collection).

a fire-pan. It is 6 cm high and 21 cm in diameter, having a rim 1.7 cm in width. There is a rectangular opening (5.6 cm by 2.4 cm) in the side-wall for the exit of ashes. The bottom of the pan is made convex so that the ashes

90

PLATE XIX.

Pottery Brazier.

will fall gradually from the centre towards the wall of the
vessel. The whole pan, with the exception of the under side
of the bottom, is glazed a light green, the greater part of
which has dissolved into a silver and gold color. The rim
is covered with a network of fine crackles. Such braziers,
but of somewhat larger size, are still in use in many parts
of China. As a rule, they are made of iron, but frequently
also of clay, as I noticed in Nanking and on a winter journey through Shantung Province.

Plate XIX represents a brazier that consists of a cylindrical
jar (10.5 cm in height and 14.3 cm in diameter) and a separate part (shown above the jar in the illustration). The latter
fits exactly into the opening of the jar, its prominent rim
(1.2 cm wide) resting closely over the edge of the vessel.
In the bottom of this upper part — which is 6.5 cm deep,
and gradually narrows below, looking, when inverted, like a
truncated cone — are six holes (five around a central one,
one of which is not visible in the illustration), each about
1 cm in diameter. This upper part is entirely glazed on both
sides; but the shade of green is lighter than that on the jar,
which is not only much darker but deeper. The light spaces
in the illustration represent spots where the color has changed
to a golden iridescence. This vessel seems to have served
as a sort of brazier or portable stove, the charcoal being
placed in the upper part, the embers dropping through the
perforations to the bottom of the jar, from which they could
easily be removed by lifting the coal-holder.[1] Among cognate
heating-devices in modern China, I know of nothing that
exactly corresponds to this type.

---

[1] The *Su hua* (colloquial) name by which this piece was called in *Hsi an fu* is
*tai ti kuan* 代 屉 罐.

The piece of pottery represented in Fig. 1, Plate xx, is apparently a charcoal-stove, consisting of a globular bowl that continues below into a narrow cylindrical foot. It is 13.5 cm high, and has a diameter at the opening of 14.2 cm, and through the foot of 10.1 cm. Projecting upwards from the rim are three triangular prongs (about 1.5 cm long), presumably for holding in place a cooking-vessel. This I conclude from the fact that in central China a portable red earthenware cooking-stove is still in use which has three similar prongs on the rim for the purpose stated. In the foot there is a dome-shaped hole (4 cm wide and 3.5 cm high) for the escape of ashes. On the inner side of the bottom is a simple spiral in raised lines, partially surrounded by a half-circle.[1]

### COOKING—VESSELS.

In Fig. 2, Plate xx, we see a bowl-shaped vessel of Han pottery with oblique handle terminating in an animal's head, much resembling the cooking-vessel found on the stove shown in Fig. 19. To obtain a clear understanding of this type, we must first discuss two related bronze types of the same period. These are the *chiao tou* 鐎斗 and the *tiao tou* 刁斗. *Tou* 斗 is a dry measure, generally called a "peck," the capacity of which has varied much, and still varies, in

---

[1] The term *lu* 爐 is somewhat vague, as it means an apparatus for burning and heating in general, including stoves, furnaces, braziers, and incense-burners. It is not possible therefore in every case to define what is meant in particular. For example, I find in the Chronicles of *Shun t'ien fu* (Peking), section epigraphy 金石, Book 128, p. 27 a, a passage relating how, in 1681, a grave inside of the *Hsi an mên* (one of the gates of the capital) was accidentally opened, and one clay stove (*wa lu*), one clay jar, and two tombstones, were discovered in it. One of the last-named had an inscription that revealed the date 794. This would be interesting evidence of the burial of stoves in graves during the *T'ang* dynasty, if we were only able to say definitely what the general expression *lu* is meant for in that passage, — very likely a single brazier or coal-pan, as, in case a censer or some more elaborate piece were intended, the author would have determined it by some additional restriction.

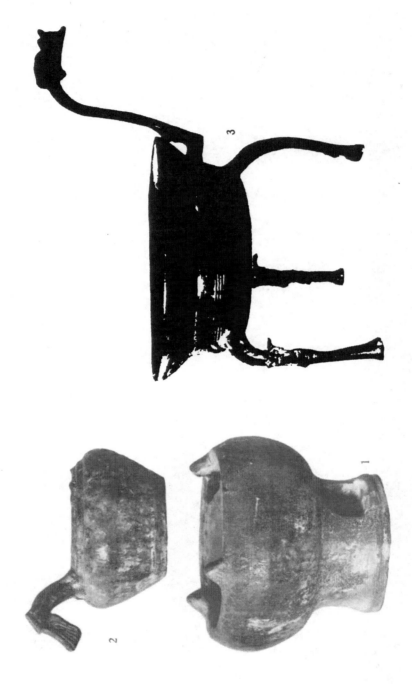

Fig. 1. Pottery Stove; Fig. 2. Pottery Cooking-Vessel; Fig. 3. Bronze Tripod Cooking-Vessel.

different parts of the country, the most usual now containing a little more than ten litres.[1]  By measuring the capacity of those vessels of the Han time which bear an inscription stating their capacity (see chapter on inscriptions), I have come to the conclusion that one *tou* of the Han period generally contained about 8000 cubic cm (8 litres); and accordingly one *shêng* (one-tenth of a *tou*), four-fifths of a litre.  Even then, a great variety of these measures must have been in vogue, and from many passages we are compelled to infer that there were *tou* of considerably smaller dimensions: for example, in a passage of the "*Shih chih*" 史 記,[2] where the consumption of seven or eight *tou* of wine appears permissible; or when *K'ung Ying-ta*,[3] commenting on a passage in the "*Li ki*," regards one *tou* of meat and one *tou* of wine as the average daily ration of a man.[4]

By the term *chiao tou*, two kinds of copper or bronze vessels are understood: (1) a vessel provided with three feet and a handle, and serving to cook food in; (2) a cooking-pan, used in camps by soldiers for preparing their food in by day, and for striking the watch by night.  The latter vessel is also called *tiao tou*.  To avoid confusion, I have restricted the term *chiao tou* to the tripod cooking-vessel, and *tiao tou* to the cooking-pan without feet.  This is regarded by Chinese authors as the principal difference between the two, as is expressly stated in the "*Tung t'ien ch'ing lu*,"[5] —

---

[1] See the tables in the article Currency and Measures in China (Journal of the China Branch of the Royal Asiatic Society, Vol. XXIV, Shanghai, 1890, pp. 90–94).

[2] Quoted by COUVREUR, Dictionnaire chinois-français, 1890, p. 856 a.

[3] A.D. 574–648 (see GILES, Biographical Dictionary, p. 404).

[4] See COUVREUR, Li ki, Vol. I, p. 547.

[5] 洞 天 清 錄, a work by CHAO HSI-KU of the thirteenth century (see WYLIE, Notes on Chinese Literature, 2d ed., p. 167; M. COURANT, Catalogue des livres chinois, Vol. I, p. 67).

that the *tiao tou* have no feet, while the *chiao tou* are provided with feet.[1]

Let us first consider the *chiao tou* made of bronze. Fig. 3, Plate xx, shows a specimen obtained in *Hsi an fu*. The total height up to the head of the animal is 24.3 cm; up to the rim of the vessel, 16 cm; the height of the feet being 11.2 cm. The diameter of the mouth is 19.8 cm; the depth of the vessel, 7.8 cm. The copper material is covered with black, reddish, and green spots. According to the verbal explanation of a Chinese archæologist at *Hsi an*, the animal forming the handle is "the scaly dragon" (*chiao* 蛟); and the vessel was used like a ladle, for scooping water, the long neck of the *chiao* serving as handle. This type comes near to that depicted in the "*Hsi ch'ing ku chien*," under Fig. 25, except that the spout is absent. The animal's neck and feet are curved in a different manner. The neck is joined to the vessel by means of two small parallel pieces, but the whole is made in one cast. The mouth of the monster is wide open. The feet are rounded out, not convex as in the illustrations of the *Ch'ien lung* Catalogue, and the lower ends are evidently worked into hoofs. Around the body of the vessel are four parallel raised lines, the so-called "girdle-

---

[1] 刀斗無足。鐎斗有足。 Quoted in *Hsi ch'ing ku chien* (Book 35, p. 7 b) and in other books on archæology. In the same passage, the *Hsi ch'ing ku chien* further remarks that the *chiao tou* is a cooking-vessel ( 溫器 ), and differs from the *tiao tou* in that the latter is used for boiling in the day-time and is beaten at night ( 與刀斗畫炊夜擊者異 ). "*Chi tiao tou*" thus came to have the meaning, in the *Ch'ien Han shu*, "to announce the watches or hours of the night." The earliest definition of the *chiao tou* as a cooking-vessel which I can find is given by YEN SHIH-KU (A.D. 579–645, see GILES, Biographical Dictionary, p. 938), who says, as quoted in the *Kin shih so, kin so*, Vol. 3, "The *chiao tou* is a cooking-vessel, and resembles a pan, but without having the shape of one" ( 焦斗溫器也。似 銚而無像 ).

ornament." Through the centre of the bottom a short nail has been driven, with two big flat heads visible on either side, for what purpose I do not know.

The "*Hsi ch'ing ku chien*" (Book 35, pp. 7–15) gives nine cuts of this type, all ascribed to the Han time, six

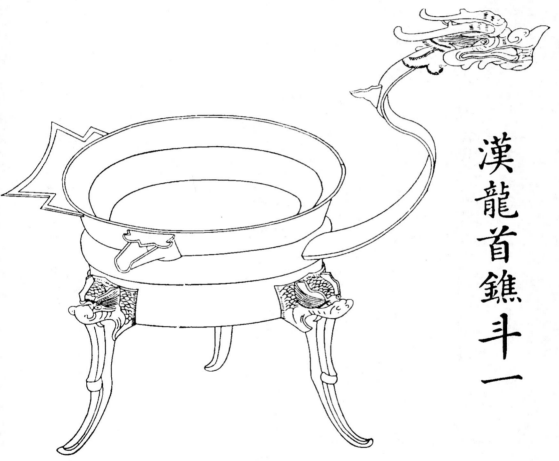

漢龍首鐎斗一

Fig. 25. Bronze Cooking-Vessel (from *Hsi ch'ing ku chien*).

of which are reproduced here in Figs. 25–30. There are four with dragon-heads (*lung shou*), i.e., the handles are of that shape, Figs. 25–28; one with the head of a wild duck (*fu shou*), Fig. 29; one for the head of which a *ju-i*

has been substituted, Fig. 30; one called *hsiung tsu*, from
the three feet representing bears; and two plain ones (素).
All have long high feet except the piece in Fig. 28, which
has very low feet, seemingly shaped into elephant-heads.
The pieces shown in Figs. 25–27 and in Fig. 29 have a curi-

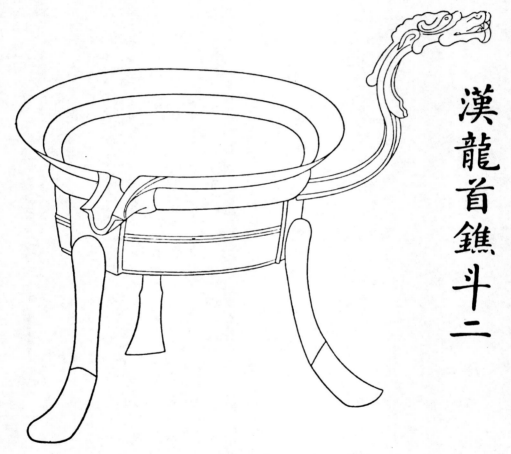

Fig. 26.   Bronze Cooking-Vessel (from *Hsi ch'ing ku chien*).

ous spout (called in the description *liu* 流) on the side, just
under the rim, the edge being continuous except in that in
Fig 26, where the spout is cut out of it.  In Figs. 25 and 27
the head of the dragon has two horns; and the upper portion

of the feet, at the place where they join the vessel, are
worked into dragon-heads. The types Figs. 25, 27, 29, and

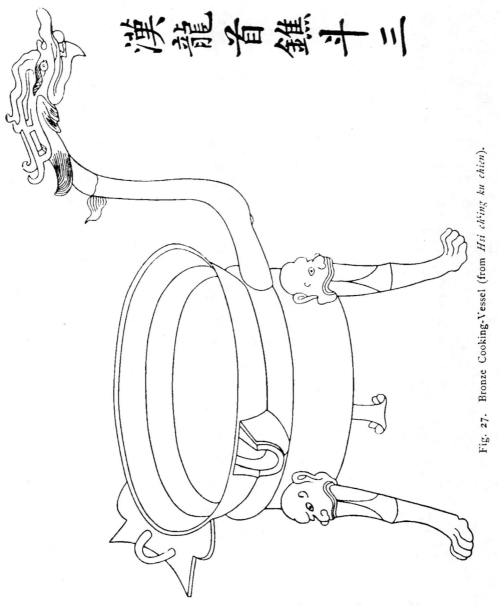

漢 龍 首 錐 斗 三

Fig. 27. Bronze Cooking-Vessel (from *Hsi ch'ing ku chien*).

30, have a peculiar tail-piece attached to the side opposite
the handle, in shape almost reminding one of the old spade

7

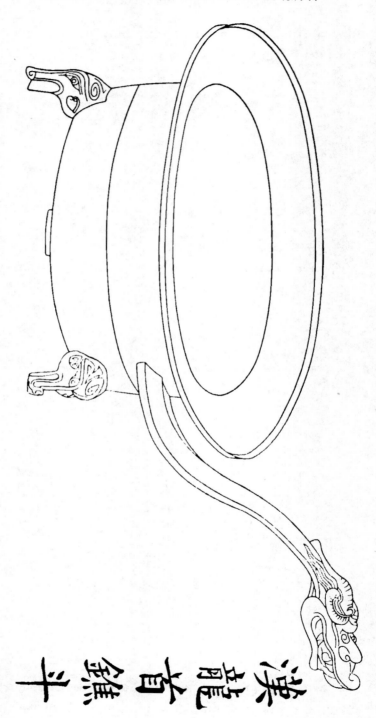

Fig. 28. Bronze Cooking-Vessel (from *Hsi ch'ing ku chien*).

money. Those in Figs. 27 and 29 have a round handle standing up on this projection, while in Fig. 30 there is such

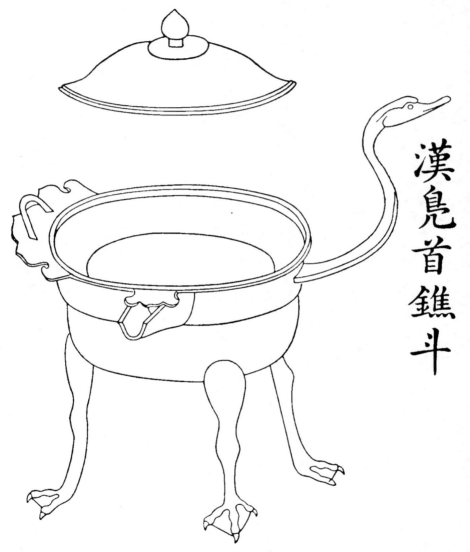

Fig. 29. Bronze Cooking-Vessel (from *Hsi ch'ing ku chien*).

a handle on the rim, thus facilitating the convenient handling of the vessel when removing it from the fire. The spout on some varieties would seem to indicate that they were used for the cooking of liquid food. The "duck-headed" *chiao tou* (Fig. 29) is provided with three bird-like webbed feet, **each** with three claws: it is the only piece having a cover on which is a knob in the highest, central part. The *"ju-i"* type

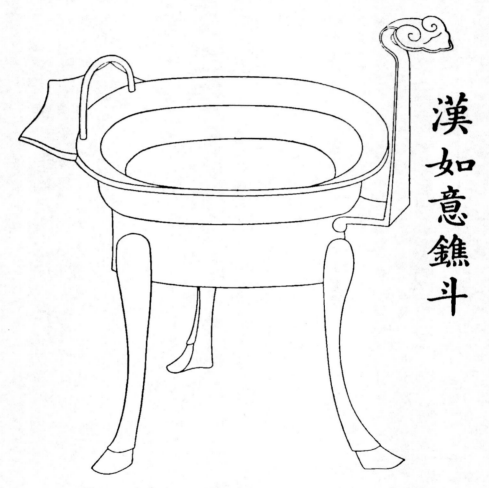

Fig. 30.   Bronze Cooking-Vessel (from *Hsi ch'ing ku chien*).

(Fig. 30) has extraordinarily long feet, slightly curved, **and** terminating in what appears to be animal's hoofs.

The *"Po ku t'u"* (Book 20, p. 9) illustrates a *chiao tou* of a style similar to No. 3 of the *"Hsi ch'ing ku chien"*

PLATE XXI.

Figs. 1, 2. Han Bronze Cooking-Pans; Fig. 3. Han Bronze Boiling-Kettle.

(Fig. 27), and gives among the measurements of this vessel also a statement regarding its capacity, which is wanting in the *Ch'ien lung* Catalogue; viz., that it holds one *shêng* 升.[1] This might show, either that, as regards these vessels, one *tou* of the *Han* time was equivalent to one-tenth of a *tou* of the *Sung* time (which is not very likely), or that the original meaning of *tou* as used in the phrase *chiao tou* was then forgotten, and *tou* simply meant a certain type of vessel. The latter conclusion is the more probable, as Hsǔ Shên 許慎, author of the dictionary "*Shuo wên*" (A.D. 100), is quoted in the same passage of the "*Po ku t'u*" as defining the *tou* as an object for heating (以 謂 斗 之 可 以 温 物 者 耳).

We now come to the other type of *chiao tou*, usually designated as *tiao tou*.

Fig. 1, Plate XXI, shows a specimen of a *tiao tou* in our collections that was likewise obtained at *Hsi an fu*. It is of very thin dark-red copper, as is peculiar to the work or the Han time, and without ornamentation, wherefore we may assume that it was the genuine article of the soldier's everyday life. Its total length is 34.4 cm. The diameter of the bowl is 19.2 cm; its depth, 8 cm. The handle, which is cast from the same piece as the bowl, is 14.7 cm long, flattened on one side, and rounded off on the other, its cut almost resembling the letter D. It broadens towards the outer end, where the flat side is 3.7 cm wide, while towards the bowl it is only 2.5 cm in width. The end of the metal handle has a socket 4.4 cm in length into which to fit another wooden handle. The handle and inner side of the bowl are

---

[1] In K'ANG-HSI's Dictionary, in a partial quotation of the passage in question, under 鐎, there is a mistake in writing 三 升 instead of 一 升.

very smooth; while the outer side is rough and coarse, probably due to long exposure to the fire in cooking.

There are also *tiao tou* coming down from the Han period, with ornaments engraved on the inner side of the bottom. Such pieces are extremely scarce nowadays; and I have seen only one (in the possession of Mr. E. T. A. Newman, district inspector of Chinese posts, and an intelligent collector) in *Hsi an fu.* On the bottom of this bronze are engraved eight fishes and two coins intertwined; and it was said to have been the first and only one which ever came to light in that place. In the ninth book of the "*Liang lei hsien i ch'i t'u shih,*"[1] an engraving of a similar specimen, and a special cut of the decorations on the bottom, are given. The design consists of four fishes, and four coins between them, grouped around a square seal in the centre which reads, "May it long be serviceable to sons and grandsons!" (*Ch'ang i tzŭ sun.*)[2] The author remarks, "This vessel has no feet, and is assuredly a *tiao tou* of the Han time; those contained in the books on archæology all have feet, but not many of these *tiao tou* are to be found."[3]

Another specimen of the same kind, pictured in the "*Chi chin chi ts'un*"[4] (Book 4, p. 9), is described by the author merely with the words "resembling a ladle" (似勺). A pair of fishes is engraved on the bottom, and there is a datemark in Han characters over the cut, which reads "made

---

[1] 兩罍軒彝器圖釋, published in Suchow.

[2] 長宜子孫。

[3] 此器無足。定爲漢時刁斗。金石書中所載類皆有足者。此刁斗不多見也。

[4] 吉金志存, published in 1859 by LI KUANG-T'ING 李光庭.

in the third year of the period *Chien chao*" (建昭三年造);
i.e., 36 B.C. The genuineness of the piece in question and
of its inscription, however, as is the case with most of the
pieces described in that book, seems very doubtful.

The "*Hsi ch'ing ku chien*" (Book 26, pp. 48, 49) has
engravings of two such *tou* with decorations of coins, but
without fishes. The one has a design showing coins of the
Han period; the other is graven with the coin *pu ch'üan*
布泉, issued by the Emperor *Wu* of the *Northern Chou*
dynasty in A.D. 561.[1]

There is a third kind of bronze *chiao tou* closely resembling the previous *tiao tou*, but posed on three low feet.
It represents, so to speak, a missing link between the *chiao
tou* proper, described above, and the *tiao tou*. Such a piece
is represented in Fig. 2, Plate XXI. It is from the collection
of Mr. Ōkazaki, correctly defined by him as a Han *chiao
tou* (Japanese, *soken*); according to his statement, used "to
boil sauce." The diameter of the mouth is 24 cm; height
up to the handle, 15.2 cm; height to the rim, 12 cm; depth
of bowl, 10 cm; height of feet, 3.6 cm. The handle describes
a short graceful curve, and is flattened at the end. It is
plain, but its shape agrees with that of the other pieces.
The feet are very low, and shaped at the upper part into
animal faces, and below into hoofs, as in the piece on Plate XX,
Fig. 3. The whole bronze is covered with a fine green
patina; and owing to this fact, as well as to its beauty of
shape and execution, and its excellent state of preservation,
it ranks high among the known specimens of this type.

---

[1] *K'ao ku t'u* (Book 10, p. 7 a) illustrates a *tiao tou*, under the heading *tou*, with
the head of a hydra and level bottom (螭首平底斗). The handle is straight
and horizontal, as in other pieces, but terminates in a head similar to those of the
*chiao tou*. The bowl is of trapezoidal outlines.

Two bronze vessels of the same type are pictured in the *Kin shih so* (*kin so*, Vol. 3), one of which bears an inscription on the handle revealing the date 65 B.C. (first year of the period *Yüan k'ang*).

If we now turn to the pottery of the same period, we recognize in it a variety of the *chiao tou* and *tiao tou* of bronze: and we are struck by the fact that in it is represented a transitional form of cooking-vessel between those two types. In Plate xx, Fig. 1, we meet with a small round vessel covered with an extremely delicate layer of a light-green glaze, having no feet (in conformity with the *tiao tou*), but provided with a dragon handle, the head and neck exactly corresponding to those of the *chiao tou*. The handle issues from the rim. The height of the jar up to the crest on the dragon's head is 10.5 cm; up to the rim, 7.2 cm; the diameter of the mouth being 8 cm. The vessel is made of light-reddish clay, and the glaze covers the inside of the bowl. The same type will be found again in a vessel which was placed on a miniature cooking-stove (Fig. 19), accurately fitting one of the holes. In this fact we may recognize a most valuable confirmation of the Chinese accounts that this vessel served for cooking-purposes. From the find in question, actual evidence is adduced that this type was the cooking-vessel in the age of the Han dynasty. The specimen in question is 6.9 cm high, with a diameter of 16.5 cm. The clay wall is 1 cm thick. The handle is attached under the rim. The glaze on the handle was covered with a layer of hardened loess, which after its removal, to judge from the general shape and the crest on the head, evidently presented a dragon-like appearance. The glaze, which also covers the inside of the bowl, is of a dark olive-green hue of unusual deepness and beauty, with glittering iridescent spots scattered over it. The loess incrustations cover the rim and upper parts, thus proving that this piece must have been long buried.

The same name (*chiao tou*) is applied also to a type of bronze vessel entirely different from those discussed before. I can find no reason why the Chinese archæologists should classify these two distinct types together, unless it was the fact that both were cooking-vessels. Fig. 3, Plate XXI, illustrates one of a pair of these bronzes obtained at *Hsi an fu*.[1] It is 13.9 cm high, and rests on three feet in the form of erect-standing bears. The handle (5.4 cm long) is of square cut, and attached to the middle part of the body, standing out from it horizontally. Its end forms a socket 3.4 cm deep for the reception of another, wooden handle. On the top surface of the short bronze handle is an animal's head in relief, consisting of two disproportionately large ears, eyes, nose, and prominent jawbones. The vessel has a spout in the form of a snake or dragon, the head of which forms the lid of the spout, and is joined to it by a hinge. It opens in such a way that the monster is then represented with open mouth, two tusks being visible in the lower jaw. The scales and legs, and what are probably the wings, of the animal, are brought out in low relief along the surface of the spout. The lid of the bowl part is circular, and also hinged. At the centre of the lid an ear is joined, not unlike the knob on metal mirrors, no doubt for the attachment of a cord by which to lift it up. The knob is surrounded by a circle. In the zone formed by this and three outer concentric circles are geometric designs.

---

[1] The colloquial name given to me in *Hsi an* for this type is "tortoise vase" (*kuei hu* 龜 壺). The feet were interpreted to me as squirrels (*sung shu*)!

There can be no doubt that this *chiao tou* represents the prototype of the more recent teapot. The two specimens of this kind of *chiao tou* in the Museum's collections are far superior in decoration and workmanship to the one depicted in the "*Po ku t'u*" (Book 20, p. 7) and the three in the "*Hsi ch'ing ku chien*" (Book 35, pp. 13–15). In the description of this piece in the "*Po ku t'u*" we meet with a curious feature. The design on the handle, "on which many figures are represented," remains unexplained, as its underlying purpose in antiquity (or the idea of the ancients) is now lost (與柄多爲物象而乏古意). This is a very interesting fact, considering that the "*Po ku t'u*" was edited between 1107 and 1111 by the art-historian WANG FU, and that at that time the original meaning or symbolism connected with many ornaments was already lost.

## LADLES.

On Plate XXII four spoons or ladles of Han pottery are represented. In Fig. 1 is shown a bowl-shaped ladle with solid handle. Its greatest length is 17.4 cm; maximum width of bowl, 9.3 cm. The handle curves convexly upward. The sides and upper surface of the handle have been flattened and smoothed by means of a stick. There are three incisions across the surface of the handle. The whole ladle has a dull-green glaze, much worn off, and there is a spurmark on the under side of the bowl.

The specimen next illustrated (Fig. 2) measures in its entire length 10.6 cm, the maximum width of the bowl being 6.6 cm. Its solid handle rises from the bowl in a curve, continuing thence straight out from the bowl. On the upper surface of the handle is a black-glazed knob. The handle

PLATE XXII.

Figs. 1-4. Han Pottery Ladles; Figs. 5, 6. Modern Porcelain Ladles.

is rectangular in section, and the under side of the bowl has been flattened out in narrow band-like sections (five in number). It has an olive-green glaze.

The ladle represented in Fig. 3 of this plate is 12.1 cm long, and the maximum width of bowl 6.8 cm. It is in imitation of a half-gourd. The handle is on the same level as the bowl, and is hollowed out. It has a light-green glaze, with blue flecks intermingled.

Fig. 4 represents the same type as Fig. 3. Its total length is 12.2 cm, and greatest width of bowl 6.7 cm. It is covered with a brilliant deep-brown crackled glaze. For the purpose of comparison, two modern spoons of white porcelain, with painted decorations, are added on this plate (Figs. 5, 6). The bowls are oval in shape, and the handles are hollowed out as in the Han gourd-spoons. That in Fig. 5 is 12.1 cm long, the greatest width of the bowl being 4 cm; while the other (Fig. 6) measures 13.9 cm in its entire length, and the bowl has a maximum width of 4.3 cm.

The "Annals of the Later Han Dynasty" mention among the mortuary objects of the Han time a gourd-ladle holding one pint.[1] In the "*Po ku t'u*" (Book 16, pp. 4 a and b) two bronze ladles of the Han time are illustrated under the term "gourd-ladle" (*p'ao tou* 匏斗). The text says, "The two implements are ladles resembling a bottle-gourd after it has been cut into halves (i.e., the half of a bottle-gourd). Among the materials for the eight musical instruments,[2] the

---

[1] 匏勺一容一升。 See DE GROOT, The Religious System of China, Vol. II, p. 403. The half of a gourd-shell for scooping up water is still a common kitchen implement in China. The character *p'iao* 瓢 ("calabash, a ladle made from the half of or in the shape of a calabash") is written also *p'iao* 瓢瓦, thus showing that such ladles were also made from clay.

[2] See MAYERS, Chinese Reader's Manual, p. 340.

gourd takes the first place, and embodies in its form the nature of heaven and earth. Nowadays the shape of the ladle is derived from the gourd, an idea inherited by the ancients."[1]   The length of the one ladle is given as 1 foot, 1 inch, and 5 *fên* 分, which, reckoning the Chinese foot as 36 cm, would make 41.6 cm; and that of the other, as 1 foot, 1 inch, equal to about 39.6 cm.   This shows that they are more than three times the length of the gourd-spoons of our Han pottery.   The ladles described in the "*Po ku t'u*" are stated to have a capacity of 3 *ko* 合.   Much larger ladles are mentioned in a passage of the "*Shih king*,"[2] where the word *tou* 斗 with the meaning of "ladle" occurs for the first time.   According to the commentators *K'ung Ying-ta* and *Chu Hsi*, these ladles had a handle three feet in length, a diameter of six inches, and a capacity of five pints (升).

Since *tou*, as we have seen, means not only a ladle, but also a measure of capacity, and in composition with other words denotes vessels which are neither the one nor the other, it is very hard to define, in many passages in litera-ture, what is understood by *yü tou* 玉斗 (a *tou* of nephrite) and by *t'ung tou* 銅斗 (a *tou* of copper or bronze).   The former, for example, are mentioned in the "Annals of the Han Dynasty"[3] as presents sent by the Duke of *P'ei* 沛公, the later Han Emperor *Kao tsu*, to *Fan Ts'êng* 范增

---

[1] 右二器皆斗也。如匏而半之。樂之八音匏居一焉。以象天地之性。今斗取象於匏。斯亦古人遺意歟。

[2] 酌以大斗。 "He fills the cups with the great ladle" (in *Shih king, Ta ya*, Book 2, Song 2, edition of COUVREUR, p. 355).

[3] The passage is quoted in *T'u shu chi ch'êng* Vol. 1118, *k'ao kung tien* Book 196, *tou pu chi shih* p. 1 b.

(278–204 B.C.),[1] who angrily smashed them to pieces. Couvreur interprets them in this case as jade ladles.[2] According to the cyclopædia "*Ts'ê fu yüan kuei*,"[3] such *yü tou* were presented in A.D. 561 to the Emperor *Wu* of the *Northern* or *Later Chou* dynasty. The bronze *tou* play an important part under the reign of the Emperor *Wang Mang* (A.D. 9–22), being conferred by him on the three prime ministers after their death, one being placed on their tomb, and one inside of it. Some passages relating to this event have been translated by DE GROOT,[4] who here explains *tou* as "peck," while COUVREUR[5] regards it as "a tablet of bronze, on which the dipper[6] was represented." Neither view appears to be correct, for in the "*Pi shu lu*"[7] it is related that the prime minister *Han Yü-ju* 韓丞相玉汝 possessed in his private collection a bronze *tou* in the shape of a *ladle* coming down from the time of *Wang Mang* (王莽時銅料一狀如勺), according to the then usual foot-measure, 1 foot 3 inches long, and with an inscription on the handle.[8] The two words 料 and 勺 in that inscription leave no doubt that the bronze *tou* of *Wang Mang* were *ladles*.

---

[1] See GILES, Biographical Dictionary, p. 218.

[2] COUVREUR, Dictionnaire classique de la langue chinoise, p. 408 a.

[3] Composed A.D. 1013 (see that passage in *T'u shu chi ch'êng*, l.c.).

[4] The Religious System of China, Vol. III, p. 1132.

[5] COUVREUR, Dictionnaire classique de la langue chinoise, p. 408 a.

[6] The dipper is called "the ladle of the north" (*pei tou* 北斗) from its resemblance to a ladle. The confusion of one author goes so far that he reverses the whole matter, and says that the peck or bushel derives its shape from the dipper, and contains ten pints (斗者取象於北斗受十升, quotation from a book *Kuang chi chu* 舫記注 in *T'u shu chi ch'êng*, l.c., *tou pu hui k'ao*, p. 3 a).

[7] 避暑錄, a book of the *Sung* dynasty (see E. BRETSCHNEIDER, Botanicon Sinicum, Vol. I, p. 180).

[8] See *T'u shu chi ch'êng* (l.c.), *tou pu tsa lu*, where all passages relating to the bronze *tou* of *Wang Mang* are collected.

The "*Hsi ch'ing ku chien*" (Book 38, pp. 40–42) illustrates three curious ladles of the *T'ang* period under the name *hu tou* 糊斗 ("ladles for gruel"[1]), but gives no description of them beyond their measurements. They have deep oblong bowls not unlike those of tobacco-pipes. One has a straight handle, while the handles of the other two are fancifully curved. The only information in literature concerning these ladles is found in a passage of the "*Tung t'ien ch'ing lu*,"[2] which is interesting, as the author seems to presume that they were developed from bronze vessels that served the purpose of porridge-ladles, and, as he shows, they were made in his time also of porcelain, the porcelain ones being cheaper than the bronze ones. The porridge-ladles which he describes, however, as having three feet, do not agree with the above-mentioned ladles engraved in the "*Hsi ch'ing ku chien*," which have no feet at all. The text runs as follows: "There is a small kettle (*yu*) of old bronze as big as a fist [with a handle] to lift it up. Above there is a cross-bar with chains (or ropes), and there is a lid. The vessel contains pap, which is thus guarded from being eaten by rats. There is an old bronze, originally an earthen jar, with a belly of the form of a wine-cup, which is supported below by a square standard. Its body is thick and heavy, and it is not known what use the ancients made of it. Nowadays they make pap-ladles as good as the old bronzes. They have three belts or rings around the long barrel.[3] Below

---

[1] *Hu* is a gruel or porridge made of wheat-flour, rice, or millet.

[2] 洞天淸錄 written by CHAO HSI-KU in the thirteenth century. See WYLIE, Notes on Chinese Literature (2d ed., p. 167). The passage is quoted in *T'u shu chi ch'êng*, l.c., *tou pu hui k'ao*, p. 2 b.

[3] *Ku t'ung* means literally "to hoop a barrel." The bowls of the porridge-ladles represented in the *Hsi ch'ing ku chien* might indeed be compared to a barrel. The bowl of the first there pictured is divided into three zones or fillets, two of which

they have three feet, two inches high.   The ladles are very
suitable for the holding of porridge.   Among the porcelains
there are long jars of *Kien-yao*,[1] black outside and white
inside; and long jars of *Ting-yao*,[2] with large belly, and
decorated with designs of garlic and rushes.   There are also
square ladles of *Ko-yao*,[3] like a *hu* (a corn-measure of five
or ten bushels).   In these a crossbar is arranged.   They can
be filled and used like a porridge-ladle.   Those of copper
(or bronze) are more convenient [for scooping up water] in
preparing a bath, but their price is higher than that of the
porcelain ones."[4]

In the "*San li t'u*" ("Illustrations of the Three Rituals"[5])
four kinds of ceremonial ladles (*shao* 勺) are pictured and

---

are decorated with geometrical designs.  I think that the author intends to denote by
*ku* the raised lines or rings brought out in the metal, which he compares to the hoops
of a barrel.

[1] Porcelains of *Kien chou* in Fuhkien (see JULIEN, Histoire et fabrication de la
porcelaine chinoise, p. 18).

[2] Porcelains of *Ting chou* in Chihli (see JULIEN, l.c., p. 61; HIRTH, Ancient Chinese
Porcelain, p. 141).

[3] Porcelains of the *Elder Chang*, one of the two kinds of celadon (see JULIEN,
l.c., p. 70; HIRTH, l.c., pp. 49 et seq.).

[4] 有古銅小提卣一如拳大者。上有提梁索股。
有蓋。盛糊可免鼠竊。有古銅元㙅。肚如酒杯
式。下乘方座。且體厚重。不知古人何用。今以爲
糊斗似宜有古銅，三箍長桶。下有三足高二寸
許。甚宜盛糊。陶者有建窰外黑內白長罐。
定窰元肚并蒜蒲長罐。有哥窰方斗如斛。中
置一梁俱可充作糊斗。銅者便於出洗。價當
高於磁石。

[5] That is, the *Li ki*, *I li*, and *Chou li*, a small work composed by a scholar of
the *Sung* dynasty, NIEH CH'UNG-I 聶崇義, from Honan Province, in the years
A.D. 954–960.  The modern editions are prefaced by a Manchu *Na-lan Ch'êng-têh*
納蘭成德, 6176.

described. They are all made of lacquered wood, and ornamented.[1] I do not find any particular references to common
ladles and spoons in ancient literature. I have no doubt
that the two gourd-shaped ladles, and the other two ladles
of Han pottery with curved handles, represent the forms of
ladles and spoons for every-day use of that period. A confirmation of this view is found in the fact that exactly such
spoons with curved handles are represented also on the Han
bas-reliefs (*Kin shih so, shih so*, Vols. 1 and 3).[2]

## BOWLS AND DISHES.

On Plate XXIII are displayed four dishes. Fig. 1 shows
a large round shallow dish 20–20.1 cm in diameter, coated
with a dark-green crackled glaze, which to a great extent
has dissolved into gold and silver colored portions. On the
lower side the glaze covers the slanting sides of the dish.
Fig. 2 represents a dish of the same type, only half as
large (diameter, 10.6–10.8 cm). It is ornamented with two
concentric grooves. The glaze, originally light green, has
changed into an almost whitish silvery tint. This dish seems
to have been twice fired, as both the surface and the under
side show small spur-marks of red clay. Both Figs. 1 and
2 strongly resemble the modern porcelain fruit-dishes *kuo tieh*
菓 碟, and were probably used as such, or for a similar
purpose, also in those ancient times. Figs. 3 and 4 are two
curious oval-shaped bowls or cups about 4 cm in height.

---

[1] See *T'u shu chi ch'êng* Vol. 1118, *k'ao kung tien* Book 198, *shao pu*.
[2] See Fig. 22. Jars with ladles emerging from them are frequent on the bas-reliefs.
The editors of the *Kin shih so* once remark: 有 一 器 如 甖 。器 中 似
有 勺 。四 人 視 之 。 "There is one vessel like a pottery jar, in which there
is an object resembling a ladle; four men are looking at it."

PLATE XXIII.

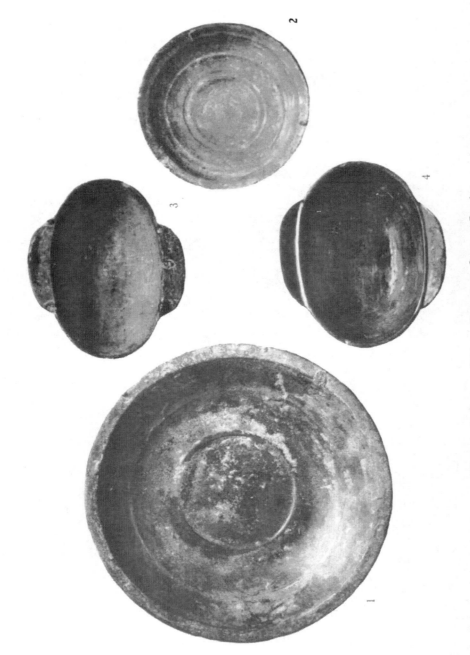

Figs. 1, 2. Pottery Dishes; Figs. 3, 4. Pottery Bowls.

That in Fig. 3 is 12.5 cm long and 7.4 cm wide; in Fig. 4, 14 cm long and 8.5 cm wide. Both pieces have earlike processes, in Fig. 3 in horizontal position, in Fig. 4 slightly turned upward. Those in Fig. 3 are 1 cm thick.

In the appendix to the "*Ch'üan pu t'ung chi*" 泉布統誌 (a standard work on numismatics), in which, aside from coins, some curious antiquities are pictured and discussed, there is an illustration in front and side views of a "bronze dish, not yet investigated, of the Sixteen States [A.D. 265–420]" 十六國未改銅器, with the same side-arms, called "ears" in the description (邊分兩耳). The author emphasizes that this piece is remarkable for its small size, being only 1 inch 寸 wide and 4 *fên* 分 ($\frac{4}{10}$ of an inch) high; and he compares it in dimensions with the goblet *shang* 觴 in the "*Po ku t'u*," which is 3 inches long and 2½ inches wide. According to the opinion of his contemporaries, such large pieces were numerous, but this was the only one of such small size. "Genuine ones of this kind," he continues, "are difficult to obtain. As the "*Po ku t'u*" originated under the *Northern Sung*, this must be an object of a time prior to the *T'ang* dynasty. Only a few of the present generation have any explanation to offer regarding its use. Some say that it was a vessel in which writing-brushes are washed off. I do not know if this is the case or not. That will have to remain a subject for further investigation."[1]

The modern wine-cups of pottery are extremely tiny; and

---

[1] 近世所見相若。而大者居多。惟此獨小。眞謂難得。夫博古一書起自北宋。則是唐以前之物也。而世人多有未解其所用者。或云筆洗。未知是否。故留以存改云爾。

8

the smallness of the above bronze cup, which is a counterpart of the pottery types of the Han time, corroborates my view that the latter were probably used as wine-cups.[1]

In the "*K'ao ku t'u*" (Book 8, p. 11 a) is illustrated an oval-shaped cup of nephrite (玉杯) with two rounded side-arms of a form exactly identical with those of our pottery cups. A cut representing the same type, only differing in its ornamentation, is found in the "*K'ao ku yü t'u*" 考古玉圖, Book b (下), p. 13 b.

The "Annals of the Han Dynasty" mention that in the ninth autumnal month of the sixteenth year of *Wên ti* (164 B.C.) the emperor obtained a jade cup on which was engraved the sentence "May the life of the ruler of men be lengthened!" and he ordered a general feast throughout the empire for changing the title of the dynastic period.[2] This new title of the reign (*Hou yüan*) was established in the next year (163 B.C.),[3] and the jade cup in question was evidently made for the celebration of this event.[4]

Plate XXIV represents a sacrificial table of light-red (unglazed) clay.[5] It consists of a long rectangular slab 55 cm in length, 34.5 cm in width, and about 1 cm thick, resting on four legs 8.6 cm high. The legs are surmounted by a

---

[1] The view of modern Chinese antiquarians is that they served as bowls in which to wash the face, but this can hardly be accepted.

[2] 漢書文帝本紀十六年秋九月得玉杯。刻曰。人主延壽。令天下大酺明年改元。 *T'u shu chi ch'êng* Vol. 569, *shih huo tien* Book 326, 玉部彙考 II, p. 13 b.

[3] See now P. MATHIAS TCHANG, Synchronismes chinois, p. iv, Variétés sinologiques (Shanghai, 1905), No. 24.

[4] Compare also CHAVANNES, Les mémoires historiques de Se-ma Ts'ien, Vol. III, pp. 459, 460.

[5] The burial of pottery tables 瓦案 is mentioned in the Annals of the Later Han Dynasty (see DE GROOT, The Religious System of China, Vol. II, pp. 403, 708).

PLATE XXIV.

Sacrificial Table of Unglazed Pottery with Three Bowls.

square capital of irregular sides, the length of which varies from 3 to 3.8 cm. They form an ornament which somewhat resembles the "squatting bear" (see p. 57). The eyes are visible, but the head is not clearly outlined.

On the sacrificial table will be noticed three bowls which are said to have been found in the grave with it. The one to the left is a flat bowl 22.1 cm in diameter and 5 cm high. It is glazed a dark green, with large decomposed portions both inside and outside. The shape of the middle piece comes very near to that of the modern porcelain rice-bowl. It is 7 cm high, has a rim at the base 7 mm high, and a diameter of 13 cm. The rim at the top is marked off by a deep groove. The glaze, which covers only the upper half of the outer surface and the entire inside, has a very peculiar tint, which is found in no other piece. It is a yellowish-white, almost ivory color, the same as occurs in a certain class of porcelains, and is covered with a fine network of crackled lines. Still more curious is the fact that the narrow bottom rim of the bowl has a light-green tinge. The third bowl, that on the right, bears a certain degree of resemblance to the type *fu* 鍑, which we saw is derived from a corresponding type in bronze. It is 8.9 cm high, the diameter of the mouth being 6.8 cm, and that of the bottom only 3.2 cm. The form of the bowl might be defined as made up of two obtuse cones united at their base. Where they join, there is a prominent rim 7 mm wide. Only the part above this rim is green glazed, the inside and the surface of the under part having no glaze. This latter part is about 8 cm in length, measured vertically from the central rim down to the foot. The upper portion is 2.7 cm long, and is entirely smooth; while on the lower part there are sixteen deep grooves running in parallel circles around it.

Fig. 1, Plate xxv, shows a rectangular clay vessel standing on four triangular feet. It is 7.9 cm high, and 18.9 cm long by 13.4 cm wide at the opening, the bottom being only 10.9 cm by 7.6 cm; so that the bowl gradually narrows towards the bottom, the walls slanting inward. It has a broad edge, averaging 1.2 cm in width. The greater part of the glazing has disappeared, and the ornaments in flat relief on the four sides are so nearly effaced that they cannot be clearly described. It appears certain, however, that they represent plant-ornaments, three floral patterns being discernible on each long side (regarding this ornament, see chapter on plant-ornaments). Just above the feet, on the short sides, are spirals. This rectangular vessel is apparently an imitation of a bronze type styled *fu* 盙 or 簠. The form of the latter character tends to prove that the vessel was originally a basket. GILES explains *fu* as a "basket, square outside and round inside, used to hold boiled grain in State worship." Fig. 2, Plate xxv, represents such a bronze *fu* belonging to the time of the *Ming* dynasty, but an excellent and doubtless faithful copy of an ancient original. It is one of a pair in the Museum's collections, and was obtained by me at *Hsi an fu*. The shape of this vessel corresponds to our pottery piece, with the exception of the two round handles at the sides (the upper ends of which form animal heads) and the hollow square base with sides slanting outward, to which the four feet are attached. On the sides, spiral ornaments are brought out in relief, and along the four slanting sides of the base are a pair of snakes facing each other. The height of this vessel is 9.6 cm; length, 16.9 cm; width, 13.9 cm. It is cast in one piece, except the two handles, which are soldered on, and coated with a light-green patina.

The *fu* is a type which existed as early as the *Chou*

PLATE XXV.

Fig. 1. Rectangular Pottery Vessel; Fig. 2. Rectangular Bronze Vessel.

PLATE XXVI.

Fig. 1. Han Pottery Urinal; Fig. 2. Modern Porcelain Urinal.

dynasty. In the *"Hsi ch'ing ku chien"* (Book 29) sixteen pieces of that period are illustrated, six of which have inscriptions. From a passage in the *"T'u shu chi ch'êng"* Vol. 1117, *k'ao kung tien*, section 缶部彙考 p. 3, which is a quotation from the *"San ts'ai t'u hui,"* it follows that such *fu* were made of pottery in *T'ung chou*, province of Shensi, in the period *Hsien p'ing* of the *Northern Sung* (A.D. 998–1003).

## URINALS.

A comparison of the piece of pottery shown in the next illustration (Fig. 1, Plate XXVI) with modern types proves it to be that indispensable article of the bed-chamber, the urinal, and we can but admire this evidence of refinement and progress in hygienic matters among the people of the Han time. It is 15.6 cm high, its maximum length being 16 cm. It has an oval-shaped body, narrowing at the top, over which rises a curved handle. On the upper side is a small nozzle, globular below, and cylindrical above. A row of small round knobs, suggesting a string of beads, encircles the handle and nozzle. On both sides of the body of the vessel the same ornamentation is brought out in low relief. It consists of six rings, the three in the upper row overlapping one another, the middle one being linked to the one underneath it.[1] Below there are two symmetrical lines running to either side in compound curves, from the ends of which tassels depend, thus indicating perhaps that this is a textile design. The glaze is a sombre green, having changed on one side of the body and along the bottom into a peculiar

---

[1] A frequent ornament, known as "connected rings" 連環 (see, for example, *Ku yü t'u pu*, Book 38, pp. 5 b, 6).

sea-blue intermixed with violet spots, seemingly the result of some underground chemical action. Back of the handle there is a big wart with a small slanting part left unglazed. The round shape of this Han urinal is frequently seen in the modern vessels used in Peking.

In Fig. 2, Plate XXVI, we see represented a specimen of the modern urinal in white and blue porcelain. It is 17.8 cm high and 18.3 cm long. The square-cut nozzle is covered with a loose lid perforated for the passage of a string, which at the same time is put through a hole in the handle on the back of the vessel. This object is used by high officials when travelling, for which purpose a special case of redwood (32 cm high) is made to hold it. This case contains, besides the compartment for the urinal, three drawers for toilet requisites. It has also a sliding door which can be locked, and a handle on top for convenience in carrying.

In an interesting paper entitled "Latrines of the East," reprinted from the "American Architect" of March 18, 1893 (p. 6), EDWARD S. MORSE gives a rather grotesque sketch of a modern Chinese urinal, and says: "Square urinals made of stone-ware are occasionally seen. These are used by old people, and I was told that they also served as pillows or head-rests." In some respects this statement seems erroneous. These urinals of porcelain or common pottery are not of occasional occurrence, but are universally used throughout China, and are indispensable requisites of every well-regulated household. They occur not only in square form, but also in round, almost globular, or ellipsoidal shapes. Their use is by no means restricted to old gentlemen, but they are employed by old and young alike; and, what is perhaps not unnecessary to remark, they are, as their narrow nozzle shows, limited in use to the male sex, the potter having been entirely heedless

PLATE XXVII.

Covered Bowl-shaped Pottery Vessels.

of the fair sex in this regard. Inquiries on this delicate subject, which I sometimes made of my Chinese friends, failed
to reveal any special contrivance of the sort for females.
Women urinate into a shallow bowl like a wash-basin, or
any other suitable vessel they can lay hold of. In the kilns
near Peking these urinals are still extensively manufactured,
and are in plain red-burnt ware, and also glazed in green,
brown, and yellow. Very neat miniature ones are turned out
for the particular amusement of little boys. Artistic pieces,
involving a higher degree of workmanship, come from the
renowned factories of *Po shan* in Shantung, several specimens
of which are in our collection.

The *Po shan* urinal, which is in the shape of a cat, the
lid forming the animal's head, is quite famous. In *King
têh chên* the same type is manufactured in white and blue
porcelain. Since, among the numerous types of clay pillows,
we find one also in the shape of a cat, the handle to which
of course is wanting, I believe that Professor Morse's in-
formants may have confounded these two objects, which are
entirely distinct.

### TAZZAS OR SACRIFICIAL VESSELS.

The two round-shaped covered vessels illustrated on Plate
XXVII represent a distinct type, called *tou* 豆 . One of these
vessels (Fig. 1), including the cover, is 13.5 cm high; the
bowl proper, 9.1 cm. The mouth of the bowl has a diameter of 13.2 cm; the cover, of 15.1 cm; the foot, of 9 cm.
An interesting feature regarding this type is the unity of
form of the two parts taken as a whole, and of each taken
separately. The vessel is composed, so to speak, of two
perfect bowls, one inverted over the other, and would seem

to be the prototype of the modern covered porcelain teacup. The one part is not imaginable without the other; whereas in other types of covered jars the vessel retains its peculiar form even without the cover, which is added merely for utility's sake. The bowl, from about the middle part, gradually narrows towards the base. The surface of the upper portion has been flattened in narrow horizontal sections or stripes of irregular width (on an average 1 cm wide) by means of a stick. Further, it is covered with a large number of very fine parallel grooved lines, which are deeper, and wider distances apart, in the lower portion. On the under side of the bottom also a number of concentric circles are deeply engraved. The inner edge of the flat rim of the bowl is surrounded by another low vertical rim curving inward, over which the cover fits. Both cover and bowl are moulded separately and were turned on the wheel. The top of the cover is flat, though somewhat convex, decorated with grooves in concentric rings, and has an upright circular rim around the edge. The outside of the vessel is coated with an extremely thin layer of a fine dark-brown glaze, which on some parts has been exfoliated. A few spots have changed into a gold color.

The other vessel (Fig. 2, Plate xxvii) is 9.7 cm high; with cover, 13.9 cm. The diameter of the mouth is 14.4 cm; that of the cover, 17.2 cm; and that of the bottom, 9.4 cm. The difference in shape between this and the previous vessel lies mainly in the fact that this is somewhat broader and has a distinct curved rim (about 1 cm high) around the base. For the rest, the description of Fig. 1 on this plate suits this piece as well. The glaze is a light green, but for the most part merges into a whitish and on the cover into a grayish color, as the white spots in the illustration indicate. Hardened earth incrustations adhere to the inner side of the cover.

The type represented by these two vessels is called *tou* 豆[1] or also *pien*[2] *tou* 邊豆. The latter name was colloquially used for them in *Hsi an fu*. In the native archæological works on bronze, we find a number of ancient bronzes, commencing with the *Chou* dynasty, described under the same name. They present several varieties of form. One consists of a deep spacious round bowl resting on a narrow base. They are usually with cover, but sometimes also without it. Another variety consists of a round shallow tray resting on a standard supported by a circular base. As the above-described pieces of pottery (Plate XXVII) have no stems, one might feel inclined to think that they do not belong to this group; but in the "*Hsi ch'ing ku chien*" (Book 29, p. 30) there is a type without stem corresponding exactly to the Han pieces in question. It is defined as a "*tou* with a girdle-ornament of the *Chou* dynasty" 周帶紋豆. On p. 40 of the same book is illustrated a *Chou tou* without stem or cover. The "girdle-ornament" must be understood to denote the engraved parallel lines on the body, which we saw also as grooves on the pottery. On most of the other pieces pictured in the "*Hsi ch'ing ku chien*" they appear as the usual decoration, and fill the entire surface of the vessel. The two ears on the vessel shown in Fig. 36 are well adapted to the bronze material of which they are made, but were naturally omitted in the pottery, as they would be liable to break off. Moreover, the ears are not essential to

---

[1] GILES, "A vessel of wood for holding food at sacrifices, feasts, etc.;" COUVREUR, "Vase de bois dans lequel on offrait de la viande cuite aux esprits."

[2] GILES, "A splint basket to contain fruits offered in sacrifice;" COUVREUR, "Vase de bambou dans lequel on offrait des fruits, des viandes etc. aux esprits."

the type of the *tou*, as there are also bronze *tou* which lack them.[1] Aside from the ears, the shape of this bowl tallies with that of the Han bowls above, and so does the cover. On the bronze cover in Fig. 36 (and this is the case with all covered bronze *tou*) we notice a hollow cylindrical knob or button (頂) surmounted by a round flat projecting rim. In the pottery this piece has shrunk into a mere edge set on the top of the cover, as probably it was too delicate a task for the potter to reproduce that elaborate ornament in clay, and presumably also as it would not appeal to the artistic sense. The connection of our Han pieces with the type of the bronze *tou*, therefore, seems quite justifiable.

To show the other varieties of *tou*, three more objects are here reproduced (Figs. 1–3, Plate xxviii). The first of these (Fig. 1) represents a bronze *tou* of the *Ming* time, illustrating the bowl with stem, the base of which is bell-shaped. The cover has been lost. It is a good imitation of the ancient *Chou* type, which is not unlike our rummers, which first appeared on the Rhine in the sixteenth century. It is 16.7 cm high.

In Fig. 2 of this plate is shown a piece of rough gray clay, unglazed, which comes down from the end of the *Chou* dynasty (see p. 14). It represents the tray-shaped variety without cover, and with slender stem terminating in a round hollow base. It is 17.6 cm high.

The next piece (Fig. 3, Plate xxviii) is a modern specimen from Peking, made of papier-maché. It is lacquered a deep black, with gold stripes around the rim, stem, and middle. Its height is 28.2 cm. This piece is extremely interesting

---

[1] See *Hsi ch'ing ku chien*, Book 29, p. 33; *Po ku t'u*, Book 18, pp. 14–16; *K'ao ku t'u*, Book 5, pp. 15 and 17.

PLATE XXVIII.

Fig. 1. Bronze Tou of Ming Period; Fig. 2. Pottery Tou of Chou Period; Fig. 3. Modern Papier-Maché Tou.

as proving that types going as far back as the age of the *Chou* are nowadays manufactured in a different material for actual use. These papier-maché *tou* are made in Peking in sets of five, together with a set of *wu kung* 五供,[1] for placing in front of the household shrine or before the altars of temples, to hold cakes, fruit, and other offerings. This is one of the vessels adopted into the Buddhistic cult from the ancient national religion of China.

The Chinese distinguish three kinds of *tou*, according to the material of which they are made. The "*Êrh ya*" gives the following definition: "The wooden *tou* is called *tou*, the bamboo *tou* is called *pien*, the pottery *tou* is called *têng*."[2] In the modern worship of Confucius these three vessels — *tou, pien,* and *têng* — are still used, and except the *têng*, which is of bronze, are made of the same ancient materials and are of cylindrical form.[3] For this purpose the *tou* is made of wood varnished with lacquer and gilt. The decoration is described as follows: "Around the body of the vessel is the pattern of 'hanging clouds,' and below it a scroll-work design. The upper part of the stem[4] has a 'wave' pattern and the 'metal plaque' pattern; the lower part of the stem, the 'axe' pattern. The cover has the 'wave' pattern; and the top is

---

[1] That is, the five objects for sacrificing, — a censer in the centre, with a candlestick and a flower-vase on either side of it.

[2] 木豆謂之豆。竹豆謂之籩。瓦豆謂之登。

[3] There is only one vase of pottery among the sacrificial objects of the Confucian cult; viz., the vase 大尊.

[4] *Chiao* 校 "leg of a table or of any other object" (COUVREUR, Dictionnaire classique de la langue chinoise, p. 451 b), which meaning is not given in Giles. It is evident from the context of the above passage, however, that the term *chiao*, with regard to the *tou*, applies only to the stem or leg, while the broad base below is styled 足 "foot."

made up of a knot of twisted rope."[1] The latter appears as a substitute for the knob of the bronze vessels.

The *pien* used in the Confucian temples are made of plaited bamboo, lacquered over, and adorned with silk taffeta. The inner side is colored red.[2] The bowl has the round form.

The *têng* is made of copper (or bronze). Around the rim of the mouth is a pattern in scroll-work; and along the middle of the body, the "thunder" scroll (the key-pattern). On the stem (literally, "pillar") there is the figure of the glutton (*t'ao t'ieh*); and on the base, the pattern of "hanging clouds."[3]

The type of the *tou* seems to have existed as early as the *Hsia* dynasty (2205–1818 B.C.). The "*Li ki*" has the following passage: "[The *tou* which the princes of *Lu* (1115–467 B.C.) used for offerings of meat were] the unornamented *tou* of the *Hsia*, the nephrite *tou* of the *Yin* (1766–1154 B.C.), and the *tou* decorated with reliefs of the *Chou* (1122–255 B.C.)."[4] *Tou* of nephrite are mentioned in another passage of the same chapter of the "*Li ki*."[5] The original material for the

---

[1] 豆考。先師廟用木柴漆塗金。通高五寸五分。深二寸。口徑四寸九分。校圍二寸。足徑四寸七分。蓋高二寸二分。頂高五分。腹爲垂雲紋回紋。校爲波紋鈑紋。足爲斂紋。蓋爲波紋。頂用絹紐。See 皇朝祭器樂舞 Book I, p. 8, WU CH'ANG edition, 1871.

[2] 籩考。先師廟編竹柴以絹飾。裏色尚紅。Ibid., p. 7.

[3] 登考。先師廟制用銅。口爲回紋。中爲雷紋。柱爲饕餮形。足爲垂雲紋。*L.c.*, p. 3.

[4] 夏后氏以楬豆。殷玉豆。周獻 (for 娑) 豆。See COUVREUR, *Li ki*, Vol. I, p. 740.

[5] *L.c.*, p. 731. The *K'u yü t'u p'u* of LUNG TA-YÜAN (Book 92, pp. 3 and 5) illustrates two old *tou* of nephrite in imitation of *Chou* models. The first of these has a cover, on which the knob of the bronzes is omitted, and in its stead is a carved rosette-like ornamentation.

*tou* seems to have been wood. This is the opinion expressed in the "*Li ki*" and the "*Êrh ya*," while the "*K'ao kung chi*" 考工志 asserts that the *tou* are made by the potters.[1] As the last-named treatise was added to the *Chou li* only under the Han dynasty, and as there is strong suspicion that the *Chou li* is a production, not of the *Chou*, but of the *Han* time,[2] it seems justifiable to conclude that *tou* vessels of pottery originated not earlier than the age of the *Han;* and this view, based on literary evidence, agrees with the archæological finds thus far made.

The trio *tou*, *pien*, and *têng*, is mentioned as early as the "*Shih king*" in the following passages : —

(1) 大雅生民。卬盛于豆。于豆于登。 "We put [the dishes] in the *tou* and [sauces] in the *têng*."[3]

(2) 國風伐柯。籩豆有踐. "The *pien* and *tou* are arranged in a row [for the performance of the marriage-ceremonies]."

(3) 小雅常棣。儐爾籩豆。飲酒之飫。 "Thou hast arranged thy *pien* and thy *tou*, and wine is drunk to satisfaction."

(4) 小雅伐木。伐木于阪。釃酒有衍。籩豆有踐。兄弟無遠。 "They cut down trees on the hillside.

---

[1] See C. DE HARLEZ, *San li t'u*, "Tableau des trois rituels" (Extrait du Journal Asiatique, Paris, 1890), p. 27.

[2] A. WYLIE, Notes on Chinese Literature (new ed.), p. 5; CHAVANNES in Bulletin de l'École française d'Extrême-Orient, Vol. III, p. 237.

[3] MAO 毛 comments on this passage: 木曰.豆。瓦曰登。 "The wooden ones are called *tou*, those of clay are called *têng*." And CHU HSI says: 木曰豆以薦菹醢也。瓦曰登以薦大羹也。 "The wooden ones are called *tou*, and serve for offerings of salted vegetables and hashed meat; those of clay are called *têng*, and serve for offerings of great soups." K'ANG HSI's Dictionary, quoting the same tradition under 登, has this addition to that sentence: 祀天用瓦豆陶器質也·

There is plenty of warmed wine.  The vases *pien* and *tou* are arranged in a row; the brothers are present."

(5) 小雅楚茨。君婦莫莫。爲豆孔庶。 "The consort of the prince places gently and reverently a great number of *tou* [for the ancestral sacrifice]."[1]

(6) 小雅賓筵。籩豆有楚。殽核維旅。 "The *pien* and *tou* are prepared; meats and nuts are served in them [at a banquet connected with archery]."[2]

(7) Same song as under No. 6. 亂我籩豆。屢舞僛僛。 "[The drunken guests] throw our *pien* and *tou* into disorder. Repeatedly they totter in the dance."

We see from these passages that *tou* and *pien* were trays used not only in the ancestral cult, but also at marriage-feasts and drinking-bouts.

As regards the further uses of the *tou*, they were employed at the Imperial Court, according to the "*Li ki*," to serve pickled vegetables in (*chü* 菹), especially those made from aquatic plants, and hashed meat (from land animals) preserved in vinegar.  In a kind of *tou* called *chia tou* 加豆 products of the soil — i.e., vegetables growing on dry land and preserved in salt — were kept.  In the ancestral temples the vases *pien* and *tou* contained both aquatic and land products.[3]  The "*Shuo wên*" defines the term *tou* as "an old vessel to eat meat from" (古食肉器也).

Most interesting is the account of the introduction of the vases *pien* and *tou* from China into Corea in the first part of the twelfth century before Christ.  The tradition is connected with *Chi tzŭ* 箕子, one of the foremost nobles under

---

[1] Compare *Li ki*, Vol. I, p. 732 (COUVREUR edition).

[2] According to tradition, the meats were served in the *tou*, and the nuts were served in the *pien*.

[3] See COUVREUR, *Li ki*, Vol. I, p. 598.

*Chou hsin*, the last emperor of the *Yin* dynasty, who, for protesting against the evil ways of his master, was thrown into prison, and on being released by the victorious *Wu wang*, founder of the *Chou* dynasty, in 1122 B.C., retired to Corea.[1] Now, the following item is contained in the "*Tung i ch'uan*," as quoted in "*T'u shu chi ch'êng*" Vol. 1120, *k'ao kung tien* Book 208, *pien tou pu chi shih:*[2] "In days of old, having made an end of the land of *Ch'ao hsien* (occupying the peninsula of *Liao tung* and the northwest of Corea), *Wu wang* appointed *Chi tzŭ* to this territory. There *Chi tzŭ* taught women and men purity and sincerity by means of [the Chinese] civilization, and taught them to take their food and drink from the vessels *pien* and *tou*."

The *pien* and *tou* used in eating by the people of Corea, are further mentioned in the Chinese accounts of *Ch'ao hsien;*[3] and their early introduction into Japan is seen from the description of this country in the "Annals of the Later Han

---

[1] See GILES, Biographical Dictionary, p. 121.

[2] 東夷傳。濊本朝鮮之地。昔武王封箕子於朝鮮。箕子教以禮義婦人貞信。飲食以籩豆。 See also *T'u shu chi ch'êng* Vol. 1335, *pien i tien* Book 13, *Ch'ao hsien pu hui k'ao* pp. 1 et seq. See further, CHAVANNES, Les mémoires historiques de Se-ma Ts'ien, Vol. IV, p. 230; ibid., Vol. V, p. 486. Chavannes doubts the authenticity of this account, as the name of *Ch'ao hsien* appears for the first time in Chinese history under the reign of *Ts'in Shih huang ti* at the end of the third century before Christ. He explains this anomaly by admitting that the Chinese, on their arrival in Corea, found there a tradition relating to a wise prince who had promulgated an admirable code in eight articles. They hastened to identify this Corean legislator with the Viscount of *Chi*, to whom the composition of the nine sections of the treatise on political philosophy called *Hung fan* is attributed. Thus the Viscount of *Chi* was made to have immigrated into Corea, though the whole remainder of his biography contradicts this statement. However this may be, the existence of the tradition that the vessels of the type *tou* were brought from China to Corea, as confirmed by their ancient occurrence in the latter country, cannot be doubted.

[3] A. PFIZMAIER, Nachrichten von den alten Bewohnern des heutigen Corea (Sitzungsberichte der Wiener Akademie, Wien 1868, p. 463). For the fact that they are traced back to the elements of culture introduced by *Chi tzŭ*, see ibid., p. 506.

Dynasty," the first in Chinese literature, and altogether the oldest document regarding Japan. There it is said that the sexes are not separated [as in China] when taking meals, that they eat with their hands, and make use of the *pien* and *tou*.[1]

In fact, among ancient Corean mortuary pottery[2] we find striking counterparts of the Chinese *tou* type, the round bowl-shaped form as well as the flat tray-like one with cover and ears, and also without them.[3] The same types likewise occur in the mortuary pottery of Japan of the so-called "protohistoric" age.[4]

---

[1] 男女無別飲食以手而用籩豆。 *T'u shu chi ch'êng* Vol. 1338, *pien i tien* Book 33, *Ji pên hui k'ao* I p. 1 a. Compare ibid., p. 4 b, the account of the *Tsin Shu*, in which the sentence occurs, 飲食用俎豆 "For their meals they use low tables and trays." The *Sui shu* (Book 80, p. 11), however, says of the Japanese: 俗無盤俎。藉以檞葉。食用手餔之。 "As a rule, they have no dishes or trays, but avail themselves of leaves of the *chieh* (*chiai*) tree; they use their hands in eating." According to B. H. CHAMBERLAIN (*Ko-ji-ki*, or Records of Ancient Matters, Yokohama, 1883, p. xxx), cooking-pots and cups and dishes — the latter both of earthenware and of leaves of trees — are mentioned in the *Kojiki*, as also chopsticks for eating the food with. The latter have doubtless been introduced from China. Regarding the above passage, see also E. H. PARKER in China Review, Vol. XVIII, p. 219 a.

[2] An historical allusion to Corean burial-pottery is found in the history of the *Wei* dynasty (*Wei shu*), where the funeral ceremonies of the country of *Wo tsü* 沃沮 in Corea are described; and it is said that they make kettles of pottery, which they fill with hulled rice, and suspend in a slanting position at the side of the entrance to the coffin (又有瓦鑞置米其中。編縣之於槨戶邊。 — *T'u shu chi ch'êng* Vol. 1338, *pien i tien* Book 30, *Wo tsü hui k'ao* p. 2 b).

[3] See P. L. JOUY, The Collection of Korean Mortuary Pottery in the United States National Museum (Report of the National Museum, 1888, pp. 589–596), where these vases are styled "tazza" (compare particularly Plates 84 and 85); see also E. S. MORSE, Catalogue of the Morse Collection of Japanese Pottery (Cambridge, 1901), plate to p. 30, Nos. 13, 15, 17, 19, 24, 26, 34.

[4] See E. S. MORSE, Catalogue of the Morse Collection of Japanese Pottery (Cambridge, 1901), plate to p. 36, Nos. 139, 142, 144. The identity of these and of other pieces with the corresponding type of Corea is noted likewise by the author on p. 34; but it is a noteworthy fact that this type is absent in the prehistoric shell-mound pottery of Japan, which proves it to be a later introduction.

Pottery Jars.

## JARS AND VASES.

The jar represented in Fig. 1, Plate xxix, has a foot 6 mm high and a short upright neck or rim 1.1 cm high. The section between this rim and the raised line surrounding the middle of the vessel is about 6 cm wide, slopes gradually outward and downward from the top, curving slightly; while the portion below the raised line gradually narrows towards the foot, also curving convexly. The jar was originally glazed a lustrous dark brown; but the glaze has mostly disappeared, and that from the upper portion entirely. The latter part is divided longitudinally by two raised lines into two sections. In each of these is an inscription in seal characters enclosed in a rectangle. These are the only characters found, thus far, which are not incised, but brought out in relief on the clay, presumably by means of a stamp. The authenticity of this writing is therefore beyond doubt. For further information as to this inscription, see chapter on inscriptions. These characters are also covered with glaze, as is apparent from the right-hand upper portion of the main seal, where it is still fully preserved.

This jar is very probably an imitation of a bronze type met with in the Han time. It is called *fu* 鍑 or 鍑. In the "*Hsi ch'ing ku chien*" (Book 31) four specimens of such *fu* are illustrated, and the second on p. 27 tallies with our specimen of pottery.

Fig. 31 represents such a bronze *fu* of the Han time from our collection. It is 8.6 cm high, and has a diameter at the mouth of 8.6 cm. The bottom is round, and along the middle surface of the body is a projecting rim 1 cm in width. On the sides are two tiger-heads in low relief, holding movable rings.

9

Fig. 2, Plate xxix, represents a jar with globular body that gradually narrows towards the bottom. It is 18.4 cm high, 16 cm in diameter at the mouth, and 7.8 cm across the bottom. It is covered with a curious brownish and bluish partially lustrous glaze, partially incrusted with thick masses of hardened carbonate of iron and clay, which seem to have pressed on the vessel during long burial under ground. Another curious feature of this piece is that its surface and

Fig. 31.   Bronze Bowl-shaped Vessel of Han Time.

interior are covered with etchings from the fibres of plant-roots which seem to have struck or grown into this jar.[1]

The jar represented in Fig. 3 of this plate (from the Williams collection) is the only one of its kind which I have come across, and is remarkable for the gracefulness of its form. The shoulders are bulging, and from here downward the vessel gradually narrows. The rim (2.5 cm wide) is bent outward and downward. The piece is 38 cm high, with a diameter of 21 cm over the opening. The circum-

[1] According to an investigation kindly made by Professor R. W. Tower, Department of Physiology, American Museum of Natural History.

Large Pottery Jars.

PLATE XXXI.

Fig. 1. Pottery Ewer dated 52 B.C.; Fig. 2. Bronze Ewer of Ming Time.

ference of the neck is 62.7 cm, and that of the base 56 cm. The jar is not glazed, but is covered with a thin dull fawn-colored paint scattered over with white or gray flecks. On the inner side, below the rim, dark finger-shaped stripes are visible, indicating that the paint, like the glaze, flowed down in streaks. Regarding the inscription of this jar, see chapter on inscriptions. The engraved lines of the characters have been whitened on the jar that they may show up better in the illustration.

On Plate xxx two large heavy jars with extraordinarily thick clay walls are represented, both of the same type. The jar shown in Fig. 1 is 25.7 cm high, with a diameter at the opening of 12 cm and at the bottom of 19 cm. The neck is 1.7 cm high. There are three long spur-marks on the under side of the bottom. The glaze is light green and crackled, with here and there light-blue and lilac spots. The jar seen in Fig. 2 of the same plate is 34 cm high, with a diameter at the top of 17.7 cm, and at the bottom of 21.1 cm. The neck is 2.6 cm high and 1.9 cm wide. There are five spur-marks on the bottom. The glaze is light green, partially effaced, iridescent, and interspersed with blue spots.

Of the foregoing three specimens (Figs. 1 and 2, Plate xxx; Fig. 3, Plate xxix) no bronze prototypes are known to me.

Fig. 1, Plate xxxi, represents a small ewer bearing an inscription which furnishes the date 52 B.C., regarding which see chapter on inscriptions. The ewer is 17.2 cm high and 6.1 cm in diameter at the mouth. The foot is 1 cm high and 5.6 cm in diameter. The vessel is made of reddish clay, and unglazed. The shape is very peculiar. I do not know of a similar piece in either old or modern Chinese ceramics. The most striking features are the narrow compressed spout and the ear-shaped handle, and, further, the parallel grooves which cover the entire surface of the body, and look as if smoothed off by a narrow bamboo strip.

My first supposition — that this piece may have been made in imitation of the Greek ceramic type *oinochoe*, to which it bears a certain resemblance — I have now entirely given up, as, first of all, in all similar Greek pieces the handle proceeds from the rim,[1] which is never the case in Chinese pottery; and, secondly, because it is possible to explain this vessel as being connected with or related to the type of the bronze *i* 匜, an example of which is given in Fig. 2 of this plate. This is a bronze ewer of the *Ming* time — 15.5 cm high, 6.8 cm in diameter at the mouth, and 9.2 cm in diameter across the bottom — with a long wide-open spout. That the above pottery jug is of Chinese make, there is no doubt, as is conspicuously shown by its inscription in the Han style of writing; and that it was by no means an exceptional piece, but a type frequently used, is seen from the phrase "Number five" at the end of the inscription, which proves that these vessels were made in a certain series of numbers.

The reason for grouping under one type the two jars shown in Plate XXXII is evident. The main characteristic common to both of them is the short contracted neck, looking almost as if cut off at the rim. Such a type is also met with among the ancient bronze vessels, and is described in the *"K'ao ku t'u"* (Book 5, p. 24 b) under the name *shou huan p'ou* 獸環瓿. The jars called *p'ou* served for holding pickled minced meat.[2]

---

[1] See, for example, FURTWÄNGLER, Griechische Keramik (Berlin, 1883, Plate XXXIII). In modern Chinese pottery, also, this sort of handle is unknown. It occurs, however, in earthenware teapots from Tibet, two of which are figured on Plate XV of The Ethnology of Tibet by W. W. ROCKHILL (Report of the U.S. National Museum for 1893, Washington, 1895), opposite p. 704.

[2] *Hsi hai* 醯醢 (see *San ts'ai t'u hui*, section on implements 器用, Book 1, p. 34).

PLATE XXXII.

Pottery Jars with Relief Bands.

To judge from the engraving in the above-mentioned book, the one referred to is formed into the same shape, so far as the base, belly, neck, and mouth are concerned, as the pottery pieces in question, and it is also provided, on its sides, with two tiger-heads[1] with rings. On p. 23 of the same book of the "*K'ao ku t'u,*" another *p'ou* is illustrated called *lung wên p'ou* 龍文瓿, from the dragon-heads with rings, on its sides, in place of the usual tiger-heads. This vase has a projecting rim, while the previous piece has simply the plain short neck cut off in the same way as in the pottery. As regards the shape of this type, I have no doubt that the pottery one represents in this case the prius,[2] or primary stage, and the bronze the secondary.

The pottery jar shown in Fig 1 of this plate is 19.9 cm high, and has a diameter at the mouth of 9.6 cm, and at the bottom of 12.5 cm. The rim at the base is 2 cm high. A band of reliefs, nearly effaced, runs around the vessel at the shoulder. The glaze is a light green, that has mostly changed into a gray earth-color interspersed with many golden specks.

The second piece on the same plate (Fig. 2) is 26.5 cm high, with a diameter at the mouth of 9.8 cm, and at the bottom of 14.2 cm. The rim below is 2.2 cm high. The shape of this vase agrees with that of the previous one. The explanation of the reliefs is given later on. It is coated with a fine dark-green glaze showing a network of fine crackles.

Another vase of this type is in the Museum's collections (Cat. No. $\frac{70}{11805}$). It is 31.3 cm high; 9.8 cm, diameter of

---

[1] See p. 140.

[2] This is also proved by the Chinese character *p'ou* formed with the radical meaning "clay" ( 瓦 ).

the mouth; and 16 cm, diameter of the bottom. The standard differs from those of the two preceding ones in its
height (6 cm) and its shape, which is that of a truncated
cone, and is therefore more sharply set off from the belly
of the vase. The three triangular prongs of the spur still
adhere to the bottom; and from their regular shape, and
the fact that they have been preserved intact and in full
size, it may be inferred that they were left intentionally, to
serve the function of feet. They raise the vase 0.5 cm.
There are no reliefs; but a band bordered by double-grooved
lines is marked around the body, and encloses two tiger-heads
on the sides. The fronts of these heads are each made up
of three triangles, and the eyeballs are clearly expressed by
high knobs. The light-green glaze has partially changed
into a brilliant silver and gold iridescence, which renders this
piece, in this respect, one of the finest in our collection.

For purposes of comparison, a bronze *p'ou* from the Ōkazaki
collection is reproduced in Plate XXXIII. The vessel is 26.7
cm high, and the diameter of its mouth 24.5 cm. The base
is of the same shape as that last described; but the form of
the body and neck varies to some extent from that of our
pottery types, while the latter agree perfectly with the illustrations in the "*K'ao ku t'u*." This proves again, as in
many other cases, that the production of one type of vessel
may result in a number of variations. The neck is much
wider than in the pottery, but just as short and contracted.
There are three animal heads with curved horns in high
relief on the upper part of the body, which divide the band
of engraved ornaments into three sections, each again sub-
divided into two symmetrical portions; so that the same
motive, composed of spirals on a background filled up with
meanders, is repeated six times on the band. Below this

Plate XXXIII.

Bronze Jar of the Early Chou Period.

PLATE **XXXIV.**

Pottery Jars Decorated with Reliefs (Figs. 1, 2) and Grooves (Figs. 3, 4).

band is another, somewhat broader, band of almost the same description, but with some slight variations; and a similar decorative band is laid around the base. The engraving is extremely clear and deep, and of rare beauty. According to the communications of Mr. Ōkazaki, this bronze piece was discovered by a farmer in a village near Peking, in a well on his estate. Long exposure to the action of the water has produced a grayish color on the surface of the bronze, which has become very brittle, like those buried in the ground. Mr. Ōkazaki sets the date of this jar as the time of the *Early Chou*, and I think there is no reason to discountenance his judgment.

In the thirty-fourth book of the "*Hsi ch'ing ku chien,*" twelve *p'ou* of the *Chou* and eighteen of the *Han* dynasty are illustrated. That shown on p. 17 comes nearest to the Ōkazaki bronze referred to above. These vessels vary extensively as to details of formation; but in their main features as expounded above, particularly in the lowness of the neck, they fairly coincide. There is one piece (p. 39) with a nozzle (*liu* 流), and a base formed to look like a rope. The undecorated (素甋) pieces of the Han time (pp. 39–42) display unmistakable ceramic forms.

There is a class of pottery styled by the Chinese *kuan* 罐 ("jar, pot"), a large number of pieces of which have been found in the Han tombs. As they are of very uniform shape and character, only four are selected for illustration here (Plate XXXIV). These jars are bulging in their upper part, which is usually decorated with a band of reliefs. They have no neck, but only a short vertical rim 0.6–0.8 cm in width. From the band downward they gradually narrow towards the foot, the diameter of which somewhat exceeds that of the mouth. They are all glazed a green in various tinges.

The figures on the reliefs are the same as those on the *hu* vases described later on, but most of them are so badly effaced that they cannot be studied. The one illustrated in Fig. 1 of this plate is 12.8 cm high, and has a diameter at the mouth of 6.7 cm, and at the bottom of 7.6 cm. It has a distinct straight foot 1 cm in height. It is glazed a light green, interspersed with silver and gold iridescence. A second jar of this type (Fig. 2, Plate xxxiv) measures 13.8 cm in height. It has a diameter at the mouth of 7 cm, and that of the bottom is nearly the same. There is no separate foot, and the glaze is dark green. The next jar (Fig. 3 of the same plate) is 12.5 cm high, with a diameter at the mouth of 6.3 cm, and at the bottom of 6.8 cm. There is no band of reliefs, but the shoulders are decorated with one groove encircling the jar, and there are two grooves around the bottom. The glaze, for the most part, is exfoliated. The last of the four jars figured on this plate (Fig. 4) is 12.9 cm high. The diameter of the mouth is 6.1 cm, and that of the bottom 1 cm greater. There is no relief-band, but it is decorated with grooves in a manner similar to that on the preceding piece. It is glazed a leaf-green, which has dissolved to a great extent, showing iridescent portions, and interspersed with dark-blue spots.

Under the name *hu* 壺, Chinese archæologists include a large number of bronze vases which vary widely in shape, and are utilized for different purposes. In the "*Po ku t'u*," where they are treated of in Books 12 and 13, fifty-three pieces are described, three of which belong to the *Shang*, seventeen to the *Chou*, and thirty-three to the *Han* dynasty. The "*Hsi ch'ing ku chien*" illustrates vases of the type *hu* in Books 19–22, all together a hundred and seventy-three pieces, — one of the *Shang*, seventy-five of the *Chou*, and ninety-seven of the *Han* period. The best synopsis of the subject is found in the "*T'u shu chi ch'êng*" Vol. 1117, *k'ao kung tien* Book 189, *hu pu hui k'ao*, where cuts of the principal types are also given from the "*Po ku t'u.*" Vessels shaped like baskets, gourds, bottles, and many others,[1] are included in this group; but here we have to deal with the *hu* merely in so far as our pottery vases are concerned, which represent only one of its many variations. There is clear evidence that all these vases have been modelled after bronze ones; that their principal shapes have been derived from them, although, with the great variability in form of the single parts, the potters may have been led to freer changes in some details. The best proof for this statement is found in the different modes in which the ornamental animal heads on the sides of bronze and pottery vases are executed.

---

[1] Among them are even a watering-pot with a rose (*p'ên hu* 噴壺: see *Hsi ch'ing ku chien*, Book 22, p. 24) and a hot-water bottle in the shape of a drooping calabash, used to warm hands and feet (see *Po ku t'u*, Book 13, p. 3: 觀其形 制有類垂匏。上為之口可以貯湯。蓋溫手足之 器也). To the group of *hu* belongs also the so-called "pilgrim's bottle," which HIRTH (Fremde Einflüsse in der chinesischen Kunst, p. 66) traces back as far as the Han dynasty on the ground that in the illustrations in the *Po ku t'u* (Book 13, pp. 12 and 13) he thinks he finds the *Urmuster* of this vessel. Basing his conclusion on this argument, P. REINECKE (Zeitschrift für Ethnologie, Vol. XXIX, 1897, p. 161) affirms that the Chinese pilgrim's bottle is derived from a Greek type, and represents one of the influences of the Greek art of Bactriana and Ferghana opened to China since the travels of Gen. *Chang Ch'ien*. This statement is untenable; for the *Hsi ch'ing ku chien* (Book 20, pp. 23 and 25) illustrates two bronze pilgrim's bottles from the age of the *Chou* dynasty, which excludes the idea of any Greek influence in this type by way of Central Asia. The former of these vases shows obviously that its shape was developed quite naturally from the round globular *hu* vases which had then long existed in China: it presents itself as a transitional form from the globular to the quadrangular vase, and can be but a link in the chain of genuine autochthonous Chinese development. If there is any resemblance to a Greek type, it is merely incidental, as hundreds of such incidental resemblances may also exist between forms of vessels of the entire Old and New Worlds.

Fig. 32 represents a bronze vase of the Han time after

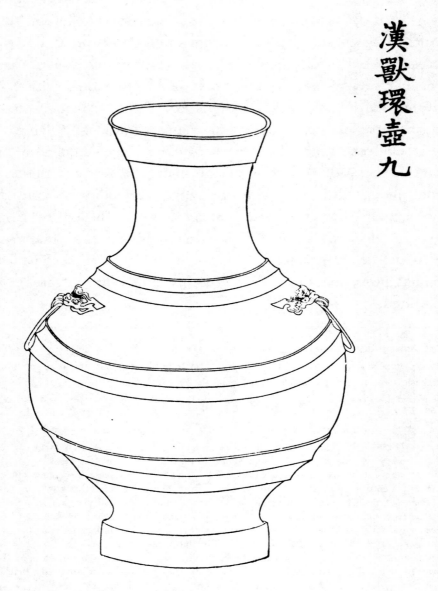

漢
獸
環
壺
九

Fig. 32.   Bronze Vase of Han Time (from *Hsi ch'ing ku chien*).

the "*Hsi ch'ing ku chien*," which will suffice to make the

漢獸鑪一

漢獸鑪二

Figs. 33, 34. Bronze Censers in Form of Quadrupeds (from *Hsi ch'ing ku chien*).

identification of the bronze with the pottery type intelligible. It is defined as a "vase with animal rings" (*shou huan hu*).[1]

Fig. 35 represents a quadrangular bronze wine-jar (方壺)[2] in the Museum's collections, and was obtained by me in *Hsi an fu*. It bears a remarkable inscription proving it to have been made in the *Shang fang*,[3] the court atelier of the Han, in the year A.D. 12.[4] It is introduced here chiefly to show the animal's head on the side, through the nose of which is a movable ring by which to handle the vessel. On the vases of pottery we meet exactly the same animal or tiger head in the same varied ornamental and conventional designs

---

[1] *Shou* is a generic term for "animals." No other explanation for this animal head is offered in Chinese archæological books. The Chinese of the present day, at least generally in *Hsi an fu*, call it a tiger's head; and I have adopted this term in the description of the pottery vases, — as a mere term, however, without pretending to know anything of the original signification of this ornament, speculations on which, in my opinion, would be fruitless, for under the *Chou* dynasty it had already developed into a conventional design, which was merely copied by artists of the *Han* time. The type of the ornamental *shou* on the sides of vessels was perpetuated on porcelain and nephrite vessels, and, up to the present day, particularly on snuff-bottles of porcelain, clay, and glass. The two illustrations in Figs. 33, 34, reproduced from the *Hsi ch'ing ku chien*, Book 38, pp. 49 and 50, represent bronze censers of the Han time in the shape of "quadrupeds" (獸), the lid being worked into the animal's head, the vessel into its body, which is supported by four legs with four claws. The first figure, besides, shows another head, on the side of the body under the hinge of the cover. There is a certain resemblance in the conventional style of these two heads and the ornamental "tiger" heads on vases, although the tusks extant in the former distinguish them from the latter; but whether the two have the same phase of development or a common starting-point remains doubtful. The above two types are interesting, and deserving of closer study, in view of the strong resemblance which they bear to the Greek Gorgon faces, which, as is known, appear in the art of India as *Rākshasa*, and were thence adopted with Buddhism in Eastern Asia, where they come up as terrifying emblems on shields in China and among the Dayak (see A. R. HEIN, Die bildenden Künste bei den Dayaks auf Borneo, Wien, 1890, pp. 43, 46).

[2] The quadrangular bronze vases occur as early as the *Chou* dynasty (see *Hsi ch'ing ku chien*, Book 19, pp. 3, 39, 40; Book 20, pp. 22, 32, 33). For those of the Han time, see Ibid., Book 21, pp. 9, 18, 56–63; Book 22, pp. 7, 8.

[3] Regarding this institution, see F. HIRTH, Fremde Einflüsse in der chinesischen Kunst, pp. 12, 13; CHAVANNES, Les mémoires historiques de Se-ma Ts'ien, Vol. III, p. 477.

[4] This inscription is reproduced and translated in the chapter on inscriptions.

as on the corresponding bronzes, with the only difference that the real, loose ring of the latter has changed on the pottery into a dead-ring brought out in low relief. It has here a mere symbolical function, which can be derived only

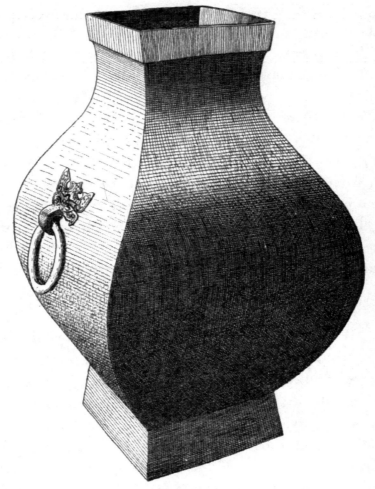

Fig. 35. Bronze Quadrangular Wine-Jar of Han period.

from the imitation of practicable metal rings; and this fact proves, first of all, the dependence of the pottery vases on the bronze type. Special references to types coinciding as to shape and ornamentation will be given at the places where these points are discussed.

A considerable number of these pottery vases have been found, which are decorated with bands showing figures, animals, and decorative designs in relief. No such reliefs have ever been discovered, to my knowledge, on the analogous bronze vases, which are embellished only with engravings made in the cast. These relief bands must therefore be claimed as a feature peculiar to this pottery, and exclusively to this particular pottery of the Han dynasty, as they are never found again in any of a subsequent period. If anything proves that this pottery is a genuine production of the Han period, it is the representations on these reliefs, as will be shown from the style and execution of the designs, and from the motives of art, which partly arose and partly reached their climax during that epoch, and which bear a striking resemblance to the art displayed on the Han bas-reliefs of Shantung, although undoubtedly they must be older than those. The importance of these reliefs, however, goes far beyond the pale of special Chinese archæology. We shall point out the connection of this art with the Old-Turkish and Scythian cultures, the art of which is partially based on sources and motives derived from Mycenian civilization.

As regards the use of the vases in question, those of bronze were employed for holding wine, from which it might be inferred that those of pottery served the same purpose. This seems more than likely from the fact that the burial of wine-vases in the grave (eight for the emperor) is expressly mentioned in the "Annals of the Later Han Dynasty,"[1] and also for the reason that the modern wine-jars (酒壺) of

---

[1] See DE GROOT, The Religious System of China, Vol. II, p. 403.

PLATE XXXV.

Pottery Vase Decorated with Medallions.

China are all made of pottery, although of much smaller dimensions than the old ones.

As to the technique of the reliefs, I am inclined to believe that the negatives of the single figures were formed in separate clay moulds, by means of which the positives were made of wet clay and then stuck on to the vase. This is particularly the case with the ornamental tiger-heads, in many of which the places where they join the body of the vessel are still plainly visible. The most instructive example of this kind is the vase shown on Plate xxxv, in the possession of Mr. Marsden J. Perry of Providence, R.I. It is 52.5 cm high, and has a maximum diameter of 35 cm. To the body of this vase eighteen medallions — partly forming animal heads, partly ornamental figures, each separately moulded — are attached. I think, too, that the same process was applied to the figures on the bands of reliefs, for some of them are easily separable from the vase, or have even been lost; and in one case several figures are placed upside down (see Plate L, Fig. 1). This latter may be explained as due to a mistake of the potter, which was possible only in case he made use of separately moulded single figures. Finally, this is the very process of making reliefs on pottery in the China of to-day.[1] Figs. 2 and 3 on Plate xxxvi illustrate two such clay stamps (diameter 5.6 cm and 6.5 cm respectively). They were obtained from a kiln near Peking. These ornaments in relief are chiefly used for the decoration of teapots and large basins, to which they are applied as above explained. Fig. 1 on the same plate shows a small wheel (from the same kiln) 14 cm in length, with the negative of a decorative design incised along its circumference. This, when pressed

---

[1] See the remarks of D'ENTRECOLLES in DU HALDE, A Description of the Empire of China (English edition), London, 1741, Vol. I, p. 342.

against the vessel while it is being turned on the wheel, leaves its impression in the form of a continuous raised band.[1]  I have no doubt that this instrument was applied also in the Han time, and that certain geometric ornamental rows (as, for example, the triangular dentil ornaments on the vase shown in Fig. 3 on Plate XLIV) were made by means of it.  On the accompanying plate (XXXVI) I have included two modern clay stands (Figs. 4 and 5) moulded by hand, — one of rectangular form (12.5 cm long and 5.8 cm wide), and one consisting of three spurs or prongs (prongs 6 cm, 6.5 cm, and 7.5 cm long), on which vessels are placed during the firing-process.  The latter form was utilized in the manufacture of pottery during the Han time, as we find a number of large vases and also other pieces, to the bottom of which the remains of the clay stand, in triangular form, still adhere.[2]  The occurrence of such so-called "spur-marks" I have indicated in every case.

According to the extent of the ornamentation, the vases may be classified into three groups : —

1. Plain pieces, without any decorations in relief, the only ornamentation being grooved lines around the belly or neck, or both.[3]  To this class belong the vases shown on Fig. 2,

---

[1] Such wheels of clay or metal were known also to the ancients: see H. BLÜMNER, Technologie und Terminologie der Gewerbe und Künste bei Griechen und Römern (Vol. II, p. 112), where an illustration is also given.

[2] "Spur-marks are generally to be found on the backs of Japanese [porcelain] plates, but are not often met with on Chinese [porcelain] plates.  They are the remains of, or marks left by, the small pillars of clay which were employed to support the plate while in the kiln, and which stuck to the glaze, and had to be broken off when the plate was taken out of the oven.  These spur-marks are generally found on the back of the plate.  Some are more marked than others, and they vary in number." — W. G. GULLAND, Chinese Porcelain (2d ed., London, 1899), p. 117.

[3] Such pieces with no decorations, and even without grooves (but with two round ears on the sides), date back to the *Chou* dynasty.  See the engraving in *Hsi ch'ing ku chien* (Book 19, p. 19), and one with two grooves (Ibid., Book 20, pp. 10, 11). For those of the Han time see the same work, Book 21, pp. 20, 21.

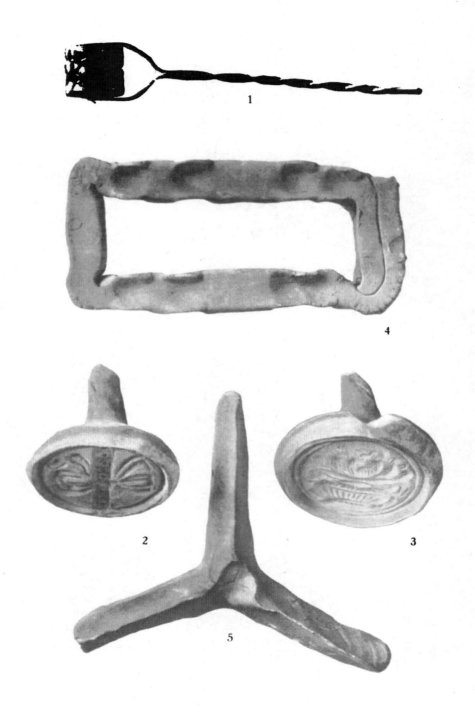

Fig. 1. Clay Stamping-Wheel; Figs. 2, 3. Clay Stamps; Figs. 4, 5. Clay Stands.

PLATE XXXVII.

Pottery Vases with Relief Bands.

PLATE XXXVII.

Pottery Vases with Relief Bands.

PLATE XXXIX.

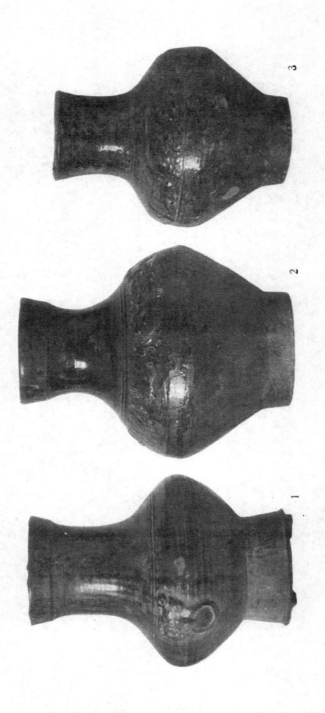

Pottery Vases.

PLATE XL.

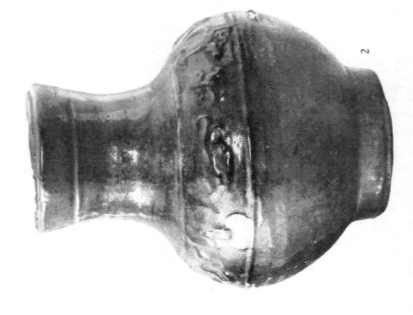

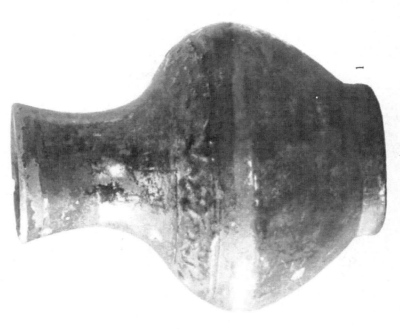

Pottery Vases.

Plate XLI; on Fig. 2, Plate XLII; and also on Figs. 1 and 2, Plate XLIV.

2. Vases having, besides grooves, two ornamental tiger-heads with rings in low relief, one on either side of the belly[1] (see Fig. 4, Plate XXXVIII; Fig. 1, Plate XXXIX; Fig. 4, Plate XLI; Plate XLII; Figs. 1, 2, Plate XLIII; Fig. 3, Plate XLIV; Plate XLV).

3. Vases decorated with bands of reliefs in addition to lateral tiger-heads and grooved lines. These pieces form the majority. (See Figs. 1–3, Plate XXXVIII; Figs. 2, 3, Plate XXXIX; Plate XL).

As regards the formation and variation in shape of this last type of vase, I distinguish the following sub-types: —

(1) With bulging, almost globular belly; with low base, clearly set off; and with constricted neck, broadening out towards the mouth. To this class belong the vases shown in the plates mentioned below, all of which have bands of reliefs.

Plate XXXVII, Fig. 1. — 30.2 cm high; diameter of mouth, 10.6 cm; of bottom, 10.5 cm. Dark-green glaze, with groove around neck.
Fig. 2. — 24.7 cm high; diameter of mouth, 8.3 cm; of bottom, 10.2 cm. Dark-green glaze; two grooves around neck.
Fig. 3. — 24.6 cm high; diameter of mouth, 8.4 cm; of bottom, 10.1 cm. Dark-green crackled glaze.

Plate XXXVIII, Fig. 1. — 22.6 cm high; diameter of mouth, 8 cm. Leaf-green glaze.
Fig. 2. — 29.5 cm high; diameter of mouth, 11.1 cm; of bottom, 11.8 cm. Yellowish-brown glaze.

Plate XXXIX, Fig. 3. — 33.3 cm high; diameter of mouth, 11 cm; of bottom, 14.6 cm. Glaze light and dark green with brown spots intermingled.

Plate XL, Fig. 1. — 17.9 cm high; diameter of mouth, 6.5 cm; of bottom, 7.5 cm. Yellowish-brown crackled glaze.
Fig. 2. — 18.5 cm high; diameter of mouth, 6.5 cm; of bottom, 8.1 cm. Greenish-yellow crackled glaze.

---

[1] Corresponding bronzes existed likewise in the *Chou* dynasty (see *Hsi ch'ing ku chien*, Book 20, pp. 9, 12, 17, 18, 21).

10

(*a*) **Same shape as (1), but with neck divided into two parts, — the lower one somewhat hourglass-shaped, the upper one like a high rim or lip protruding over it a little.[1] To this type belong the vases shown in the following plates.**

Plate xxxviii, Fig. 3. — 36 cm high; diameter of mouth, 11 cm; of bottom, 12.2 cm. Light-green glaze, much worn.
    Fig. 4. — 34.4 cm high; diameter of mouth, 12.5 cm; of bottom, 14.2 cm. Light-green glaze, much worn.
Plate xxxix, Fig. 2. — 31 cm high; diameter of mouth, 10.6 cm; of bottom, 11.5 cm. Brilliant dark-green glaze.

(2) **With ellipsoidal belly; base higher than in (1), and somewhat trapezoidal in outline; neck long and tubular, and similar to that in (1a), in having an overlapping vertical rim. This type is represented on the following plates, one of which, however (Plate xli, Fig. 1), does not belong wholly to this type, it being a combination of (1) and (2), having the belly and base of (1) and the neck of (2).**

Plate xxxix, Fig. 1. — 34.8 cm high; diameter of mouth, 12.5 cm; of bottom, 15.4 cm; base, 5.7 cm high. Brilliant green glaze with fine crackles. Three large spur-marks on bottom; two double-grooved circles around belly.
Plate xli, Fig. 1. — 36.4 cm high; diameter of mouth, 12.5 cm; of bottom, 15.7 cm; base, 1.9 cm high. Dull-green crackled glaze with yellow and gold spots. Four spur-marks on bottom.
Plate xlii, Fig. 2. — 37 cm high; diameter of mouth, 12.2 cm; base, 6.8 cm high. Leaf-green glaze, parts of it exfoliated. Potter's mark on inner side of neck (see chapter on inscriptions).
Plate xliii, Fig. 1. — 40 cm high; diameter of mouth, 15 cm; of bottom, 15.6 cm. Deep-green glaze. No. 1484 of British Museum, London.
Plate xliv, Fig. 1. — 46.4 cm high; diameter of mouth, 15.2 cm; of bottom, 19.4 cm. Rim, 1.6 cm wide. Olive-green crackled glaze with many gold flecks. Two spur-marks on bottom.

---

[1] In the corresponding bronzes a band of ornaments is sometimes laid around this part. See the exact counterparts to type (1a) in the *Hsi ch'ing ku chien* (Book 19, p. 17; Book 20, p. 7), which are from the *Chou* dynasty. For bronze types of the *Han* time, see Ibid., Book 21, pp. 28, 29, 32, 35, 36, etc.

PLATE XLI.

Pottery Vases.

PLATE XLII.

Pottery Vases.

PLATE XLIII.

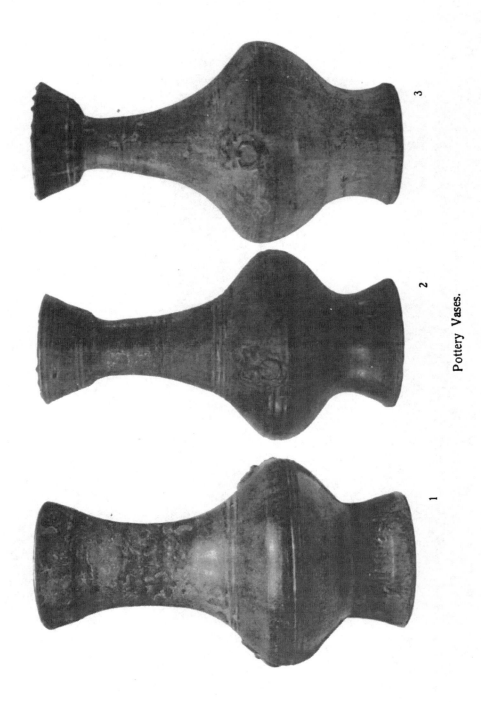

Pottery Vases.

PLATE XLIV.

Pottery Vases.

(*a*) With ellipsoidal belly; base high, and shaped in the form of a truncated cone; neck slender, and gradually and gracefully narrowing towards mouth, and surmounted by a bowl-shaped top-piece of trapezoidal outline, harmonizing with the shape of the base. Compare with the specimens shown on the plates mentioned below.

Plate XLI, Fig. 3. — 43.1 cm high; diameter of mouth, 13.2 cm; of bottom, 16.2 cm; base, 8.9 cm high. Dull-green glaze with yellow and black stripes running vertically.

Plate XLIII, Fig. 2. — 43.7 cm high; diameter of mouth, 12.5 cm; of bottom, 16.7 cm. Decorated with two rows of parallel raised lines 6–7 mm long. Yellowish light-green glaze with gold, blue, brown, and black spots. For inscription on this vase, see chapter on inscriptions.

Fig. 3. — 43.5 cm high; diameter of mouth, 12.7 cm; of bottom, 15.5 cm. Very light-green glaze with iridescence; light-blue specks on the neck; network of fine crackles. Ten accidental warts on rim.

(*b*) Belly and base coincide with those of (*a*) under this type; neck broad, tubular, almost cylindrical, with slightly concave outlines. For examples, see plates mentioned below.

Plate XLI, Fig. 2. — 36.8 cm high; diameter of mouth, 12.6 cm; of bottom, 15.6 cm; rim, 1.2 cm wide. Dark-green glaze, partially changed into silver gloss.

Plate XLII, Fig. 1. — 38.2 cm high; diameter of mouth, 14.1 cm; of bottom, 14.7 cm. Dark-green glaze with brown, black, and gold flecks. Spur-marks on bottom.[1]

(3) With belly bulging more width-wise than length-wise; base low, as in (1); neck similar in shape to that of (2), the top-piece being larger, and projecting still more over the neck. See the following plates.

Plate XLIV, Fig. 2. — 42.4 cm high; diameter of mouth, 14.7 cm; of bottom, 20.6 cm; rim, 1 cm wide. Olive-green crackled glaze, in natural preservation. Only one tiger-head with ring, the other is wanting.

Fig. 3. — 38.6 cm high; diameter of mouth, 15.2 cm; of bottom, 21.5 cm; rim, 1.8 cm wide. Three spur-marks on bottom, which is entirely covered with glaze. Interior of bowl-shaped neck filled with a lustrous yellow glaze, which is the only example of this kind known to me.

---

[1] For corresponding bronze, see *Hsi ch'ing ku chien*, Book 22, p. 2.

Plate XLV, Fig. 1. — 52 cm high; diameter of mouth, 16.3–16.6 cm; of bottom, 20.8 cm; maximum circumference, 113.8 cm; rim, 1.8 cm wide; foot, 4.3 cm high. Glaze around the neck, light leaf-green; around the middle and lower parts, dull green and iridescent. Worn on some portions of the base, and the whole glaze crackled. There is one groove around the neck, and there are two triple grooves around the body, enclosing tiger-heads. In the rings is a human figure in low relief, holding a stick or pestle in the right outstretched arm, and with a square vessel in front of it resembling a mortar (perhaps the well-known motive of the hare in the moon pounding drugs, but the figure is not clear).

Fig. 2. — The vase shown on this plate, which is in the possession of Mrs. E. C. Bodman of New York, is one of the largest and finest of its kind. It is 48.5 cm high; the diameter of the mouth being 15.3 cm, and that of the bottom 21.9 cm. The rim is 1.9–2 cm wide. The greatest circumference of the belly is 109 cm; that of the base 69.7 cm, and of the neck 42.9 cm. The base is 4.3–4.6 cm high, and the bottom measures the same in thickness. There are two grooves on the lower part of the belly 2 cm apart (the lower one 4 mm wide, and the upper one 5 mm), and a double groove 7 cm above the latter. Farther up, a raised line separates the belly from the neck, which gradually narrows towards the mouth, having a protruding lip. A very remarkable feature regarding this vase is that its interior, and even the inner side of the bottom, are glazed completely, the latter exhibiting a brilliant silver lustre. On the upper portions of the neck, inside, a deep bright-green color is preserved, which must be considered as the original appearance of the glaze. This color has also kept well on the outside of the lip; but the nearer we approach the base, the more we see that the process of dissolution of the glaze has advanced. It is worthy of note that the two sides of the vase, defined by the two vertical lines drawn over the tiger-heads, present a distinct aspect as regards the appearance of the glaze. That on the one half shows, on the whole, its original green hue; while on the other half a general pale-yellowish and a silver gloss prevail. It must be supposed, therefore, that it was on this latter side that the vase lay in the ground, or that just this particular half of the vase, for some unknown cause, was more exposed to the action of subterranean moisture and gases. The whole outer surface is covered with small particles of earth, presumably loess, which adhere very firmly. It seems to contain iron, as appears from the numerous brown flecks of rust scattered over these earth incrustations. A small iron nail, the head of which is visible on the surface, has been driven into the side of the base — whether incidentally, or for what purpose, I am unable to state.

(4) With belly and neck like those of (2); base in the

PLATE XLV.

2

1

Pottery Vases.

shape of a truncated cone, and comparatively high. The bottom of the vessel is between the belly and the base, and not on a level with the base-line, as in all previous pieces; so that this base would be defined more correctly as a standard (Plate XLI, Fig. 4). This is the only piece of its kind known to me. It is 38.4 cm high; the diameter of the mouth varying from 9.5 cm to 9.6 cm, and that of the base of the standard (10 cm high) being 19.5 cm. The rim at the mouth is 0.8 cm wide; that below, 1.1 cm. The nose of one of the two lateral tiger-heads is perforated, but not the other one. The vessel is covered with a crackled leaf-green glaze extending over a large portion of the interior side of the neck. The body is decorated with eleven circles glazed a light yellow. Among the "marvellous objects of good foreboding" (瑞 應) engraved on the bas-reliefs of the Wu family, we see a jar, explained in the inscription accompanying it as being of silver, with a base drawn as a large trapezoid,[1] apparently conforming to the truncated-cone base of this piece of pottery.

### RELIEFS ON VASES.

The subjects represented in the reliefs on vases, and, as we shall see hereafter, also in those on hill-jars, are, generally speaking, but manifold variations of one and the same main theme, that may be defined as animal pictures and hunting scenes. Hunting was a favorite pastime of the ancient Chinese. Several odes in the "*Shih king*" describe grand hunts. The sportsmen of those days pursued the wild goose and duck, the boar, wolf, fox, hare, badger, deer, tiger, and rhinoceros.[2]

It was not under the Han dynasty that hunting-scenes

---

[1] See *Kin shih so*, section *shih so*, Vol. 4.
[2] JAMES LEGGE, The Chinese Classics, Vol. IV, Prolegomena, p. 148.

were first conceived of as a subject of art; but they had their beginning in the *Chou* period, as we may judge from a unique and costly monument in bronze, a tazza (*tou* 豆) of the *Chou* dynasty, an engraving of which is preserved in the "*Hsi ch'ing ku chien*," Book 29, p. 26 (reproduced in Fig. 36). Its height, according to the text (the measurements being recalculated from the statements in Chinese feet), is 24.4 cm; its depth, 10.8 cm; the circumference of the mouth 19.2 cm, that of the foot 12.4 cm; and its weight, 63 ounces. No further description is given in the "*Hsi ch'ing ku chien*." This bronze vessel is defined as a "*tou* with a hundred animals" (*pai shou tou*), from the representation of a number of animals and hunting-scenes engraved on it. In this respect it is of primary importance for the study of the reliefs on the Han pottery. It shows, first of all, that the representation of animals, and particularly animals in motion, was a favorite subject of earliest Chinese art, and that it had already adopted at that time a fixed, stereotyped, conventionalized expression. Two typical scenes, which are several times repeated, represent especially the prototype and exact counterpart of a type frequently occurring on the Han reliefs. For brevity's sake we might define these scenes as "the hunter and the animal." In the one marked *a* in the illustration (once on the cover, and twice on the body, of the vase) we see a man-like creature holding a spear which he is thrusting into an animal.[1] In one case he holds the spear over his head with both hands, directing it right into the monster's jaws.

---

[1] This and the other scenes represented on this *Chou* bronze refute the following statement made by Dr. S. W. BUSHELL in his book Chinese Art (Vol. I, p. 89): "The human figure never occurs in these primitive bronzes, . . . and we see only occasional sketches of animals, such as tigers and deer. The artist, in fact, neglects the *ordinary animal world* to revel in a mythological zoölogy of his own conception, peopled with dragons, unicorns, phœnixes, and hoary tortoises."

No attempt has been made to sketch this hunter's face, and only a tuft of hair is outlined on the head.[1] Further, con-

周百獸豆

Fig. 36. Bronze Tazza (from *Hsi chʻing ku chien*).

sidering this point, his wild appearance, and his sometimes

---

[1] This, however, must not be imputed at the outset to lack of ability of the artist, or looked at in the light of the hackneyed phrase "primitive art:" it seems rather to

weird representation on the Han pottery, one might feel tempted to see in this strange figure a demon-like being or a deified hero, and I have therefore several times called it "the demon," in descriptions of the pottery reliefs, as a brief term for this otherwise undefinable type.[1] On the *tou* in question we see exactly the same type, but without the spear, in the scenes marked *b*, which occurs three times on the edge of the cover and once on the vase itself. This variation without weapons will be met with likewise on the reliefs of our Han pottery: the coincidence even extends so far that the position of the arms — one of which is curved almost like a spiral, while the other is stretched forward towards the animal opposing him — is the same in the works of Han ceramics. In *b* on the body of the vessel, the hunter is accompanied by a dog, which is leaping up to him. It would lead us too far, and be beyond the present scope of our work, to submit all the figures here represented to a minute analysis. I may be permitted to call attention, however, to the five running hydras engraved in the round knob of the cover, and to the strutting birds with outstretched wings on the main part of the cover, both of which types we shall meet again on our Han reliefs. I may mention also the two animals (marked *c* in the illustration) which represent undoubtedly a tapir, that is found again on the Han bas-reliefs of the *Hsiao t'ang shan*.[2] From an archæological

---

have had its natural cause in technical conditions, considering the small dimensions of the whole vessel, and accordingly the narrowness of the space at the artist's disposal, which did not allow him to go into details.

[1] In doing so, I do not by any means pretend to say that this figure is really and unquestionably a demon.

[2] See CHAVANNES, La sculpture sur pierre en Chine, p. 76; WILLIAMS in The Chinese Repository, Vol. VII (1839), pp. 46, 47; PAUTHIER and BAZIN, Chine moderne, p. 587; F. DE MÉLY, Les lapidaires chinois (Paris, 1896), p. 170. It is curious to note that the engraving of the tapir in the *Pen ts'ao kang mu* and other works on

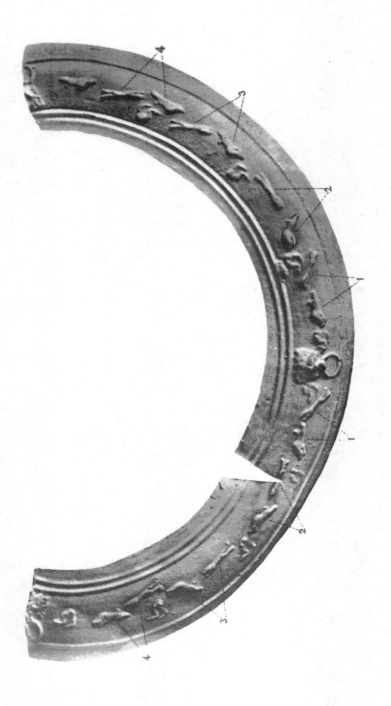

Relief Band on Vase.

view-point the application of spirals in the bodies of animals (see particularly on the base of the tazza) is highly interesting.[1]

The bands shown in the following illustrations are made from photographs of plaster casts taken from the reliefs.

That in Plate XLVI is divided by the two tiger-heads into two symmetrical sections. To the left of the central tiger-head is a greyhound in flying gallop, with only two legs represented, hunting a wild boar in the same pose. The dog's body and head are directed upward, the boar's downward, the outlines of both animals forming an obtuse angle. The boar is well characterized by its short clumsy body, its long protruding snout, the shape of its head, and its short erect tail. The eyes of animals are visible in many cases as thickened spots in the clay. This pair is followed by three other pairs of boars and dogs, but each in a different position. In the second pair the dog's head is turned downward, as if descending a steep path, while the boar seems to be running upward. In the third pair, both animals are somewhat more distant from each other. The legs of the dog, galloping *ventre à terre*, form almost one line with its paunch, to indicate that it has reached its utmost degree of speed; while the boar, at half right angles to the ground, appears to be making its final struggle to escape. The fourth representation shows us the greyhound making its last leap towards its adversary, as if ready to catch him at the next bound. These four scenes are intended probably to represent the four consecutive stages of the same chase; i.e., they

---

natural history, far from giving a drawing of the animal after nature, reproduces almost minutely the ancient stereotyped form as we find it already established in the *Chou* and *Han* time. The tapir is called *mo* 貘, and is believed to eat iron: hence it is designated in Manchu *selekje* (*sele,* "iron," + *je-,* "to eat"); in Mongol, *selekcen;* in Tibetan, *gas lcags zan* ("eater of cleft iron").

[1] Compare my remarks in Globus, Vol. LXXIX, pp. 69, 70.

describe the story of the same dog pursuing and finally reaching the same boar in four scenes, which thus become a real "moving picture." Between these four scenes will be observed four curious figures, besides one to either side of the tiger-head, evidently used to separate the four stages of the chase, one from another. They resemble a pair of human feet; but what they were really intended to represent — whether merely an ornamental figure or something else — cannot be made out. They may have stood as an abbreviation for the hunter; but this, of course, remains mere guess-work. If we turn now to the other side of the band, we remark along the upper edge of the relief the same ornament which is applied to the lower border on this side. In general, the subject of this other half is symmetrical with the one described, in that here are also four pairs of hunting-dogs and boars; but whereas the feet-like figure mentioned appears on both sides of the central tiger-head, it occurs on but one side of the opposite tiger-head. On the other side of it, where a larger space is left, is a single galloping boar, which we must imagine to be outside of the chase, and there merely as a space-filler. Starting from the tiger-head in the centre of the illustration, we see, to the right, the single pairs in the same postures of motion as the corresponding ones on the other side. The distances between the animals are somewhat wider, as this entire section of the band is longer.

The next band of reliefs (Fig. 1, Plate XLVII) is very artistically composed. There are the two main sections formed by the two tiger-heads, as on the other vases; but there is a further division of each of these sections into halves by means of a large wave, as on some of the hill-jars (see Index), consisting of several curved lines starting from the lower border-line, and rising in a bold sweep almost to the upper

PLATE XLVII.

Relief Bands on Vases.

edge of the band. On the upper ends of the cast we see the under border-lines terminating in wave-lines, which exist also on both sides of the central tiger-head, but are omitted in the illustration, the cast being cut short here. These wave-lines are continued above into spirals, which confine the figures next to the tiger-heads in a separate field; so that now the entire band results in a sixfold division. To the right of the central tiger-head with ring is a tiger drawn in profile, with wide-open mouth, in flying gallop, only two of its legs being represented. Behind it is another tiger trotting, its head turned backward. Two hydras (*ch'ih* 螭), one above the other, follow, with the same distinction, that the one is represented galloping, the other trotting. On the summit of the wave we observe two animals, which I think represent monkeys,[1] the one in an attitude like that of climbing down a tree, the other in a squatting posture. In the next field are three animals, — a dragon-like monster above, very indistinct; below, a bird with outstretched wings and large tail-feathers, stalking (perhaps a wild goose or swan); and a lion (or tiger) behind it. This animal is separated from the adjoining man, with animal-like demoniacal grimace and wide-open mouth, by a spiral above, continuing into an undefinable animal and a wave-line below. The same figure is represented to the left of the central tiger-head in our illustration, and again in the field next to the latter, in a scene depicting the struggle with a lion, where this figure is thrusting a spear into the lion's jaws. The "demon's" feet are straddled apart (sprawled out), his arms wide apart.

---

[1] The monkey occurs in Chinese art as early as the *Chou* dynasty. In full figure, this animal is found on the lids of the *Chou* bronze vessels called *ho* 盉 (*Hsi ch'ing ku chien*, Book 31, pp. 57, 59). It is frequently seen on the roofs of houses on the Han bas-reliefs. Regarding later representations of monkeys, see Index (monkey).

The lion is represented in a very realistic way, the shape of its big head, its well-formed mane, the carriage of its tail, its claws, leaving no doubt as to its identity. A large flying bird is over it. The monkeys over the wave, and the two hydras, and a tiger looking backward, comprising the next field, tally with that described on the other side. The following galloping tiger is likewise identical with its counterpart in the other section; but, instead of the decorative spiral there, we see here, over its tail, a squatting monkey. The vase bearing this relief is 36.8 cm high; diameter of mouth, 12.5 cm; diameter of foot, 14.5 cm; height of foot, 3 cm. The band (6.4 cm wide) is bordered above by two raised circles; below, by one circle. The glaze is a grayish green, without brilliancy, and is wanting in a few places.

In the description of the bronze *tou* of the *Chou* dynasty above (p. 150), I pointed out the occurrence of the demon-like figure which we here meet on the Han pottery. There as well as here this figure plays at the same time the rôle of hunter. What its mythological bearing or real significance is, is not revealed. A similar figure occurs repeatedly on the bas-reliefs of the *Wu tsê shan;* and a selection of three of these, drawn after the "*Kin shih so*," is shown in Fig. 37. There it is always single, and by itself, placed behind a chariot or a rider on horseback.[1] Unfortunately, none of them is accompanied by an explanatory inscription, as is the case with most of the figures on these bas-reliefs. A striking analogon to this demon is found on two unpublished bas-reliefs of the Han time, rubbings of which I obtained in the city of *T'ai an*. Both represent the same subject, with some variations. The one stone slab is said to stand in the open

---

[1] One of these is referred to by CHAVANNES (La sculpture sur pierre en Chine, p. 62) as "a fantastical animal which is merely an ornament pure and simple."

air, six *li* north from *I chou fu* in Shantung.[1] The other
is stated to be in the Confucian temple of *Chi ning chou*,
in a room inaccessible, and closed to the public.[2] The two
bas-reliefs are much effaced and decayed. The animal in
front of the figure is hardly traceable. The outlines of the
figure, which is in profile, are still clear. It has the left
arm stretched backward, and the right arm uplifted as if to
strike a blow. The artist may have made a correct observation of this natural position of a man about to make an
assault, but evidently he was not able to draw correctly

Fig. 37. Demon-like Figures on Han Bas-Reliefs of the *Wu tsê shan* (from *Kin shih so*).

what he had seen or imagined. The right arm forms almost
a half-circle, the hand being just above the head. There is
no sign of an elbow to either arm. This figure occurs on
the reliefs of the Han pottery, in most cases one arm (sometimes both) held akimbo, but even then outlined in the same
semicircular curve. A demon of nearly identical type is found
also on several metal mirrors of the Han time (see *Kin shih*

---

[1] This is the assertion of the Chinese rubbing-man, for which I must decline all
responsibility, not having visited the place myself. I acquired, all together, seventeen
rubbings of Han bas-reliefs said to be in or around *I chou fu*.

[2] I was in *Chi ning chou* myself for several days, and inspected all monuments in
the Confucian temple there, but did not find the bas-relief in question, probably for
the reason stated above by my Chinese authority. I have several other rubbings of
bas-reliefs alleged to be in the same secret room.

*so*, section *kin so*, Vol. VI, mirrors Nos. 37 and 42: the latter mirror is in the Museum's collections, Cat. No. $\frac{70}{13983}$).

The band illustrated in Fig. 2, Plate XLVII, belongs to a small jar of the type *kuan* (see p. 135). The jar is 12.1 cm high; the diameter of the mouth being 6.8 cm, and of the foot 7.5 cm. It is made of brownish-red clay covered with a yellowish-green glaze. The band is 4 cm wide. Glaze and relief figures have suffered much from long burial under ground. The band is not divided by tiger-heads. Starting from the upper right-hand side of our illustration, we see the demon with pointed head[1] and open mouth, shaped almost like a bird's beak, rushing at a boar. His arms are rounded, the left one akimbo. His body is spotted with tiny black dots. The boar is very true to nature: its mighty head with pointed muzzle, its short erect tail, the shape of its body and feet, are well brought out. In the following scene we meet with the same type of demon, only in this case the right arm is uplifted, and he is making for a galloping tiger. Next to this one there is a trotting hydra, then a large bird walking with wings wide outspread, perhaps a wild goose or swan; next a boar galloping full swing towards an open-mouthed tiger which seems to threaten to devour it. The next field is occupied by a galloping winged hydra, in the delineation of which the artist has bestowed great care. The rapidity of its gallop is carried out with a certain boldness combined with gracefulness of form. Above its tail we see a small quadruped with head formed *en face* and short tail, the right fore-leg uplifted, while the other three feet stand on the ground as if starting to walk. It is probably

---

[1] Thus it appears on the pottery; but we shall see hereafter that this triangle seems to represent a cap, which is the well-known Scythian cap derived from Siberian art.

Relief Band on Vase.

a sheep. Under the head of the hydra are four snakes
standing erect, whose heads and eyes are plainly visible on
the pottery. In front of the hydra is a hunter in the attitude
of running, his left arm stretched forward towards the monster,
while he holds in his left hand an oval object that might be
taken for a weapon. The man's right arm is drawn backward, and his right hand is touching his pointed cap.

The band seen in the next illustration (Plate XLVIII) is the
decoration on a globular vase 29.7 cm high; the diameter
of the mouth being 11.1 cm, and of the bottom 12 cm. Its
short neck with straight rim curves outward somewhat. About
2 cm below the rim is a deep dark groove. A band of
reliefs runs around the body, bordered above by a groove
and two raised circles connected by rows of parallel oblique
lines, below by one raised circle. The foot of the vase is
2.8 cm high. The glaze is a yellowish green. A black
color has been profusely applied under this glaze in spots
or lines to express various ornamental motives. The two
tiger-heads that divide the band into two sections are worked
out more elaborately than on the other pieces. The ears
are formed into two superposed spirals dotted with black
spots. The forehead is indicated by a half-circle, and brows
are marked over the circular eyes. Inside of the nose-ring
there is an ornamental figure which cannot be identified.
The same figure appears outside, to the right of the central
tiger-head. The lower border of the band is filled with a
wave-line. The small knobs interspersed here and there for
the ornamental purpose of separating the single fields — an
artistic expedient analogous to that on certain metal mirrors,
from which it might probably be derived — are a remarkable
feature of this relief. On either side of the conventional
heads with rings is a realistic tiger. The one on the right

is in flying gallop, with but two legs represented; the other, trotting; while on the opposite side both tigers are trotting, and one of them has the head turned backward. The stripes of the tigers are indicated by lines in black. Over the tiger in the right section there is a bird-like creature hard to recognize. A deer follows, with a knob on either side of it; then a large flying bird (wild duck or goose) with outstretched vertical wings, so many of which appear on the Han bas-reliefs; under it a small animal that I cannot determine. The next scene portrays the struggle between the demon and the lion. The hunter or demon has an animal's head, with mouth open as if letting forth a yell. His left arm is resting on his hip, while with his right hand he is pushing a weapon against the lion's head. The hair of the lion's mane is colored black. Longitudinal stripes run over its body. In the next field we see a rider on horseback looking backward, and shooting with bow and arrow. His horse is in flying gallop. This is a conventional type which very frequently occurs on the reliefs of this pottery, with only two variations. In the one the horse gallops over a plain, and is accordingly drawn horizontally; in the other the galloping horse is represented rearing up in a leap to the top of an ornamental wave. On one of the bas-reliefs of the *Hsiao t'ang shan* (Plate XXXVIII in the work of CHAVANNES) a battle between archers on horseback is represented, the horses being in full gallop. None of the archers there, however, are shooting backward. As to the motive of the flying gallop in ancient Chinese art, compare the admirable paper of SALOMON REINACH, "La représentation du galop dans l'art ancien et moderne" (Extrait de la *Revue archéologique*, Paris, 1901, pp. 83 et seq.), who has made a most careful study of the horses on the bas-reliefs, accompanied with very instructive sketches. Compared with the latter, the flying gallop appears on the pottery reliefs still more exaggerated and carried to the extreme, the legs of the animal being stretched out farther, and the legs of each pair being much closer together, so that they sometimes dwindle into one in the artist's representation, and are almost on the same horizontal plane with the outline of the paunch.[1]

It is further the great and indisputable merit of S. Reinach to have established beyond doubt the fact (p. 85) that the horses of the Han bas-reliefs — characterized by the enormous development of the breast, neck, and croup, by the massive proportions of the body and the slenderness of the feet — represent a peculiar breed of noble horses imported from Bactriana into China at the time of the Han dynasty, as is well attested by the records of Chinese history. Only the horses in the battle scene of the *Hsiao t'ang shan* (quoted above) form an exception, and represent the ordinary (but originally also imported) small breed of horses now found throughout China (the now usually so-called "Mongolian pony"). The horses on the reliefs of our pottery must all be attributed to the latter race, as comparison with those of the bas-reliefs proves, not one example of the large Bactrian horse being found there. One might think from such a conclusion that the vases bearing these reliefs must or could have therefore originated in the time before the arrival of Bactrian horses in China; i.e., before 138 B.C., the date of the first journey of Gen. *Chang Ch'ien* to the West; but

---

[1] This motive of art makes a passage in CHUANG TZŬ more intelligible. "I judge of horses as follows," remarks the philosopher (in the translation of GILES, p. 312), "their straightness in running must be that of a line." This saying is only imaginable, if, indeed, the philosopher had actually seen an artistic representation of the conventional motive of the flying gallop, which I am therefore inclined to infer must have existed in China at least in the third century before Christ.

11

for myself, I should not attribute much importance to such negative evidence.

On the left section of the band of reliefs (Plate XLVIII) we see to the left of the trotting tiger (already referred to above) the same type of mounted archer shooting backward at a tiger just ready to pounce upon him. Under the horse is a galloping greyhound hunting a hare, likewise in full-swing gallop. The hare is characterized unmistakably by his long upright ears and short tail. Above the hare there is a creature resembling a snake, and over the latter a quadruped, not unlike a monkey, walking on all-fours. Farther along, in the upper row, are three flying birds, and a trotting hydra below.

The next illustration (Fig. 1, Plate XLIX) shows a decoration on a vase measuring 35.7 cm in height; the diameter of the mouth being 13.5 cm, and of the foot 15.9 cm. The foot is short, measuring but 3.3 cm in height. The band is 7.4 cm wide, and shows one circle near its upper border, and another near the lower border. The tiger-heads divide the relief into two parts, each of which is separated into three fields by triple wave-lines enclosing the field below and reaching nearly to the top of the figures. These lines are glazed in light color that they may the better stand out from the green-glazed background. Concerning the tiger-heads with rings there are two remarkable features, — first, the ornamental treatment of the head-line; second, what appears elsewhere as the nose, to hold the ring, is here shaped like a paw with four distinct claws. To the right of the central tiger-head in our illustration is a galloping tiger in profile, the two front-legs and one hind-leg of which are represented. The mouth is wide open, and the tail carried erect. From the side of the wave next to it a snake or dragon-shaped head is emerging. The second field is occupied

PLATE XLIX

Relief Bands on Vases.

by the "demon" running at a lion, but he holds no weapon in his hands. The animal is lifting its right paw upward against him. Over the wave following are two snake-like creatures difficult to recognize. The third field is taken up by a large winged galloping hydra. Over it are two running creatures, each with one arm uplifted: they are probably monkeys. To the left of the central ornamental tiger-head is a trotting tiger, a counterpart to the galloping tiger on the other side. Its head is represented *en face*, with two eyes, nose, mouth, and two erect ears, distinctly marked. Each of its feet has four claws. Its tail runs in a double curve, and is striped with parallel black lines under the green glaze. In front of this tiger we see the type of the "demon" in a queer attitude, as if, startled at sight of the tiger, he were about to cry out and fall backward. Unfortunately there is a crack in the vase, which passes just over his body. His whole appearance is that of an animal. His feet are set with three claws. He has a large rounded tail behind. His fingers, as seen on one hand, bear long nails (compare Fig. 37). A spiral running downward is attached to the back of his head, the significance of which I do not know. His large wide-open jaws are turned upwards. In the wave separating this field from the next one we observe also an animal *en face*, but so effaced that its identification seems hopeless. Over the top of the wave is a well-designed boar in full gallop, the victim of the mounted archer in front of it. The outlines of the horse are extremely good, but the head is too large in proportion to the body. A sort of tassel is attached to the under side of the horse's chin. In the wave concluding this field is a galloping greyhound, and over it a somewhat indistinct bird with tail-feathers. As to the third field, we find there (to the right in the illustration) a galloping tiger of the same type as that mentioned above to the right of the central tiger-head. In this case the tiger's stripes are well indicated by black lines. The vase has a brilliant leaf-green crackled glaze, and black spots showing under it.

The band illustrated in Fig. 2, Plate XLIX, is divided into halves by the two tiger-heads. On either side of these is a tiger, the corresponding pairs being marked *a* and *b* respectively in the illustration. Although the general posture of the two animals is identical on both sides, neither their measurements nor details are just alike. The two corresponding tigers (*a* in the figure) agree fairly well. There is a slight difference in the curve of the tail. Both are in flying gallop, and at the same time ready to leap, as indicated by the uplifted fore-legs; both have the mouth wide open; both fore and hind legs are represented. In the two tigers *b* a remarkable difference is to be found, in that the one near the central ornamental tiger-head is represented galloping, with but two legs visible, while the other is trotting. Only in the former has an attempt been made to represent the tiger's stripes by dark-green lines under the light-green glaze. With both the tail is erect, but only that of the galloper is curved. Between these pairs of tigers a hunting scene is represented in each section. On the right half of the illustration we see the hunter on horseback, setting his horse in full gallop. The horse is generally well and clearly shaped, the head extended, the nostrils emphasized. The rider is looking backwards, and shooting with bow and arrow at a boar behind him. The right hand is just pressing the arrow against the bow, the string being drawn back against the man's breast. The six animals following him are arranged in two rows, one above the other. In the upper row there are three

Relief Bands on Vases.

boars in full gallop. Below there is first a trotting dog, apparently the hunter's hound. Then follow what appear to be two deer, the first stretched out in full gallop with only two legs represented, the second running. On the other half we notice to the left of the tiger in the lower row a dog very similar to that on the corresponding side, then two boars facing each other, the smaller one running preparatory to leaping, the other larger one going at full speed. Above these are two hydras, and another bigger one by itself on the left. These creatures have long lank bodies, a long narrow curved neck, dragon heads with wide-open mouths, a long tail ornamentally curved and ending in a half-circle. Above the last-mentioned monster there is a man, indicated only by thin raised lines. His head is very indistinct. He seems to be kneeling on the ground, and to be holding a lasso in his hand, as if about to insnare the head of the animal in front of him. The vase is 30.5 cm high; diameter of the mouth, 10.6 cm; diameter of foot, 10.9 cm; height of foot, 2.2 cm. It has a short stout neck with straight rim, and is glazed a brilliant leaf-green.

The band shown in Fig. 1, Plate L, is from the vase represented on Plate XXXII, Fig. 2, and is 4.8 cm wide. Peculiar to the designs on this piece is the tortoise enclosed by each of the tiger-rings. It is also a singular fact that three of the animals depicted on this band are represented upside down: namely, a winged hydra with horns; another winged hydra with long neck, looking back and biting into its recurved tail; and a third hydra without horns and wings, with tail out straight, and head turned. I cannot imagine what induced the artist to invert just these animals and leave the others in their normal position, unless it may be that the animals were separately moulded, and the molds for these three were by accident wrongly placed. The other animals represented here do not differ much from those on the other pieces. Twice we find the motive stag and boar superposed, the boar turned to the side opposed to the stag; between these two pairs, a tiger. There are three greyhounds hunting stags, two of them unfortunately much effaced, but the other so happily drawn, with its long pointed head, big breast, and thin loins, that it is unmistakable.

The vase shown in Fig. 2 of Plate xxxvii is decorated with the band of relief figures here represented (Fig. 2, Plate L). The band is 5 cm wide. To the left of the central tiger-head there is an animal of lion-like shape, with an object resembling a spear in its forehead (cf. p. 155), being attacked by an archer on horseback. The same type of animal, with a curved weapon sticking in its head, occurs on the upper right-hand side of the band, in a struggle with the "demon," who occurs in exactly the same position to the left of the horseman on the left of the central tiger-head. On that side, farther to his left, are a boar in flying gallop, a boar with a fish and a snake above it, a trotting tiger with head *en face*, and an animal turning its head backward and stretching its tongue out. The same type is found again to the right of the central tiger-head, from which it is separated by a fish, and is followed by the mounted archer shooting at a tiger behind him. There is a fish again over the lion-like figure on the right end of the band.

The vase to which the next band (Fig. 1, Plate LI) belongs is 36.7 cm high, with a rim 1 cm wide. The diameter of the mouth is 12.6 cm, and of the foot 15.7 cm, the height of the latter being 1.7 cm. Around the middle of the neck are two parallel grooved lines (see Fig. 1, Plate XLI). The glaze is a dark green with yellowish spots. The band of

PLATE LI.

Fig. 1. Relief Band on Vase; Fig. 2. Cover to Pottery Tripod Kettle shown in Fig. 38.

reliefs is 6.9 cm wide. A curious circumstance regarding the representations on this relief is that the archers do not ride horses, but are mounted on hydras. These animals have an extremely long, slightly curved neck, a dragon-shaped head, a thin body, and long legs. The animal to the right of the ornamental tiger-head *a* is in the position which S. Reinach designates as "*cabré allongé;*" i.e., the hind-legs rest on the ground, while the fore-legs are raised high in the air.[1] The two feet of the rider are visible under the monster's belly. His figure is much effaced, but it would seem that it was never shown in much detail, only the outlines of the head (without eyes or any further features), upper part of body, and arms being expressed. His head is looking backwards, which can be inferred merely from the position of the bow. The left hand is pressing the arrow against the arm of the bow, while the right one was intended by the artist to be pulling back the string. He failed in this attempt, however, through ignorance of the laws of perspective, since this arm appears stretched to the right side of the figure. Under the front leg of the hydra we see what might be taken for a man lying wounded or dead on the ground. In attempting to express this subject the artist was utterly helpless. A tiger is galloping at full speed towards this man, as if ready to devour him. A trotting tiger is represented over this one, all four of the legs being shown. Under the hind-legs a wave line begins, and continues to the tiger-head *b*. The running tiger above mentioned seems to be the target of the hydra rider following, which type does not differ from the preceding one, except that the animal

---

[1] REINACH (l.c., p. 90) observed only one case of this position in China on the bas-reliefs, which, however, may be explained differently, as will be shown later on; here also, I think it is merely incidental.

has a larger muzzle. Next follows the dancing demon in front of a tiger, whose head seems to be represented *en face*. This type of demon varies somewhat from the other designs. The left arm (behind the body) is curved somewhat into a spiral, as if the artist had attempted to place it akimbo. The right arm is raised high up, as if threatening the encountering beast. The right leg is drawn backwards as if kneeling, and the left foot is uplifted. To the right of the tiger-head *b* we notice a tiger and a hydra with rider, in the same attitude. The archer's manner of sitting on the animal's back differs somewhat from the other representations in that the feet are stretched forward as far as possible, and the body leaned backward. He seems to be shooting at a running tiger behind him, whose head is *en face*. The next field, which is continued on the left of our illustration, is occupied by a fourth rider and three animals. What the figure of the riding animal was really meant to represent in this case is hard to decide, as it seems to be halfway between a hydra and a courser. The head and neck are those of the former, and there even seems to be a horn-like process in front of the head. Along the neck a horse's mane is clearly outlined; and the legs with distinctly marked hoofs are evidently borrowed from the same animal. The tail, on the other hand, belongs to the hydra. The under animal is a hydra; the next two, a running and a galloping tiger, about which there is nothing new to be said.

In the next three plates, portions of relief bands are illustrated directly from photographs of the vases on which they occur.

On Plate LII is shown a portion of a relief band on a vase glazed a dull green. The vase is 42.6 cm high, and has a diameter at the mouth of 13.8 cm, and at the bottom of 16.1 cm. The band is 7.3 cm wide. Below, on the left

PLATE LII.

Portion of Relief Band on Vase.

PLATE LIII.

Half of Relief Band on Vase.

(not visible in the illustration), is a stag in *galop volant*, the antlers indicated by faint yellow lines under the glaze. A greyhound is following it in close pursuit. Above these two is a tiger with head *en face*, the eyes, brows, and stripes on the body, being indicated in black. Above the tiger is a small dog, and another below it to the right. Then follows a large snake, over which is a hydra on which a man is standing with left foot raised. Below, in front of the hydra, is again a dog, and next a wave on top of which are two undefinable animals. Farther along is a trotting tiger or leopard, the four legs all represented; and below this a bird, apparently perching on a rock, its beak bent forward grasping a round object, the outline of which is glazed yellow (perhaps a bird of prey standing over its booty).[1] The last figure in this field represents a large running tiger, its stripes being indicated by deep blackened grooves. Above it is one snake; and below it, two snakes.

The next relief, the figures of which are unusually high, is from a vase 26.1 cm high, with a diameter at the opening of 8.6 cm, and across the bottom of 10.1 cm, and with a deep-brown lustrous glaze. The upper edge of the band (Plate LIII), which is 3.8 cm wide, is bordered by a raised zigzag line. The first relief, beginning on the left of the illustration, is the demon-like creature, with legs wide apart, the left arm stretched out behind, and the right reaching the muzzle of the sheep or goat standing in front of him. A standing bird with large round conspicuous breast, pointed beak, and long pointed tail-feathers, follows next; and beyond it, a tiger with head *en face*, and an animal with two horns,

---

[1] On one of the bas-reliefs of the *Hsiao t'ang shan* we see, on the roof of a house, a bird of prey rushing down on a hare, in about the same posture (see CHAVANNES, La sculpture sur pierre en Chine, p. 74 and Plate XXXVI).

prancing, the two front legs bent upwards, its muzzle reaching the thigh.

The vase, half of the band of which is shown in Fig. 1, Plate LIV, is 35.2 cm high, with a diameter at the mouth of 11.1 cm and at the bottom of 14.5 cm, and has a yellowish-brown glaze. The band of reliefs is 6.7 cm wide. Inside of the tiger-rings is a running quadruped with head *en face* and straight tail. Then follow a wild goat or chamois, the right leg gracefully raised as if starting to walk; a large flying bird above it, with a long tuft of feathers attached to the back part of the head; a galloping boar, and below it a double wave-line; a huge angry wave with a quadruped on either side of it resembling those in the tiger-rings. A galloping tiger is jumping up against the other side of the wave, and behind it is a trotting lion or tiger (the head effaced), while above the tail of the latter is a small animal. Then come a raised wave up which a chamois is leaping, and a man standing with his right foot on a hydra, his left foot raised, with knee bent (not in the illustration).

Fig. 38 represents a tripod vessel of pottery which is undoubtedly a close imitation of the ancient tripod caldrons called *ting* 鼎 .[1] It has a cover, and on either side a somewhat high ear or loop-handle. It is 17.7 cm high up to the handles, and 24.9 cm wide between the handles. The body of the vessel is somewhat globular, the centre being indicated by a knob on the lower part outside. Its three feet (4.2 cm high) form an isosceles triangle, the base of which is 7.5 cm long, and the sides each 8.5 cm. These are moulded in the shape of elephants; and the heads, on

---

[1] The burial of pottery tripod vessels is expressly mentioned in the Annals of the Later Han Dynasty (see DE GROOT, The Religious System of China, Vol. IV, pp. 403, 708).

PLATE LIV.

Fig. 1. Half of Relief Band on Vase; Fig. 2. Clay Disk.

which there is no indication of a face by the usual ornamental lines, adhere to the bottom of the vessel. The long trunk is distinctly curved; and the nostrils are indicated by two shallow cavities on the under side of the vessel's foot, between which a narrow raised and unglazed part is left that must have stood on the spur in the furnace. The loop-handles are

Fig. 38. Pottery Tripod Kettle (from specimen in Clarke collection).

7 cm high and 4.5 cm wide, and are turned over towards the outside. The perforations in the two handles have received somewhat different treatment. The one is regularly pierced in the form of a narrow staple; the other shows the same general form on the outer side, while the inner side is blind, so to speak, being covered with a thin wall of clay,

leaving only a small irregular hole in the middle, due no doubt to an accidental break.

The cover (17.8 cm in diameter), represented from a cast in Fig. 2, Plate LI, is the only ornamental part, the decoration being in low relief. The lines are glazed a light yellow under the dark-green glaze. The arrangement of the ornaments strongly reminds one of that on the metal mirrors of the same period, the points in common being the principle of arrangement (in circular zones), the knob in the centre, and three knobs in the second zone, which divide it into three sections. These three knobs are displayed in the form of an isosceles triangle having a base 8.8 cm and sides 9.3 cm long. All together, there are four zones. In the centre, around the high pointed knob, is a four-leaved rosette arranged crosswise. Whether it is intentionally composed of leaves, or is merely a decorative emblem, remains an open question.[1] Animals in motion are represented in the next zone, and so distributed that four (three boars and a stag, all galloping) fill the base of the triangle, six (one boar, two tigers, one effaced, the demon, and tiger *en face*) one long side, and three (hydra, unicorn, and tiger) the other long side. The

---

[1] An analogous leaf-cross appears on a clasp (*kou* 鉤) of the Han time, the engraving of which will be found in the 兩罍軒彝器圖釋, Book 11, p. 8 a. On the disk of a Han tile designed in the *Ts'in Han wa tang wên tzŭ* (秦漢瓦 當文字) of *Ch'êng tun* (程敦), Book 2, p. 5 b, four similar heart-shaped leaves are grouped around a central knob, but not connected with it, as on the lid above. In the *Hsi ch'ing ku chien* (Book 39, p. 27), exactly the same floral design is shown on a metal mirror attributed to the Han period, and explained in the descriptive text as 作菱花形 "having the shape of blossoms of the water-chestnut," *Trapa natans* L. (see BRETSCHNEIDER, *Botanicon sinicum*, Part II, p. 219; Part III, p. 440). It therefore seems evident that the Chinese regard this pattern as a plant-ornament. In the same way a somewhat different design on a Han tile is explained (see FORKE, Mitteilungen des Seminars für orientalische Sprachen, Vol. II, Sect. 1, p. 100, and Plate 12, Fig. 148).

following zone (not visible in the illustration) is made up of a waved band ornament, and divided off by a circle from the outer zone forming the edge, and decorated with a zigzag line, the so-called "hill-ornament" (山紋). The whole vessel is coated with a leaf-green glaze, which also covers the inside of the bowl. The under side of the lid shows only a glazed spot in the centre (perhaps accidental). The glaze on the outside is darker than that on the inside, and the original color is in a better state of preservation. The fact that the inside is covered with a layer of loess, brilliant with colors of the rainbow, indicates long burial of the piece under ground, presumably in such a way that the lid was taken off or placed beside it; while the outside, protected by stones or the walls of the coffin, was less exposed to the action of earth or other agencies. It would also seem that a falling heap of earth had exerted a strong pressure on the open (uncovered) vessel, as is evident from the breaking-off of the two ears, which were subsequently attached again.

Because of the similarity in principle of ornamentation, an otherwise undefinable piece of pottery may be described here. It is a flat disk of gray clay unglazed (Fig. 2, Plate LIV), 9 cm in diameter, thick in the centre, and gradually thinning out towards the periphery. The signification of this piece is unknown to me.[1] On both sides the same decoration

---

[1] My friends in *Hsi an fu* called it *ya hsiu* 押袖 "[a thing] to be put into the sleeve," and explained it as an instrument to be turned to and fro on the palm for strengthening the hand-muscles, in the same manner as iron metal balls made especially in *Pao ting fu* are used nowadays. There is probably little or no ground for this interpretation. Very likely this clay piece was made in imitation of some metal object. These balls have been made the subject of a special paper by W. JOEST, Allerlei Spielzeug (Internationales Archiv für Ethnographie, 1893, Vol. VI, pp. 163–173). The observation made therein (p. 165) by VON BRANDT, that they are manufactured only in *Pao ting fu*, is correct. An additional note to this paper by G. SCHLEGEL appeared in the same volume (pp. 197, 198). Schlegel's identification of those iron balls

is brought out in low relief, but with this difference, that the star on one side has six points, while that on the other side is composed of seven triangles.

The notable feature of the ornamentation is the mode of arrangement in zones, which is the same as that on the metal mirrors and bronze drums of the Han period. In the first outer zone there is a double circular row of small round beads or knobs, divided off by a circle. The next zone is occupied by the star figure. Inside of the triangles, and in the spaces between them, are spiral curves and rows of beads. The third zone, surrounded by two concentric circles, is taken up by a circular row of beads. In the centre is a svastika-like figure, the four ends of which are curved inward in a somewhat spiral-like shape; beads also are scattered here and there, and four spirals are attached to the circle surrounding the central figure.

## HILL—CENSERS.

Among the bronzes of the Han period illustrated in the "*Hsi ch'ing ku chien*" there is a sort of censer called *po shan lu* 博山鑪 or 博山香爐 — i.e., "brazier or stove of the vast mountain" — from the cover being formed into the shape of a hill,[1] which is the principal characteristic of

---

with the term *ling* 鈴 is inaccurate. They are never called *ling*, but *kang ch'iu* 鋼毬 "steel balls," in the colloquial language of Peking. GILES (p. 244) gives 手球 or 鐵球. They are known also in Tibet, frequently used by Lamas, and styled in Tibetan *lcags ril*. The word *ling* denotes a bell, and nowadays especially a small bell carried by dogs, horses, donkeys, etc., on the collar. The various quotations given by Schlegel have therefore nothing to do with the subject in question.

[1] 蓋象山形。故仍謂之博山鑪。 *Hsi ch'ing ku chien*, Book 38, p. 46 b. The hill-decoration on covers seems to have been in vogue down to the T'ang dynasty. The *Hsi ch'ing ku chien* (Book 39, p. 56) shows a *hsün lu* 薰鑪 of the T'ang time with a cover worked into hills, having the appearance

these vessels. Some have loose covers, others have the lids made fast. Aside from these, they consist of a bulging bowl resting on a stem usually fitted into a dish below.[1]

The purpose of the under dish is, as explained in the "*Hsi ch'ing ku chien*," to receive the particles of ashes dropping from the burning incense. In this connection the *Ch'ien lung* Catalogue criticises the view expressed in the "*Pu pi t'an*" (an appendix to the "*Mêng ch'i pi t'an*," middle of the eleventh century), which says that, to keep the brazier, when hot, from burning the mats, a dish is made for putting water in to soak its foot; but the "*Hsi ch'ing ku chien*" retorts, that, as the dish of the censer under consideration has a perforation in the bottom, it cannot hold water, and serves only the purpose stated above.[2] It is questionable, however, whether, in that passage of the "*Pu pi t'an*," the *po shan lu* is understood, as this term is not employed, or whether it is not simply a brazier placed on a separate tray.[3]

On the other hand, the "*K'ao ku t'u*" (composed between 1086 and 1094), Book 10, p. 15 b, gives the following

---

of a relief map, but not attaining the height or the effect of the Han pieces. In a discussion of the *po shan lu* in the *Kin shih so* (*kin so*, Vol. 3), a passage from a song of the poet LI T'AI-PO 李太白 (eighth century) is quoted, in which mention is made of a golden *po shan lu* cast by a famous bronze-founder of *Lo yang*, to show that this kind of vessel was highly valued at the time of the *T'ang*.

[1] 此器分三層。蓋爲山形。下爲承盤。 — *Hsi ch'ing ku chien*, Book 38, p. 43 b.

[2] 補筆談謂防罏熱灼席則爲盤薦水以漸其趾。今按盤底有孔非可以盛水者。惟云承火炧之墜則得之矣。 — Ibid., p. 43 b. The perforation mentioned in the text appears somewhat exaggerated on the accompanying illustration.

[3] Indeed, in looking up in the original the full passage in question, I find that the *Pu pi t'an* does not speak there at all of *po shan lu*, but generally of bronze censers (*t'ung hsiang lu*), and also adds expressly that the dish serves, moreover, to receive the ashes, in the same words as the *Hsi ch'ing ku chien*, which claims this sentence as its own.

explanation: "The censer resembles a vast mountain (or island) in the midst of the sea. Below, it has a dish brimful of hot water to soak and steam the fragrant herbs, resembling the eddying rings of the ocean waves. Many generations are in possession of this vessel, the shapes of which vary in size."[1]

How are these contradictory statements to be reconciled? Both are correct, for each of these explanations refers to a different variant of the same type of vessel. The confusion is caused by the "*Hsi ch'ing ku chien*" overlooking this fact, and closely linking its quotation from the "*K'ao ku t'u*" with that from the "*Pu pi t'an*," all unaware of its own inconsistency in rejecting the definition of the latter book (regarding the filling of the dish with water) and at the same time admitting the assertion of the "*K'ao ku t'u*." The bronze *po shan lu* pictured in the latter work is coverless (compare the similar type in nephrite in Figs. 40, 41), and the mountains are moulded in the bowl in plastic shape, so that naturally there is no other expedient but to burn the aromatic substances in the lower dish. The interpretation heralded by the "*K'ao ku t'u*" is therefore correct. The five hill-censers in the "*Hsi ch'ing ku chien*," however, form another variation of this type; and all are composed, as set forth above, of a round-shaped bowl and a separate perforated hill-shaped cover. Accordingly this type is suited for burning

---

[1] 香爐象海中博山。下有盤貯湯使潤氣燕香以象海之回環。此器世多有之。形制大小不一。 *T'u shu chi ch'êng* Vol. 1124, *K'ao kung tien* Book 236, *lu pu hui k'ao* p. 2, quoting this passage, and repeating the engraving of the *K'ao ku t'u*, writes 四 instead of 回, probably from misreading the variant 𦥯 (for 回), which is in the text of the *K'ao ku t'u. Hsi ch'ing ku chien* has 回 correctly.

the incense in the bowl, the smoke issuing through the holes of the cover. For this reason the receiving-dish could easily be dispensed with in these pieces, as, of the five in the "*Hsi ch'ing ku chien*," two are without it, one being provided with a low circular stand or base only, and the other posed on three feet.

According to the unanimous judgment of Chinese archæologists, the type of the hill-censer originated under the Han dynasty, and was the first censer ever made, as the following texts will show.

Lü Ta-lin 呂大臨, author of the "*K'ao ku t'u*" (l.c.), makes the following statement: "It is an old affair of the Han dynasty. When the rulers are married, they bestow the *po shan* censers. The book '*Tsin tung kung chiu shih*' says, 'When the heir-apparent is in mourning, he has one *po shan* censer.'"[1]

Lung Ta-yüan (end of twelfth century) expresses the view that the *po shan lu* was originally a vessel of the Han palaces, whence it came, and that the people of the Han dynasty made it for a long time.[2]

"*T'u shu chi ch'êng*" (Vol. 1124, *k'ao kung tien* Book 236, *lu pu hui k'ao* p. 2 a) quotes from a book, "*Ku ch'i p'ing*" 古器評 ("Commentary on Old Vessels"), thus: "At the time of the Emperor *Wu* of the *Liang* dynasty, in the second year of the period *Chung ta t'ung* (A.D. 530), which

---

[1] 按漢朝故事。諸王出間則賜博山香爐。晉東宮舊事曰。太子服用則有博山香爐一云。 *San ts'ai t'u hui* (section 器用, Book 2, p. 38), reproducing the *po shan lu* figured in the *K'ao ku t'u* with its text, writes 出閣 for 間. The *Ko chih ching yüan* (Book 58, p. 1 a) quotes this passage from the *Tsin tung kung chiu shih* as follows: "When the heir-apparent is in mourning, he has a *po shan lu* of bronze; when he marries, he has one plated with silver."

[2] See the extract from the *Ku yü t'u pu* on p. 190.

12

corresponds to the cyclical signs *kêng hsü, Wang Chung-ta* —
of whom there is no historical record, and whose genealogy
nobody can trace — possessed a *po shan lu.* This vessel
was unknown during the three generations [i.e., *Hsia, Shang,*
and *Chou* dynasties]. CHANG CH'ANG[1] says, in his book
'*Tung kung ku shih,*' 'When the heir-apparent marries, he
arranges two *po shan* censers, and this was the beginning
of the latter.' The cover of the vessel [of *Wang Chung-ta*]
was shaped into a hill, and the vapor ascended up through
it. Below, a hydra (*ch'ih*) and a tiger, with heads raised,
were formed to support it, and it was provided with a dish.
Although very ingenious, still, compared with the old pieces
[of the Han time], it was far from equalling them."[2]

The question regarding the origin of censers in connection
with the *po shan lu* is discussed in another book as follows:
"By means of southernwood and mugwort[3] the ancients com-
municated with the spiritual beings, and did not burn incense.
Therefore they had no censers [literally, "incense stoves"].
The vessels which nowadays are called censers are all made
after models derived from sacrificial vessels in the ancestral

---

[1] Died 48 B.C. (see GILES, Biographical Dictionary, p. 48). Thus the type of the
hill-censer must have existed in the first half of the first century B.C. Another evidence is the fact that *Liu Hsiang* (80–9 B.C.; GILES, l.c., p. 501) is reported to have
possessed a *po shan lu* (see *Hsi ch'ing ku chien,* l.c.; and *K'o chih ching yüan,* Book
57, p. 1 a).

[2] 按中大通二年太歲庚戌梁武帝也。王仲達
於史無傳、莫能考其世系、博山爐。不聞於前
代。惟張敞東宮故事云。皇太子納妃、貝博山
香爐二。豈始於此乎。是器蓋爲山形。通氣於
上。下作螭虎仰首以承之復載以盤。雖若甚巧
然較之古制殆未可彷彿也。

[3] *Hsiao* and *ai,* both species of *Artemisia* (see E. BRETSCHNEIDER, Botanicon
Sinicum, Part II, pp. 247, 252; Part III, p. 145).

temples of the ancients. The censer in the shape of a 'goblet' (*chio*)[1] took its pattern from the goblet of the ancients. The 'lion'[2] censer is made after the ancient tazza with a single stem. The censer in the form of a ball[3] is copied from a cauldron of the ancients, and there are many others of this kind. There are also new casts, but in shape, resembling forms that existed in times of old, with the only exception of the *po shan lu*, which was used in the palace of the heirs-apparent of the Han time. It is from this type that the manufacture of censers was first commenced. There are counterfeits among them which can be discriminated by the color (patina) of the object."[4]

The "*Hsi ch'ing ku chien*" (Book 38, p. 43 b) emphasizes the fact that the real beginning of the *po shan lu* was in

---

[1] An ancient type of libation-cup, said to resemble an inverted helmet posed on three feet, appearing from the *Shang* dynasty. There are three specimens in our collection.

[2] *Suan i*, a conventional type of lion frequent in art (see DE GROOT, The Religious System of China, Vol. III, p. 1324; and chapter on lion, of this paper).

[3] *Hsiang ch'iu*, a sphere composed of two halves, and in open-work carving. A specimen of such a censer of the *Ming* period is in our collection.

[4] 古以蕭艾達神明而不焚香。故無香爐。今所謂香爐皆以古人宗廟祭器爲之。爵爐則古之爵。㹇狼爐則古踽足豆。香毬則古之鬵其等不一。或有新鑄而象古爲之者。惟博山爐乃漢太子宮所用者。香爐之制始於此亦有僞者當以物色辨之。 This quotation is given in *T'u shu chi ch'ing*, Vol. 1126 (*k'ao kung tien* Book 250; 古玩部, *hui k'ao* p. 4), without source; and in Vol. 1124 of the same work (Book 236, *lu pu tsa lu* p. 2 a) as derived from the *Tung t'ien ch'ing lu* 洞天清錄, which is a work of the thirteenth century (see WYLIE, Notes on Chinese Literature, p. 167). The *Ko chih ching yüan* (Book 58, p. 1) has the same extract from the *Pai shih lei pien* 稗史類編, except the last clause regarding the counterfeits. There is one variant in the text of the latter cyclopædia: "a tazza with stem of carnelian" (瑪足豆) in lieu of "single-footed tazza" (踽足豆).

the Han time, as proved by the three books, "*Tung kung ku shih*"[1] 東宮故事, "*Tung t'ien ch'ing lu*"[2] 洞天清錄, and "*Hsi ching tsa chi*"[3] 西京雜記.

"The people of the *Chou* dynasty burned *Artemisia*. The people of the *Han* time commenced to make *po shan lu*, and what they burned in them were only fragrant species of orchids (*lan hui*)."[4]   In the same source it is further related that importations of new aromatics took place under the Han Emperor *Wu* (140–87 B.C.) in consequence of the far-reaching expansion politics of this monarch and the newly opened commercial relations of China.   Then for the first time she received from Annam Baroos camphor and cloves (literally, "fowl's tongue");[5] and, as a result of her intercourse

---

[1] Composed by CHANG CH'ANG 張敞, who died in 48 B.C. (see the above extract, p. 178, on the *po shan lu* of the *Liang* dynasty, in which the passage referred to by the *Hsi ch'ing ku chien* is evidently contained).

[2] A work of the thirteenth century (see A. WYLIE, Notes on Chinese Literature, 2d ed., p. 167).   Compare p. 93.

[3] Sixth century (see A. WYLIE, l.c., p. 189).   There are two passages in this work relating to the *po shan lu*, which I take from the *T'u shu chi ch'êng*.   The one runs, "*Ting Huan* of *Ch'ang an* made a nine-storied *po-shan* censer.   He engraved on it queer birds and curious animals in countless numbers.   All these supernatural and strange creatures could turn round of themselves" (長安丁緩作九層博山香爐。鏤為奇禽怪獸窮。諸靈異皆自然運動).   The other passage says that *Chao* (called *Fei yen*, "The Flying Swallow," from her wonderful dancing), when she was empress, was presented by her younger brother, who was in the palace *Chao yang*, with a golden five-storied *po-shan* censer (趙飛燕為皇后。其女弟在昭陽殿遺飛燕以五層金博香爐).   The former courtesan was made empress in 16 B.C. by the Emperor *Ch'êng ti*.   She resided in the gorgeously fitted up palace *Chao yang*, a part of the *Wei yang kung*, the famous palace of the Han in *Ch'ang an*, erected 200 B.C. (see A. FORKE, in Mitteilungen des Seminars für orientalische Sprachen, Vol. I, Sect. I, p. 112).

[4] 周人炳蕭。漢人始為博山爐而所焚惟蘭蕙。— *Ko chih ching yüan*, Book 57, p. 1a, 周祈名義考。

[5] See E. BRETSCHNEIDER, Botanicon Sinicum, Vol. III, p. 459.

with the countries of the West, Parthian incense[1] and attar
of roses made their first appearance on the Chinese market,
while orchids and *Artemisia* were no longer used.[2] Undoubtedly it was these new aromatics of foreign countries
which made the necessity felt of producing a suitable vessel
in which to burn them, and which set before the metal-founder
and the potter of that time the task of finding a new and
adequate form in which to mould it. What their productions were, we shall presently see. The facts of history agree
perfectly with the view of Chinese archæologists, as quoted
above. Owing to the lack of frankincense in China during
the times before the Han, there was no need of an incense-
burner. It remained for the Han to supply this need after the
import of foreign products; and the result was the formation
of a new type of vessel, the *po shan lu.* This was the only
*ingenious* invention for the purpose of fumigating, and it
found no successor. How the artistic idea underlying the
production of this work sprang from the *Zeitgeist,* we shall
attempt to point out after surveying the specimens at our
disposal. All other vessels of subsequent epochs employed
as censers must be traced back to types of vessels developed
in connection with the ancestral worship of the *Shang* and
*Chou* dynasties, as our Chinese authorities justly remark.[3]
The most favorite form of these, now found in Buddhistic,

---

[1] See E. BRETSCHNEIDER, in Notes and Queries on China and Japan, Vol. IV,
1870, p. 145. Identified with gum benjamin or benzoin (F. PORTER SMITH, Contributions
towards the Materia Medica and Natural History of China, p. 36).

[2] 自武帝通南越中國始有龍腦雞舌等香。通
西域始安息薔薇水等香而蕙蘭與蕭不復用
矣。— *Ko chih ching yüan,* l.c.

[3] Dr. BUSHELL (Chinese Art, Vol. I, p. 93) also makes the correct observation,
"The bronze incense-burners of later times are often modelled in the lines of the
ancient ancestral vessels."

Taoistic, and Confucian[1] temples, is derived from the ancient *ting* 鼎, a bronze cauldron with three or four feet, originally devoted only to meat-offerings. The fact of a type of vessel sanctified for millenniums within the strict boundaries of rigid religious observances, suddenly changing its object under outside currents of influence, but still retaining its original shape, is of paramount ethnological value, since it proves a higher degree of tenacity of forms, and greater change-ability of the ideas embodied in them: the forms survive, while the ideas vanish or alter.

A few words on the designation *po shan lu*. In the *Kin shih so* (section *kin so*, Vol. 3) an inscription engraved on the lower dish of a *po shan lu* is reproduced, derived from the "*Chi ku lu*" 集古錄 of OU YANG-HSIU (A.D. 1007–1072). It will be found in Book 1 of the latter work, p. 20 b. The censer in question was in the collection of *Liu Yüan-fu*[2] (a friend of *Ou Yang-hsiu*), who was for a long time district magistrate of *Ch'ang an*, and availed himself of this opportunity to accumulate a great many antiquities of the *Ts'in* and *Han* periods "hidden in ruined estates and destroyed mounds, and frequently brought to him by ploughmen and herd-boys."[3] The inscription on the *po shan lu* shows that it came from *Lien cho* 蓮勺宮, the palace of the Emperor *Hsüan*, and was made on the day *chi ch'ou* 巳丑 of the first month of the

---

[1] The *Ch'ü fu hsien chi* 曲阜縣志 describes and illustrates by good cuts all religious objects in the marvellous temple of Confucius in his birthplace, the city of *Ch'ü fu*, among them a *hsiang ting* 香鼎 of cloisonné enamel made in 1732 (Book 9, p. 5), and another censer of bronze (Book 14, p. 2).

[2] 劉原甫. In the *Kin shih so*, erroneously written 父.

[3] 多古物奇器埋沒於荒基敗冢往往爲耕夫牧豎得之。 — *Chi ku lu*, Book 1, p. 21 a.

PLATE **LV.**

Figs. 1, 4. Pottery Hill-Censers of Han Time; Figs. 2, 3. Bronze Hill-Censers of Han Time.

third year of the period *Wu fêng* (55 B.C.) by *Hsü An-shou*
徐安守 of *Wei ch'êng* 渭城 (i.e., *Hsien yang hsien*), a
native of *Ting ch'ang* 定昌 (i.e., *P'ing yang fu* in Shansi).
The designation "*po shan lu*" does not occur in the inscription;
but the vessel is called in a roundabout way 銅一斗鼎下槃;
i.e., "a bronze *ting* or cauldron of one peck capacity, with
a tray below."[1]    Accordingly the name "*po shan lu*" seems
not to have existed in the Han time, nor is there any other
testimony to this effect, and, to judge from the sources where
it is found, it may not be older than the sixth century A.D.
Apparently it was first a colloquial expression.[2]    The authors
of the *Kin shih so*, after expressing their satisfaction over
the discussion of the *po shan lu* given in the "*K'ao ku t'u*,"
which they call "very clear" (按考古叙論甚明), attack
a certain writer, *Hsieh* 薛氏,[3] for re-forming the name of the
vessel into "the vast mountain, as it were, in the sea, with
a tray below" (象海中博山下槃), an expression which
is somewhat severely criticised as "very stupid and confused"
(語甚蒙混).

Fig. 1 on Plate LV shows a hill-censer of Han pottery.
It is 17.4 cm high.    The bowl is 4.5 cm deep, and has a
diameter across the mouth of 6.5 cm.    The dish beneath

---

[1] This definition shows that the hill-censer of the *Lien cho* Palace was a type like
the above bronze censer.    Furthermore, the editors of the *Kin shih so* remark that the *po
shan lu* obtained by *Liu Yüan-fu*, and that illustrated in the *K'ao ku t'u* from a private
collection of a certain *Li* from *Lü chiang* in Anhui (廬江李氏), are not
necessarily identical, but that they closely approach each other, which means in our
language that they are two varieties of the same type; but they did not see that the
differentiation of the two must be based on the argument stated above.

[2] *Kin shih so* quotes from a book, *Ku ch'i chi* 古器記, of the above
mentioned LIU YÜAN-FU, who says, 俗謂之博山爐。

[3] Probably HSIEH SHANG-KUNG of the *Sung* dynasty, author of a book containing
inscriptions on bronzes (see BUSHELL, Chinese Art, Vol. I, p. 77).

measures 51.8 cm in circumference. This vessel has the shape of a goblet, resembling somewhat the type of a *tou* 豆 or rummer, narrowing below into a short foot or base, which is connected with a flat round dish having a rim 1.1 cm in width. This measurement denotes at the same time the thickness of the clay of the whole piece (except the cover). The bowl has a rim of the same width, surrounded on its inner edge by a narrow straight rim 0.5 cm wide, to hold the cover, which fits over it. The cover is of conical shape and 6.8 cm high. It is moulded into a hill encompassed by a double row of waves. The upper part is perforated by four holes arranged in a square. Cover, bowl, and plate are coated with a deep, brilliant dark-green glaze in an excellent state of preservation, and veined with a network of fine crackles.

Fig. 2 on the same plate represents the exact counterpart in bronze of the foregoing piece of pottery. Presumably it is likewise a genuine relic of the Han time. It is 16.5 cm high, the diameter of the plate being 13 cm, and the height of the cover 6 cm. The cover is worked into the motive "hill surrounded by water," with twelve perforations along the lower margin, and seven in the upper part. The four upper ones are displayed in the same way (in the form of a square) as in the corresponding pottery. The stem of the bowl is riveted to the dish below. It differs from its companion piece in the slender shape of its standard, which is set off distinctly from the bowl; while in the former the outlines of the bowl gradually merge into the low foot.

This bronze censer corresponds to *po shan lu* No. 1 in the "*Hsi ch'ing ku chien*," but it has no perforation in the dish, like that one, which seems to be an exceptional case, for the *raison d'être* of which I cannot account (see above,

p. 175). The engravings of hydras, or *ch'ih* dragons, on the lower part of the stem in the *Ch'ien lung* bronze, are missing in ours. The former has the stem smooth, while the middle of the latter bulges at the centre, which is the case with the *po shan lu* No. 5 of the *Ch'ien lung* Catalogue.

Fig. 3, Plate LV, illustrates another variety of a bronze hill-censer of the Han time in our collections. It is 9.5 cm high, the diameter of the dish being 13.5 cm. The bowl rests on three short feet riveted to the round plate below. The lid is fastened to the bowl by means of a hinge. Its outer zone appears in open-work of extremely thin metal, which is characteristic of the Han period. Only the raised part on the top is worked up into high relief, forming one central hill surrounded by six waves, which, in their turn, are encircled by another outer row of six larger waves. Between the waves of this row are six round perforations. The whole object is covered with a dark-green patina. Generally this piece corresponds to No. 3 of the "*Hsi ch'ing ku chien*," except in its tripod character, which is found there as No. 4.

Fig. 39 represents a bronze censer of the Han time in our collection, which might be defined as a combination of a *hsün lu* 薰鑪 and a *po shan lu;* but it differs from that in the previous figure by having a handle 11.4 cm long and 0.9 cm wide attached to the bowl, and in lacking the receiving-dish. Its total height is 8.8 cm, and total length 20 cm, the diameter of the mouth being 7 cm, and the width of its rim 0.6 cm. The lid is of the same workmanship and ornamentation as that of the foregoing specimen, except that in place of the hill in the centre is the full figure of a bird with outspread wings and long curved tail-feather. On the fifth *po shan lu* illustrated in the "*Hsi ch'ing ku chien*" (Book 38, p. 47), the lid is surmounted also by a bird (presumably a cock). The handle is cast in the same mould as the vessel, and rests on a short leg 1.8 cm long. Its end is shaped into an animal's head. Along the upper side of the handle is an engraved inscription, reading "In the seventh month of the third year of the period *Yen kuang* (A.D. 124), the work was made by *Chang Shun*" (see Inscriptions). An exact type of this vessel is not pictured in any of the standard Chinese books on bronzes. It is of special interest

Fig. 39.   Bronze Censer of Han Time, with Bird Decoration.

from the fact that it has been adopted, though in less elaborate work, as one of the utensils employed in the Buddhistic cult in China and Japan. In the Japanese book "*Shoso Butsu-zō-zu-i*" 諸宗佛像圖彙 (Vol. 5, p. 16 b) there are two illustrations of it under the title "*shiu-ro*" 手爐 ("hand-censer").[1]

The piece of pottery represented in Fig. 4, Plate LV

---

[1] See also the first painting in the 78th number of the Kokka, portrait of *Shōtoku Taishi* 聖德太子 (572–621), who holds such an implement in his hand. Regarding this personality, see E. PAPINOT, Dictionnaire japonais-français des noms principaux de l'histoire et de la géographie du Japon (Hongkong, 1899), p. 190; CL. E. MAITRE, in Bulletin de l'École française d'Extrème-Orient, Vol. III, 1903, p. 586.

(17.5 cm high), in its outward shape is akin to the type of the preceding *po shan lu*, although the hill formation on the cover is wanting. The bulging bowl rests on a high stem fixed in a deep circular bowl. The loose cover is dome-shaped, and has two circular rows of eight perforations each. The decoration consists of an eight-pointed, star-shaped figure in the centre on the top, around which run two circular rows of squares with their diagonals. The perforations are in the angles of these squares. The glaze, only partially preserved, is a very light green, and the surface has a brilliant silvery gloss.

In the "Illustrated Book of Ancient Jades"[1] (*Ku yü t'u p'u*), Book 79, pp. 1—4, two ancient hill-censers of nephrite are pictured and described. The description of No. 1 (Fig. 40) runs thus: "*Lu* ["stove", "brazier"] with connected cover;[2] the total height is 7 inches, 1 *fên;* the circumference of the body, 1 foot. Below there is a receiving-plate which is 1 inch 8 *fên* high and 1 foot 2 inches in circumference. The color of the nephrite is bluish green, and the carving is without blemish. Mountain-peaks rising in several ranges form the cover. The lower part of the censer is supported by the leaves of the aquatic plant *Hsing*[3] 荇 [*Villarsia nymphoides, Gentianaceæ*]. On the body the design of 'leather balls' (*p'i ch'iu*) is carved. On the lower part of the stem (*tsu fei*, 'calf of the leg') eight seal characters are engraved, reading 'May Heaven make your sons and grandsons rich,

---

[1] See p. 49.

[2] *Lien kai*, here rendered by "connected cover," is an untranslatable term, as *kai* may mean anything that covers. The hill formation of this vessel covers the bowl, according to the Chinese point of view, but it is certainly not a cover in our way of thinking.

[3] There are six such leaves, as also in the corresponding bronze type of the *K'ao ku t'u*, which explains them under the same name.

古玉博山爐一

Fig. 40.  Nephrite Hill-Censer (from *K'u yü t'u p'u*).

古
玉
博
山
爐
二

Fig. 41. Nephrite Hill-Censer (from *K'u yü t'u p'u*).

noble, prosperous, and proper!'[1]    This is a phrase for imploring blessings.    In the receiving-plate below, in the censer, are stored up fragrant herbs in hot water, like the great ocean.    The mountain-peaks on the cover have the form of the Three Isles of *P'êng-lai.*[2]    The *po shan* is primarily a vessel of the Han palaces, and comes from them.    For a long time the people of the Han dynasty have made it."[3]

The other piece (Fig. 41) in the same work (p. 3) is the same in size and shape as the preceding.    It differs from it in the following points.    The color of the jade is "sweet-blue" or "sweet-green" (甘青), the exact equivalent of which I cannot define.    On the front of the brazier is an animal's face (獸面), agreeing with those on the Han bronze and pottery; and a row of knobs ("connected beads," 連珠) encircles the upper part.    The inscription on the foot reads,

---

[1] The same inscription is on the bronze of the *K'ao ku t'u*, with the reading 與 *hsing* for 成 *ch'êng*.

[2] See MAYERS, The Chinese Reader's Manual, Nos. 559, 925; PARKER in China Review, Vol. XVIII, p. 58.    They are called also "Islands of the Three Genii" (三仙山). Their names are *P'êng lai*, also called *Fang yo* 防丘 and *Yün lai* 雲來; *Fang chang* 方丈; and *Ying chou* 瀛洲.    G. SCHLEGEL (*T'oung Pao*, 1895, Vol. VI, pp. 1–64) has devoted a lengthy dissertation to these islands, which he locates near the province of Hizen, Japan.    This conclusion, however, is not acceptable, as we have here a mere geographical myth, all Chinese accounts of these islands being of a purely legendary character (see also CHAVANNES, Les mémoires historiques de Se-ma Ts'ien, Vol. II, p. 153, note).

[3] 右爐連蓋。通高七寸一分。身圓徑一尺。下有丞盤高一寸八分。圓徑一尺二寸。玉色翠碧。無瑕琢。山峯層叠爲蓋。爐下以荇葉承之。身上琢刻皮毯之文。足胕刻以篆文八字曰。天成子孫富貴昌宜。乃祝嘏之詞。爐下承盤貯湯薰香象大瀛海也。蓋上山峯象蓬萊三島也。博山本漢宮之器。其來已久此亦漢人所制者。

"Life as long as the mountains, felicity as great as the sea!"
(壽山福海).[1]

The type of these two nephrite hill-censers corresponds to the bronze piece of the "*K'ao ku t'u*," which is reproduced also in the "*San ts'ai t'u hui*," "*T'u shu chi ch'êng*," and *Kin shih so* (*kin so*, Vol. 3). Both agree in the moulding and carving respectively of the hills in the bowl, — a feature which distinguishes them from the specimens described before, and constitutes them a peculiar variation of this type. They have also the six leaves grouped around the stem in common.

The three Isles of the Blest — *P'êng lai, Fang chang*, and *Ying chou* — are first mentioned under the reign of *Ts'in Shih Huang ti* (221–210 B.C.) as the abode of the immortals. The emperor despatched there several thousand young girls and boys across the eastern sea in search of them.[2] It was believed that the drug preventing death grew there; that all beings there, birds and quadrupeds, were white; that the palaces and gates were made of gold and silver.[3]

The belief in the existence of these islands received a new impetus under the Han Emperor *Wu* (140–85 B.C.). Since about 133 B.C. the first notions of alchemy have sprung up in China; to wit, that cinnabar was transmutable into gold in the furnace, and that immortality could be attained by him who should eat and drink out of vessels made of such

---

[1] This is a common adage frequently written or printed on slips of red paper and pasted on the gates of houses in Peking. It is found also on amulets. CHAVANNES (Journal Asiatique, 1897, pp. 148, 149) has made known a bronze medal on which the same characters, transcribed in the Mongol 'P'ags-pa writing, occur. See also CHAVANNES, in Journal Asiatique, 1901, p. 208.

[2] See CHAVANNES, Les mémoires historiques de Se-ma Ts'ien, Vol. II, pp. 152, 190; MAYERS, The Chinese Reader's Manual, pp. 163, 197.

[3] See CHAVANNES, l.c., Vol. III, p. 437.

gold.[1] *Li Shao-chün*,[2] magician and adept, whose influence held sway over the emperor, persuaded him to sacrifice to the furnace, and to send an expedition over the sea in search of the Fortunate Isles. In 110 B.C. the emperor travelled to the east, and, arriving at the shore of thė sea, looked off in the distance, hoping to see Mount *P'êng-lai*.[3] In 104 B.C. he betook himself to the bank of the *P'o hai* to sacrifice from a distance to the inhabitants of *P'êng lai*. He still hoped to be able to penetrate their marvellous hall.[4]

At that time he built the palace *Chien chang* 建章宮, about twenty *li* northwest from the modern *Hsi an fu*. There was an artificial lake there with three islands in it, intended to represent the three Isles of the Blest,[5] "imitating what there is in the ocean, — sacred mountains, tortoises, fishes, etc." The magicians who went over the sea in search of Mount *P'êng lai*, says SSŬ-MA CH'IEN, never brought any proof of their words. The emperor gradually grew weary of the strange propositions of the adepts, but could not free himself entirely from the bonds connecting him with magic, and still continued to hope that he would really find what they promised.[6] The conception of these islands of happiness, remote in the sea, must have deeply impressed the imagination of the people of that period, as is also proved by the fanciful descriptions given of them in later books.[7] If we now take

---

[1] See CHAVANNES, l.c., Vol. III, p. 465. W. A. P. MARTIN derives the alchemy of Europe from that of China (see his essay, Alchemy in China, in his book The Lore of Cathay, pp. 44–71).

[2] See CHAVANNES, l.c., Vol. III, pp. 463–466; MAYERS, l.c., p. 122.

[3] See CHAVANNES, l.c., Vol. III, p. 504.

[4] Ibid., Vol. III, p. 513.

[5] Ibid., p. 514. SSŬ-MA CH'IEN makes in this passage four islands, adding that of *Hu liang* 壺梁. See also A. FORKE, Ch'ang an im Altertume (Mitteilungen des Seminars für orientalische Sprachen, Vol. I, Sect. 1, p. 115).

[6] See CHAVANNES, l.c., Vol. III, p. 519.

[7] Translated by DE GROOT, Les fêtes annuellement célébrées à Amoy (Paris, 1886), Vol. I, pp. 166–168, and by SCHLEGEL in the article referred to above.

into consideration that the censers called *po shan lu* originated in the Han period; that, as shown above, they must have existed in the first part of the first century before Christ (very likely when *Wu ti* was still reigning); and if we recall that LUNG TA-YÜAN (the author of the "*Ku yü t'u p'u*") compares the hills of the *po shan lu* with Mount *P'êng lai*, and the water-filled receiver with the great ocean; and that LÜ TA-LIN (the author of the "*K'ao ku t'u*") employs a similar comparison in reference to the water in the dish, — I think it seems justifiable to assume that the artistic idea personified in the *po shan lu* might have been an attempt at representing Mount *P'êng lai*, the abode of the blessed, enveloped by the waves of the ocean. It must be borne in mind, however, that there is no contemporaneous tradition of the Han epoch that confirms such a view, and that *Lung Ta-yüan* mentions the *P'êng lai* only by way of likeness, and does not say that it was the real intention of the artist of the Han time to bring out the idea of the *P'êng lai* in these censers. Even if he should have expressed himself clearly to this effect, it would not prove with mathematical certainty that it was so objectively: for the author, who wrote about 1174, lived more than twelve hundred years later than the time in which the idea of the art works in question was conceived of; and whatever he says, or might have said, is certainly nothing but subsequent reflections of a late epigone on things the real origin of which was then lost, — reflections, on the other hand, which could have been suggested or inspired by the remains of a well-founded oral tradition handed down from generation to generation.

The presumption that such a tradition regarding antiquities in metal, nephrite, and other substances, was still current in the time of the *Sung*, and was incorporated in the great

13

catalogues of ancient art, like the "*Po k'u t'u*," "*K'ao ku t'u*," and "*Ku yü t'u p'u*," becomes necessary, — a *conditio sine qua non*, — since otherwise the judicious agreement of the chronological classification of the antiquities made out in these books with the facts of history and with the development of Chinese culture, would be inexplicable. The subjective suggestion of *Lung Ta-yüan* might therefore be the outcome of some living tradition, and bear a certain foundation in an objective fact, especially as the simile of the boiling water in the receiving-dish to the ocean seems to be so far-fetched and so forced that to our mind no normal imagination would ever conjure it up. In presupposing that this vessel embodied the idea of Mount *P'êng lai*, and that the water in the dish was fancifully believed to mean the sea surrounding it, this exaggerated figure of speech becomes intelligible. Nevertheless I should not lay any particular stress on this side of the argument, but should attach far greater importance to the striking historical coincidence of the production of the *po shan lu*, which we have seen was then a new type of art, simultaneously with the culminating point of the idea of the Fortunate Isles, which had then undoubtedly reached its climax, and held the minds of the people in suspense. Certainly this new and peculiar art motive, the hill-shaped cover, must have had some reason for its existence, and have been suggested by some idea. This idea surely must have arisen in the minds of artists of the day, as the names of those artists are expressly recorded. All historical data referring to this vessel make it a Chinese invention of the time; and its whole make-up bears every mark of Chinese genius, and excludes all thought of any foreign influence. Certainly, therefore, in searching for the idea which might have crossed the vision of the artist, we are justified in looking for it in the domain of coeval folk-lore, where our only recourse is to the legend of the Isles or Mountains of the Blest, a most prominent feature in the religious faith and yearnings of those days. It was undoubtedly more than a speculation of some religious adventurers and charlatans: it was a deep-rooted belief, nourished on the soil of Taoism, expressing the human desire for a better land, for a better immortal life beyond the grave.

It was to these longings of their contemporaries that the artists attempted to give life; and we shall see in the next section that the hill-censer did not remain the only production of these aspirations, but that the idea of the mountainous Isles of the Blest inspired the potters of the time to much more magnificent works, in which the art of the Han epoch had its greatest triumph. These were undoubtedly mortuary jars; and it might be concluded justly that the burying of hill-censers and hill-jars in the grave had also a symbolical signification, and implied the mourner's wish that his beloved deceased might reach the land of bliss and attain immortality on the Fortunate Isles.[1]

We read in the annals of the father of history, SSŪ-MA CH'IEN, so brilliantly translated and commented on by Professor CHAVANNES, that the Emperor *Wu* caused special utensils to be manufactured for the sacrifices *fêng* and *shan;* and that when

---

[1] I wish to call attention here to a passage in the *Shi i chi* 拾貫記 of the fourth century, quoted by SCHLEGEL in the article of the *T'oung Pao* referred to, on p. 10. There the Three Islands are designated also as the "Three Vases" (三壺), because, it is said, they have the shape of a vase (此三山形如壺). This is apparently a very inadequate comparison; and it occurs to me that it may have been the representation of these islands or hills on censers and on pottery jars which later caused the islands themselves to be conceived of as vase-shaped. Mount *P'êng lai*, as the resort of sages and hermits, subsequently became a favorite subject among Chinese and Japanese painters (see a kakemono by *Ōkyo Maruyama* in the Kokka, No. 149, October, 1902).

he showed them to the assembled literati, several among them declared that they did not conform to those of antiquity.[1]   Perhaps hill-censers and hill-jars were included in this series.   However this may be, that interesting passage proves sufficiently well that an innovating spirit had come over the art of this time, that the forms of vessels handed down by tradition were copied slavishly no longer, but that new patterns also, new styles deviating from the accepted rules, were invented.   Such a reformer of art was *Ting Huan* 丁緩, who is credited, as already stated, with having made a nine-storied hill-censer engraved with quaint birds and animals.   Unfortunately all that is known about this interesting personage is very little, and limited to the brief account in the "*Hsi ching tsa chi*,"[2] which, notwithstanding, reveals an unexpected feature in the art life of the Han period, — the influence of the artist on the public, and the subjection of patterns to the changes of fashion.   *Ting Huan*, it is recorded, was a clever craftsman of *Ch'ang an*.   He used to make lamps[3] composed of seven dragons and five phœnixes, between which hibiscus and lotus flowers were interspersed, — marvellous works.   Further, he made the censer styled "reclining on the mattress," called also "the censer in the bedclothes."   This practice originally turned out to have been a custom in the houses, but this fashion was afterwards abandoned, when *Huan* commenced to alter it by making a mechanical contrivance[4] of rings which turned round four

---

[1] CHAVANNES, l.c., Vol. III, p. 498.

[2] Quoted also in the *Ch'ang an chi*, Book 32, *i shu ch'uan*.

[3] In my edition of the *Hsi ching tsa chi*, printed in the collection *Han Wei ts'ung shu*, there is the reading 蒲 燈 *p'u têng* ("lamps made of rush"); but in the numerous books which extract this passage I find only 滿 燈, without exception.

[4] I do not know of what this contrivance really consisted, as nothing more is known about it, and, through not understanding the matter, I might mistranslate the

times, so that the body of the censer was constantly on a level, and could be placed on the bedding. Therefore he gained celebrity. Then follows the passage about his *po shan lu* already cited. Moreover, our account continues, he made a fan consisting of seven wheels, each ten (Chinese) feet in diameter, connected with one another, and attached to a man who turned it round, so that all in the house were shivering with cold.[1] No doubt, the man who antici-

passage in question. In the *T'u shu chi ch'êng* Vol. 1124, *k'ao kung tien* Book 236, *lu pu hui k'ao*, this *wo ju lu* is given as a special type of stove, introduced by the words 以 銅 爲 之 花 文 透 漏, which are followed by the exact wording of the *Hsi ching tsa chi*. In the section *i wên* there is a lengthy poem on this stove by a certain CH'ÊN CHIAO 陳 樵 of the *Yüan* dynasty. The placing of a brazier on the bed, in which also aromatic substances were burned, for warming it, has nothing extraordinary about it, considering the fact that other means also for this purpose are usual in China. This is the "hot-water old woman" (*t'ang p'o* 湯 婆), or "foot old woman" (*chiao p'o* 脚 婆), also "tin female slave" (*hsi nu* 錫 奴), colloquially "hot-water wife" (*t'ang fu jen* 湯 夫 人), an earthenware bottle filled with hot water, which HUANG T'ING-CHIEN (A.D. 1050—1110: GILES, Biographical Dictionary, p. 338), a celebrated poet of the *Sung* dynasty, wittily eulogized: "For a thousand cash you may buy a foot-woman who will sleep with you night after night until day-break." The direct opposite is the "bamboo wife" (*chu fu jen* 竹 夫 人), in poetry also "dark female slave" (*ch'ing nu* 青 奴), from a dark kind of bamboo used for it, called "Dutch wife" in the hobson-jobson of the East, — a hollow frame of bamboo used in bed to insure coolness during hot summer nights, mentioned as early as in the dictionary *Shuo wên* (A.D. 100), and much praised by the poets, sarcastically and ungallantly lauded by the poet SU SHIH (1036—1101) as the "speech-less bamboo wife." It is mentioned also in a novel of the *Liao chai chi i*, Book 11, p. 12. Compare *T'u shu chi ch'êng*, l.c., Book 237, *t'ang p'o pu* and *chu fu jên pu*, and *K'o chih ching yüan*, Book 54, pp. 8, 9, where these subjects are treated with due seriousness. See also GILES, A Glossary of Reference, p. 76, and on the "Dutch wife" in Japan (Transactions of the Japan Society of London, 1893, Vol. I, p. 46).

[1] 丁 緩 長 安 巧 工。常 爲 滿 燈 七 龍 五 鳳 雜 以 芙 蓉 蓮 藕 之 奇。又 作 臥 褥 香 爐 一 名 被 中 香 爐。本 出 房 風。其 法 後 絶 至 緩 始 更 爲 之。爲 機 環 轉 運 四 周 而 爐 體 常 平 可 置 之 被 褥。故 以 爲 名。．．．． 又 作 七 輪 扇。皆 徑 丈。相 連 續 一 人 運 之。滿 堂 寒 顫。

pated two thousand years ago the modern idea of the electric fan was a skilful mechanic of no small individuality, who knew his business and his public, and understood how to influence its taste and fashions. This intimate glimpse into the studio of an artist of that age affords, at any rate, a better understanding of how such a new motive of art as the hill-shaped censers and jars could then spontaneously spring up in an ingenious mind bold enough to throw off the fetters of the past, and to imbue itself with the thoughts of the time.

There is possibly another lesson which might be learned from a consideration of this type of art works. We have repeatedly brought up the question as to which material in a vessel the priority is due, — pottery or bronze. As this group is well defined in its chronology, and even in the localities where it arose (which seem to centre around the ancient *Ch'ang an*, the then capital of the Han), the question is more easily answered in this case. The hill-censers of bronze, nephrite, and pottery, must have all been made in the course of the same period, and I am therefore inclined to think that no one of these materials can claim chronological precedence over another. I am rather under the strong impression that all three originally arose almost simultaneously in a restricted area, as the individual productions of a small, well-defined school of artists under the reign of the Han Emperor *Wu*. Once introduced into the realm of art, they were of course imitated also in later periods.

### HILL—JARS.

From the graves of the Han period a large number of cylindrical jars have been brought to light, with conical covers, on which the motive "hills surrounded by waves," or

Pottery Hill-Jars.

"a mountainous island in the midst of the sea," is worked
out most elaborately with much predilection for the subject
and to a great degree of perfection. All of these jars are
posed on three low feet shaped into the form of squatting
bears. All their bodies are decorated with bands of low
flat reliefs; the subjects represented, and their style, being
very similar to those on the large globular vases; but, as
the figures on the jar reliefs are generally of larger dimensions than those on the vase reliefs, they afford a still better
basis for a study of this ornamentation. All these jars are
glazed in the various shades of green; none of the other
colors, like brown, yellow, etc., being applied to them (at
least so far as known to me). The purpose for which vessels of this type were used cannot be made out, as no references
to them are found in Chinese literature.[1] There can be no
doubt, however, that they were mortuary pottery, for which
a special reason may be seen in the signification of the hill
motive explained on the previous pages.

To illustrate the outward appearance of these hill-jars,
two of them are shown on Plate LVI. The form of the cover
is the one most usually found on these vessels; i.e., it has
a hill in the centre consisting of three peaks, which are shown
in Fig. 1 from the side, and in Fig. 2 in front view, the
third peak being hidden behind the larger central one. The
full description of Fig. 1 will be found on p. 204 in connection with the band of reliefs separately illustrated there.
The jar represented in Fig. 2 of this plate is 14.5 cm high,
having a depth of 12 cm and a diameter of 19 cm. The
band of reliefs is 7.7 cm wide. Of the figures visible in
the illustration, we note a monkey walking on all-fours at

---

[1] In *Hsi an fu* they are called *liu-li hua lien* 琉璃花連.

the top of an ornamental wave, and two quadrupeds, probably tigers, facing each other, the one trotting, the other
in flying gallop; and, what is very curious, the lower outline
of this animal describes nearly a half-circle. The cover, as
already mentioned, has three peaks in the centre, which are
surrounded by four large ornamental waves of conventional
design, and rising above the edge of the cover. On the
front peak is represented a human or rather demon-like
figure, walking-stick in hand, in the act of ascending a hill, a
type which frequently occurs on these lids.

Plate LVII represents another of these jars on which the
hill motive is brought out in a somewhat different way. Its
total height is 21.4 cm; up to the rim, 14.3 cm; the diameter
being 18.1 cm. It has a dark-green glaze. The band of
reliefs around the body of the jar consists of two discon-
nected sections separated by narrow plain spaces. In each
section a wave-line runs along the lower border of the relief,
rising in two large high waves that extend nearly to the
upper border. In the spaces between these waves one animal
is formed, making three for each subdivision. On the left
of the illustration is a tiger in profile, with head *en face*.
The figure on top of the wave is probably a monkey, the
animals on either side of which are in the act of ascending
the wave. A small animal, presumably a goat, occupies the
next field. This animal is delineated in yellow-glazed outlines. Of the other animals represented (not seen in our
illustration), a stag with six large antlers, running at full
speed, and a hare sitting upright (the figure of the "demon"
rushing at it), are notable.

The cover, an excellent work of art, has a triple row of
plastic waves, — in the outer row, four pairs; in the second
row, four waves emerging behind the interstices formed by

PLATE LVII.

Pottery Hill-Jar.

PLATE LVII.

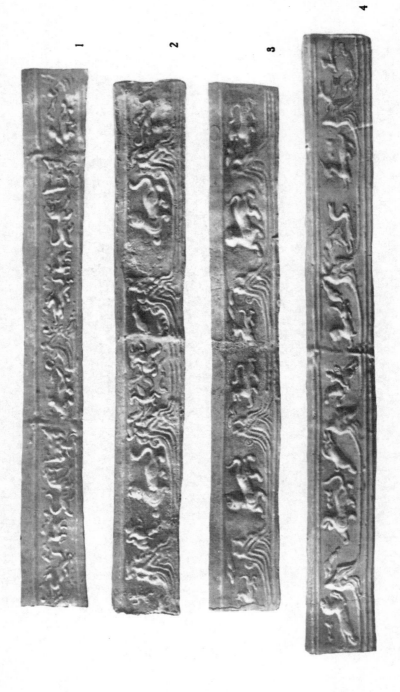

1

2

3

4

Relief Bands from Hill-Jars.

the double waves of the first row. In the centre, four waves surround a hill. Some waves are rounded, others are pointed. The latter have a small round knob on the top, which is glazed black. The rounded projections forming the waves and the hill have a leaf decoration on their surface, the veins of the leaves being expressed in dark-yellow lines (perhaps this is intended for a fern). All spaces in front of the outer row of waves, those on the waves of the back row, and those between the topmost and second rows, are filled up with numerous figures of men and animals. It is a densely populated island. Of the figures which are discernible, I notice three men running one behind the other, a hydra, a monkey walking on all-fours, large birds, two monkeys on their backs with their extremities upstretched and with long tail coiled at the end, a stag, a bird on the wing, a galloping tiger.

Plate LVIII illustrates four bands of reliefs covering the bodies of hill-jars. For the purpose of illustration, plaster casts were taken from them, as also from the previously described vases, and reproduced unfolded. The latter were then photographed, all details being carefully compared with the original objects, and corrected accordingly, or touched up in the photograph of the cast.

The jar the band of which is represented in Fig. 1, Plate LVIII, has a total height of 24.1 cm, and a depth of 10.8 cm. Up to the rim it measures 15.2 cm in height, the diameter being 19.6 cm. The width of the band is 5.9 cm; the circumference, 62 cm. The cast is 68 cm long. The band is divided into two symmetrical halves by the two tiger-heads with rings (this is the only occurrence of them on this type of jar) derived from the large globular vases. Each half is divided again by a wave into two sections of unequal size.

Our illustration starts on the left with a running monster with one long horn on its forehead. Above it, on either side, are two snake-heads. The next figure is that of an animal resembling a wild goat, a snake being over it. There are black spots on its body under the green glaze. A trotting tiger with a snake in front of it in vertical position terminates this section. Beyond the conventional tiger-head with ring is the frequently occurring type of trotting tiger, with head *en face* and right fore-paw uplifted. The animal to the right above it is not clearly discernible. Beyond the following wave we observe exactly the same representations as on the left. On the cover of this jar are represented four animals, — a hydra (*ch'ih*), clearly recognizable; a deer with what appears like a bird over it; a snake caught by a large bird of prey swooping down on it from above (*ni fallor*); and a galloping quadruped.

In the three following reliefs the artist availed himself of the same motive for the arrangement of the figures. We see here in low relief large high waves similar in form to those moulded in the clay on the covers of these jars; and four such waves serve the purpose of dividing the band into four ornamental divisions. They are connected one with another by two parallel lines running along the lower border, and terminating (Figs. 2 and 3) in spirals at the points where they reach the waves.

Fig. 2, Plate LVIII, represents the band on a hill-jar which appears to have been much exposed to subterranean moisture, resulting in a partial obliteration of the green glaze, which has changed in places into a silvery-white gloss. It had also a destructive effect on the relief-work, producing numberless small pores in the clay. The band is 7.3 cm wide, 62 cm in circumference, and the cast of it is 67 cm long. Small animals are seen above and along the sides of the waves, probably monkeys, but most of them are much worn. Two of the four fields, which are non-contiguous, offer a certain parallelism in their subjects. In the first field on the left we see the "demon" facing a tiger — the head of the latter *en face*, its body in profile, and its right fore-paw uplifted — which we met on the previous relief, and which is found again in the third field of this relief. Here, to the left of the tiger, the space has been washed off, and is now filled up with clay; but, on the ground of the parallelism of the fields, we might infer from the former design that this place likewise was taken up by the "demon." On the tiger's feet, four claws are discernible. The second field contains three figures, — a large flying bird with a long crest or bunch of feathers on the back of its head, outspread wings, erect tail-feather, and long legs; the "demon;" and a boar in oblique position, leaping down from the top of the wave. In the fourth field is a winged galloping hydra with an animal on its back and the "demon" in front of it, which last is seen on the left of the illustration.

The jar to which this band belongs has a total height of 22.5 cm, and of 15.1 cm up to the rim. It is 11.1 cm deep and 19.3 cm in diameter. The cover, like those of all other jars of this type, is of conical shape and plastically worked into large sea-waves, four all together. The top is formed of three parallel wave-bands. Between these latter and the border wave-lines are four spaces, each filled with the representation of an animal galloping at full-swing. These are a stag with large antlers, a boar with colossal head, a lion, and a tiger.

In the next figure (Fig. 3, Plate LVIII) is represented a band from a hill-jar 24.4 cm high, 11 cm deep, 15 cm from base to rim, and with a diameter of 19.5 cm. The band itself is 6.7 cm wide and 63 cm in circumference, and the cast is 67 cm long. The green glaze is much decomposed, and has a gold and silver shimmer. Beginning on the left of the illustration, we find a running hydra, a crawling snake above it, two small creatures on the wave (no longer recognizable), a tiger with head *en face*, the two fore-feet on the ground, the two hind-feet raised as if running down a path. On the next wave a bird seems to be perching, with head turned upward and long pointed beak, a snake over it. A tiger running at a trot, head in profile and mouth open, and a hydra, occupy the next field; then follows a tiger *en face*, of the same execution as that before described. Beyond the wave is a running tiger in profile. On the cover are represented a unicorn, a tiger, a hydra, and a deer with a large bird over it.

The last band shown on Plate LVIII (Fig. 4) is from a hill-jar (see Fig. 1, Plate LVI) 27 cm high, including cover, 13 cm deep, and 16.3 cm from base to rim, the diameter being 22.5 cm. The width of the band is 7.3 cm, and its circumference 72 cm. The cast measures 78 cm. This jar, which is covered with a light-green glaze (the spur-marks showing on the under side of the bottom) is the largest, and the finest in execution of relief-work, of this type. The composition of the relief is remarkable in that the artist has made skilful use of the ornamental waves, causing his animals to rear up in the act of leaping against them, thus attaining an impressive effect of lively description. It looks as if these animals were about to plunge over the wave, — a sort of steeple-chase race. This motive is not found on any of the Han bas-reliefs of stone. On the left we observe a tiger rearing up in the act of leaping, while a boar in flying gallop seems to be just on the point of springing over it from the other side. Then follow a trotting tiger, its right fore-paw raised from the ground, with a snake over it, and a huge boar in *galop volant* at half right angles to the lower border, its fore-legs almost reaching the crest of the waves. In the same position is the horse on the other side, with the usual type of archer astride it. He is followed by a running tiger at which he is aiming, and a boar rushing down from the top of a wave. On this wave is a bird with outspread wings. The next field is occupied by a galloping and a trotting hydra facing each other. Over the hydra to the right is a bird on the wing; over the left, the outlines of an undecipherable animal. In the last section a large stag with mighty antlers is breasting the wave, his fore and hind legs being almost in the same straight line. He is followed, as it would seem, by a deer galloping; below him, a long coiled snake.

The cover is correspondingly worked up most elaborately (see Fig. 1, Plate LVI). The three peaks in the centre stand out very prominently, the central one rising 2 cm high. The waves, also, are more plastic than on the other examples. The island is populated not only with animals, but also with men. In two fields we notice a mountaineer. In the one there is the figure of a man below, at the foot of the hill (see Plate LVI, Fig. 2, where it is distinctly visible), the left foot stretched backward, and the right far forward and bent, as if just in the act of ascending a steep mountain-path. In both hands he holds a staff crooked at the upper end. His head is animal-like, and not dissimilar to that of the "demon." On his back there is a long pointed object, which might be a burden. In the field next to this one we see exactly the same type of mountain-climber just at the point where he has nearly reached the top of the hill. There is a snake under his right foot. What was a staff before, appears here as a bird with long neck and head; but it is held by the man in the same way as the staff.[1] Under this man is a tiger with uplifted right fore-paw and lolling tongue. Another interesting type is a squatting goat with horns curved downward, and with what appears to be a beard under its chin.

On Plate LIX are shown two jars which in shape, and in being posed on three feet, are identical with those previously described. They differ from them, however, in lacking the plastic hills and waves on the covers. These are flat, but also decorated with reliefs.

The jar shown in Fig. 1, Plate LIX, is 23.5 cm high, 15.2 cm deep, and with a diameter of 25.3 cm, the rim being 1.3 cm

---

[1] It may be justifiable to identify this bird-staff with the well-known pigeon-staves prevalent at that time. According to MAYERS (The Chinese Reader's Manual, p. 86), a custom prevailed under the Han dynasty of bestowing upon persons above the age of eighty a jade-stone staff (*yü chang* 玉杖) upon which the figure of a pigeon was engraved, the pigeon being believed to have peculiar powers of digesting its food; and a wish for similar strength on the recipient's part was also thus symbolized. Such staves were also surmounted by the full figure of a pigeon, and were made of bronze. There is a rough engraving of such a staff in the *Chi chin chi ts'un* 吉金志存 of LI KUANG-T'ING 李光庭 Book 4, p. 26 a, and a better illustration of one may be found in the *Hsi ch'ing ku chien*, Book 38, p. 19. See also STEWART LOCKHART, A Manual of Chinese Quotations (Hongkong, 1903), p. 270. Neither Mayers nor Lockhart in quite accurate. The passage in the *Hou Han shu*, to which their statement alludes, runs as follows (quoted in *T'u shu chi ch'êng* Vol. 579, *ch'in ch'ung tien* Book 29): 鳩部紀事 p. 2 b): 後漢書禮儀志。仲秋之月縣道皆案戶比民年始七十者授之以玉杖餔之糜粥。八十九十禮有加賜玉杖長尺端以鳩鳥爲飾。鳩者不噎之鳥也。欲老人不噎。 "In the month of mid-autumn the districts and provinces made the regulation that the people, in single families and groups of five families, who had commenced their seventieth year, should receive a staff of nephrite [to enable them] to eat their gruel; the custom for octogenarians and nonogenarians should be to add the ornament of a pigeon to the top of the nephrite staff being one foot long. The pigeon is indeed the bird which does not choke, thus indicating the wish that the old men may not choke."

Cylindrical Pottery Jars.

wide. The cover is divided into four circular zones, the outer one being the lowest; while, of the remaining zones, one is raised about two mm above the other. The second zone from the outer edge is curved convexly, while the outer and the two inner ones are each level. In the inner circle are the remains of a knob (partly broken off), through which appear in relief two dead-rings side by side. They are derived naturally from the cover of a metal vessel, where such rings were actually applied as handles. In the second zone is a large group of animals, but most of them so worn away that they have lost their individuality. The "demon" with the lion, stags, boars, hydra, may be traced in a general way. The next zone is occupied by six compound spirals. Along the outer zone is a fourfold zigzag pattern.

The band around the body of the jar is 9.2 cm wide, and measures 79.3 cm in circumference. The animals on the relief are arranged in three rows, and are more numerous than on the other pieces, but unfortunately are so much rubbed that many details are lost. Two ornamental tiger-heads holding rings form the boundary between two sections, which are each divided again by a raised vertical line. In some cases the same animal, in the same position, is repeated in the three rows; for example, the tiger, the boar, and the stag. There are several lions with well outlined manes.

The other jar on this plate (Fig. 2) is 18 cm high, 12.5 cm deep, with a diameter of 19.1 cm. The feet are not bear-shaped, but represent a very curious round face, with two large distinct eyes and well-marked eyebrows, and a broad flat nose. They are not unlike the representations of elephants on ancient bronzes found as early as the *Chou* dynasty. There are no reliefs on the body of the vessel, which is ornamented merely with three double grooves. The cover, which, like the preceding one, has three spur-marks on it, is divided into four circular zones for purposes of decoration, one zone being somewhat higher than the other. The middle part is formed by a flat circle 8.2 cm in diameter, with a knob in the centre not unlike those on the metal mirrors. A dead-ring which passes through a supposed perforation in the knob is brought out in relief. There are two isolated warts near the periphery, which may be accidental. The zone around this central portion is taken up by a row of triangles with a small light-glazed dot inside of each. In the spaces between these isosceles triangles are also similar dots (one in each) and smaller triangles with sides parallel to those of the larger ones. The third zone is occupied with fanciful ornaments consisting of combinations of curves, waved lines, and dots. In the outer zone, along the edge, there is a simple row of triangles. The glaze is a brilliant leaf-green, that on the inner side of the cover being a dark mottled brown. The glaze on the outer side of the cover is crackled. The lines and dots forming the ornaments in the two middle zones of the cover are glazed yellowish-white and brown.

Plate LX represents a jar of the same type, but without cover. It is 18.5 cm high, 13.5 cm deep, and 21.2 cm in diameter. Its original light-green glaze has entirely changed into a marvellous silvery iridescence with light-blue flecks, covering the whole vessel. The ears of the bears forming the feet are very large and deep.

This is the most artistic and unique piece as regards the choice and execution of subject for the relief. It is an entirely new and surprising revelation of the art of the Han time. The entire surface of the body is taken up by a gorgeous mountain landscape moulded in deep grooves and with firm and bold strokes in the clay. The representation

Cylindrical Pottery Jar with Relief Work.

PLATE LXI.

Pottery Censer with Perforated Lid.

differs, however, from all others in this pottery in its unconventionality, breathing of the freer air of natural observation. Wonderfully shaped pointed cliffs rise from the lower border. On the right of the illustration a chamois is standing on the summit of a peak, the feet closely pressed together on the narrow space, the back arched,[1] head and neck eagerly stretched forward, looking below, as if trying to reach a plant beyond the abyss, — in a most natural attitude, — the most admirable and artistic conception of that period. On the other side is a man sitting on a rock, with legs crossed and hanging down. What the wave lines below and to his right are intended to mean, is hard to make out. There are, further (not visible in the illustration), a large standing bird with long neck, curved and stretched forward, and five distinct feathers on its head, and a large carnivorous quadruped, its fore-paws touching a peak.

Plate LXI represents a distinct and unique piece, but one that resembles the hill-jars in certain features as to the formation of the lid. Its total height, including the lid, is 24.3 cm; excluding the lid, or up to the rim of the jar, 14.6 cm. Diameter of mouth is 10 cm; diameter of bottom, 12.5 cm. The clay walls are 1 cm thick. In the formation of the lid, four triangular pieces have been cut out and bent outward. These bear incisions which have the appearance of veins in the leaf of a tree, and which, in the same way, we found on the waves of hill-jars. Upward from these four openings the lid narrows, forming a constricted neck, above

---

[1] Stags with arched bodies appear on two Han bronze vases in *Hsi ch'ing ku chien*, Book 21, pp. 14, 15. In this case, however, this motive is of merely ornamental character, which is clearly evident from the whole conception of the deer, having a long vertical tail curved into a spiral (!), which otherwise appears in this form in connection with hydras.

14

which are four square overhanging projections surmounted
by a round top-piece. Four triangles are cut out of it, and
bent outward in the same manner as below. There was
probably once a knob on the top, which is now broken off.
Thus, all together, there are eight openings in this lid. This
vessel, therefore, seems to have served for the burning of
fragrant herbs. The jar is covered with a dark-green glaze,
and the greater part of the surface is overlaid with hard earth
incrustations. The glaze on the lid has changed in many
places into a silver and gold iridescence.

A bronze type corresponding to that of the hill-jar does
not seem to exist: at least, I have never come across any,
nor is there any figured or described in any of the Chinese
archæological books. There is only one Chinese bronze
vessel (illustrated in No. 114 of the "Kokka," and here reproduced in Plate LXII) whose type may be said to come very
near to our pottery jars, and, even in that, though their most
essential characteristic, the hill-shaped cover, is wanting. It
has, however, the same cylindrical shape; it is likewise posed
on three feet formed in the figures of squatting bears; it is
provided with two tiger-heads holding loose rings, in the
characteristic Han style; and its surface is decorated with
figures and ornaments in relief, among which bold waved
lines are prominent, and strongly remind one of those in the
pottery jar Plate LX. The Japanese descriptive text designates
it as *kodōro* 古銅鑪 ("an old bronze brazier"), being in
the possession of a Mr. Ueno 上野氏, and as an extremely
precious and rare piece. It is annotated with the remark
that the ornamentation is different from anything depicted
in the "*Po ku t'u*" and "*Hsi ch'ing ku chien*," and breathes
of an Indian flavor 印度風, resembling the pedestal of the
bronze statue of Yakushi 藥師 (Buddha Bhêshajyaguru) in

PLATE LXII.

Ancient Cylindrical Bronze Jar.

the temple of Yakushi in Yamato;[1] but since no date is given, it is said there that its age cannot be ascertained, but that it may belong to the period of the Six Dynasties 六朝 (i.e., roughly, the time between the *Han* and *T'ang* dynasties). The Buddhistic character of the figure covered by the ring cannot be denied; but, whatever the period of the piece in question may be, it betrays eloquently the genuine style of Han art, and the style which we noticed in the pottery hill-jars of this epoch.

---

[1] A fine collotype of this work will be found in the Kokka, No. 96, Plate v.

# V. — NOTES ON ORNAMENTATION.

### INFLUENCE OF SIBERIAN ART AND CULTURE ON ANCIENT CHINA.

THE pottery of the Han dynasty, from the view-point of technique and form, is doubtless a genuine and original production of Chinese culture. So far as this pottery represents imitations of actual objects, we have seen that these objects are true constituents of Chinese civilization, and existed in the period to which the pottery is to be ascribed. So far as it embodies vessels of a great variety of shapes, we have noticed that a great number of these can be traced back to forms in bronze and nephrite which were developed either during the age of the Han, or, for the most part and to the greatest extent, in ages much earlier, on Chinese soil. Of the remaining pieces which do not betray any affinity to forms in other substances, it may be asserted with perfect safety that they do not display any traces suggestive of foreign influence, but appear as thoroughly Chinese ceramic productions. Indeed, this entire pottery forms a well-defined group, which, in its character as pottery, bears no resemblance to any other known groups of ancient pottery in Asia, nor to that of Siberia or of Turkistan.[1] We may well exclude Corea

---

[1] The pottery of Turkistan described by Dr. Hoernle represents excavations from Khotan, from various sand-buried sites in the Takla Makan desert, and offers, according to Dr. Hoernle, very curious analogies to objects of Greek art of an early date, attributable to the influx of Græco-Buddhist art into Khotan, which was early in intimate connection with India and Buddhist culture. All these ceramic productions are made of an extremely hard, burnt clay, which has no glaze, but only a "gloss." In going over carefully the specimens published in Dr. Hoernle's report, I cannot trace

212

and Japan, since the old pottery of Corea seems to depend largely on that of China, and since it is an established fact that pottery was carried from Corea to Japan, and that Corean potters were working in the latter country. As to one type of Corean-Japanese pottery, we are able to show by means of historical records that it was transported from China to Corea (see p. 127).

Entirely different, however, is the question regarding the origin of some of the artistic motives with which this pottery is decorated. The whole style of this art, dominated by the conventional design of the "flying gallop" as its fundamental leading motive, bears such a striking resemblance to that of ancient Scytho-Siberian art, that a connection between the two must be presumed almost *a priori*. At the outset it should be stated that the representation of the lion, entirely foreign to the Chinese, can be considered but as a loan.

The motive of the shooting archer on horseback[1] is absolutely alien to the Chinese mind, and derived from Turkish art. This is proved by two things, — first by the historical

---

the slightest Chinese influence in them as regards either their technique and shapes, or their representations and styles of decoration; nor do I believe that this field of ceramics has ever exerted any influence on that of China. On the finds of pottery in Turkistan, compare A. F. RUDOLF HOERNLE, A Report on the British Collection of Antiquities from Central Asia, Part I (extra number to the Journal of the Asiatic Society of Bengal for 1899), Calcutta, 1899, pp. xii, xxxi, with frontispiece, and Plate XIX; Ibid., Part II (reprinted from the Journal of the Asiatic Society of Bengal, Vol. LXX, Part I, extra No. 1, 1901), Calcutta, 1902, pp. 42–55, with Plates VIII–XIII; A. F. R. HOERNLE, A Note on the British Collection of Central Asian Antiquities,- presented to the XIIth International Congress of Orientalists in Rome, Oxford, 1899, pp. 19–23 (this is a reprint of the same note published before in the Indian Antiquary, 1900, pp. 99–101); SVEN HEDIN, Durch Asiens Wüsten (Leipzig, 1899), Vol. II, pp. 16, 35 ff; M. AUREL STEIN, Preliminary Report on a Journey of Archæological and Topographical Exploration in Chinese Turkestan (London, 1901), pp. 30, 31, with Plates I and II; M. A. STEIN, Sand-buried Ruins of Khotan (London, 1904), pp. 242, 302.

[1] On one vase only (illustrated in Fig. 1 Plate XXXVIII) do I find a pedestrian, a man standing erect on the ground, and shooting with a bow or crossbow at a tiger in front of him.

testimony of *Ssŭ-ma Ch'ien*, who tells us that the tactics of mounted archers were derived by the Chinese from Turkish peoples towards the end of the fourth century before Christ; and secondly, by ancient Siberian art, in which we find just the same motive, in a much earlier period, of a composition identical with that of the Han time. Consequently, if the Chinese have received this matter itself from the Turks, if the Turks themselves had found the expression of an artistic representation of it, it is inevitable and logical to conclude that the Chinese obtained both the matter and the motive from the Turks.

I will first outline the historical background of this important event. A supposition which I myself had long considered is at once confirmed by a close scrutiny of Professor CHAVANNES',[1] in the fifth volume of his translation of *Ssŭ-ma Ch'ien's* Annals, just published, one of the most splendid and lasting monuments produced by French scholarship. I feel I cannot do better than repeat the actual words of the great sinologue, to which I can but readily subscribe: "The princes of *Ts'in* have occupied the present province of Shensi since the eighth century before our era; the princes of *Chao*, whose star rose later, ruled, from the fifth century before Christ, in the centre and in the north of the present province of Shansi. These two families, whose territories were contiguous, had sprung, as indicated by *Ssŭ-ma Ch'ien*, from the same ancestor. They found themselves placed in a position intermediary between the states in the south and east, properly Chinese, and the nomadic peoples of probable Turkish race in the north and west. To which of these two ethnical groups were they attached? No doubt, they had been deeply sub-

---

[1] Le voyage au pays de "Si-Wang-Mou," Appendix II, in Les mémoires historiques de Se-ma Ts'ien (Paris, 1905), Vol. V, p. 484.

jected to the Chinese influence which finally was radically
to transform them; but a certain number of facts lead us
to think that they arose from this immense ocean of the
Turkish tribes, whose incessant ebb and flow have in the
course of centuries, now covered, now abandoned, northern
China. Here are some of the facts. The princes of *Ts'in*,
up to the middle of the fourth century before our era, were
not considered as forming a part of the 'Middle Kingdoms:'
they were treated as barbarians. Likewise, in 307 B.C., King
*Wu-ling* of *Chao* formally adopted the dress and manners
of the warlike peoples of the north, — a measure which the
historian presents as inspired by political reasons, but which,
however, was presumably nothing but a return to the ancient
customs. In the country of *Ts'in*, in 678 B.C., the detestable
practice appears for the first time of interring a deceased
prince with his most faithful servants. This custom is ascribed
to the *Hsiung nu* by *Ssŭ-ma Ch'ien*,[1] and we know from
Herodotus (IV, 71) that this was the usage also of the Scythians.
In the country of *Chao* we see a prince making a drinking-
cup of the skull of his dead enemy; we find again the same
particular with the *Hsiung nu*, who are of Turkish race.
All these concurrences[2] go to show that the countries of

---

[1] See CHAVANNES, l.c., pp. 50 and 543; further, F. HIRTH, Die Insel Hainan nach
Chao Jua-kua (reprint from Bastian-Festschrift, Berlin, 1896, p. 29). The Issedones,
identified by Tomaschek with the Tibetans, gilded, according to Herodotus (IV, 26),
the skulls of their fathers, which they used as sacred objects (ἀγάλματι χρέωνται: others
translate "sacred image," "sacred ornament," "precious vessel;" see H. CARY, Herodotus, London, 1901, p. 246), and to which they offered yearly great sacrifices (compare W. TOMASCHEK, Kritik der ältesten Nachrichten über den skythischen Norden,
in Sitzungsberichte der Wiener Akademie der Wissenschaften, 1888, Vol. 116, p. 749,
where the whole classical and modern literature regarding this subject in Tibet is
brought together). A Tibetan engraving of a hermit holding a skull-cup may be seen
in L. A. WADDELL, Lhasa and its Mysteries (New York, 1905), p. 239. For the same
custom among other peoples, see R. ANDREE, Ethnographische Parallelen und Vergleiche (Stuttgart, 1878), pp. 133–136.

[2] Many more instances might be enumerated. The great individuality of *Ts'in
Shih huang ti* is strikingly non-Chinese. His deep inward hatred towards everything

*Ts'in* and *Chao,* neighbors and kinsmen of each other, belonged to the great family of Turkish peoples." The introduction of the Turkish-Siberian costume by King *Wu-ling* (325–299 B.C.) of *Chao* is described by *Ssŭ-ma Ch'ien*[1] in a highly dramatic way, reproducing the lengthy deliberations and speeches of the prince and his ministers concerning this revolutionary innovation, which the king, according to the historian, found necessary in order to get the better of the numerous surrounding barbarous tribes. He adopted their clothing and the practice of shooting with the bow on horseback, and taught this latter art to his people.[2] The former was the inevitable consequence of the latter measure, for the Chinese garments were spacious, loose, and flowing, and an awkward obstacle to riding and shooting, whereas the Siberian costume was narrow and tight-fitting.[3] *Wu-ling,* accordingly, to use a modern term, was the first to organize the tactics of mounted infantry[4] in China, borrowed from the Turks,

---

Chinese, his aversion to the traditions and manners of antiquity, his contempt for the rites of the *Chou* dynasty and for the whole Confucian school, point in the same direction. It is by no means incidental that all emperors of China who impress us by an outspoken individual personality — like *Kubilai, K'ang hsi,* and *Ch'ien lung* — were aliens and of Mongol and Tungusian extraction; whereas Confucianism finally led to the annihilation of individuality, and created types of men all alike under the pressure of ethical ritualism; and in this point the Chinese came to be the antipodes of the Greeks. E. H. PARKER also (China Review, Vol. XIV, p. 224) justly remarks, "Most of the really noble characters in Chinese history appear to have been Tartars."

[1] CHAVANNES, Les mémoires historiques de *Se-ma Ts'ien,* Vol. V, pp. 69–84.

[2] Ibid., p. 73.

[3] This agrees with the results of the archæological investigation in southern Siberia (see W. RADLOFF, Aus Sibirien, Leipzig, 1884, Vol. II, p. 98). Compare also J. H. PLATH, Nahrung, Kleidung und Wohnung der alten Chinesen (München, 1868), p. 61.

[4] It will be remembered, from the history of the British-Boer war, what a great sensation the mounted infantry of the Boers then generally created, and how much attention and study were devoted to it by the European armies, and particularly by the general staff officers of Berlin. During and after the expedition of the European allies to China, this mode of tactics was put into practical use, curiously enough, against a people which had availed itself of the same tactics for more than two thousand years. There is nothing new under the sun. There is not the slightest doubt that the ancient Siberian peoples, Tungusians, Turks, and Scythians were the originators of this

which, as expressly stated by *Ssu-ma Ch'ien*,[1] did not exist
there before, and to establish squadrons of mounted archers
for the protection of his frontier against the designs of in-
vading nomads.    In earlier times, only war-chariots had been
in use in Chinese armies.

We now come to discuss the motive of the archer on
horseback in ancient Siberian art.    The most striking counterpart to the Chinese design is met with on an urn dug
up from a grave-mound near Krasnoyarsk, and figured on
Plates III and IV of the famous work of PHILIPP JOHANN VON
STRAHLENBERG, "Das Nord- und Östliche Theil von Europa
und Asia" (Stockholm, 1730), also reproduced in RADLOFF's
"Siberian Antiquities"[2] (Appendix, pp. 27, 29).    On this urn,
hunting-scenes are depicted.    The hunters are all mounted
on horses, and the horses are represented in the motive of

---

manner of fighting, and became the teachers of the Chinese in this art.    It was by
means of these tactics that the Manchu conquered the Chinese Empire.    It does not
make any principal difference, of course, whether the horseman, when riding, shoots
with a bow or a carbine: the former, however, is more difficult, if it is a question of
hitting the goal.    The test of the candidates for the military examinations, who at full
gallop were obliged to hit the target three times with three arrows, — a practice
abolished since recent years, — was the last survival of this ancient custom (compare
ETIENNE ZI, S. J., Pratique des examens militaires, Variétés sinologiques, No. 9, Shanghai,
1896, pp. 7, 15, 16, 25, and notice also the Chinese engraving on p. 17, representing
the shooting candidate on flying horse, in many traits not unlike the ancient Han
designs; for the same subject in a Japanese engraving, see Globus, 1904, Vol. LXXXV,
p. 158).    The motive of the archer in flying gallop shooting backward seems to have
occurred in the art of Japan since the eighth century (see a very interesting example
from that period in the work Histoire de l'art du Japon, ouvrage publié par la commission impériale du Japon à l'exposition universelle de Paris, 1900, Paris [1900],
p. 67, Fig. 31).

[1] See CHAVANNES, Les mémoires historiques de Se-ma Ts'ien, Vol. V, p. 47 note, p. 81.

[2] Сибирскія Древности. Матеріалы по Археологіи Россіи, No. 5 (St. Petersburg,
1891).    It is not clearly stated in Strahlenberg what the material of this urn is (see
pp. 353 and 356, where he speaks in a general way of large and small urns of clay
found in tombs); but Radloff (Aus Sibirien, Leipzig, 1884, Vol. II, p. 131, where a
description of the scenes on this urn is given) calls it "a silver goblet from the Yenisëi,"
and attributes it to the peoples of the later iron period of that region.    Compare also
the riding archer killing a boar, on a round metal plaque among the antiquities of the
iron age of Perm in Arpelin, Antiquités du Nord Finno-Ougrien, (Helsingfors, 1877),
Part II, p. 142, No. 610.

the flying gallop, except one, which is trotting. One rider holds a hawk on his left fist. Another aims with bow and arrow at a huge bird just flying over him,[1] and turns his head back so that it almost rests on the horse's mane, face upward. The last rider shoots with bow and arrow while turning backward at an animal supposed to be behind him, in exactly the same position as noticed on the Han pottery. His body is half in profile, his head *en face*. He has on the pointed Scythian cap.

There are pictographs on the Yüs River which, in their style and representations, agree with the scenes on this urn.[2] There we see also hunters astride of coursers, who, armed with bow and arrow, pursue deer, stags, elks, and foxes. One of the archers hunts such animals, and flies his arrow by standing high in the saddle. Radloff emphasizes the fact that the pictographs of this period never represent pedestrians, but exclusively riders on horseback, which proves that the peoples of this period were a tribe of horsemen, — a fact confirmed by the large number of horse-skeletons, harness, and stirrups, found in their tombs, and by the accounts of the Chinese annals, which describe the Turks as having been an equestrian people from the oldest times.

Among the pictographs on the rocks near Suliek, on the Yenisei, Aspelin describes also an archer turning back in the saddle in the manner of the Scythians, and thrusting his arrow in a direction opposite to the head of the horse.[3]

---

[1] Compare with this motive the engraving of the summer chase in the *Érh ya* (Book 2, p. 22 a), on which an archer on horseback turns back to aim at a bird flying over him in the clouds. In the picture above this one, representing the spring chase, we see the replication of the Han motive, a mounted archer shooting backward at a stag.

[2] See W. Radloff, Aus Sibirien (Leipzig, 1884), Vol. II, p.    .

[3] Un des cavaliers se retourne sur la selle à la manière des Scythes et lance sa flèche dans une direction opposée à la tête du cheval (I. R. Aspelin, Types de peuples

The same type of archer astride a courser, and shooting backward, with head *en face*, sometimes in profile, is met with again in the art of the Sassanidian dynasty of Persia (A.D. 226–636), and I am convinced that there also it was derived from the province of Scythian or Old-Turkish art. An example of this motive on a Sassanidian silver bowl of the Eremitage in St. Petersburg may be seen in REINACH's "La représentation du galop," p. 61, Fig. 85 (or "Antiquités de la Russie méridionale," p. 425).[1] A still more surprising instance, in which the archer is adorned with the Scythian cap,[2] is represented on one of the most precious relics of Sassanidian textile art, now preserved in the Archi-Episcopal Museum of Cologne.[3]

---

de l'ancienne Asie Centrale, Journal de la Société Finno-Ougrienne, Helsingfors, 1890, Vol. VIII, p. 133). — Unfortunately, there are no collections whatever of Siberian antiquities in New York; neither is the vast amount of Russian literature on the subject accessible here, and what I have of it myself is very little. If this chapter could have been worked up in St. Petersburg, it would probably stand on a more solid basis of facts, and thus be more forcible. I must therefore ask the indulgence of the reader for its many shortcomings. — The hunters on the representations of South-Russian antiquities are armed with lances (KONDAKOFF, TOLSTOI et REINACH, Antiquités de la Russie méridionale, pp. 225, 226). In the same way, lancers are met with on the Siberian petroglyphs (Inscriptions de l'Yénissei recueillies et publiées par la Société finlandaise d'archéologie, Helsingfors, 1889, p. 13; I. R. ASPELIN, Types de peuples de l'ancienne Asie Centrale, l.c., p. 131), on the bas-reliefs of the *Hsiao t'ang shan*, and frequently in later Chinese and Japanese art (see, for instance, the porcelain plate on Plate VII of REINACH, La représentation du galop). Mounted archers in profile, shooting straight forward, occur likewise in Scythian, Siberian, and Chinese art; but I preferred to make the other type alone, as being more characteristic, the basis of the investigation.

[1] The best and largest reproduction is found in Compte-Rendu de la Commission Impériale Archéologique pour l'année 1867. Atlas. St. Petersburg, 1868, Plate III.

[2] Or, rather, helmet. See E. DROUIN (Sur quelques monuments sassanides, Journal Asiatique, 1897, p. 446), who clearly recognized the Scythian costume on the Persian bas-reliefs: "Le casque est conique terminée en boule, sans visière, vrai casque tartare que l'on retrouve plus tard sur les dessins de l'époque mongole . . . . Le cavalier de droite porte un casque pointu et un costume qui le désignent comme un scythe."

[3] This marvellous silk-weaving was discovered on Sept. 22, 1898, at the opening of the shrine of St. Cunibert, in the church of the same name at Cologne. It was published with a good heliogravure by Canon A. SCHNÜTGEN in the Zeitschrift für Christliche Kunst (Düsseldorf, 1898), Vol. XI, Columns 225–228. In the same volume (Columns 361–368) F. JUSTI published an additional article, "Die Jagdscene auf dem sassanidischen Prachtgewebe," and interpreted the archer as the Persian prince Bahram Gor

The pointed, so-called "Scythian" cap, which seems to be represented several times in the archer and in the figure of the "demon" in the Han—pottery, as may be inferred from the triangular pointed representation of the head (see p. 158), goes back to a very remote antiquity among the tribes of Siberia: it is found as early as in the bronze period,[1] as is proved by a bronze plaque from the Altai, representing a hunter with two dogs, and by the statuette of a miner,[2] both of whom are dressed in this peculiar head-gear. It is met with again in the Siberian iron period. Radloff[3] discovered, in a tomb on the Buchtarma, the figure of a rider cut out of gold-leaf, who seemingly wore on his head a pointed cap almost as long as half of his upper body. One of the riders in the pictographs of the Yüs River was also clad in such a cap. In a similar way, remarks Radloff, the Chinese describe the caps of the Haka. On the silver bowl mentioned above, three of the riders are represented with this cap.

---

(A.D. 420—438). See also K. WOERMANN, Geschichte der Kunst aller Zeiten und Völker, Vol. I, p. 484. Regarding another representation of Bahram Gor, see DROUIN in Journal Asiatique, 1896, p. 350. The churches of Cologne were rich in Sassanidian textiles, those of St. Ursula and St. Gereon now being in Berlin. — The same·motive of shooting-archer may be seen on a Persian brass plate (see F. SARRE, Ein orientalisches Metallbecken des 13. Jahrhunderts, Jahrbuch der königlichen preussischen Kunstsamm-lungen, 1904, Heft 1, Fig. 2), on which two series of hunters with horses in flying gallop, one of eight riders (p. 6) and another of sixteen riders (p. 8), are represented. From the Mongolian characteristic of the figures and the representations of Chinese motives on this vessel, Sarre is inclined to infer that it had its origin in the second half of the thirteenth century, when the Chinese world of forms had gradually obtained a most marked influence in Persian art, and is perceived in all its productions (pp. 15, 16). The author, however, does not appreciate the conventional Scythian type of the archer shooting backwards, nor the motive of the flying gallop, both of which make me think that this vessel is older than its supposed date. See also a Chinese weaving, made after a Sassanidian design with the same motive, in O. MÜNSTERBERG, Japanische Kunst-geschichte (Braunschweig, 1904), Vol. I, p. 116; and compare with the latter the pas-sage in Zeitschrift für die Kunde des Morgenlandes (Göttingen, 1837), Vol. I, p. 187.

[1] See W. RADLOFF, Aus Sibirien, Vol. II, p. 98.

[2] Both are illustrated in RADLOFF, l.c., Plate IV, Figs. 1—3; the miner also in ASPELIN, Antiquités du Nord Finno-Ougrien, Part I, p. v.

[3] See l.c., p. 138.

As to the motive of the flying gallop,[1] it has already been convincingly demonstrated by S. REINACH, that Chinese art derived it from Siberian art. According to the ingenious investigations of this French archæologist, this conventional motive appears neither in Assyria nor in Egypt, neither in the classical art of Greece nor in that of Etruria or Rome, nor in European art of the middle ages, the Renaissance, and the present age up to the time of the French Revolution (p. 11), when, in 1794, it appears for the first time in a popular engraving in England (p. 113). More than a millennium anterior to our era, however, it appears in the domain of Mycenian art, then in Scythian and Siberian art, in the Caucasus, in the Persia of the Sassanidæ, and in China and Japan. The latter two countries present an uninterrupted repetition of this motive up to the present day. From Mycenian art it migrated, through still little-known intermediary agencies, into the territory of the Scythians in the north and northeast of the Black Sea, and spread farther to Siberia, and from there to China towards 120 B.C.,[2] to Persia towards A.D. 220. Neither the art of the Achæmenidæ nor that of the Arsacidæ, which are ramifications of classical Greek art combined with Assyrian imitations (p. 78), furnishes one example of the flying gallop; while that of the Sassanidæ (since A.D. 226) shows a great number (p. 60) which must be traced back to the models of Siberian metal plaques. This theory of the migration of the flying-gallop motive is based first on the supposition that it does not correspond

---

[1] It occasionally occurs also on bronze vessels of the Han time. The *Po ku t'u* (Book 13, p. 15 a) contains the engraving of a flat vase ( 扁 壺 ), on the front side of which a deer in flying gallop is represented.

[2] The oldest evidence was found by REINACH (La représentation du galop, pp. 94, 95) in a horse represented on a square coin issued under that date.

to any real movement of the actual galop,[1] as proved by the kinemato-photographic reproductions of the horse-gallop, in which that motive adopted by art does not occur, and can therefore have been fixed only once; and, secondly, on the undeniable historical fact that cultural relations and connections existed between the areas of Mycenæ, Scythia, the Caucasus, Siberia, Persia, and China. Even if it could be proved that the flying gallop, as represented by the art of these regions, should really thus appear to the observing human eye (and this is not the case, according to frequent observations of my own), so that, on the ground of such a physiological phenomenon, the independent origin of this motive in different regions and times might be concluded, — I think, notwithstanding that, the very fact that the uniform representation of this motive is restricted to just those cultural provinces which have indubitable historical connections and mutual influences, and is entirely lacking in other vast regions and periods, is sufficient to force upon us inevitably the supposition of a mutual interdependence. In a word, this question is not a matter of physiology, but of history and archæology. From the fact that the Chinese have copied the mounted Turkish archer, we may safely conclude also, that, since the horse of this pattern was represented in flying gallop, they received also the latter motive from the same source.

On p. 209 we noticed on a hill-jar (Plate LX) a chamois standing on a peak, with its feet closely pressed together. I am very much inclined to see in this motive the free creation of an individual artist of the Han period, on account of the peculiar naturalistic action in which the animal is conceived, — stretching down its neck as if searching for food. If, on the

---

[1] "Le motif du galop volant ne peut avoir été imaginé qu'une fois, parce qu'il ne répond pas à la réalité et n'est qu'un symbole." — REINACH, La représentation du galop, p. 83.

other hand, we now consider the dependence of some themes of old Chinese art on that of Siberia, it cannot be denied that also a certain prototype of this particular motive is anticipated in that field, which might have given the inspiration for the Chinese production.  In the so-called Siberian copper or bronze age, copper bells in the shape of hemispheres have been found, to which stonebucks (probably argali) or stags are attached.  In the Eremitage of St. Petersburg there are five such bells from the Altai, some about 20 cm high. The purpose which they served can hardly be surmised.[1] The stags show the same curiously shaped antlers, consisting of three or more circles side by side, as on the Siberian bronze plaques; while the argali is represented with one horn curved back and downward, of natural appearance.  On several pieces, both animals have the feet closely pressed together, resting on the topmost point of the bell, and in this respect they agree exactly with our Chinese design; but the point in which they differ from it is that the neck and head are straight, and that the body is not arched.  It is therefore not a case of harmonious coinciding: at all events, it might remain an open question whether there is any connection in this case.  Since, however, contact between the two arts is established, it might be well to call attention to the two somewhat analogous conceptions, to be kept in mind for further research, if ampler material should be available.[2]

---

[1] W. RADLOFF, Aus Sibirien, Vol. II, p. 89.  Illustrations of these objects may be seen in ASPELIN, Sur l'âge du bronze altaico-ouralienne (Congrès international d'anthropologie et d'archéologie préhistoriques, compte-rendu de la 7ième session, Stockholm, 1876), Vol. I, Fig. 47, p. 576; ASPELIN, Antiquités du Nord Finno-Ougrien, Part I, p. 68, Nos. 305, 306; RADLOFF, l.c., p. 96 with Plate v, Figs. 4 and 5; F. R. MARTIN, L'âge du bronze au musée de Minoussinsk (Stockholm, 1893), Plate 33, Fig. 4; PALLAS, Reise durch verschiedene Provinzen des russischen Reiches, Vol. III, p. 386 and Plate VII; see also R. ANDREE, Die Metalle bei den Naturvölkern (Leipzig, 1884), p. 125.

[2] For other relations of ancient Chinese and Siberian antiquity, see the paper of P. REINECKE, Über einige Beziehungen der Altertümer Chinas zu denen des skythisch-

The intimate political and social connection of the Turkish peoples with China, from her earliest history down to the middle ages, is now a well-established fact, since the discovery and decipherment of the Old-Turkish inscriptions of Siberia, and the fundamental investigation made by Professor Hirth,[1] Professor Chavannes,[2] and O. Franke,[3] concerning

sibirischen Völkerkreises (Zeitschrift für Ethnologie, Vol. XXIX, 1897, pp. 141–163). This paper is based partially on an inadequate method. To establish the possibility of a comparison between types of vessels of different cultural provinces, it is a fundamental necessity that these types agree in the principal features of their formation: coincidence in only one or in a few characteristics does not constitute a relationship. I utterly fail to see any similarity between the Chinese tripod vessel (*ting*) of Silver Island and the Scythian bronze kettle from Hungary, which Reinecke groups together on pp. 148 and 149, except, perhaps, in the very accidental feature of the loop-holes; but the Chinese *ting* always has either three or four feet, while the Scythian kettle has only one stem; and this makes a principal difference, which Mr. Reinecke too easily overcomes by calling this the "simplification of the foot" in the Siberian specimens (on p. 159). If this Scythian type *must* resemble a Chinese one, it is perhaps somewhat similar to a *tou*. The *tou*, however, is never provided with loop-handles: consequently the Scythian kettle is neither a *ting* nor a *tou*, but an entirely different and distinct type. Besides, W. RADLOFF (Aus Siberien, Vol. II, p. 88) remarks expressly that the Siberian peoples of the bronze period did not know of tripod vessels, but put the kettles themselves into the fire. The author goes altogether too far in his individual explanations and in finding fault with the Chinese writers, who, according to him, "mostly interpret the objects wrongly" (p. 157), who give a "rather arbitrary and at all events not serious explanation" of their bronze axes (p. 155), as seems to him to be also the case with the bronze rattles of the Han time, in explaining which the Chinese authors are utterly helpless; and "their surmise that they were theatre-requisites (sic!) is very insufficient and improbable" (p. 152). No such assertion was ever made by any Chinese authors, who interpret this instrument sensibly and clearly enough as a musical rattle used in the military orchestra of the Han time as a signal for stopping the sound of the drums. That in the so-called "knife-money" the form of ancient knives is rather faithfully copied (p. 153) is a statement not convincingly proved. Its mere name, "knife-money," given to the coin from its resemblance to the shape of a knife, does not prove that its shape was objectively derived or developed from that of a knife. On the contrary, the ancient Chinese knives, a specimen of which I possess, were different from those which might be artificially reconstructed on the basis of the knife-money.

[1] "Nachworte zur Inschrift des Tonjukuk," in W. RADLOFF, Die alttürkischen Inschriften der Mongolei, 2d series (St. Petersburg, 1899); Sinologische Beiträge zur Geschichte der Türkvölker, St. Petersburg (Academy of Sciences), 1900; Über Wolga-Hunnen und Hiung-nu (Sitzungsberichte der philos.-philol. Classe d. K. bayer. Akademie, 1899, München, 1900, Vol. II, 2, pp. 245–278).

[2] Documents sur les Tou-kiue (Turcs) occidentaux, St. Petersburg (Academy of Sciences), 1903; Notes additionnelles sur les Tou-kiue (Turcs) occidentaux (*T'oung Pao*, 2d series, Vol. V, 1904, pp. 1–110).

[3] Beiträge aus chinesischen Quellen zur Kenntnis der Türkvölker und Skythen

the history of the Turks, based on Chinese sources. There is no doubt, after the archæological researches made by W. Radloff in southern Siberia, that the ancient Turks had been in the possession of a comparatively high civilization; and, in the present state of our knowledge, the question as to which cultural elements the Turks received from China, and which China received from the Turks, seems to me of primary importance. Since we have to deal here principally with the art relations of the two peoples, I wish to refer to two further points, — one, the representation of Turks in Chinese art; the other, the interchange of metal vessels between Turks and Chinese, which took place as early as the Han time.

The representation of foreign tribes in art seems to have been carried out in China at an early date. The oldest mention of such productions which I can find seems to be that of the twelve bronze statues which *Ts'in shih Huang ti* had founded from metal obtained from weapons taken from his vanquished enemies. They are said to have represented the barbarians *I* and *Ti*, who were of extraordinary size. Each statue weighed a thousand Chinese pounds.[1] On the bas-reliefs of the *Wu tsê shan*, CHAVANNES[2] observed two bizarre beings having enormous heads, and reminding one of the barbarians described by *Wang Wên-k'ao.*[3] The five personages represented on one of the bas-reliefs of the *Hsiao t'ang shan* (*en face*, wearing official hat and ceremonial girdle), designated

---

Zentralasiens (aus dem Anhang zu den Abhandlungen der königl. Preuss. Akademie der Wissenschaften, Berlin, 1904).

[1] See ED. CHAVANNES, Les mémoires historiques de Se-ma Ts'ien, Vol. II, p. 134; F. HIRTH, Chinesische Ansichten über Bronzetrommeln (Leipzig, 1904), p. 17; A. FORKE, Ch'ang an im Altertume (Mitteilungen des Seminars für orientalische Sprachen, Vol. I, Sect. I, p. 110).

[2] See La sculpture sur pierre en Chine, p. 65 and Plate XXXI.

[3] Ibid., p. xxvi.

15

by the characters *hu wang* 胡王, may be claimed presumably as representatives of Turkish tribes.[1]

The oldest representation of Turks now extant, however, seems to have been handed down on the engraving of a

有柄温爐 廬江李氏

Fig. 42. Stove of Han Time, showing Full Figures of Turks (from engraving in *K'ao ku t'u*).

stove of the Han time, preserved in the "*K'ao ku t'u*" (Book 10, p. 14), here reproduced in Fig. 42, on which four Turks in full figures appear as supporting the stove, as they are explained also in the descriptive text of LÜ TA-LIN,[2] no

---

[1] See ED. CHAVANNES, La sculpture sur pierre en Chine, p. 76 and Plate XXXVIII.

[2] 四胡人貢之.

doubt on the ground of a well-founded ancient tradition. Even without this, however, we should unmistakably recognize in these characteristic figures representatives of Turkish tribes from their entirely un-Chinese physiognomy, which the artist took every pains to express. Their heads are big and round; a circle of spiral-formed curls outlines the head and face; a bushy mustache curls up at the ends under the large curved nose; and the eyebrows are much exaggerated. They are clad only with an apron. Breast and abdomen are massive and protruding. Their arms and ankles seem to be adorned with bracelets and anklets.

In a painting of *Ku K'ai chih*, L. BINYON[1] sees in the picture of a nobleman a Turk by blood, as it would seem from his features. "Certainly there is nothing Chinese in his full eyes, his slightly aquiline nose, and his beard," he says. This opinion seems very doubtful to me as long as this scene is not identified with the story underlying it, though another subject represented on this painting was identified by Chavannes.[2] The figure in question has nothing in common with the usual style in which Turks are depicted in Chinese art.[3]

The fact that the Turks received works of art from China in the Han period will be seen from the following passage. In the city of *Yüeh yü ti*, in the country of Western *Ts'ao* (*Ishtikhan*), where the people worshipped the gods *To-si*, they used a set of gold utensils, on the left of which there

---

[1] A Chinese Painting of the Fourth Century (The Burlington Magazine, 1904, Vol. IV, No. 10, p. 42 a).

[2] *T'oung Pao*, 1904, pp. 323–325.

[3] Compare also the engraving illustrating the four representatives of foreign tribes in the *Ērh ya* (Book 2, p. 31 a), which is reproduced in PAUTHIER, Chine (Paris, 1853), Part I, Plate XII (see p. 57). On Chinese sources of the middle ages containing pictures of foreign peoples, see F. HIRTH, Wiener Zeitschrift für die Kunde des Morgenlandes, Vol. X, pp. 227, 228, 232, 236, 237.

was an inscription saying that this was a present made by the Chinese Emperor in the Han epoch.[1] On the other hand, the fact that metal vessels were sent from Turkistan to China at the end of the second century before Christ, will be recognized from an intensely interesting passage in the "Annals of the Liang Dynasty," biography of *Liu Chih-lin*.[2] "*Liu Chih-lin* cherished antiquity, and was fond of queer objects. In *Ching chou*[3] he collected dozens of ancient vessels. Among the manifold things which he had, there was a vessel resembling a bowl, capable of holding one *hu* [five or ten pecks], in the upper part of which there were characters inlaid with gold unknown to the contemporaries. Moreover, he presented to the heir-apparent four kinds of ancient vessels. The first was a pair of ornamented bronze jars for holding wine,[4] with two ears [handles] and an inscription inlaid with silver, reading, 'Made in the second year of the period *Chien p'ing* [5 B.C.].' The second was a pair of ancient engraved goblets inlaid with gold and silver, with an inscription in seal characters, reading, '*Ts'in Jung ch'êng hou shih Ch'u chih sui ts'ao.*'[5] The third kind was

---

[1] See CHAVANNES, Documents sur les Tou-kiue (Turcs) occidentaux (St. Petersburg, 1903), p. 139.

[2] Quoted in *T'u shu chi ch'êng* Vol. 1126, *k'ao kung tien* Book 251, 古玩部 紀事 p. 2 a. The passage will be found in the *Liang shu*, Chap. 40, p. 6 b, and agrees in its readings with the quotation in the *T'u shu chi ch'êng*.

[3] A prefecture in Hupeh Province.

[4] 銅鴟夷榼 (*Ko chih ching yüan*, Book 51, p. 2 b, where the same paragraph from the *Nan shih* 南史 is quoted) writes 鵝. *Ch'i i* is a leathern skin or pouch to hold wine, but the term is apparently applied also to metal vessels serving the same purpose, as is evident from the passage above and the other quotations in the *Ko chih ching yüan*, l.c.

[5] According to a suggestion of Professor Conrady, "Made in the year when the Marquis of *Jung-ch'êng* in *Ts'in* arrived in *Ch'u*." A marquisate of *Jung-ch'êng* is mentioned in CHAVANNES, Les mémoires historiques de Se-ma Ts'ien, Vol. III, p. 156.

a foreign wash-basin,[1] which had an inscription on it reading, 'In the second year of the period *Yüan fêng* (109 B.C.) presented by the country of *K'iu tzŭ* (Kucha).'[2] The fourth kind was a wash-bowl of old make, with the inscription, 'Made in the second year of the period *Ch'u p'ing* (A.D. 191).'"[3]

It seems that the gift of this Turkish wash-basin under the reign of the Emperor *Wu* was important enough for the vessel to be engraved with an inscription in China commemorating the date of its arrival. This fact becomes most important for the consideration of the cultural relations of the Chinese to the tribes of the Tarim valley, particularly viewed in the light of the researches made by Professor HIRTH,[4] who, on the basis of a passage in the "*Ch'ien Han shu*," established the fact that the inhabitants of this country were acquainted with the art of metal-founding, and ascertained from the later accounts of the "*Wei shu*" that the country produced copper, iron, and lead. From a fourth-century

---

[1] *Ko chih ching yüan*, Book 52, p. 7 b, quoting this clause from the *Nan shih*, has 罐 for 灌. Perhaps it was a jar out of which to pour water for preparing a bath.

[2] F. HIRTH, Nachworte zur Inschrift des Tonjukuk, p. 114, in W. RADLOFF, Die alttürkischen Inschriften der Mongolei, II (St. Petersburg, 1899).

[3] 梁書劉之遴傳。之遴好古愛奇。在荆州聚古器數十。百種有一器似甌可容一斛。上有金錯字。時人無能知者。又獻古器四種於東宮。其第一種鏤銅鴟夷榼二枚。兩耳。有銀鏤銘云。建平二年造。其第二種金銀錯鏤古鐏二枚。有篆銘云。秦容成侯適楚之歲造。其第三種外國澡灌一口。銘云。元封二年龜茲國獻。其第四種古製澡盤一枚。銘云。初平二年造。

[4] Sinologische Beiträge zur Geschichte der Türkvölker, I, Die Ahnentafel Attila' nach Johannes von Thurócz (St. Petersburg, 1900), p. 244.

source brought to light by Hirth,[1] it is evident that these people melted iron by means of mineral coal, which they obtained from the same mountain where their iron-mines were exploited. Hirth is therefore inclined to think that the Chinese transcription of the name of this place — occurring in the three forms *Kui-tzŭ*, *K'iu-tzŭ*, and *K'üt-tzŭ* — represents an old Turkish word corresponding to Teleutic *kyza*, with the sense of "furnace of a smith." The first mention of the name *Kucha* in Chinese literature is made in the "*Ch'ien Han shu.*" The deduction of Hirth, that the metal industry of this region, facilitated by the richness of the mountains in coal, played a part at the time of the first occurrence of this name (i.e., in the first century before Christ), is now conspicuously confirmed by the above passage in the "*Liang shu,*" which announces the existence of a metal vessel in China from this old locality of copper and iron foundries, as handed down to posterity in the inscription of this piece.

In connection with the Turkish-Chinese relations, a number of ethnological questions will have to be studied, particularly such as relate to the horse and its harness, saddle, stirrups,[2]

---

[1] F. HIRTH, l.c., p. 245.

[2] According to F. HIRTH (Verhandlungen der Berliner Anthropologischen Gesellschaft, 1890, p. (209)), stirrups were known in China at least around the year A.D. 477, and may then have already existed centuries before. From the bas-reliefs of Shantung it becomes evident beyond doubt that stirrups and saddles were in general use during the Han period (see particularly the large engraving of the horseback-rider in *K'in shih so, shih so* Vol. 3 [p. 26 a], where the heavy stirrups with broad base, as used nowadays, are clearly outlined). The Chinese stirrups are of exactly the same type and shape as those of the Siberian iron age (see W. RADLOFF, Aus Sibirien, Vol. II, p. 133, and Plate X, Figs. 1 and 2), and in all probability were derived from the Turks, as were also saddles and horseshoes. It is known that VICTOR HEHN (Kulturpflanzen und Haustiere in ihrem Übergang aus Asien nach Griechenland und Italien, 7th ed. by Schrader and Engler, Berlin, 1902, pp. 21, 623) claims that the Turks were the domesticators of the horse, from which centre it was propagated to the Euphrates, further, to the Nile and into Europe. I think that, by a close study of the history of the horse in ancient China, considerable light may be thrown on this subject, and I am under the impression that presumably the problem in the sense of Hehn may thus be solved.

and horseshoes; the art of riding; the game of polo,[1] which originated in Turkistan; and the great problem of the origin of the Chinese iron industry.[2] One of the most interesting elements of culture which China has borrowed from Turkish tribes, and to which I wish briefly to allude here, is falconry. O. SCHRADER,[3] from a study of the history of falconry in ancient Europe, has arrived at the result that Turkistan must be considered to be the mother-country of falconry, whence it was carried to the Occident during the first invasions in the Migration of Peoples. A study of the same question in Chinese literature, the working-up of which I must reserve for a special paper, leads me to the same solution of the problem with regard to the countries of Eastern Asia. The whole method of hawk-training, as laid down in detail in the Chinese and Japanese falconers' books, coincides in such a striking manner with the same practice followed in Europe, and also by the Persians and Arabs, that it must needs be attributed to a common source of origin.[4] This can be sought only in the vast steppes of Central Asia and in the culture of the ancient Turks.

The principal witnesses to this fact are the early represen-

---

[1] A paper on the history of this game is in course of preparation.

[2] F. HIRTH (Chinesische Ansichten über Bronzetrommeln, Leipzig, 1904, p. 18) is inclined to think that the northern Chinese, whom we find first in the possession of iron, received their iron industry from Turkish tribes. — An interesting parallel to the *K'un wu* sword of *Mu Wang* (HIRTH, l.c., p. 21 note, "Hunnenschwert") is the σιδήρεος ἀκινάκης of Herodotus (IV, 62), in which the Scythians worshipped the image of the god of war.

[3] Reallexicon der indogermanischen Altertumskunde (Strassburg, 1901), p. 211.

[4] To mention only one of many instances: the hood, a leather cap for blindfolding hawks in order to tame them, was unknown to European falconers before the Crusades. It was introduced by the German Emperor Frederick II, who adopted it from the Syrian Arabs. In his famous treatise, De Arte venandi cum avibus (1245), the first which appeared in the West, he calls the hood *capellus*, and credits its invention to the Arabs. The use of the hood ( 皮套 ) has been well known to Chinese falconers since times of old, and is still prevalent in China. A specimen of one is in the collections of the American Museum of Natural History.

tations of falconry found in Siberian art. I refer to a find made between the villages of Reshnetikof and Krasnoyarsk, district of Tara, not far from the Irtysh. To a fragment of skin two figures of silver are attached, representing two riders on horseback, each holding a large falcon on his right fist, while with their left hands they are seizing hold of the bridle.[1] A silver goblet from the Yenisei, belonging to the later iron period, displays scenes of hunting, several riders chasing on horseback, one shooting at a bird flying over his head, another one having a hunting-bird on his left fist.[2]

Nothing would be more valuable for the history of falconry than the discovery of ancient pictorial representations of it among the antiquities of Turkistan. Indeed, I. KLEMENTZ[3] has made such a find on a wall-painting in a cave near Turfan, which represents two figures of men. One of these seems to be carrying a falcon; but the picture has suffered so much damage that nothing definite about it can be said.

In China, hawks, eagles, and other large birds of prey, are early mentioned in the "*Shih king*" and in the "*Li ki,*" particularly in poetical comparisons; but in classical literature no mention is made of falconry or of the training of birds for the chase, which seems to have come up not earlier than the Han dynasty, and soon developed into the favorite sport and pastime of emperors and noblemen, reaching its climax

---

[1] A. HEIKEL, Antiquités de la Sibérie occidentale, p. 84 and Plate XXIV, Fig. 2 (Mémoires de la Société finno-ougrienne, Vol. VI, 1894).

[2] W. RADLOFF, Aus Sibirien (Leipzig, 1884), Vol. II, p. 131; illustrated in PH. J. v. STRAHLENBERG, Das Nord- und Östliche Theil von Europa und Asia (Stockholm, 1730), Plate IV, B. See also the hunter on horseback with falcon, on a round metal plaque among the antiquities of the iron age of Perm, in ASPELIN, Antiquités du Nord Finno-Ougrien (Helsingfors, 1877), Part II, p. 151, No. 641.

[3] Nachrichten über die von der Kaiserlichen Akademie der Wissenschaften zu St. Petersburg im Jahre 1898 ausgerüstete Expedition nach Turfan (St. Petersburg, 1899), Part I, p. 45 (Plate VII). For a fine illustration representing modern falconers of Turkistan, see J. E. HARTING, Bibliotheca Accipitraria, A Catalogue of Books, Ancient and Modern, relating to Falconry (London, 1891), plate opposite p. 192.

under the Mongol rulers whose hawk-hunting expeditions are so picturesquely described by WILLIAM RUBRUCK and MARCO POLO.[1]

The oldest representation of falconry in China is found on one of the Han bas-reliefs of the *Hsiao t'ang shan*.[2] A man on foot holds a falcon on his right fist;[3] and a greyhound is hunting a stag in front of him. The next in point of time are two wood engravings in the dictionary "*Êrh-ya*,"[4] which may be stated to present a rather faithful copy of the illus-

---

[1] The two greatest sovereigns of the present dynasty, *K'ang-hsi* and *Ch'ien-lung*, maintained the warlike habits of the Manchu by frequent hunting-expeditions to the north of the Great Wall. They proceeded at the head of a little army, by which the game was enclosed in rings, and thus exposed to the skill of the emperor and his grandees. The Jesuit Father Gerbillon, who accompanied the Emperor *K'ang-hsi* on several of these expeditions, relates that a portion of the train consisted of falconers, each of whom had charge of a single bird. According to the official work Statutes and Regulations of the Present Dynasty (大清會典, WUCHANG edition, Book 4, p. 33 a), the court office of falconer is connected with that of the keeper of the imperial hunting-dogs (養鷹狗處), and is in charge of a high official who has administration over the breeding and training of the animals. He has under him a staff of twelve officials, composed of two commandants of Manchu forces, five subaltern officers of the Guards wearing the plain blue feather, and five sub-assistants. These are responsible for all affairs in connection with the training of the imperial falcons and eagles. Three Manchu secretaries, for clerical work and interpretation from Manchu into Chinese, are associated with this board.

[2] See CHAVANNES, La sculpture sur pierre en Chine, p. 77 and Plate XXXVIII (below on the left).

[3] In Japan the hawk is carried on the left hand, as usually among the ancient Turks, the Arabs, and in Europe. In China, where falconry is practised nowadays in Chihli and Shantung, the same old custom of holding the bird on the right prevails, while the left governs the bridle. In modern Turkistan, however, it seems that the falcon is carried on the right, to judge from the illustration in Harting's Bibliotheca Accipitraria, l.c. — The *Man* tribes in the south of China seem to carry or have carried the falcons on their left hands, as I see from the description of a bronze statue (in *T'u shu chi ch'êng* Vol. 1126, *k'ao kung tien* Book 250, *ku wan pu hui k'ao*, p. 5 b), seven feet high, representing a *Man* riding on a lion, which a certain *Yeh Sên* 葉森 acquired by exchange in 1320. According to the description, there is a hole in the back of the figure, through which water can be poured, which flows out of the lion's mouth. The *Man*, with dishevelled hair, is lifting a falcon on his left hand, and making signs to it with his right to control it.

[4] Book 2, p. 23 b. The lower one of the two is reproduced in G. PAUTHIER, Chine (Paris, 1853), Part I, Plate 10 (see text, pp. 55, 56).

trations to this work extant in the fourth or sixth centuries.[1] At all events, it may lay just claim to the honor of being the oldest graphic book-illustration of falconry in the world; the oldest English (and altogether European) representation, reproduced in the excellent book of JOSEPH STRUTT, "The Sports and Pastimes of the People of England" (new ed., London, 1898, p. 88, Fig. 3), being from an Anglo-Saxon manuscript of the end of the ninth century or the beginning of the tenth, preserved in the British Museum. While the oldest Chinese book on falconry seems to come down from the *Sui* dynasty (A.D. 518–617),[2] the first European print on the subject is the German book of ANON, printed in Augsburg in 1472,[3] followed by the earliest English print made at St. Albans in 1486, a copy of which was recently exhibited in the J. Pierpont Morgan collection of ancient books at the Columbia University Library.

Another important question which now seems to me to be connected also with the history of Turkish culture, and to which I can here allude only in a few words, is the origin of furniture in China. The fact that the Chinese are the only peoples in Asia who use raised chairs, settles, and high tables, has been frequently noted, and the great resemblance of Chinese furniture[4] to our own must strike every observer.

---

[1] See E. BRETSCHNEIDER, Botanicon Sinicum, Part I, p. 35.

[2] J. E. HARTING, Bibliotheca Accipitraria, A Catalogue of Books, Ancient and Modern, relating to Falconry (London, 1891), p. 207; after H. SCHLEGEL and WULVERHORST, Traité de fauconnerie (Leiden, 1844–53).

[3] Ibid., p. 45.

[4] A brief, fairly good description of articles of Chinese furniture is given in the New International Encyclopædia (New York, 1904), Vol. VIII, p. 349 b. See also E. DESHAYES, Le mobilier des anciens Chinois, conférence du Musée Guimet, 1905. Of great importance for the history of furniture in China is the third chapter in *I Tsing*'s memoirs (J. TAKAKUSU, A Record of the Buddhist Religion as practised in India and the Malay Archipelago by I-Tsing [Oxford, 1896], pp. 22–24), from which it follows that the Buddhist monks in China, imitating the practice of their brethren in India, used to sit on chairs at meals.

There is no doubt that in the times of their classical literature they did not sit on chairs at tables, but simply squat-
ted on mats spread on the ground.[1]  Also on the bas-reliefs
of the Han time, where so many scenes of home life are
depicted, we observe the same mode of sitting and a conspicuous absence of chairs and tables.  This should confirm
the supposition that the latter came into vogue only in post-
Christian times, and were due to foreign influence.

In the "*San ts'ai t'u hui*," section on implements 器用,
Book 12, p. 14 a, a chair with back surmounted by a headrest, side-arms, and a footstool in front of it, is depicted.
It somewhat resembles a folding-chair.  The descriptive text
runs thus: "The *Sou shen chi* (A.D. 320) says, 'The *Hu*
[Turkish] couch is a piece of furniture of the *Jung* and *Ti*.'
The *Fung su t'ung* [second century A.D.] says, 'The Han
Emperor *Ling* (A.D. 168–188) was fond of the costume of
the Turks [*Hu*].'  *Ching shih* made Turkish couches, which
were the first.  The present *Tsui wêng*[2] made chairs of bamboo
and wood, but all different, and all are derived from the
idea of the Turkish couches."[3]

It may at first seem strange that the chair should be
derived from Turks of Central Asia; but the Turks were
then a highly cultured people, who at that time might well
have obtained chairs and tables from the west, and they

---

[1] Compare J. H. PLATH, Nahrung, Kleidung und Wohnung der alten Chinesen
(Abhandlungen der bayerischen Akademie der Wissenschaften, I Cl., München, 1868,
Vol. XI, Part III, p. 219, or p. 27 of the separate reprint).

[2] *Tsui wêng* is a sobriquet given to *Ou-Yang Hsiu* (1007–1072), but whether he
is meant in this case, or somebody else, seems doubtful.

[3] 搜神記曰。胡床戎翟(=狄)之器也。風俗通曰。
漢靈帝好胡服。景師作胡床。此蓋其始也。今
之醉翁諸椅竹木間爲之。制各不同。然皆胡床
之遺意也。

may have acted as mediators between the west and China with regard to these objects. But this whole question will have to be thoroughly investigated, though the chief obstacle in the way is that no reliable historical work on this subject exists regarding the west of Asia and Europe.[1] In the light of the above Chinese passage, M. AUREL STEIN's discovery of a decorated wooden chair and arm-chair in a ruined dwelling-house of Turkistan[2] seems most remarkable to me, and a living witness in corroboration of the Chinese view.

### THE MOTIVE OF THE LION.

During the antiquity of China, lions were unknown, and no mention of them is made in so-called "classical" literature. There is an old term, *suan i*,[3] which occurs once in the "*Mu T'ien tzŭ ch'uan*"[4] (third or second century before Christ), and in the dictionary "*Êrh ya*"[5] (in its present shape from

---

[1] See O. SCHRADER, Lexikon der indogermanischen Altertumskunde, p. 344.

[2] M. AUREL STEIN, Sand-buried Ruins of Khotan (London, 1904), pp. 354–357.

[3] 狻猊 or 麑.

[4] 穆天子傳。狻猊日走五百里。 "The *suan-i* runs daily five hundred *li*" (see *T'u shu chi ch'êng* Vol. 584, *chin ch'ung tien* Book 59, *shih fu chi shih*, p. 1 b). LI SHIH-CHÊN, author of the *Pên ts'ao kang mu*, has copied this sentence in his description of the lion. Giles translates *suan* by "a lion from Tibet," though nothing in Chinese literature seems to corroborate such a statement; nor have there ever been, nor are there now, any lions in Tibet. The Tibetan word for "lion" (*seng-ge*) was borrowed from the Sanskrit *siṁha*.

[5] *Suan i* is there defined 如虎竊猫。食虎豹。 "Like a tiger which has lost its hair; and it devours tigers and panthers." Commentary: 即獅子也。 出西域。 "This is the lion; it comes from the Western countries (Central Asia)." The explanations of the later works, like *Shu i chi* (sixth century), *Yu yang tsa tsu* (eighth century), *Êrh ya i* (twelfth century), and even *Pên ts'ao kang mu* (sixteenth century), as quoted in *T'u shu chi ch'êng*, may here be passed over in silence, as they offer nothing remarkable regarding the question of the history of the lion in China, and their statements are derived principally from the scholiasts of the *Han shu*. The passage in the *Pên ts'ao kang mu* is translated by WILLIAMS, The Chinese Repository, Vol. VII, 1839, p. 595.

the third century A.D.), and is explained by the commentators to mean "lion." Whether by the passage in the former book the lion is really intended, cannot be decided; but it appears rather doubtful, considering the fact that lions were not introduced into China before the end of the first century of our era. It is certainly very possible that the inexplicable *suan i*[1] was identified as the lion at later epochs. According to the compilation of historical notes in the "*T'u shu chi ch'êng*," in the chapter on lions,[2] they are first mentioned by their present name (*shih tzŭ*) in the biography of *Pan Ch'ao* (A.D. 31–101), in the "*Hou Han shu*." They were sent as tribute, together with an animal called *fu pa*,[3] by the king of the *Yüe-chi*, when he asked for a Chinese princess in marriage. This event took place in A.D. 87 (in the winter, in the tenth month), under Emperor *Chang*. Under the reign of Emperor *Huo* the country of Parthia (*An hsi*) sent an envoy in the winter of A.D. 101 to present a lion (or lions);[4] and in A.D. 133 (sixth month), under Emperor

---

[1] Undoubtedly this word is not Chinese, but represents the sounds of some foreign word, which, I am inclined to think, is Sanskrit *çvan* ("dog"). This is confirmed by the dog-like representations of the Buddhistic lion. The Turkish-Mongol-Manchu designation for the lion is *arsalan, arslan* (W. RADLOFF, Versuch eines Wörterbuches der Türksprachen, p. 327; H. VÁMBÉRY, Die primitive Cultur des turko-tatarischen Volkes, pp. 185, 204). From the latter word, pronounced also *arlan*, W. SCHOTT (Skizze zu einer Topographie der Producte des chinesischen Reiches, Abhandlungen der Berliner Akademie, 1842, p. 372) derives the Chinese *suan-ni*.

[2] See Footnote 4, on preceding page.

[3] 符拔 or *t'ao pa* 桃拔, *t'ien lu* 天鹿 ("celestial stag"), in the *Sung shu;* and *fu pa* 扶拔 or *p'i hsieh* 辟邪 in the *Êrh ya i*. According to the *Han shu*, it occurs in the country of *Wu i* 烏弋國 in Turkistan, and was sent as tribute also from Parthia (安息) in A.D. 87, on which occasion it is described as having the shape of a female unicorn, but without horn (形似麟而無角). According to the *Êrh ya i*, however, it resembles a stag, has a long tail and one horn (似鹿長尾一角者); and such is the creature depicted in the *T'u shu chi ch'êng* (l.c.), *t'ao pa pu hui k'ao*.

[4] On the lions of *Ta Ts'in*, see F. HIRTH, China and the Roman Orient (Leipsic

*Shun*, the country of *Su le* (Kashgar) presented a lion. The scholiasts of the Han Annals describe the animal as follows: "The lion resembles a tiger, and is yellow; it has side-whiskers, and the soft hair at the end of its tail is of the size of a grain-measure" (*tou*). *Yen Shih-ku* says, "The lion is the animal called *suan i* by the '*Êrh ya;*' the character 猊 also [i.e., as well as 髥] means 'hair on the sides of the jaws.'"[1]

The first statues of lions of which we know are mentioned under the year A.D. 147 in an inscription on one of the pillars belonging to the tomb of the *Wu* family, where it is recorded that a certain artist, *Sun-tsung*, had made the lions at a cost of 40,000 "cash."[2] Unfortunately these lions have been lost.

There are two engravings representing lions in the "*T'u shu chi ch'êng*," — the one called *suan i*, a reclining lion with its cub, in which only the legs and claws are fairly delineated; the other, called the "lion of Libya,"[3] resembling a bear more than a lion, the head surrounded by rows of hair curled in small spirals, and the hair parted along the spine. The other lion has a parting on its head, from which

---

and Munich, 1885), pp. 219 et seq. — In the *Hsin T'ang shu*, the character 獸 ("quadruped") is used for "lion" (E. BRETSCHNEIDER, Notes and Queries on China and Japan, Hongkong, 1870, Vol. IV, p. 105).

[1] 師子似虎正黃。有髥猊尾端茸毛大如斗。師古曰。師子即爾雅所謂狻猊也。猊亦頰旁毛也。 The first clause is found in the *Tung kuan chi* 東觀記, written about A.D. 170 (E. BRETSCHNEIDER, Botanicon Sinicum, Part I, p. 205), and is also made use of in the *Pên ts'ao kang mu*.

[2] See CHAVANNES, La sculpture sur pierre en Chine, pp. xii, xl; S. W. BUSHELL, Chinese Art, Vol. I, p. 39.

[3] 利未亞州獅. The name "continent of Li-wei-ya [Libya]" for Africa was introduced by the early Jesuits. Under the name "Libyan lion" the animal is noticed in the *K'un yü t'u shuo* 坤輿圖說, published by Ferdinand Verbiest (see WYLIE, Notes on Chinese Literature, p. 58; H. CORDIER, L'imprimerie sino-européenne en Chine (Paris, 1901), p. 59).

the hair falls down on both sides in long curls. The *suan i* pictured in the "*San ts'ai t'u hui*" is still more fabulous and conventional, with four bands fluttering from its body, as on the sea-horse 海馬 in the same book, and a much exaggerated bushy tail with five spirals in the centre. But the standard work on natural history, the "*Pên ts'ao kang mu*," even beats this one by its grotesque figure of the lion, which is essentially that of Buddhist art,[1] with two flames over its head, and its tail represented as a flame. In fact, the Chinese have never had a correct conception of the lion,[2] nor have their artists ever drawn a natural sketch of a lion from life, but merely copied the fanciful conventionalized types of lions introduced into China from India with Buddhism.[3]

The type of the conventionalized Indian lion existed possibly as early as the Han time. At least, in the "*Ku yü t'u p'u*," Book 73, p. 7, a book-weight of nephrite in the shape of a coiled lion (古玉蟠獅書鎮) is illustrated,

---

[1] "The conventionalized lion of Indian art betrays its anterior Asiatic character particularly in the arrangement of its mane. A series of lion-like animals appears in art as early as the Aśoka period. Especially the gates of the Stūpa of Sāñchī show such representations. Afterwards these conventionalized lions became still more baroque. The so-called śardulas (North India) and yālis (South India) of the later Indian art are overloaded with shaggy hair and petty curls." — A. GRÜNWEDEL, Buddhistische Studien (Veröffentlichungen aus dem Kgl. Museum für Völkerkunde, Berlin, 1897, Vol. V, p. 70); see also A. FOUCHER (L'Art gréco-bouddhique du Gandhāra, Paris, 1905, p. 215), who also derives the lion of Indian art from Assyria.

[2] When the Chinese pilgrim *Sung Yun* (A.D. 518) saw two young lions at the court of Gandhāra, he wondered that the pictures of these animals, common in China, were not at all good likenesses (YULE, Marco Polo, 3d ed., Vol. I, p. 399).

[3] In No. 152 of the Kokka (January, 1903) there is a colored woodcut representing a pair of lions (*karashishi*) after a painting of *Eitoku Kano* (1543–90). In the descriptive text it is said, "The present subject, being drawn from pure imagination, is not supposed to represent lions as they would appear in nature. In fact the lion is treated by artists in this part of the world as if it were a semi-fabulous creature; so that a production of this kind should not be judged from the standpoint of mere accuracy, but from that of artistic conception." But what do "pure imagination" and "artistic conception" mean, if the question is merely that of reproducing a fixed type derived from a foreign art copied, and copied many hundred times, in painting, in stone, in bronze, and in iron?

which is ascribed in the text to the Han time (觀其制作之工亦漢人物也). The cheeks, head, and tail of this lion are covered with a mass of spirals, and it is characteristic of the Chinese notion of this animal that it is observed in the accompanying description that "hair, mane, teeth, and claws are just like the living specimen" (毛髮牙爪宛然如生).[1]

---

[1] Worthy of special note are the lion pillars in the province of Shensi, which I observed on my journeyings from *T'ung kuan* as far as *Hsi an fu*, and which were noticed also by A. FORKE, Von Peking nach Ch'ang an und Lo yang (Mitteilungen des Seminars für orientalische Sprachen, Berlin, 1898, Vol. I, Section I, p. 72), with regard to the place *Chih shui*. These stone pillars, of square section, with bevelled corners (usually about 1.5 m high, and sometimes from two to four grouped together), are placed before the façades of houses, about one metre distant from them. If there are two pillars in front of a house, they are arranged just in front of its front corners. A single pillar stands in the middle. They are scattered all over the streets of the city of *Hsi an* in great numbers, likewise so in *Lintung*, in *Weinan*, and in the two suburbs on the east and west sides of the latter place. Several such pillars are surmounted only by a round knob or a pear-shaped object, the point placed towards below. Aside from the lions of decidedly Buddhistic character (one of these stone lions is in our collection), they are adorned also with the figures of monkeys and what appears as the Indian stage-fool. The monkey occurs either single, or riding or squatting on the back of a lion. In the former case it is sitting in an upright posture, holding between its forepaws an undefinable object at which it is gnawing. Another type has its forepaws crossed over its breast. The monkey on the lion holds an object in its right paw, while with its left it grasps hold of the lion. Compare the figures of monkeys found in Turkistan: A. F. R. HOERNLE, A Note on the British Collection of Central Asian Antiquities (Oxford, 1899), p. 21; Idem, A Report on the British Collection of Antiquities from Central Asia (Calcutta, 1902), Part II, pp. 47, 48, 55; M. A. STEIN, Preliminary Report on a Journey of Archæological and Topographical Exploration in Chinese Turkestan (London, 1901), pp. 30, 31. The mime is either single, sitting on the top of the stone pillar, or riding on a lion's back, sometimes stretching both his hands into its jaws. Forke saw also a bird perching on his shoulder. His characteristic attribute is the wide, usually conical and pointed, fool's-cap, whose tip falls down in front of him. Some are provided with side-laps drawn over the ears. On other representations this cap runs up in spiral-like windings, thus affording nearly the appearance of a snail-house. His face bears the melancholy expression of good-natured humor. According to the tradition of the local population, the beginnings of these stone pillars go back into the age of the *T'ang* dynasty, which seems to me an acceptable view. They are doubtless nothing more than replicas of the Indian *stambhas*, and in particular the Indian lion-pillars known as *simhastambha* (A. GRÜNWEDEL, Buddhist Art in India, London, 1901, p. 20; Idem, Kunst des alten Indiens, in Baumann's Kunstgeschichte, p. 548). It thence follows clearly that also the figure of the fool must be of the same origin, as is corroborated also by the same representation of the fool found by me on the reliefs of the stone *pai lou* (memorial

How is it, then, that we meet with a rather natural representation of a lion on the reliefs of our Han pottery? Was this a type introduced into China from a foreign art, or was it conceived of by Chinese artists themselves of the Han period? Before answering this question, which now naturally arises, we should consider the curious fact that designs of lions occur also on works of art of another kind belonging to the Han epoch, — on metal mirrors.[1] Professor Hirth has made the subject of special investigation a certain group of ancient mirrors, decorated with patterns of vine and grapes among which birds and quadrupeds are interspersed. On some of these he observed "lions with unmistakable mane,"[2] and remarks, further,[3] "that the metal mirrors, the principal

---

archways) in the province of Honan, whose architectural and decorative style bears an essentially Indian character. That the prototype for the mime was given in Indian art, will be recognized from a fragment of pottery found in Turkistan, which, according to A. F. R. HOERNLE (A Note on the British Collection of Central Asian Antiquities, p. 20), exhibits "a man habited in what strikingly resembles the mediæval court-fool's dress." The migration of the Indian mime to China, and his representation in Chinese art during the *T'ang* time, are certainly important for the history of the mime and the dramatic art connected with him, in view also of the fact that the Annals of the *T'ang* Dynasty make express mention of Indian jugglers having arrived in China (F. HIRTH, Chinesische Ansichten über Bronzetrommeln, p. 56). Another interesting question related to these lion and fool pillars is whether the Buddhistic lion-dances (W. GRUBE, Zur Pekinger Volkskunde, Berlin, 1901, p. 111; Mrs. BRYSON, Child Life in China, London, 1900, p. 76; J. T. HEADLAND, The Chinese Boy and Girl, New York, 1901, pp. 143, 144), still current in all Buddhistic countries, one of which I witnessed in Peking, are connected with the mime (in Tibet the lion-dancer is introduced by a harlequin mummer — L. A. WADDELL, The Buddhism of Tibet, London, 1895, p. 539); and, if so, whether the mime as lion-dancer might not have given the impetus to the artistic motive of the mime riding on the lion's back. And further, should it be merely incidental, or should it not rather be an idea borrowed from India, that the European harlequin frequently appears with the mask of a lion (OTTO DRIESEN, Der Ursprung des Harlekin, ein kulturgeschichtliches Problem, Berlin, 1904, pp. 144, 169–170)?

[1] In the *Kin shih so* (*shih so*, Vol. 6) there is an illustration of a brick (the first of the bricks there, following the tiles) of the Han time, with the date of the first year of the period *Ching ning* (33 B.C.) on the one side, and a much conventionalized animal's head on the other, explained as a lion-head (獅 首 形). I can see nothing lion-like in this design.

[2] F. HIRTH, Über Fremde Einflüsse in der chinesischen Kunst (München, 1896), p. 14.

[3] Ibid., p. 26.

16

ornament of which is formed by the grape sacred to Dionysius, contain, besides, a series of other attributes of this god derived from the animal world, particularly the lion, which is plainly to be recognized by its mane on several mirrors." The mane of these lions is always sketched in the conventional style of spirals; the head is clumsy and misdrawn, but the outlines of the body, and the bushy tail, are not badly executed.[1] At all events, it is evident that this lion is not related to the Indian Buddhist figure,[2] but must come from another source, for the reason also that there is no question of Buddhist influence in that period.[3] On the other hand, the lion of the metal mirrors is a type entirely different from that on the Han pottery, which — in the representations of its outlines, its mane, and its mode of motion — can be compared only to the lion of Mycenian art.[4] I have no doubt that this pottery is older than the grape mirrors, and that the source to which the type of the pottery lion must be traced back is materially and chronologically quite different from that which furnished the lion pattern for the later mirrors. It cannot be imagined that the type of the pottery lion arose in China, since there is the greatest probability of its having arrived in China at a much earlier date than any living lion. Of course, although history furnishes us the date A.D. 87 as the first in which lions were officially brought as tribute from Central Asia, we need not cling over-scrupulously to that year as the very one in which the Chinese first made their acquaintance with the lion. It may very well be, that,

[1] To judge from a great number of such mirrors collected by me in *Hsi an fu.*

[2] The view of O. MÜNSTERBERG (Japanische Kunstgeschichte, Braunschweig, 1904, Vol. I, p. 107), that "the sacred lion *shishi* [i.e., the Buddhistic lion] arose from the lions with spiral-like, curly manes on the metal mirrors," is erroneous.

[3] Buddhistic lions appear on metal mirrors in the *T'ang* period; compare the *T'ang* mirror of "dancing phœnixes and lions (*suan i*)" in the *Po ku t'u,* Book 20, p. 6.

[4] See particularly the lions on the famous dagger-blade of Mycenæ.

owing to their close connection with the Turkish and other tribes of Central Asia, many individuals had seen or heard of lions long before, without such facts having been recorded in history; but all that would not favor a supposition that the Han artists conceived a type of lion of their own. Even supposing that in the Han period living specimens of this animal were transported into China in a much larger number than history permits us to compute, it is hardly credible that it would have come under the notice of artists for original life-like portrayal;[1] for we can almost establish a law that, at all times and in all countries, artists are apt to copy, not nature, which is simply a schooled phrase, but models which they find ready made.[2] During the *T'ang*,[3] *Yüan*, and *Ming* dynasties[4] the tribute gifts of lions from Central Asia and

---

[1] There is, however, one exception. *Wei Wu-t'ien*, a painter of the *T'ang* dynasty, is recorded as having painted a lion sent as tribute by a foreign nation in 756 (see A. PFIZMAIER, Kunstfertigkeiten und Künste der alten Chinesen, Wien, 1871, p. 56; H. A. GILES, An Introduction to the History of Chinese Pictorial Art, Shanghai, 1905, p. 49); but this was of no consequence for the history of art.

[2] The fact that painters do not work after nature, but create pictures living in their minds or notions, symbolical of the works of nature, could not be better illustrated than by the following story regarding the great *T'ang* painter *Wu Tao-tzü*. About 750 the Emperor conceived a longing to see the scenery on the *Chia ling* River in Ssüch'uan, and sent *Wu Tao-tzü* to paint it. *Wu* came back with nothing in the way of sketches; and when the Emperor asked for an explanation, he replied, "I have it all in my heart." Then he went into one of the halls of the palace, and in a single day he threw off a hundred miles of landscape (see H. A. GILES, An Introduction to the History of Chinese Pictorial Art, p. 43). Numerous other examples of a similar nature could be cited, e.g., the many Chinese painters who could handle the brush only under the influence of wine in a moment of bacchical ecstasy.

[3] See, e. g., CHAVANNES, Documents sur les Tou-kiue (Turcs) occidentaux, p. 135.

[4] S. WELLS WILLIAMS (The Middle Kingdom, New York, 1901, Vol. I, p. 317) remarks that the last instance of a live lion brought as tribute to China was in 1470, from India or Ceylon. This statement can hardly be correct, for it is known that the Emperor *K'ang hsi* received lions from Europeans, which in one of his writings led him to the belief that the lion originates in Europe (Mémoires concernant les Chinois, Paris, 1783, Vol. IX, p. 225). This fact seems to have become the idea underlying the artistic representation in China of Europeans leading a lion by a halter (A. FORKE, Von Peking nach Ch'ang an und Lo yang, Mitteilungen des Seminars für orientalische Sprachen, Berlin, 1898, Vol. I, Section 1, p. 47). The Chinese Chronicle of Macao

Persia to China increased, and were much more frequent than in the Han epoch; and during those periods a great number of the masters of the brush flourished; and not one of them ever thought of painting a lion after nature. Why? Because they were content with the existing conventional type of lion then generally adopted in art; and there was no tradition of a natural type of lion then known, and available to the artists as a stepping-stone, to justify them in such a revolutionary move. If the artists of the Han time had modelled their own lion true to nature, it would surely have been perpetuated through subsequent ages: since there is not the slightest trace of such a natural type in the vast province of Chinese and Japanese art, it is impossible to credit the Han period with such a creative invention. It cannot have been otherwise than that the natural life-like design of the lion on the Han pottery reliefs was an import which for a brief time exercised its influence and was fashionable, but soon sank into oblivion, — and quite naturally, since this animal was not familiar to the mass of the Chinese people, and thus lacked popularity. The conventionalized lion of India, following in the trail of Buddhism, had a much better chance to endure, as here the religious idea upheld its representation in art, and its numerous connections with the legends woven around Buddha and the saints deeply appealed to the popular mind. Both lions were derived in China from a foreign art. The one had to recede as being a mere temporary inorganic scion in the social milieu; the other persisted on the ground of social justification in the religious life of the nation.

---

(澳門記畧, Book F, p. 36 a) mentions elephants, rhinoceroses, and lions as being possessed by the Portuguese. Also from Samarkand, lions were despatched to China as late as 1489 and 1490 (BRETSCHNEIDER in China Review, Vol. V, p. 126 b).

I have no doubt that the prototype of the figure of the lion on the Han pottery reliefs found its way to China through the same channels as the design of the archer on horseback; i.e., through the medium of Scythian and Old-Siberian art. The occurrence of the lion on works of Scythian art is very frequent; and as to Siberia,[1] we have many examples of it on the famous gold plaques in the Eremitage, on which, as in China, the lion is represented, particularly in hunting-scenes.[2] The type of the Scythian and Siberian lion is undoubtedly derived from Mycenian and Greek art, and thus the transplanting of it to Chinese soil is historically and logically accounted for. And last, not least, the historical fact that lions were sent to China in the beginning by the Turks,[3] is strong evidence also for the derivation of the artistic motive of the lion from the same region — in the same manner as we observed also in the tactics and in the motive of the mounted archer.

Further, the great predilection for the boar and the tiger on the reliefs of the Han pottery cannot be merely incidental. The boar in flying gallop is a favorite in the art of Mycenæ and on the Siberian gold plaques in the Eremitage;[4] and the tiger, according to REINACH[5] almost unknown to Greco-Roman art, is a frequently treated subject in ancient Scythia and Siberia.

---

[1] E. DESOR, in a letter to Sir John Lubbock (Journal of the Anthropological Institute, 1873, Vol. III, p. 175), speaks of the handle of a knife found at Krasnoyarsk adorned with the figure of a *lioness* or tigress; but this case seems rather doubtful, as he adds that this animal is provided with something like an elephant's trunk.

[2] Numerous examples in KONDAKOFF; TOLSTOI et REINACH, Antiquités de la Russie méridionale, pp. 299, 307–310, 384, 387, 399; and see Index under *lion*.

[3] According to the *K'uang yü chi* (seventeenth century), lions are also imported into China from Khotan and Samarkand (see W. SCHOTT, Skizze zu einer Topographie der Producte des chinesischen Reiches, Abhandlungen der Berliner Akademie, 1842, pp. 371, 372).

[4] Antiquités de la Russie méridionale, p. 395.

[5] La représentation du galop, p. 80.

## PLASTIC FIGURES OF ANIMALS.

Fig. 43 represents a bird. It is 10.2 cm high, and measures 12.5 cm from beak to tail. A small piece has been chipped off from the lower side of the head. The eyes are clearly expressed, the spine and wings indicated by grooves. The bird's body rests on a hollow base. It might be supposed that this was not a separate piece by itself, but that it had

Fig. 43. Pottery Bird (from specimen in Clarke collection).

been stuck on, possibly, to the top of a lid, by way of decoration. Compare, for example, the raised phœnix on the lid of a bronze *lien* 匲 in the "*K'ao ku t'u*," Book 10, p. 19.[1]

On the other hand, it appears that fowl and dogs made of pottery were placed on graves to watch them, as may be concluded from the saying 陶犬瓦鷄 (now employed

---

[1] 蓋有立鳳爲飾。

PLATE LXIII.

1

2

in the sense of a useless thing or person), said to have originated with the Emperor *Hsiao I* (*Yüan ti*, A.D. 508–554) of the *Liang* dynasty, who urged the people not to put those clay objects on his grave, since, he said, the clay dogs would not keep guard at night, nor the clay cocks crow at dawn.[1]

Another bird of clay, apparently a cock, is represented in Plate LXIII, of the same type as the previous one, only in this the tail-feathers are more extended, and more elaborated in detail. It measures 15 cm from the tip of the bill to the tip of the tail. On the same plate is shown a duck made of pottery (23.7 cm from tip to tip). This bird was a favorite with the Han artists, and was frequently cast in bronze, particularly as a support for lamps, a beautiful example of which I saw at a Chinese collector's in *Hsi an fu*. Both of these are in the possession of Mr. Marsden J. Perry of Providence, R.I.

### JOTTINGS ON THE RACES OF DOGS IN ANCIENT CHINA.

The representation of dogs on the reliefs of the Han pottery and on the Han bas-reliefs of Shantung naturally gives rise to the questions, What race of dogs were they intended to portray, and what breeds of dogs then existed in China? This difficult problem becomes the more attractive, as it has already been taken up several times by naturalists and by specialists in the field of dog-history, who have made partial attempts to anticipate results the basis for which must first be laid by means of philological

---

[1] See GILES, Biographical Dictionary, p. 281; I. H. STEWART LOCKHART, A Manual of Chinese Quotations (Hongkong, 1903), p. 335; C. PÉTILLON, Allusions littéraires (Shanghai, 1895), Vol. I, p. 228.

research.  To understand the following investigation, it
must be premised that the generally accepted theory of
zoölogists at present is founded on the assumption of a
polyphyletic origin of the now living races of dogs; i.e., that
there are several groups which must be traced back to different primeval wild ancestors, the domestication of which
must accordingly have taken place in different parts of the
world and at various periods.[1]  The most important of these
stocks or breeds are the wolf-dog, the shepherd-dog, the
pariah-dog, the greyhound and sporting-dog, the mastiff, the
American dog.  The most remarkable group, from the view-
point of cultural history and archæology, is represented by
the mastiffs.  All of these are supposed to go back to one
wild genus, the *Canis niger*, or black Tibetan wolf; and
Tibet is considered the original home of all varieties of
mastiffs, to which belong the so-called "mastiff of Tibet"
(*Canis molossus tibetanus*), the Old-Assyrian hound, the *molossus* of the ancients, the St. Bernard, Newfoundland, bull-dog,
and pug-dog.[2]  In Assyrian sculptures, particularly in the
Palace of Assurbanipal (seventh century before Christ), we
find representations of powerful dogs which have been identified as the Tibetan mastiff.  Indirectly the same animals
are represented also on the sculptures of Sânchî of ancient
India (third century before Christ) in the lions with dog's

---

[1] "Ziehen wir die prähistorischen und die kulturgeschichtlichen Thatsachen zu Rate,
so deutet alles darauf hin, dass die zahmen Rassengruppen in ganz verschiedenen,
zum Teil weit auseinander liegenden Kulturkreisen entstanden sind, eine Wanderung
der Rassen dagegen erst später erfolgte." — C. KELLER, *Über den Bildungsherd der
südlichen Hunderassen* (*Globus*, 1900, Vol. LXXVIII, p. 107 b).

[2] According to C. KELLER.  However, TH. STUDER (Die prähistorischen Hunde in
ihrer Beziehung zu den gegenwärtig lebenden Rassen, Abh. d. schweizer. paläontolog.
Gesellsch., 1901, Vol. XXVIII) presumes that the pariah-dogs, greyhounds, and mastiffs
of Tibet form one race, and go back to *Canis tenggeranus* Kohlbrugge, which existed
until recently on the Island of Java.

heads. This representation, remarks GRÜNWEDEL,[1] reminds one of the treasure-guarding griffins of Ktesias, which I think have been correctly identified by BALL[2] as the great shaggy Tibetan dogs. They are the prototype of the so-called "Corean dog."

Turning now to our own subject, we find in all recent authors on the Tibetan dog the surprising statement that the first mention made of it occurs in Chinese literature. C. KELLER, in his fundamental book "Die Abstammung der ältesten Haustiere" (Zürich, 1902, p. 74), records it as a plain fact that the oldest accounts of the Tibet mastiffs are contained in Chinese literature in the "*Shu king*," according to which, in 1121 B.C., such a dog, trained as a bloodhound for the pursuit of men, was offered as a gift to the Emperor of China, and that frequently nowadays Tibetan dealers still bring such dogs into the Chinese Empire. MAX SIBER, in his "Der Tibethund" (Winterthur, 1897), — a finely illustrated monograph on the subject, giving an almost exhaustive series of quotations from modern European literature, — is somewhat more explicit on this point. He says (p. 15) that in 1121 B.C. a people called *Liu*, and living in the west of China, sent to *Wu-wang*, the Emperor of China, a dog of Tibetan race called in Chinese history *ngao*. According to the Chinese historians (thus he translates the French *commentateurs*) of the "*Shu king*," this dog was four feet high, and trained to hunt men, as was customary at one time in middle America and the West Indies. As his authority he quotes a note of G. PAUTHIER, in his "Le livre de Marco Polo"

---

[1] GRÜNWEDEL., Buddhist Art in India (London, 1901), p. 51.

[2] Indian Antiquary, 1883, Vol. XII, p. 235. On the dogs of ancient India see CHR. LASSEN, Indische Altertumskunde (Bonn, 1847), Vol. I, pp. 299–301, and J. W. McCRINDLE, The Invasion of India by Alexander the Great (Westminster, 1896), pp. 363–364.

(Paris, 1865).[1]  In a recent dissertation entitled "Zur ältesten Geschichte des Hundes" (München, 1903, p. 58), OSKAR ALBRECHT discusses the same "fact" in still more decisive terms.  He remarks, "In fact, it is well testified to by numerous [!] historical accounts that dogs were introduced from Tibet into China.  When the Yün-thai-Tatars (*sic!*), alleged relatives of the Scythians, invaded China and were repulsed by the Chinese, their envoys brought to the Emperor with the tribute also a dog.  The literati wished to insinuate to the Emperor that this had been done to jeer him; but he observed that he felt grateful for the gift, as this animal was useful, and deserving of higher appreciation also in China. A dog imported under *Wu-wang* in 1121, under like circumstances, is explained expressly [!] as one of the Tibetan race.  This is certainly by no means the first importation of the kind, since we see the Tibetan dog introduced into Mesopotamia at a much earlier date; and now the accounts of the mutual relations existing between China and the Western countries date much farther back, and are very old, particularly with regard to Tibet.  In later times the Tibetan mastiff is still more frequently mentioned; e.g., in the annals of the Han, about 100 B.C.  This historical testimony to a constant [!] importation of dogs from Tibet into China from the oldest times up to the present [!], together with the linguistic argument, make it very probable that the Chinese dog is altogether

---

[1] The original text of PAUTHIER (Vol. II, p. 380, note) runs thus: "L'an 1121 avant notre ère (*Li taï ki sse* k. 6, fol. 5) des peuples d'un pays situé à l'ouest de la Chine, appelé Liu, envoyèrent près de Wou-wang, pour lui offrir un chien de la *race tibétaine,* appelé dans l'histoire chinoise *N'gao.* Le même fait est rapporté dans le Chou-king (chap. Liu Ngao), en même temps que le discours du prince Tchao, ayant le titre de 'Grand protecteur du royaume' (*taï pao*), fait à cette occasion.  Ce chien, selon les commentateurs chinois du chou-king, avait quatre pieds de hauteur, et était dressé, comme certains chiens en Amérique, à faire chasse aux hommes de peau différente.  Marc Pol pouvait donc dire avec raison de ces chiens mastins, qu'ils étaient grans comme asnes."

of Tibetan origin, and derived by the Chinese from the Western countries. That the mastiff imported under *Wu-wang* is designated by the special name *ngao*, is only a seeming contradiction to our supposition that the Tibetan word *khyi* and the Chinese *khiuen* are identical, for the lexicographers describe this *ngao* as an animal four feet high, but capable of being trained, and particularly serviceable as a shepherd's dog [GILES's Dictionary is here quoted as authority]. This only goes to show that several breeds of dogs, already trained in their home country for various purposes, were introduced, — now a hunting-dog, *khiuen* (GILES); and then a shepherd-dog, *ngao* (GILES). Whatever else we know concerning the dogs of the Chinese, points likewise to an importation: an autochthonous domestication cannot be proved for any of them."

Let us now examine these statements from the original Chinese sources.[1] We do not possess any historical records of any literature regarding the early domestication of animals,

---

[1] Collections of quotations regarding dogs are to be found in all Chinese cyclopædias, the most complete in the *T'u shu chi ch'ëng*, Vols. 593 and 594. So far as they are important for the history of dog races, — and there are comparatively few such, — they are extracted in the following notes. The bulk of the other accounts are stories about dogs, dog-spirits, monstrosities, etc., interesting contributions partially to the psychology of the dog, partially to the psychology of the Chinese in their relations with dogs, and valuable folk-lore material. There are many good and touching stories from the life of particularly intelligent or faithful dogs worthy of the record of a Brehm. Some of them have been translated by A. PFIZMAIER, Denkwürdigkeiten aus dem Tierreiche Chinas (Sitzungsberichte der Wiener Akademie, Wien, 1875, pp. 36–51, or pp. 34–49 of the separate reprint). — In contributing this small share to the history of the dog, I feel I must state at the outset, that, not being a naturalist, I disclaim any pretension to authority on the subject. The sole purpose by which I have been guided is to place in an objective manner before the eyes of the specialist the material available, and thus enable him to draw his own conclusions. Taking the illustrations here inserted and the reliefs of the Han pottery, he has access to all archæological sources thus far in existence. My own feeble attempts at identifying dog species must not be taken to be more than they pretend to be, — impressions of one of the laity, who will most gratefully acknowledge any competent criticism, and welcome any further contributions to the following problems from those who know better, whether *pro* or *contra* his results.

and therefore we should not expect to find such in China. The "six domesticated animals" of the Chinese — horse, ox, pig, sheep, dog, and fowl — existed in and with the nation when it appeared on the stage of history. They were there, and later historians could not explain their origin. They took them as one of the facts which cannot be accounted for, and as altogether too plain and natural to require discussion. In short, what has become a problem to our modern science was not a problem at all to them. *Huang ti* is credited with the taming of bears, leopards, panthers, lynxes, and tigers, which he employed in battle against his adversaries;[1] but the simple question of training dogs remained untouched even by legend. The statement of Albrecht, however, that the dog is altogether a foreign import in China, is unfounded, and lacks historical basis. The dog has doubtless been a constituent of Chinese culture since most ancient times, which is all that we are able to state with safety; whereas the question as to who were the domesticators of the dog in Eastern Asia must naturally remain unanswered, at least from the standpoint of history.

The Emperor *Chou Hsin* (1154–1123 B.C.) of the *Yin* dynasty, in ill repute on account of his extravagance and debauchery, maintained a great number of dogs (狗), horses, and rare objects, and filled his palaces with them.[2] This is the first mention of dogs in the "Annals of *Ssŭ-ma Ch'ien*."

The *ao* or *ngao* (獒) is mentioned in the "*Shu king*."[3] After the defeat of *Chou Hsin* (1154–1122 B.C.), the last tyrant of the *Shang* dynasty, the *Chou* dynasty opened com-

---

[1] CHAVANNES, Les mémoires historiques de Se-ma Ts'ien, Vol. I, p. 28.
[2] Ibid., Vol. I, p. 200.
[3] J. LEGGE, The Chinese Classics, Vol. III, p. 345; *Shu king*, edition of COUVREUR, p. 209.

munications and amicable relations with all neighboring nations. One of these, the inhabitants of the country *Lü*, in the west, offered to *Wu-wang* (1122–1116 B.C.) as tribute "their *ao*," i.e., dogs (the plural is to be inferred from the possessive pronoun) of their country, which are designated as *ao*. No description or characteristics of them are given.[1] This donation provoked the memorial of the grand tutor *Shih*, in which he denounces rare foreign animals and curious objects of foreign people in general, and upholds the ideals of plain home life.[2] Of the situation of the *Lü* country we are ignorant. We only notice that the *Lü* formed a tribe of the Western Barbarians (西戎). LEGGE (*l.c.*) remarks in a note that "*Lü* was the name of one of the rude tribes lying in the west, beyond the 'nine provinces' of the empire, and that *ao* is the name of a kind of hound. It was, according to the '*Êrh ya*,'[3] 'four

---

[1] The whole report on this dog is limited, indeed, to this one sentence of six characters: 西旅底貢厥獒.

[2] See also PLATH, Die Beschäftigungen der alten Chinesen, p. 49.

[3] The dictionary *Êrh ya* (Book 3) offers the following explanations under the heading "dog." There are three designations for a bitch, — *tsung* 獏, *shih* 師, *ch'i* (*k'i*) 獥: a bitch bearing three puppies, two puppies, or one puppy respectively. These names are identical with those of pigs under the same conditions. A dog whose hair has not yet grown is called *k'ou* (未成毫狗). A dog with long snout is called *hsien* (or *lien*); one with short snout, *hsieh hsiao*. (長喙獫。短喙猲獢。) Compare J. LEGGE, The Chinese Classics, Vol. IV, p. 192. Both are hunting-dogs, and already mentioned in the *Shih king* (edition of COUVREUR, p. 134). *Chao* 獥, according to the *Êrh ya*, is a dog of great strength; and *mang* 尨 is identical with 狗. The *Shuo wên* (A.D. 100) gives the following names of dogs, and their definitions: —

獫黑犬黃頤也。 *Hsien* (or *lien*) is a black dog with yellow chin.

尨犬多毛也。 *Mang* is a dog with plenty of hair.

獟犬吠不止也。 *Hsien* is a dog that barks incessantly.

獒犬知人心可使也。 *Ao* is a dog that knows man's heart, and can accordingly be utilized.

feet — ancient feet, that is — high.' The *Shuo wên* describes it 'as knowing the mind of man, and capable of being employed.' From an instance of its use, quoted in the 集傳 from *Kung-yang*, it was evidently a bloodhound."

There is nothing at all in the Chinese accounts to show positively that this dog was a Tibetan mastiff, although it might be freely left to everybody's personal opinion to suppose that it was one. To derive this dog directly from Tibet, however, or only to ascribe it to Tibetans, to say nothing of the unfounded assertion that we have here the first real historical mention of the Tibetan *molossus*, seems out of the question, for the reason that in that period there was no such country as Tibet, nor was even a name like that of the Tibetans in existence. From the little we know about the situation of the *Lü* country, and the classification of the *Lü* among the *Jung*, we can only gather that the *Lü* were one of the numerous branches of Turkish tribes.[1]

---

犺健犬也。 *K'ang* is a strong dog.

狄赤犬也。 *Ti* is a red dog.

狡少犬也。 *Chiao* is a young dog.

匈奴地有狡犬巨口而黑身。 In the country of the *Hsiung nu* (Huns), there are *chiao* dogs with huge mouth and black body.

獿獀南越名犬也。 *Nao-sou* is a *Yüeh nan* (Annamite) name for dog.

[1] Dr. LANGKAVEL, in an article in the journal Der Hund (Vol. IX, pp. 38 and 169), which is not accessible to me, but is quoted by MAX SIBER (Der Tibethund, p. 15), relates that at the time of Marco Polo, in the thirteenth century, Tibetan grandees presented to China dogs as large as donkeys (*liu-ngao*), which were able to overcome tigers. According to him, the Chinese Annals of the Han (in the text, *die chinesischen Annalen des* [!] *Han 142–87 vor Christus* [!]) describe the Tibetan dogs as being as big as donkeys. There, also, they are called "red," which is evidence that then red ones had already occurred, besides the frequent black and red ones. This seems to be a somewhat confused mixture of the statements of Marco Polo and the *ao* in the *Shu king*. There is no account directly referring to Tibetan dogs in the Han Annals. But it may be that this note of Langkavel alludes to the account on *Ki pin* (Kashmir; see CHAVANNES, Documents sur les Tou-kiue [Turcs] occidentaux, p. 336) in the section on the Western Countries of the *Han shu*, where it is related that in the country of

In later literature the term *ao* is rarely used, and is restricted to poetry, and it is not then intended to denote the mastiff. In the modern Chinese literature on Tibet, the mastiff is, as a rule, called *ta t'ou kou* 大頭狗 ("the dog with the big head").[1] Colloquial terms are *Man kou* 蠻狗 ("dog of the barbarians"); *Tsang kou* 藏狗 ("Tibetan dog"); *ta chung kou* 大種狗 ("dog of large species").[2]

*Ki pin* large dogs are produced. The commentaries to this passage quote the *Kuang chi* of *K'uo I-kung* (*Liang* dynasty, 502–556; see E. BRETSCHNEIDER, Botanicon Sinicum, Part I, p. 164), where it is said that the large dogs of *Ki pin* are as big as donkeys, of red color, and are called in several places *yao t'ao* ("shaking a peddler's drum"). 西域罽賓國傳。罽賓國出大狗。注。郭義恭廣志云。罽賓大狗大如驢赤色數里搖鞀以呼之。(*T'u shu chi ch'êng* Vol. 593, *po wu hui pien, ch'in ch'ung tien* Book 116 犬部紀事 I, p. 9 a.) The description of this dog is therefore not contained in the *Han shu*, but in a much later commentary, and so cannot be much depended upon.

[1] See, for example, 西藏記 (published 1792), pp. 32, 33; W. W. ROCKHILL, in Journal of the Royal Asiatic Society, 1891, p. 273.

[2] Dictionnaire chinois-français de la langue mandarine parlée dans l'ouest de la Chine, par plusieurs missionaires du Ssüch'uan méridional (Hongkong, 1893), p. 209 b. The word *ao* was probably never a current term for any species of dog, but was artificially constructed *ad hoc*, after analogies. Compare the formations 驁 *ao*, "a vicious horse;" 鰲 *ao*, "a huge sea-fish;" 鼇 *ao*, "a huge turtle;" 鷔 *ao*, "a bird of ill omen;" 謷 *ao*, "a worthless fellow," — all in the second tone, and allowing the conclusion that *ao* originally implied the notion of something huge, weird, and extraordinary. — The principal Tibetan names for the mastiff are the following: (1) *bzang k'yi*, "the excellent dog" (Dictionnaire tibétain-latin-français par les missionaires catholiques du Tibet, Hongkong, 1899, p. 108 a, "molossus, bouledogue;" JÄSCHKE, A Tibetan-English Dictionary, p. 496 a, "a species of large dogs" [after Csoma]; the translation of SARAT CHANDRA DAS, A Tibetan-English Dictionary, Calcutta, 1902, p. 1109 a, "the Tibetan lap-dog," is wrong; see further H. RAMSAY, Western Tibet, Lahore, 1890, p. 33, "the *zang khee* is the real Tibetan mastiff"); (2) *rgyal k'yi*, "the royal dog" (FREDERIK SCHIERN, Über den Ursprung der Sage von den goldgrabenden Ameisen Kopenhagen and Leipzig, 1873, p. 45 [after Montgomerie], where other valuable notes on the mastiff not utilized by Siber are given); (3) *go k'yi*, which, as a term for the mastiff, I infer from the Chinese translation *Tsang k'ou* ("Tibetan dog") in the Dictionary of Four Languages (see above); from Ramsay's, l.c. *gho khee* ("watch-dog"), I conclude that *go* (thus written in the Manju-Chinese Dictionary) is intended for *sgo* ("door"), and, indeed, in JÄSCHKE's Dictionary, p. 115 b, *sgo k'yi* ("a door-guarding dog, watch-dog") is noted; (4) *gom k'yi*, "larger Tibetan mastiff" (G. SANDBERG, Handbook of Colloquial Tibetan, Calcutta, 1894, p. 168 a).

The great resemblance of the wolf to the dog is noted also by the early Chinese writers. The "*Êrh ya*" says that the wolf has the feet of the dog (豺狗足); and the "*Shuo wên*," that the wolf has the voice of a dog (豺狼屬狗聲). Most interesting in this connection is a passage in the "*Tso chuan*" given in the "*Mao shih ming wu t'u shuo*"[1] (Book 2, p. 3), under the description of the wolf, where again it is quoted after the book 疏 of HSING PING 邢昺 (932–1010: GILES, *Biographical Dictionary*, p. 296): 戎狄豺狼不可厭也。 "To the *Jung* and *Ti*, the wolf is not an object of dislike," by which is meant, apparently, that the wolf was not dreaded by these presumably Turkish tribes, and was accustomed to live in the neighborhood of human dwellings. This reminds us of JÄSCHKE's[2] remark, that the Tibetan wolves, where more numerous, — as, for instance, in Spiti, — commit ravages among the sheep, but are otherwise not much dreaded by man, and, like the wolf in general, they are easily tamed. If we compare in the above-mentioned book "*Mao shih*," etc., the picture of the wolf (Fig. 45) with that of the dog (*mang* 尨, Fig. 44), a striking resemblance between the two is noticeable as regards shape of head, mouth, and body, legs, claws, tail, and hairiness. GILES explains *mang* as a shaggy-haired dog, and even translates it by "the Tibetan mastiff;" and it cannot be denied that the traditional design of the *mang*, in spite of its shortcomings and conventionalized features, bears a speaking likeness to the pictures of the

---

[1] 毛詩名物圖說. Methodical Repertory on the Natural Science of the *Shih king*, giving illustrations of all plants and animals mentioned in it, with brief discussions on their names. Compiled by WU CHUNG-HSÜ 吳中徐 and TING SHIH-FU 鼎寶夫. No date; according to J. LEGGE (The Chinese Classics, Vol. IV, Prolegomena, p. 179), published in 1769.

[2] A Tibetan-English Dictionary (London, 1881), p. 332b.

Tibetan mastiff. The *mang* is mentioned as early as the "*Shih king*," and must therefore have been known to the Chinese at an early date. It now seems questionable whether the Tibetans are to be looked upon as the trainers of the mastiff; if not, rather ancient Turkish tribes tamed the wolf — an animal with which they were always quite familiar, and which played an eminent rôle in their tribal traditions and creation myths[1] — at a much earlier period. Of the fact that the dog in general was known in the South-Siberian bronze age, we possess well-authenticated archæological evidence in a bronze

Fig. 44. Engraving of a Dog (from *Mao shih*).

Fig. 45. Engraving of a Wolf (from *Mao shih*).

plaque representing a hunter accompanied by two dogs.[2] In this connection it is worthy of note that a fierce kind of dog, called *p'i ngan* 猈犴 (usually translated "bull-dog"), whose picture is painted on the doors of jails because of his ability as a watch-dog, is said to originate from the land of the Turks (胡地).[3]

---

[1] H. VAMBÉRY, Die primitive Cultur des turko-tatarischen Volkes (Leipzig, 1879), pp. 197, 198.

[2] W. RADLOFF, Aus Sibirien (Leipzig, 1884), Vol. II, p. 89 and Plate IV, Fig. 3.

[3] C. PÉTILLON, Allusions littéraires (Shanghai, 1895), Vol. I, p. 232; LOCKHART, A Manual of Chinese Quotations (Hongkong, 1903), p. 501. Giles translates *ngan* by "tapir" (see ZACH in China Review, Vol. XXV, p. 142). The noteworthy passages regarding this animal are collected in *T'u shu chi ch'êng*, Vol. 593, *po wu hui*

17

Other evidence pointing to the same fact is found in the peculiar "dog of the kind which is found with the barbarians *Ti*," mentioned by *Ssŭ-ma Ch'ien*.[1] And if the country of *Lü*, which sent the hound *ao*, was a branch of the Western *Jung*, everything is indeed apt to show that these large extraordinary dogs, including the mastiffs, — if the term *mang* may be identified with the latter, — came from Turkish regions.[2] It is noteworthy, too, that, according to the dic-

---

*pien, ch'in ch'ung tien*, Book 115 犬部彙考 pp. 6b–8a, and compare particularly the definitions there given, 犴胡犬也 and 犴胡地之野犬也, etc. See, also, G. SCHLEGEL, Uranographie chinoise, p. 593.

[1] CHAVANNES, Les mémoires historiques de Se-ma Ts'ien (Paris, 1905), Vol. V, pp. 28, 30, 31.

[2] It is of course quite possible that *Canis molossus* was domesticated in various territories, at various periods, and by various peoples of Central Asia, — a question which unfortunately will never be cleared up satisfactorily. ALBRECHT (Zur ältesten Geschichte des Hundes, p. 56) also seems to presume two districts for its domestication. From the standpoint of the historian, however, I think I should warn naturalists against a premature over-estimation of the age of Tibetan culture and a rash supposition that Tibet and the Tibetans should be credited with the taming of *Canis niger*. There is a great difference to be noted between Tibet and Tibetans. Of the existence of a Tibetan nation in the Tibet of nowadays we know nothing before the seventh century A.D., and of the prehistoric conditions of this country we know nothing at all. Tibetans (i.e., a great number of scattered, partially nomadic tribes without a political union) may have existed millenniums before our era: at least, the K'iang of the Chinese annals are usually asserted to be the ancestors of the subsequent Tibetans, but they did not then inhabit Tibet proper. If we find the Tibetan mastiff on Assyrian sculptures of the year 668 B.C. (C. KELLER, Die Abstammung der ältesten Haustiere, pp. 42, 73), I do not see how we can look upon Tibet as the place of origin of the Assyrian *molossus;* and, even if the Tibet of that period were suddenly brought under the search-light of history, the historical connection of Tibet with Assyria, which doubtless has never existed, could not be construed If we suppose, on the basis of the very suggestive but nevertheless hypothetical argument laid down above, that *Canis molossus* existed among the ancient Turks, — whether we may personally feel like attributing the domestication to them or to any other tribe from which they might have derived it, — the appearance of the mastiff in Assyria presents itself as a natural historical phenomenon; for it cannot be denied that there is a certain substratum of ancient Turkish culture in Assyria, whether the Sumerians may have been Turks, kinsmen of Turks, or not. If, as we tried to show, falconry was an old Turkish invention, spreading likewise to China and to Europe, I do not see any reason why falconry in ancient Assyria (B. MEISSNER, Falkenjagden bei den Babyloniern und Assyrern, Beiträge zur Assyriologie, Vol. IV, pp. 418–422; PINCHES in Proceedings of the Society of Biblical Archæology, Vol. VI, pp. 57 et seq.) should not have been of

tionary *"Shuo wên"* (A.D. 100), the land of the *Hsiung nu* (Huns) possessed a special kind of dog, called *chiao* 狡 , with

Turkish origin. The problem, on the whole, certainly remains one for future solution. Here only one suggestion. It is now a well-established and generally adopted fact that bronze, in all provinces of antiquity of Europe, goes back to Babylonia, and finally to the Sumerians, who are the only people in the world who have in their language an independent genuine word for "bronze" (see O. SCHRADER, Lexikon der indogermanischen Altertumskunde, p. 200). Only one centre of bronze exists, therefore, for the entire western part of the Old World. How is it now about the eastern half? There, on the continent of Asia, four bronze centres would be left, — Babylonia; India; the Altai, from which all Siberian bronze radiated; and China. After our experience in the countries of Europe, would it now be logical to let four independent bronze regions come up in Asia? The almost uniform alloy of Babylonian and European bronze has always been much emphasized, and taken as a strong proof of its migration from one locality. Certainly such an arbitrary composition of copper and tin can have been invented only once. It is strange to see that two important facts seem to have remained unnoticed by archæologists in this field. The one is, that the chemical investigations of old Siberian bronzes made by Struwe at the instigation of Radloff yield, on an average, the same proportions of the two metals as the Babylonian bronze; i.e., 90 : 10 (see W. RADLOFF, Aus Sibirien, Leipzig, 1884, Vol. II, p. 83). The other is, that as regards a curious Chinese bronze implement from Yünnan Province, figured and described by JOHN ANDERSON (A Report on the Expedition to Western Yunan via Bhamô, Calcutta, 1871, p. 415), to use the actual words of the author, "It is certainly remarkable that the composition of this bronze is the same as that which characterizes the bronze implements found throughout Northern Europe, the percentage being copper 90, tin 10 = 100" (after an analysis by Dr. Oldham). Many more analyses of Chinese as well as Siberian bronzes are of course required to decide this important question; and not only that, but, above all, much more intense historical and archæological investigations than have been heretofore undertaken. If the late Chinese and Indian bronzes show so many different alloys, and such deviation from the common standard, this is not to be taken in the sense of an independent origin of these bronze provinces, as has been done so frequently before, and most illogically. This fact is quite natural, since we possess thus far neither the most ancient nor *the* ancient Chinese and Indian bronzes, but only a few more or less poorly authenticated from comparatively late historical epochs. When the idea of making an alloy of copper and tin was once suggested, it was certainly an easy matter for highly civilized nations, who produced a great number of individual craftsmen and artists, to experiment and employ various other alloys of the two for various purposes in the course of centuries and millenniums. That Sumerian is somehow connected with Altaio-Siberian bronze culture, and the latter with that of China, is an *a priori* supposition, even now not to be overlooked, because the historical foundation, though even in its crudest outlines, is there given. To return now to the above problem regarding the migration of *Canis molossus*, if the identifications made by the zoölogists, of the Assyrian sculptures with this species, are correct, and if their whole tracing of the genealogy of this dog race is true, — which I am not competent to judge, although it seems plausible to me, — the Tibetan mastiff can have reached Assyria only from Turkish regions through the medium of Turkish tribes, who, according to Chinese accounts, were in all probability familiar with this dog.

large mouth and black body,[1] which characteristics are essential to the mastiff. For the rest, the definition of the *chiao* in the "*Shuo wên*" is that of a young dog (少犬). This *chiao* appears as a fabulous animal in the "*Shan hai king*." "On the nephrite hills there is an animal of a shape like a dog, striped like a panther, and with horns like an ox; it is called *chiao;* its voice is like that of a barking dog; in the country where it appears, dogs will be abundant."[2] In a book, "*Jui ying t'u*" 瑞應圖, occurs now the interesting passage that the *Hsiung nu* offered panther-dogs with pointed mouth, red body, and four feet.[3] Whether these latter animals were mastiffs or not, — red color occurs with them, in fact, — it is evident that the *Hsiung nu* possessed extraordinary dogs, which arrived also in China.

MARCO POLO, speaking of Tibet (Book II, Chap. 45), mentions its large and fine dogs, which are of great service in catching the musk-beasts; and in Chapter 46 he says, "These people of Tebet are an ill-conditioned race. They have mastiff dogs as big as donkeys, which are capital at seizing wild beasts [and in particular the wild oxen which are called *Beyamini*, very great and fierce animals]."[4]

---

[1] 匈奴地有狡犬。巨口而黑身。 See K'ANG HSI's Dictionary *sub voce*. GILES translates *chiao ch'üan* by "a black Peking dog."

[2] 玉山有獸狀如犬而豹文。角如牛。名曰狡。音如吠犬。見則其國犬穰。 — Ibid.

[3] 匈奴獻豹犬錐口赤身四足。 — *Ko chih ching yüan*, Book 87, p. 2 a. The express mention of four feet probably alludes to a dog with panther stripes, two horns, and without its two fore-feet, also called *chiao* 狡 by the contemporaries, mentioned by KUO P'O (A.D. 276–324) for the seventh year of the period *T'ai k'ang* (A.D. 286), and quoted in the commentary to the *Shan hai king* to our passage above.

[4] YULE, The Book of Ser Marco Polo (3d ed. by H. CORDIER, London, 1903), Vol. II, pp. 45, 49. In the new edition, Yule's explanation of *beyamini* from an

It is clear that Marco Polo was acquainted with the Tibetan mastiffs; but the general opinion based on his accounts, that in his time mastiffs were exported in large numbers from Tibet to China, is not correct. It is true that he mentions big mastiffs at the court of the Khan, two barons with the title *Chinuchi*, "which is as much as to say 'The keepers of the mastiff dogs,'" and two thousand men who are each in charge of one or more great mastiffs, so that the whole number of these is very large.[1] But Marco does not say a word about these dogs having come from Tibet.[2] That these dogs were Tibetan mastiffs will have to remain a supposition, which we are entitled to entertain, from a reason that seems to have escaped all the numerous commentators of Polo's book. The only possible correct derivation of the word

---

artificially constructed and non-existing word *buemini*, i.e., Bohemian, — "a name which may have been given by the Venetians to either the bison or urus," — still remains. I have given a plausible, and I still think correct, interpretation of the name in Denkschriften der Wiener Akademie der Wissenschaften, 1900, Vol. XLVI, No. VII, pp. 20, 52, where I derived it from the Tibetan word *ba-men*, the gayal wild ox, *Bos gavæus* (see SARAT CHANDRA DAS, A Tibetan-English Dictionary, Calcutta, 1902, p. 860), which is indeed the same species as is recognized by Yule under Marco's term.

[1] YULE'S Marco Polo (3d ed.), Vol. I, p. 400.

[2] Probably for the reason that such was not the case. I cannot find, in fact, any instance in Chinese history of dogs having been exported from Tibet into China. Among the objects of tribute sent from Tibet to the Emperor of China (see W. W. ROCKHILL in Journal of the Royal Asiatic Society, 1891, pp. 244 et seq.), dogs do not appear. The voluminous geographical descriptions of the provinces of Ssŭch'uan and Kansuh, which treat in special chapters of all products of these regions and of all those imported there from Tibet, and through which I specially hunted for this purpose, do not mention any dogs from this country, either; and if such a constant importation of Tibetan dogs into China went on, as Albrecht and others would have us believe, we should certainly expect to find a record of such fact in the annals of the border provinces. According to ROCKHILL (l.c., p. 273, note), mastiffs are rare in Eastern Tibet; and, according to H. RAMSAY (Western Tibet, Lahore, 1890, p. 33), the pure mastiffs are procurable only in Lhasa, very handsome, and costly to purchase. These conditions would exclude at the outset any lively border-trade in these dogs. There is no doubt that they occasionally occur in western China, particularly among the numerous Tibetan tribes of western Ssŭch'uan (see, for example, A. HOSIE, Three Years in Western China, 2d ed., London, 1897, p. 134), where they might have been since days of old, but the frequency of this occurrence must not be over-estimated.

*chinuchi*[1] is from Mongol *chinoa* (pronounced *chinō* and *chono*), "wolf," with the possessive suffix *-chi*, meaning accordingly a "wolf-owner" or "wolf-keeper." Now, one of the Tibetan names for the mastiff is *chang-k'i* (according to pronunciation, written *spyan k'yi*),[2] which signifies "wolf-dog." Since Mongol culture and literature are entirely founded on Tibetan, and the greater number of all Mongol terms are literal translations from Tibetan, there is no doubt that the translation of the word for mastiff as "wolf" came also with that dog from the Tibetans to the Mongols. On this basis, we might translate *chinuchi* as a "keeper of Tibetan mastiffs," and this seems to be the only safe evidence available for the assertion that Tibetan mastiffs were kept in China during the Mongol period; and it seems to me that this importation took place rather through the medium of the Mongols from Mongolia than by the direct way from Tibet.[3]

If it is impossible to draw from Chinese literature the direct conclusion as to the existence of the Tibetan mastiff in ancient China, — and this may be conjectured only from certain indications, — an archæological specimen might probably be more apt to reveal the state of the case. This is the full figure of a dog of Han pottery, with green glaze, which, for the most part, has dissolved into a silver iridescence. It is in the possession of Mr. Marsden J. Perry, Providence, R.I. Fig. 1, Plate LXIV, shows the side view, and Fig 2 the front view, of the dog. The figure is hollow, apparently made over a mould, and measures over all 28.7 cm.

---

[1] The numerous explanations given for it are collected in the same book, pp. 401, 402. The Mongol dictionaries of KOWALEWSKI and GOLSTUNSKI do not contain the formation *chinuchi*, but it may well be that it existed in the thirteenth century.

[2] H. RAMSAY, Western Tibet, Lahore, 1890, p. 33.

[3] According to W. LOCKHART (Proceedings of the Zoölogical Society of London, 1867, p. 43), "a noble black dog, as large as a full-sized Newfoundland, is brought to Peking from Mongolia: he is used as a sheep-dog."

PLATE LXIV.

Pottery Mastiff in Side and Front View.

It represents a large, muscular, sturdily built dog, which, in my opinion, plainly shows the bull-dog character. The representation of the tail is identical with the carriage of the tail in the Tibetan mastiff,[1] and the well-executed face leaves no doubt as to the correctness of this identification. The only deviating feature seems to be the erect ears, while, as a rule, the mastiff's ears are long and drooping. This, however, is considered by zoölogists to be the result of domestication; so that in ancient times there might well have been forms of the animal which were nearer in this respect to their ancestor the wolf. On the other hand, we should not lay too great stress on such details, but bear in mind that the piece in question is the work of a potter of the Han time, whose intention was certainly not to furnish an absolutely true, life-like portrayal of this dog, but merely to outline it in its general and most striking features. If we consider that the idea connected with such statuettes of dogs — which were placed in the grave of the dead, perhaps of a sport-loving nobleman — was doubtless to guard their masters, and keep off from their graves the evil influences of obnoxious spirits, we clearly recognize that the artist had good reasons for representing the dog in the posture of watching, listening, pricking up its ears; and I think it is more adequate to the purpose to look upon this feature rather as an intention of the artist, prompted by the subject, than as a racial quality of the animal. Further, it should not be expected that in all its particulars it will fully agree with the modern Tibetan

---

[1] "The only peculiarity that I have noticed about them [the Tibetan mastiffs] is that the tail is nearly always curled upward on the back, where the hair is displaced by the constant rubbing of the tail." — A. Cunningham, Ladák, Physical, Statistical, and Historical, London, 1854, p. 218. See, also, the illustration of a Tibet dog in G. W. Leitner, The Languages and Races of Dardistan, Lahore, 1877, Plate opposite p. 107.

mastiff and its descendants in the West. Since it is very likely that the mastiff reached the Chinese through the medium of the ancient Turks, we can but presume that it has been modified, not inconsiderably perhaps, under the influence of Turks as well as Chinese, and might have adopted some peculiar form under the hands of both these most active animal-breeders. Very curious is the harness represented on this figure, which consists of a double collar, — one for the neck, the other going around the body, — seemingly broad leather straps, with metal fittings at the ends, which are joined on the back in a massive hook for the attachment of the halter. The necessity of guiding the animal by means of such a solidly worked harness is evidently an intimation of its un-broken strength and ferocity, which the Tibetan mastiff, ac-cording to all accounts, still possesses, and strong proof for the identification of this dog with the latter or a related species. On the bas-relief shown in Fig. 49 we notice a greyhound led by a hunter, but provided only with a neck-collar, which was sufficient to hold him under control, while an additional means of force was necessary for the mastiff. It is interesting to note that also the mastiffs on the As-syrian sculptures above referred to are managed by hunters with long halters attached to the neck-collars, though the belly-band is there wanting.

The question as to whether descendants of mastiffs still exist in China, and whether traces of them are to be found in other breeds, ought to be made the subject of special investigation by a naturalist in China. I give here only a few notes from Chinese records regarding large and fierce dogs, which may perhaps suggest localities where an inquiry in this direction might be advantageously started.

"A large breed of dogs, so fierce and bold that two of

them together will attack a lion" (i.e., a tiger, in Polo's language), is mentioned by MARCO POLO in connection with the province of Kueichow.[1] These are presumably identical with the *Yü lin* dogs (鬱林犬) mentioned by the Chinese as being produced in *Yü lin chou* (Kuanghsi Province), extremely high and large, with drooping ears, and tail different from that of the common dog.[2]

In the "Annals of *Ch'êng tê fu*" (Jehol) 承德府志 Book 29, p. 25 a, occurs this passage: "The hunting-dogs are clever in seizing wild animals, and are kept in great numbers in Mongolia. These are the 'hunting-dogs higher than stags' that *Chou Po-ch'i* of the *Yüan* dynasty mentions in a poem in his 'Diary of a Journey to the Capital.'"[3]

The following passage occurs in the "*K'un lun nu ch'uan*" 崑崙奴傳: "In first-class houses there are fierce dogs that watch the doors to the halls of singing-girls. Men who are not regular customers are not allowed to enter unceremoniously. If they enter, the dogs bite them to death. Their warning is like that of a spirit, their fierceness like that of a tiger. They are dogs from *Mêng hai* in *Ts'ao chou*."[4]

---

[1] YULE, Marco Polo (3d ed.), Vol. II, p. 126.

[2] *T'u shu chi ch'êng* Vol. 593, 犬部彙考 p. 5.

[3] 田犬善捕野獸。塞外多畜之。元周伯琦上京途中紀事詩所謂獵犬高於鹿者是也。 This passage is briefly alluded to by O. FRANKE, Beschreibung des Jehol-Gebietes (Leipzig, 1902), p. 89.

[4] 一品宅有猛犬守歌妓院門。非常人不得輒入。入必噬殺之。其警如神。其猛如虎。即曹州猛海之犬也。— *Ko chih ching yüan*, Book 87, p. 2 a. *Ts'ao chou* is a prefecture in Western Shantung: such a place there as *Mêng hai* is not known to me. — Compare the similar passage in *T'u shu chi ch'êng* Vol. 594, 犬部紀事 II, p. 10 b.

Thus far the oldest representation of a dog in China (speaking from the view-point of our limited knowledge of Chinese antiquities and of the small number of those which have survived) is found on a bronze tazza of the *Chou* dynasty, reproduced above in Fig. 36 and described on pp. 150—153. It is apparently a hunting-dog, as it accompanies a huntsman. The small scale on which it is represented was naturally conditioned by the size of the vessel, which itself is only 24.4 cm high, and therefore cannot offer conclusive evidence in judging of the size of the animal. The ears of the dog stand upright. Its tail is short, and turned upward.

It seems very likely that this dog is intended to represent a wolf-dog, whose genealogy will be seen in C. KELLER, l.c., p. 54. The Chinese breed is unfortunately called by naturalists *tchau*, which is the Pidgin-English word "chow-dog;" i.e., edible dog. It is desirable that this term should disappear from scientific nomenclature. An illustration of three Chinese wolf-dogs will be found in MAX SIBER, "Der Tibethund," p. 39. A description of its skull is given by C. KELLER (p. 50), who groups this breed with *Canis palustris;* but it has been changed, he remarks, by long domestication, "as in a country with ancient culture could hardly otherwise be expected."[1]

The representations next in point of time are those on the reliefs of the Han pottery (pp. 153, 154, 162, 163, 165, 166, 169) and then those on the Han bas-reliefs.

Fig. 46, made from an original rubbing, represents a grey-

---

[1] For some notes regarding this dog, see G. WELLS WILLIAMS, The Middle Kingdom (New York, 1901), Vol. I, p. 318; J. DYER BALL, Things Chinese (4th ed., Hongkong, 1903), pp. 215—217. The notion of these two authors that this dog is related to the Eskimo dog, and "perhaps the original of the species," is a mere opinion, for which I do not see the proof. For an illustration of this dog, see R. FORTUNE, Two Visits to the Tea Countries of China, Vol. II (London, 1853), p. 120.

hound[1] from one of the Han bas-reliefs of *Wu liang* in Shantung.[2]　It is sitting on the ground in front of a cart-horse, a man standing on its left. It seems to belong to the people driving in the cart.　Its interpretation as a greyhound is also given by CHAVANNES,[3] with the remark that greyhounds are peculiar to Shantung.　Though Chavannes does not give the source for this statement, I should think we might safely accept it on the authority of a cautious and matter-of-fact scholar of his type.[4] The same figure of a greyhound, in exactly the same posture as

Fig. 46.　Greyhound represented on Han Bas-Relief of *Wu liang* in Shantung (from a rubbing).

---

[1] The modern colloquial name is 細狗, "slender dog" (COUVREUR, Dictionnaire français-chinois, p. 563).　GILES (p. 630 b) writes 西狗, which seems to be a misprint.

[2] In the work of CHAVANNES (La sculpture sur pierre en Chine) it will be found on Plate XXV, second panel.

[3] Ibid., p. 59.

[4] I looked up the Annals of Shantung Province 山東甬志 with the expectation of finding there something that would elucidate this question, but they contain only the following passage with regard to dogs (Book 24, section on Products 物產, p. 6 b): 犬。種類不同。有吠犬短喙善守。有田犬長喙而瘦駛善獵。有食犬體肥供饌近多畜。西洋種矮小潔獝可玩名哈吧狗。 "There are various kinds of dogs.　There are the barking dogs with short muzzle, excellent watchers.

above, is delineated on another bas-relief,[1] also here squatting in front of a cart whose horse has been unharnessed and is standing to one side under a tree; while a man, probably the teamster, is to the left in front of the dog.

A large greyhound that has been set on a man is represented on the bas-reliefs of *Wu liang*, and will be found on Plate XXI b, lower panel, of the work of CHAVANNES (cf. Ibid., p. 56).[2]

Figs. 47 and 48 represent hunting-hounds from two bas-reliefs of the *Hsiao t'ang shan*, drawn from original rubbings of the monuments.[3] Both reliefs are of colossal size, and the most varied scenes are united in them. On the lower section of the one, two hunting-scenes — one on the right, the other on the left side — are represented. That on the right shows eight hunters afoot, carrying nets over their

There are hunting-dogs with long muzzle and thin legs, excellent hunters. There are the edible dogs with fat body, which are served as food, and reared in large numbers. There is a kind from the Western Ocean, low, small, clean, and cunning, with which you can play; it is called 'ha-pa dog.'" *Hsi yang* ("Western Ocean") is the usual term for Europe, but is here apparently a mistake for *hsi yü* ("western countries"). The statement of Chavannes, however, is confirmed by W. LOCKHART, who states (Proceedings of the Zoölogical Society of London, 1867, p. 43) that from Shantung a beautiful black, long-haired, long-backed, long-legged terrier, very much like a black skye, is brought to Peking.

BISHOP A. FAVIER, in his book Péking, histoire et description (Lille, 1900, p. 383), remarks, "De vrais chiens de chasse, il n'y en a pas; cependant, pour la chasse du lièvre au faucon, les Chinois se servent quelquefois d'un levrier ressemblant au chien Kurde."

[1] CHAVANNES, l.c., Plate X, upper panel to the left.

[2] S. REINACH (La représentation du galop, p. 90), who gives a sketch of this dog in Fig. 132, observes that it is represented in the posture of *cabré allongé*, the only example of this motive found on the bas-reliefs. But here, I think, Reinach contradicts himself; for if, according to his definition, the *cabré allongé* is a conventional expression of the gallop in certain spheres of art, this term cannot be applied to the case under consideration, as the artist's idea was not, and could not possibly have been, to represent this dog galloping. The dog has simply raised himself on his hind-feet, ready for an assault on the man, on whom he has been set. The coincidence of this posture and the *cabré allongé* seems merely incidental.

[3] For the entire composition in which these dogs occur, I would refer the reader to the book of CHAVANNES, La sculpture sur pierre en Chine, where the two dogs in our Fig. 47 will be found on Plate XXXVIII, to the right below, and the dog in our Fig. 48 on Plate XXXIX, on the left.

shoulders, and eight dogs preceding them, two of which are
selected for illustration (Fig. 47).    There is unfortunately a

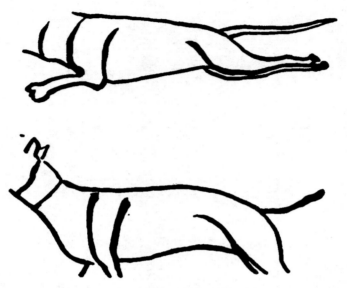

Fig. 47.    Hunting-Hounds represented on Han Bas-Relief of the
*Hsiao t'ang shan* (from a rubbing).

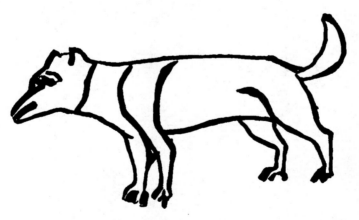

Fig. 48.    Hunting-Hound represented on Han Bas-Relief of the
*Hsiao t'ang shan* (from a rubbing).

crack in the stone here, which destroyed the heads of the
animals; but their outlines are so well and sharply drawn

Fig. 49. Figures of Hunting-Dogs on Han Bas-Relief (from *Kin shih so*).

that they must be pronounced to be among the best dog-pictures of this period. The upper animal, in flying gallop, closely agrees with the designs on the pottery vases. The difference between this type and the preceding greyhound is obvious, but the relationship of the two is also evident. According to C. KELLER,[1] the hunting-dogs are in closest genetical relations with the greyhounds, and are likewise of southern origin, first appearing on African territory. The ancient Egyptian wall-paintings, according to his investigations, allow us to recognize the gradual transformation of greyhounds into hunting-dogs. I believe we can also follow up this change, or at least observe features in common between these two types, on the Chinese bas-reliefs. The mode of carrying the tail in the greyhound shown in Fig. 44 is the same as that of the hunting-dogs in Figs. 47 and 48, and agrees with that of the Egyptian greyhounds and the Russian *barzoi* figured in KELLER (p. 64). The hunting-dogs on the bas-relief in Fig. 49 doubtless display in their long heads and extended muzzles the unmistakable greyhound character.[2] The latter relief is here reproduced from the original edition of the "*Kin shih so*" (section *shih so*, Vol. 4), and belongs to those of the village *Chiao ch'êng* 集城村, sixteen *li* in a southwesterly direction from the city of *Chia hsiang*, western Shantung. In that place there are, all together, four stone slabs[3] with relief figures of the Han time,

---

[1] Die Abstammung der ältesten Haustiere, p. 68.

[2] Compare, for example, the French greyhound in ALFRED BARBOU, Le chien, son histoire, ses exploits, ses aventures (Paris, 1883, p. 164), which in the formation of its skull, its thin loins, and in the carriage of its tail, agrees strikingly with the dogs of this Chinese bas-relief.

[3] According to the *Huan yü fang pei lu* 寰宇訪碑錄 Book I, p. 9 a. This is an excellent and useful catalogue of inscriptions, portraits on stone, and statues, covering the whole empire, in chronological order, from the *Chou* down to the *Yüan* dynasty inclusive, in twelve books. Each inscription is given with its title, character

two of which are reproduced in the "*Kin shih so*," section *shih so*, Vol. 4. From one of these our illustration Fig. 49 is taken. It occupies the lower panel of this relief in incised carving. On the right we see two pedestrians carrying hunting-nets with poles over their shoulders, two dogs running in front of them. On the left another hunter holds a greyhound by a halter. Two hares are depicted in front of the latter, and above them a dog making in the opposite direction, and a bird on the wing.[1]

Figs. 50—52 represent other dogs taken from the bas-reliefs of the *Hsiao t'ang shan*, which are remarkable for the truthfulness of their conception, the boldness of their motions and actions, and the correctness of outlines. The stag-hunt in Fig. 52 is notable for the motive of the flying gallop, expressed in the same style as on the pottery reliefs.

All the aforementioned greyhounds and hunting-dogs have smooth-haired tails; only two (on the bas-reliefs of the *Hsiao t'ang shan*) are represented with bushy tails, the hair being drawn in an ornamental and much exaggerated manner on the lower side of them.

The hunting-dog is twice represented, — in two wood-engravings in the "*Êrh-ya*"[2] (Book 2, p. 23 b, edition of

---

of its script, date, and locality; and to monuments of art is added the name of the private collection in which they are preserved. It was compiled by SUN HSING-YEN 孫星衍, Taotai of the *Yen-Yi-Ts'ao-Chi* Circuit in western Shantung (lived 1752—1818: see GILES, Biographical Dictionary, p. 691) and HSING CHU 邢澍, T. 階州, *Chih hsien* of *Ch'ang hsing* in *Hu chou fu*, Chehkiang. The preface of *Sun* is dated 1802. The book was published at *Su chou* in 1883.

[1] The explanation of the editors of the *Kin shih so* runs as follows: 下層一人率犬。二人荷罼。雉兔奔走。盖畫田獵之狀。
"On the lower panel one man leads a dog, two men carry nets over the shoulders. A pheasant and a hare are running at full speed, for it represents a hunt."

[2] Regarding the date of these engravings see p. 234.

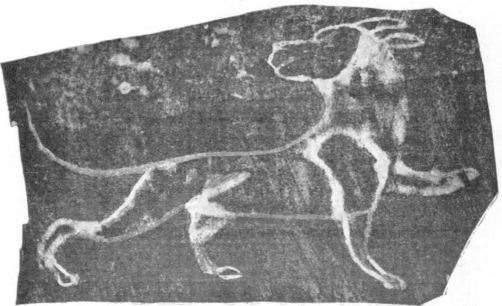

Figs. 50, 51.   Figures of Dogs on Han Bas-Reliefs of the *Hsiao t'ang shan.*

18

1801), which depict the starting-off to and the return from the chase. In the one hunting-scene the dog's ears are erect, in the other drooping. The former is turning its head back towards its master. Both have a long muzzle, a slender body, and a tail curled up at the end. Around the neck they wear a collar with a bell. Both are unmistakably types of greyhounds. The second of these hunting-pictures will be

Fig. 52.   Figure of Dogs on Han Bas-Reliefs of the *Hsiao t'ang shan*.

found reproduced in M. G. PAUTHIER's "Chine" (Paris, 1853), Vol. I, Plate XI.

In Fig. 53 is reproduced the engraving of a dog in the cyclopædia "*San ts'ai t'u hui*," section on animals 鳥獸 Book 3, p. 14 a, published in 1607 by WANG CH'I. Our illustration is taken from the original edition of this work. In the accompanying text the stereotyped division into the three kinds of dog — hunting, barking (i.e., watch-dog), and edible dogs — is referred to,[1] but it is not stated which species is to be understood by the illustration. No doubt,

---

[1] 犬有三種。一者田犬。二者吠犬。三者食犬，

it represents a hunter, as the long drooping ears indicate. It is long-haired, with a bushy tail.

Fig. 54 is a reproduction from the engraving of the dog 狗 given in the "*Pên ts'ao kang mu.*"[1] This work was compiled by LI SHIH-CHÊN in the latter half of the seventeenth century (1552–1578), the first printed edition being issued in the period *Wan li.*[2] The edition from which the above illustration is derived is from 1784. In style of drawing, it

Fig. 53. Engraving of a Dog (from *San ts'ai t'u hui*).

Fig. 54. Engraving of a Dog (from *Pên ts'ao kang mu*).

agrees with that in the "*San ts'ai t'u hui;*" it is also the same conventional type, except that here the ears stand erect,[3] and the animal is in a jumping position. The same cut was copied in the "Book on Eatable Things" (食物本草), section on animals, Book 6, p. 40 b (edition of 1803).[4]

In "*Mao shih ming wu t'u shuo,*" Book 2, p. 7, the dog

---

[1] 獸部畜類圖, Book 下, p. 48 a.

[2] A. WYLIE, Notes on Chinese Literature (2d ed.), p. 100; E. BRETSCHNEIDER, Botanicon Sinicum, Part I, pp. 54 et seq.

[3] "The hounds [of the Lolo], a species of thick-built terrier, rough-haired and mostly black, with straight legs, a coarse tail and muzzle, erect ears, tan eyebrows, and about 20 inches of height, though very useful for tracking and rousing game, will not face a beast at bay." — E. C. BABER, A Journey of Exploration in Western Ssüch'uan (Royal Geographical Society, Supplementary Papers, Vol. I, London, 1886, p. 39).

[4] The preface to the original edition is dated 1691 (see M. COURANT, Catalogue des livres chinois, coréens, japonais, 4ème fasc., Paris, 1903, p. 134).

is described under the name *lu* 盧,[1] and the same picture of it is given as in the "*Pên ts'ao kang mu*," — open mouth; pointed, erect ears; bushy tail; hairy neck, breast, and body; four paws. The *lu* is explained as a hunting-dog (田犬), and the first mention of it occurs in "*Shih king*," I, 8, 8 (ed. COUVREUR, p. 109). MAO SHIH quotes a passage from the "*Chan kuo tz'e*" 戰國策, in which it is stated that the home of the *lu*, the swiftest dog on earth (天下之駿犬也), is the ancient feudal state of *Han* 韓, which formed part of the present southern Shensi and western Honan. The feudal state of *Sung* 宋 (eastern part of Honan and northwestern part of Anhui) was famous for another kind of dog, called *ch'iao* 鵲 ("magpie"), white and black in color, while the *lu* was entirely black. Both were considered noble dogs, *liang ch'üan* 良犬.

I think that the foregoing notes and illustrations will have sufficiently proved the early existence of greyhounds in China, and form an interesting problem for naturalists, among whom this fact seems not to have been generally known hitherto. C. KELLER (l.c., p. 68) only remarks with regard to this subject, "Towards Eastern Asia, greyhounds seem to be scarcer; they are, however, employed in Burma for the stag-chase, according to oral communications of Ferars." I am of course convinced that also anatomical investigations and biological studies of living Chinese specimens will be necessary to place the question beyond doubt, and to fully identify the species. The foremost final problem will certainly be, Is the Chinese greyhound an independent race, a domestication of Eastern Asia, or will it, with all other forms of greyhounds, rank among the descendants of the ancient Egyptian species ori-

---

[1] J. LEGGE, The Chinese Classics, Vol. IV, p. 158.

ginating from *Canis simensis* (see the genealogy in KELLER, p. 70)? And if the latter should be the case, how, then, will the way of its migration from Egypt to China be explained?

Well known is the Pekingese pug or sleeve dog (袖巴狗兒; GILES, p. 480 c), so called from being carried in the capacious sleeve of its master.[1] The literary name of it is *wo* 猧, or written (and probably so originally) 猥;[2] at all events, a recent formation of word and character in conformity with 倭, therefore perhaps "dwarfish dog." The colloquial name is *ha pa*, a term doubtless of Turkish origin,[3] as is confirmed

---

[1] J. DYER BALL, Things Chinese (4th ed.), p. 217.

[2] See K'ANG HSI after *Yü pien:* 猧犬名。亦作猥。

[3] See the Dictionary in Four Languages (御製四體清文鑑), Book 31, p. 36, where the following equivalents for 哈叭狗 are given: Manchu, *Kabari;* Tibetan, *'o yo* (or *'o yog*); Mongol, *Khaba.* The Turkish form could be ascertained from the unique manuscript copy of the same dictionary in the British Museum, where Djagatai, with transcription in Manchu characters, is added as a fifth language. The Manchu word is explained by SAKHAROF (Manchu-Russian Dictionary, p. 254 a) as a "small lap-dog with short little legs;" likewise the Mongol word by GOLSTUNKI (Mongol-Russian Dictionary, p. 63 a; it is not in KOWALEWSKI). I cannot find any Turkish equivalent; but compare Turkish *Kaba* ("bushy, thick" [of hair]) in W. RADLOFF, Versuch eines Wörterbuchs der Türk-Dialekte, column 433. Tibetan *'o yo* is accordingly not a terrier, as SCHMIDT translated (in JÄSCHKE), but a puppy, as now SARAT CHANDRA DAS correctly has it. A skull of this dog in the British Museum is illustrated and described by I. E. GRAY, On the Skull of the Chinese Pug-Nosed Spaniel or Lap-Dog (Proceedings of the Zoölogical Society of London, 1867, pp. 40–43). It is said there on p. 42: "The nose of the Chinese or Japanese pug is said by some to be artificially produced by force suddenly or continuously applied: but that is certainly not the case in the skull that is in the British Museum; for the bones of the upper jaw and the nose are quite regular and similar on the two sides, showing no forced distortion of any kind, such as is to be observed in the skulls of some bulldogs; for I believe that some 'fanciers' are not satisfied with the peculiarity, and do sometimes try to increase the deformity by force." According to W. LOCKHART (Ibid., p. 43) "There are two kinds of pug in China, — one a small black-and-white, long-legged, pug-nosed, prominent-eyed dog; the other, long-backed, short-legged, long-haired, tawny-colored, with pug-nose and prominent eyes. The sleeve-dog is, according to him, a degenerated long-legged variety of pug rigidly kept on diet, and never allowed to run about on the ground. They are kept very much on the top of a *kang* or stove bed-place, and not allowed to run about on the ground, as it is supposed that, if they run on the ground, they will derive strength from the ground, and be able to grow large. Their food is much restricted, and consists chiefly of boiled rice."

by the history of this dog, which arrived in China from Turkish regions. Another colloquial term is *shih tzŭ K'ou* 獅子狗 ("lion-dog").

The first mention of these tiny pets is made in the history of the *T'ang* dynasty. "*K'iu Wên-t'ai*, king of the Uigur in Turfan, offered to the Chinese Court in A.D. 624 a pair of dogs (a male and a female) six inches high and one foot long, who understood how to drag a horse by the bridle and to carry a lighted candle in their mouths. They were said to have come from *Fu lin*,[1] and it was then for the first time that there were dogs of *Fu lin* in China."[2]

---

[1] Here probably the Byzantine Empire. Regarding the identification of *Fu lin*, see CHAVANNES in *T'oung Pao*, 1904, pp. 37 et seq.

[2] 唐 武 德 中 高 昌 王 文 泰 獻 狗 雌 雄。各 一 高 六 寸 長 尺 餘。性 甚 慧。能 曳 馬 銜 燭。云 本 出 拂 菻 國。中 國 有 拂 菻 狗 自 此 始 也。— *T'u shu chi ch'êng* Vol. 594, l.c. 犬 部 紀 事 II p. 8 b. See, further, E. CHAVANNES, Documents sur les Tou kiue (Turcs) occidentaux (St. Pétersburg, 1903), p. 103; FAVIER, Péking, Histoire et description (Lille, 1900), p. 383. — The notion of ALBRECHT, that frequent imports of dogs from abroad into China took place, is by no means confirmed by Chinese history. For corroboration of this I have perused the *K'ung hsien pu hui k'ao* 貢 獻 部 彙 考 (*T'u shu chi ch'êng*, Vol. 547 *shih huo tien*), where all tribute gifts sent to the Court are enumerated in chronological succession. Presents of dogs are there mentioned only fourteen times, which is a very low number in comparison with that of other objects; and quite naturally, since dogs could never be a very fashionable gift for an emperor. They were usually given together with horses and falcons. In connection with the above tribute from Turfan, I may mention, that, according to the *T'ang shu*, in 697 a two-headed dog (兩 頭 犬) was despatched to China from Bukhārā (安 國), and in 721 horses and dogs by the King of Kucha (on the northern border of the basin of the Tarim, Turkistan). — Mention is made also of Persian dogs in the *Chiu T'ang shu*, described as variegated, red with white spots (駿), also as red (赤), and making 700 *li* a day, then called Persian dogs (波 斯 多 白 馬 駿 犬 或 赤 日 行 七 百 里 者。駿 犬 今 所 謂 波 斯 犬 也: *T'u shu chi ch'êng* Vol. 594, l.c. 犬 部 紀 事 II p. 13 b). In the Account of Western Countries of the *T'ang* Annals (*T'u shu chi ch'êng* Vol. 1242, *tien i tien* Book 56, *T'ao chih pu hui k'ao* p. 4 b) many good dogs

I am not familiar with the history of this dog in the West, and must leave it to more competent writers to trace its species and the eventual rôle which it seems to have played in Byzantium and the west of Asia. It is most valuable, surely, to find the exact date of the first introduction of a foreign sport into China fixed here. From a piece of court gossip preserved in the "*Yu yang tsa tsu*" 酉 陽 雜 俎, written towards the end of the eighth century,[1] we see that these lap-dogs came in the middle of the eighth century also from Samarkand: "In the period *T'ien pao* [A.D. 742–755], the Emperor on a summer's day played chess with the hereditary prince of the first order. The concubine of the second rank (*kuei fei*) looked on, seated in front of the board, and, when several of the Emperor's chessmen were about to be taken, she let her puppy-dog from Samarkand (*K'ang kuo*) off on the side of her seat. The dog jumped on the chessboard and threw the men into disorder, by which the Emperor was greatly amused."[2]

The modern pug-dog preserves in its name the recollection of its Turkish provenience (see the above quotation from the "Annals of the Shantung Province"). In the description

---

(多 善 犬) are attributed to Persia. — PH. JOH. V. STRAHLENBERG (Das nord- und östliche Theil von Europa und Asia, Stockholm 1730, p. 373) tells of an import of Russian dogs into China by saying: "In China sind sie [Hunde] ziemlich rar, und wollen da nicht wohl fort [which, as Strahlenberg did not visit China, he could have known only from hearsay]. Die Kaufleute und Reisende aus Russland dahin, nehmen gemeiniglich einige mit sich, welche daselbst brav bezahlet werden, insonderheit wenn sie etwas abgerichtet, und einige Künste können."

[1] WYLIE, Notes on Chinese Literature, 2d ed., p. 193.

[2] 天 寶 中 上 嘗 於 夏 日 與 親 王 棋。貴 妃 立 於 局 前 觀 之。上 數 子 將 輸。貴 妃 放 康 國 猧 子 於 坐 側。猧 子 乃 上 局。局 子 亂。上 大 悅。— *T'u shu chi ch'êng*, Vol. 594, l.c. 犬 部 紀 事 II, p. 10 b. Compare also the note of PARKER in China Review, Vol. XIII, p. 381.

of the prefecture of *Shun t'ien* 順天府志 (Book 50, p. 28 a), where occurs the following passage regarding it, it is even called straightforward Turkish or Central Asiatic: "A small and alert class are the Turkish (*Hu* 胡) dogs, which nowadays are called *ha pa* dogs.    There is also the name *pa'rh* 叭兒 dogs.    The long-haired among them are designated 'monkey-lion dogs' (*nao* 猱, 'a long yellow-haired monkey'). The people of the locality call it *shih nung kou* 獅猥狗 (*nao, nung,* or *nang,* 'a fierce, shaggy-haired watch-dog')." It is interesting to note that the Chinese pug-dogs were introduced as far as Lhasa.    In Tibetan they are called *lags k'yi* (i.e., hand-dogs), "because it is believed that, if a human being lays hands upon a young eagle when freshly hatched, the bird is transformed into a dog of the Chinese pug breed."[1]

In Japan these dwarfy lap-dogs are called *chin,*[2] in England known as Japanese spaniels,[3] and undoubtedly derived from the Pekingese breed.    This becomes clearly evident from a comparison of Chinese and Japanese pictures of these dogs, which also formed a favorite subject of painters.    A collotype after a group of puppies painted by *Mao I* 毛益 (Chinese, thirteenth century) will be found in No. 27 of the "Kokka;" and a colored woodcut after a kakemono, representing the same subject, of *Ōkyo Maruyama,* in No. 143 (April, 1902) of the same journal.    These paintings, deserving of the study of the naturalist, will probably teach him more about the nature of these animals than any photographs could.

---

[1] H. RAMSAY, Western Tibet (Lahore, 1890), pp. 33, 35.

[2] J. J. REIN, Japan (2d ed., Leipzig, 1905), Vol. I, p. 255; I. C. HEPBURN (A Japanese-English Dictionary, Tokyo, 1894, p. 56 a) writes this word with the characters 佛林狗, in which the remembrance of the origin of this dog from the land of *Fu lin* is still preserved.

[3] J. DYER BALL, Things Chinese (4th ed.), p. 217; R. FORTUNE, Yedo and Peking, A Narrative of a Journey to the Capitals of Japan and China (London, 1863), p. 98.

The question may arise, Should all modern pugs in Eastern Asia be traced back to the ancestors imported from abroad under the *T'ang* dynasty? Mention is made of at least one similar breed, seemingly a Chinese production. The "*Ku chin shih hua*"[1] 古今詩話 has it, "In the period *Shun hua* (A.D. 990—994) [the people of] *Ho chou* (in *Chung ch'ing fu*, Ssŭch'uan Province) sent dogs as tribute from *Lo chiang* (in *Mien chou*, Ssŭch'uan Province). They were very small, of an intelligent mind, constantly tame and docile [they were sitting] at the sides of the imperial couch, and at every audience they must wag their tails and bark first; then people were respectful."[2] These *Lo chiang* dogs are further mentioned in the work "*Tung hsien pi lu*" 東軒筆錄 (written at the end of the eleventh century[3]), and cited in the same passage of the cyclopædia "*Ko chih ching yüan*." There this dog is described as being red, and as having a small tail (赤而尾小), and is considered a dreaded watch-dog whose "warning is like that of a spirit" (其警如神). This quotation refers to the period *Ch'ing li* (A.D. 1041—1048).

## PLANT—ORNAMENTS.

S. W. BUSHELL[4] thinks that vegetable forms are very rare as motives of decoration in ancient works of Chinese art. At all events, they have existed not only from the days of

---

[1] A work of the Sung dynasty (E. BRETSCHNEIDER, Botanicon Sinicum, Part I, p. 163).

[2] 淳化中合州貢羅江犬。甚小而性慧。常馴擾於御榻之側。每坐朝犬必掉尾先吠人乃蕭然。— *Ko chih ching yüan*, Book 87, p. 2 a.

[3] WYLIE, Notes on Chinese Literature (2d ed.), p. 196.

[4] Chinese Art, Vol. I, p. 89.

the *Han* time, but at least as early as the *Chou* dynasty. That plant-ornaments appear scarce in both these periods, is chiefly the fault of the material which has survived, bronze[1] and stone not appropriately lending themselves to the application of these more delicate decorations; while wood-carvings, basketry, and other works, on which they were doubtless more frequently employed, have perished. The most frequent use of plant-forms was made in the *Chou* time on nephrites, — a fact which we can ascertain partly from literary records, partly from the illustrations of ancient nephrites in the "*Ku yü t'u pu.*" On the ring-shaped jade tablets received from the emperor by the vassals of the fourth rank, four stalks of grain vere engraved; and on those of the fifth rank, four bundles of rushes.[2] Very elaborate plant-decorations of conventionalized style, and of most graceful arrangement, may be seen on many pieces figured in the "*Ku yü t'u pu;*" as, for example, in Book 20, pp. 1, 3 (ascribed to the *Shang* and *Chou* dynasties); and in Book 9, pp. 3, 5, etc. Plant-ornaments were also brought out on bronze vessels; for instance, stems of cereals on the type *chia* 斝.[3] In a great number they occur on the *Chou* bronze bells, where they require special and careful study.

As to the Han period, we found large leaves surrounding the standards of bronze and nephrite hill-censers, and fern-like leaf-decorations frequently on the covers of hill-jars; further, a rosette on the lid of a tripod jar (p. 172) agreeing with one on a Han metal mirror. I need not call attention to the several representations of trees on the Han bas-reliefs;

---

[1] A rich decoration of leaves on a Han bronze basin may be seen in *Hsi ch'ing ku chien*, Book 33, p. 55.

[2] See the engravings in COUVREUR, Dictionnaire classique de la langue chinoise, p. 587 c.

[3] Ibid., p. 409.

but the fact may be emphasized that three motives of plant-ornaments, usually attributed to much later times, occur as early as the Han time, and are in all probability much earlier still. The one is the well-known star-figure composed of six or eight conventionalized leaves grouped around a small central circle, as we found it on a Han pottery vessel (p. 116), and usually designated as a "plum-blossom" in Chinese and Japanese art. In the "*Hsi ch'ing ku chien*" (Book 21, p. 22) a bronze vase of the Han period is illustrated under the title "Han vase with plum-blossom patterns" (漢梅花紋壺). Along the one visible edge of the mouth of this vase we see nine of these conventionalized "plum-blossoms," making all together eighteen; and on the surface of the handle is one real plum-blossom true to nature, the outlines of the five petals being well drawn, two of them gracefully turned inwards, and the veins on them finely sketched. This composition shows how it was made possible for that stereotyped ornamental form to receive the name of "plum-blossom." That it really developed on Chinese soil, from the real picture of a plum-blossom, is hardly credible: for, first of all, there is nothing to show such a development; and, secondly, we find this motive in early Mycenian and Greek art, and conveyed thence to Scythian art;[1] so that its derivation from this region is not out of the question. The

---

[1] See particularly KONDAKOFF, TOLSTOI, and REINACH, Antiquités de la Russie méridionale, p. 192, where an eight-petalled leaf occurs on a vase from Kertch; p. 239, where a six-petalled leaf may be seen on a gold bracelet from Coul-Oba; and p. 319, where the same occurs on a gold diadem. Also in the art of India it is very common. I cannot agree with A. FOUCHER (L'art gréco-bouddhique du Gandhāra, Paris, 1905, Vol. I, p. 218), who sees in this pattern an Indian design which he explains as a wild rose, as Gandhāra has remained a country of roses. This ornament in his Fig. 96, on p. 221, is identical with the Chinese *mei hua*, and as regards the art of Gandhāra doubtless derived from classical art. The same ornament may be seen on a wooden chair excavated by Stein in Turkistan (M. AUREL STEIN, Sand-buried Ruins of Khotan, p. 356).

plain fact that "plum-blossom" is a mere name by which to
call this ornament, is corroborated by its occurrence (eighteen
times) on a bronze bell of the *Chou* dynasty (*Po ku t'u*,
Book 25, p. 14 a), where it consists of five petals, and is
explained as 荇葉 ("leaves of the aquatic plant *Villaria
nymphoides*").

Bamboo-drawing is a special branch of art in China, and
there are numerous illustrated books to teach the delineating
of the stems, branches, and leaves of this plant in the most
varied fanciful combinations. The bamboo in black and
white, and in color, is said to have originated under the
*T'ang* dynasty;[1] but in the Han period bamboo-drawing
seems to have already been an accomplished art, as we see
from the "Han tablet of bamboo leaves" (漢竹葉碑) at
*Ch'ü fu*, reproduced in the "*Kin shih so*," section *shih so*,
Vol. 4 [p. 52 a]: there, single pieces of bamboo stems and
leaves, combined in artistic and tasteful groups, are engraved
and freely scattered around the margin of this stone tablet.[2]

H. A. GILES,[3] speaking of the *Ming* painter *Ting Yün-
p'êng* (worked around 1600), refers to the great number of
woodcuts of pictures by this artist which have been preserved,
but no doubt all of them poorly representing the real genius
of the painter. "One of these," he continues, "is entitled
'Joined Trees,' and consists of two rugged trees, bare of
leaves, and joined about halfway up, something like the
Siamese twins, by two branches which have grown together."
This engraving is there reproduced on the same page, with

---

[1] H. A. GILES, An Introduction to the History of Chinese Pictorial Art (Shanghai,
1905), p. 139.

[2] See, besides, the bamboo drawn by *K'uan yü*, reproduced by P. LOUIS GAILLARD,
Nankin d'alors et d'aujourd'hui (Variétés sinologiques, No. 23, Shanghai, 1903), on
Plate XXIV (opposite p. 249), also pp. 41, 291.

[3] An Introduction to the History of Chinese Pictorial Art (Shanghai, 1905), p. 164.

the annotation "A charming example of *Ming* decorative treatment of a landscape subject." This would all be very well were it not for the fact that this motive occurs in just exactly the same form on a stone bas-relief of the *Han* time, which is dated A.D. 171,[1] and reproduced in the "*Kin shih so*," section *shih so,* Vol. 2 [p. 10 a],[2] that is to say, this motive existed in China 1429 years earlier than *Ting Yün-p'êng* had conceived of it; so that it can be considered neither as a characteristic *Ming* motive nor the individual invention of the latter painter, but simply, like many others, an ancient conventional motive of art handed down and copied through centuries. The agreement of the design of the *Han* time with that of the *Ming* is most striking, even in minutest details; the style of drawing, the composition and arrangement of the decayed branches and twigs, the curve of the branch uniting the two trees, and the twig growing out from the last-named, being perfectly identical; so that from the archæological point of view the *Ming* engraving must be looked upon as the copy of a copy going back at last to this form of the *Han* time. There will certainly be many missing links between the two which are still unknown to us. The motive on the bas-relief referred to is the oldest of which we know, but it will not presumably be the first then made. A glance at the whole composition engraved on this stone slab, representing several so-called "objects of good omen," shows that the other scenes depicted — viz., the white stag (白鹿); the large "yellow" dragon with a smaller dragon below; and the tree

---

[1] On its locality, and the personality to which it is devoted, see CHAVANNES, La sculpture sur pierre en Chine, p. xxix.

[2] With the explanatory inscription 木連理。

from which sweet dew descends (甘露降), with a man standing in front of it with uplifted hands to receive it (承露人) —
were all conventional motives of art at that time, and thus
accordingly also the joined trees.

A new era in the art of plant-ornamentation sprang up
in China during the epoch of the *T'ang* dynasty. Hitherto
unknown forms appear on the metal mirrors and on the
decorated sides of inscribed stone tablets (in the *Pei lin* of
*Hsi an fu*) of this period. These are derived from Persian
elements, when China had received as an inheritance the
art of the Sassanidæ.

# VI. — INSCRIPTIONS.

RECENTLY (in 1904) a book was published at Shanghai in ten volumes under the title "*T'ieh yün tsang kuei*" 鐵雲藏龜; i.e., "Tortoise-Shells collected by *T'ieh-yün*." His family name is *Liu* 劉. Six volumes contain reproductions of inscriptions, photo-lithographed from facsimile rubbings, found on fragments of ancient tortoise-shells used for purposes of divination. A large portion of these fragments discovered in *Wei hui fu*, Province of Honan, in 1899, came into the possession of Mr. Chalfant and Mr. Couling of *Wei hsien*, and is now in the Museum of the Royal Asiatic Society of Shanghai.[1] The explanatory text to these inscriptions, furnished by LIU T'IEH-YÜN in the preface to his publication, is meagre and disappointing. He explains only a number of the cyclical signs, and makes some plausible guesses at a few characters varying from the forms of the inscriptions on ancient bronzes. Mr. Chalfant informs me that he and Mr. Couling have succeeded in deciphering many more characters, and also in joining and combining many fragments, thus yielding a sensible rendering. I must therefore refer the reader to the forthcoming publication of these gentlemen.

The last four volumes[2] of LIU T'IEH-YÜN bear the special title "*T'ieh yün tsang t'ao*" 鐵雲藏陶, giving inscriptions

---

[1] See Journal of the China Branch of the Royal Asiatic Society, Vol. XXXV, Proceedings, p. xxi.

[2] The volumes are not numbered.

287

on ancient pottery.   One of these volumes (43 pages) contains facsimiles of a number of clay seals employed for sealing official letters and packages; the other three, comprising 142 pages, inscriptions on pottery of the *Chou* dynasty.   The author, however, gives no identification of any of the characters.   Nevertheless this collection, as entirely new and well-reproduced material, is extremely valuable.   Most of these pottery seals consist of one, two, or four characters, and are engraved in the clay, presumably by means of stamps. A few characters only stand out in the clay in relief.   The style of these characters leaves no doubt that this pottery must be attributed to the *Chou* time; and the resemblance of the character on our piece of pottery (Fig. 5, Plate III, pp. 13, 14, 122) to those in *T'ieh yün's* book, where it occurs several times, confirms my judgment in assigning this piece to the same period.   An attempt to decipher the ceramic seals of *T'ieh yün* is extremely difficult, not only because there are so many new forms among the symbols which cannot be identified with the characters on ancient bronzes, and require the slow work of decipherment by other methods, but also, which is the chief obstacle, because *T'ieh yün* does not reproduce the pieces of pottery at all, but merely their marks, which deprives us of the primary means of getting at the significance of the character: for it is always possible (and it seems indeed to be the case many times) that a single character may denote the type or the name under which the particular piece was then known; so that if we know the form of a vessel, and are able to classify it from the Chinese point of view, a not unimportant step might be advanced towards the elucidation of the seal.   Even if the inscription has another content, we cannot deviate from this principle of philological criticism, that an inscription must be studied in connection with the object on which it occurs, and which it apparently tends to interpret to some extent. If we were able to read all characters contained in *T'ieh yün's* facsimiles of pottery seals, we might obtain a new and not uninteresting palæographic contribution to our knowledge of *Chou* writing, but none to advance our knowledge of contemporaneous pottery, which certainly ought to be the first result of such an investigation.

Following are the inscriptions that occur on the pieces of pottery described in the preceding pages.

1. SEAL–MARK ENGRAVED BY MEANS OF A STAMP ON THE STANDARD OF A TAZZA OF POTTERY (豆) OF THE LATER CHOU (OR TS'IN) DYNASTY (SEE PP. 14, 122).

I suppose that the above character might read 鐙, which has the meaning of 登; i.e, a variety of *tou* 豆 made of pottery. Several variants of the same seal occur among the facsimiles of *T'ieh yün's* book quoted above, five of which are here selected for reproduction. They occur there on pp. 10 a, 55 b, 61 b, 75 a, 76 a, respectively.

Compare also the following two variants of the character 鐙 on Han bronzes from the "*Chi ku chai chung ting i ch'i kuan chih*," Book 9, pp. 28 a, 29 a. To demonstrate the close agreement in the style of

19

the characters and in the contents of the inscriptions on the bronzes of the Han period with those on the pottery of the same time, I shall first reproduce some inscriptions of Han bronze pieces in the collections of the American Museum of Natural History.

## 2. INSCRIPTION, ENGRAVED IN THE CAST, ON A BRONZE QUADRANGULAR VASE (方壺) FIGURED ON P. 141.

*In Modern Characters.*

始　中
建　尚
國　方
四　銅
年　二
朔(?)斗
月　鈁
工　重
造　十
　　斤

TRANSLATION. — "A bronze made in the *Shang fang*. A quadrangular vase [containing] two pecks (*tou*). It weighs ten catties. In the first month of the fourth year of the period *Shih chien kuo* [A.D. 12] the work was made."

NOTES. — Regarding the *Shang fang*, see p. 140. Regarding the meaning of 鈁, see E. v. ZACH, "Lexicographische Beiträge"

(Peking, 1905), Vol. III, p. 84. The reading of the character marked with an interrogation-point is doubtful. Professor ÉDOUARD CHAVANNES, to whom I submitted this inscription, was kind enough to make the following suggestions in a letter dated Sept. 30, 1905. He is inclined to read the character in question 年, and refers, for forms similar to the one above, to "*Kin shih tsui pien*," Chapter 5, inscriptions dated A.D. 63 and A.D. 81. The character preceding this one would read 元 (see, in the same work, an inscription of the year 36 B.C.). The difficulty which then arises is to explain the sign following the year-period *Shih chien kuo*, which Chavannes supposes might stand for 之. The translation of this paragraph would then have to be, "In a month of the first year of the period *Shih chien kuo* [A.D. 9]." The only objection to this interpretation is the unusual use of the particle 之 between a *nien hao* and the number of years, as Chavannes also justly remarks. I should therefore prefer to adhere provisionally to the explanation 朔, which I afterwards found confirmed in some degree by the accompanying form occurring in a Han inscription in the "*Li tai chung ting i chi kuan chi fa t'ieh*," Book 19, p. 4 b, and there transcribed as 朔.

## 3. INSCRIPTION ALONG THE HANDLE OF A BRONZE PO SHAN LU (HILL–CENSER) OF THE HAN TIME (SEE P. 186).

*In Modern Characters.*

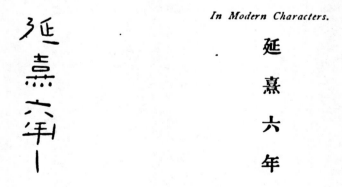

延光三年七月工張順造

TRANSLATION. — "In the seventh month of the third year of the period *Yen kuang* [A.D. 124] the work was made by *Chang Shun.*"

## 4. INSCRIPTION ON A BRONZE TRIGGER TO A CROSSBOW (NU CHI 弩機) OF THE HAN TIME.

*In Modern Characters.*

延熹六年

TRANSLATION. — "Sixth year of the period *Yen hsi* [A.D. 163]."

### 5. Inscription on a Jug of Han Pottery (see p. 131).

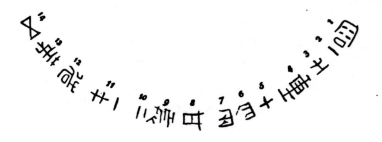

It reads from right to left as follows:

₁容 ₂一 ₃升。 ₄重 ₅十 ₆四 ₇兩。 ₈甘 ₉露 ₁₀二 ₁₁年 ₁₂造。 ₁₃第 ₁₄五。

Translation. — "[The jug] contains one pint, weighs fourteen ounces, and was made in the second year of the period *Kan lu* [52 B.C.]; No. 5."

Notes. — As the capacity of this jug is 790 cc, we may accept this as the average unit of the pint in the Han period; and since its weight is between 17.5 and 17.75 ounces English (say, roughly, 18 ounces), 1 Chinese ounce (*liang*) of that time may be taken as $\frac{7}{9}$ of an ounce English. Statements regarding the capacity and weight frequently occur on Han vessels of bronze, so that the similar practice on pottery might be thence derived. The appearance of the lines in the above inscription leaves no doubt that it was incised into the wet clay before baking. It therefore remains a problem how the makers of pottery (and the same holds good for the bronze-casters) could know in advance the exact capacity and weight of the vessel, which after baking certainly became much lighter. The only explanation which I can offer seems to be that they had ascertained these data by a great deal of experimenting, numerous examples of the same shape, size, and thickness, having been made before, and thus yielded at last the desired figures.

As to the numbering of ceramic pieces, compare the work of T. HAYASHI ("Objets d'art du Japon et de la Chine," Paris, 1902, Vol. II, p. 105), who illustrates a porcelain vase of *chün yao* with the mark — (= 1) on the bottom, and remarks, "Les vases et les coupes les plus recherchés de l'espèce *Chün yao* sont numérotés, par des incisions à la roue, depuis le chiffre 1 jusqu'au chiffre 5." See also ST. JULIEN, "Histoire et fabrication de la porcelaine chinoise" (Paris, 1856), p. XLIX.

For the more important characters occurring in this inscription analogous forms from Han bronze vessels are here added, and directly reproduced from the facsimiles of the book "*Chi ku chai*, etc."

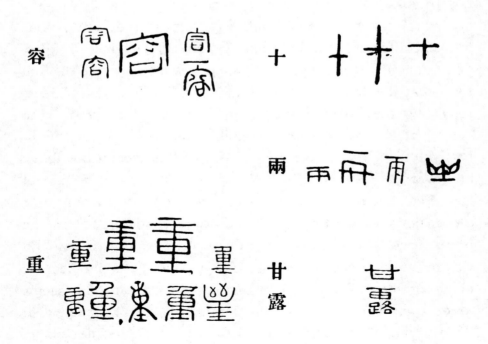

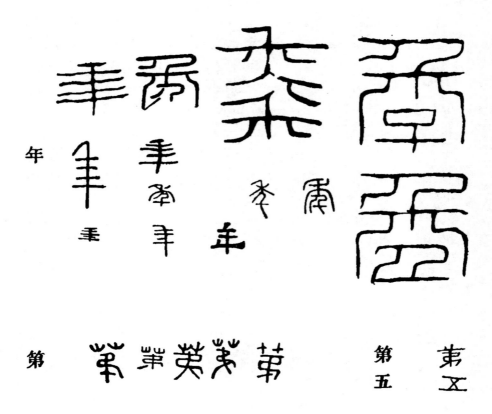

6. INSCRIPTION, CONSISTING OF TEN CHARACTERS, ON A JAR IN THE POSSESSION OF MR. R. H. WILLIAMS OF NEW YORK (SEE PP. 130, 131).

In Modern Characters.

| 9 | 7 | 5 | 3 | 1 |
|---|---|---|---|---|
| 一 | 重 | 二 | 乍。| 嚴 |
| 斤。| 十 | 斗。| 容 | 氏 |
| 10 | 8 | 6 | 4 | 2 |

TRANSLATION. — "*Yen shih* made [the vessel]. It contains two pecks (*tou*), and weighs eleven catties (*chin*)."

NOTES. — The reading of the first three characters is corroborated in a legend on a bronze vase (壺) of the Han dynasty, here reproduced after the "*Chi ku chai.*" The fact that the same artist's name appears on a bronze and on a pottery vessel shows that both arts were sometimes practised by the same hand.

I give below the identification of other characters from Han bronzes after "*Chi ku chai.*"

作

容 See under No. 5.

重 See under No. 5, where also the above peculiar form [ ] is given.

斗

斤

The above inscription seems to have been engraved by means of a burin after the baking process, since the lines are of varying depth and width; whereas, in wet clay, lines can be incised straight, firm, and of even depth (as in

Inscription 5). In examining the vessel, the points where the graver interrupted the line, and then went at it again with his tool, are plainly visible. The numeral — (No. 9), for example, is composed of three distinct parts. Further, on incising lines into wet clay, the edges coil up and become thickened, and this should be visible after baking; but this is not the case in the above inscription. The fact that it was made after baking, however, does not militate against its origin contemporaneously with the jar.

7. INSCRIPTION PRESSED BY MEANS OF A STAMP INTO THE INTE-RIOR SIDE OF THE NECK OF A GREEN-GLAZED VASE OF HAN POTTERY IN THE POSSESSION OF MR. R. H. WILLIAMS OF NEW YORK.

The reading of this inscription remains doubtful. The first two characters represent perhaps the *nien hao* 神爵 *Shên chüeh;* i. e., the period 61–57 B.C. Compare the accompanying similar forms in a Han inscription from "*Chi ku chai*," reading "fourth year of *Shên chüeh* [58 B.C.]." The stamp was pressed into the clay while still wet, before baking.

The signs in the second half of the inscription look like figures giving weight or capacity, or both; but they are so pressed together, and seem to be written in such an unusual mode (two rows, one beside the other), that, without a number of analogous examples, decipherment is hopeless.

## 8. TWO SEALS IN LOW RELIEF, OPPOSITE EACH OTHER, ON A HAN POTTERY JAR (SEE P. 129).

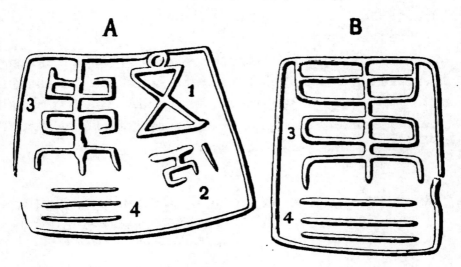

A Chinese archæologist in *Hsi an fu* proposed to read this as follows: —

A 　吾<sup>1</sup>東<sup>3</sup>
　　司<sub>2</sub>三<sub>4</sub>　　and accordingly　　B 　東<sup>3</sup>
　　　　　　　　　　　　　　　　　　　　　　三<sub>4</sub>

What seems certain is, that the two characters in Seal B are identical with the last two in A (3 and 4): whence it follows that the four charac-ters in A can be read only in the succession as numbered above. Compare the example given herewith of a four-character inscription on a Han bronze from the "*Chi ku chai.*"

On closer inspection, it will be found impossible to ap-

prove of the above reading.   Character 1 should read either
五 or perhaps 十 五.   The difficulty is in the point or circle
on the top, without which it is the common sign for 五.
I know of one example of this character in which the point
occurs on a Han bronze in "*Chi ku chai*," Book 9, p. 28 b: thus,
Ⴃ where it occurs in the composition of the year-period
*Wu fêng*, and is accordingly interpreted as 五.   On
the other hand, 乂, with a small vertical or horizontal
stroke added above, occurs on ancient coins in the sense of
十 五 (see, for instance, T. DE LACOUPERIE, "Catalogue
of Chinese Coins," London, 1892, p. 169; for a doubtful old
character representing 吾, see the same work, p. 333).   That
the character No. 2 cannot be identified with 司 will be
seen from the above bronze inscription, where the Han form
of this character occurs in a different way.   No. 3 does not
read 東, as this always requires an additional horizontal
stroke in the middle, and is without the uppermost horizontal
stroke in our character.   On account of the following numeral
三, one might be inclined to recognize in it 第; but the
latter always retains in Han inscriptions its radical, which is
there wanting.

9. MARK PRESSED BY MEANS OF A STAMP INTO THE INTERIOR
    SIDE OF A HAN POTTERY VASE IN THE POSSESSION
    OF MR. R. H. WILLIAMS (SEE P. 131).

The seal resembles the character for "man" (人);
but whether it was really intended to represent it, must remain
very doubtful.   Probably it is merely a potter's mark, without
any particular signification attached to it.

# APPENDIX I.

## HAN ROOFING–TILES.

THE roofing-tiles described on the following pages come from the ancient palaces of the Han dynasty situated west of *Hsi an*. No remains of the latter are left nowadays. The original appearance of these tiles is shown in Fig. 3 of Plate LXV, — a long half-cylinder, to the front of which a round disk is attached. The half-cylinder rested on the lower end of the roof so that the disk projected over the eaves, and was visible to the passer-by from below. This part was therefore embellished with ornamental characters in relief, usually four of them, containing a sentence of good omen or the special name of the palace to which they belonged.

The first tile of this sort was described in 1721 by LIN T'UNG,[1] and the subject soon attracted a great deal of attention from Chinese archæologists, so that subsequently quite a number of books arose concerning these antiquities. These publications are for the most part illustrated by engravings

---

[1] See WYLIE (Notes on Chinese Literature, 2d ed., p. 144), who speaks of an "old brick;" but the term 甘泉宮瓦 leaves no doubt that in this treatise the question is of a tile of the Han palace *Kan ch'üan*. FORKE (see below) does not mention this work. I have searched in vain for it in China: it is no longer in existence. From the statement of Wylie, it seems that *Lin t'ung* was also the first to convert an ancient tile into an ink-pallet.

300

after the actual rubbings from the tiles, and a discussion and explanation of the curious characters on them are given. Complete tiles were, of course, comparatively scarce: with the majority the cylindrical part was broken off. It is curious to note the idea which soon came into vogue of making these relics of the past in some way useful, and of converting them into ink-pallets. On the complete tiles these were brought out on the surface of the cylinder; on the fragmentary ones, on the back of the disk. To prevent the ink from soaking into the tile, and so gradually appearing on the surface, which would have defaced the characters, the surface was varnished with oil and made waterproof. This work having been done chiefly in the *Ch'ien lung* period, in the second half of the eighteenth century, a great number of the tiles have come down to us in the shape of varnished ink-slabs. Modern imitations[1] are numerous, but there are infallible characteristics by which the genuine ones and modern fabrications may be distinguished. First of all, the modern clay is widely different from that of the Han time, the locality

---

[1] It should be well understood that such imitations are made by the Chinese for the Chinese, since there is a market demand for such things in the country. Every Chinaman knows what they are; and it is absurd to speak, as so many foreigners do, of a "counterfeit industry" in China. If the Chinese imitate their ancient bronzes and porcelains, they are quite as much justified in doing so as we are in reproducing old Delft ware or rococo furniture, or a thousand other objects of the past. The foreigner in China is usually irritable, and prone to believe that the actions of the people are directed against his sacred person as a wilful insult. I have already refuted (in the *T'oung Pao*, Series II, Vol. IV, 1903, p. 265) the unfounded statement of Lessing, director of the Kunstgewerbe Museum of Berlin, that the Chinese manufacture bronzes for export abroad. Modern imitations of ancient bronze form a part, for example, of the dowry of every farmer's bride near Peking; and the Chinese dealer finds a ready sale for them without waiting for the haughty foreigner to enter his shop. The standpoint of the Chinese merchant is generally correct: if customers come to him for antiquities, he is right in expecting them to know something about the business, and to know how to judge. If they buy imitations in the belief that they are genuine material, it is in his estimation their own fault. The dealer does not consider it his place to lecture to his customers concerning the art of discriminating old objects from new ones.

of which is unknown, and which cannot be replaced by any of the modern products. The Han clay is of very even consistency, and hard, like stone, yielding a clear ringing tone on sounding it; while in modern imitations the clay is brittle, badly prepared, and gives no sound. This is the principal criterion by which the antiquarians of *Hsi an* are still guided. Secondly, regard is paid to the palæographic feature; no imitators, however clever they may be, being able fully to grasp and reproduce the ductus and original grace and life of the Han characters. Further, there is a general inferiority in execution manifest in all the modern works, which only too obviously betrays their date. These imitations, however, are not only the work of yesterday, but have come in since the eighteenth century, and it is still more difficult to recognize a production of that period.

Some pieces were found with the characters missing. These were then sometimes added in the *Ch'ien lung* period, while the tiles themselves may be genuine. Broken pieces were also patched together, and fragments so cleverly supplemented that the added pieces were visible only to the eye of the special expert.

The inscriptions on ancient tiles first engaged the attention, among European scholars, of CHAVANNES, who studied those of the *Ts'in* in connection with the other inscriptions on bronze and stone of this dynasty.[1] Further, they form the basis of a paper by FORKE,[2] who investigated the formation of the characters in their relations to other ancient scripts, but relied exclusively, in his research, on the illus-

---

[1] ÉDOUARD CHAVANNES, Les inscriptions des Ts'in (Journal Asiatique, 1893, pp. 517–520).

[2] A. FORKE, Die Inschriftenziegel aus der Ch'in- und Han-Zeit (Mitteilungen des Seminars für orientalische Sprachen, Vol. II, Sect. I, Berlin, 1899, pp. 58–100, with 14 plates).

PLATE LXV.

Han Pottery Tiles.

trations in Chinese books, a great number of which he reproduced. In the present publication the illustrations of these tiles are the first that have been made from photographs taken directly from specimens, which were obtained by me in *Hsi an fu.*

Plate LXV represents three perfect tiles. The inscription on the one shown in Fig. 1 reads 延年益壽 *yen nien i shou,* "May your years be prolonged, and your longevity be increased!" see FORKE (l.c., p. 70). The tile is varnished brown, and a black ink-slab (16.4 cm long) is worked out on the surface of the front part of the cylinder, which is 49.2 cm long. Adjoining this part at the back is a ribbed portion 20 cm long, followed by a somewhat lower compressed piece. The width of the cylinder is 17.6 cm, and its circumference 26.8 cm. The diameter of the disk is 17 cm, and that of the knob 1.7 cm. The clay is of uneven thickness, varying from one to two centimetres.

The characters on the next tile (Fig. 2, Plate LXV) read 甘林 *Kan lin,* referring to the *Kan ch'üan* (泉) *lin* Palace (see FORKE, l.c., p. 88). The cylinder of this tile is also varnished a light brown, and has been turned into an ink-pallet. The length of the cylinder is 45.1 cm; width, 15.3 cm; circumference, 23.5 cm. The diameter of the disk ranges from 15.3 cm to 15.6 cm.

Fig. 3, Plate LXV, presents a perfect specimen of a roofing-tile made of a coarse grayish clay, unvarnished, and without any other subsequent artificial alterations, thus showing the entire original appearance of these tiles. The inscription reads 億年無疆 *i nien wu chiang,* "Thousands and thousands of years without end" (see FORKE, l.c., p. 69). A large knob surrounded by twelve smaller ones forms the centre of the disk. The half-cylinder has a slanting piece at the back, which is 4 mm lower down, and 5 cm long. It is compressed at its lower ends to afford a stronger hold on the roof. The length of the cylinder is 52 cm; width, 16 cm; circumference, 25 cm. The diameter of the disk is 17.1–17.4 cm; that of the knob, 2.5 cm.

Plate LXVI shows six disks of tiles. That in Fig. 1 (diameter, 17.1–17.2 cm) bears the characters 廷 年 益 壽, the same as Fig. 1, Plate LXV, but with some graphic variants. The tiles with these inscriptions, according to the investigations of Chinese authors, come from the *Wei yang* Palace. This disk has been varnished a dark brown, and the back turned into an ink-slab of black slate. The ink-line drawn by me through the middle indicates that the part below it, with its two characters, is genuine; whereas the upper part, also taken from a genuine piece of a Han tile, was fitted to the lower fragment, and the upper characters on it were added afterwards, in agreement with the corresponding inscriptions on well-preserved pieces. The joining of these two parts into one piece is so skilfully done that no suture or seam is visible.

The characters on the disk represented in Fig. 2 of this plate (diameter, 18.1–18.5 cm) read 與 天 無 極 *yü t'ien wu chi*, "May your life be eternal, like Heaven!" see FORKE (l.c., p. 68), who translates "infinite like Heaven." There is a large, high, almost square knob in the centre, 2 cm in diameter. The disk is varnished a dark brown, and has been turned into an ink-slab of slate. The ink-line drawn by me in the upper left-hand portion indicates that the small segment to the left of this line was pieced in, but the joining is by no means visible on the specimen.

The tile-disk Fig. 3 shows the natural appearance of the clay, and has not been converted into an ink-pallet. It

PLATE LXVI.

Han Pottery Tile Disks.

measures in diameter 15.7 cm. Parts of the edge have been broken off, as indicated by the broken lines, which have been added. The centre is taken up by a square (sides about 5 cm long) with the seal character 便 *pien*, which stands for *pien tien* 便殿, "side-hall, side-palace" (see FORKE, l.c., p. 76). An engraving of the same specimen as ours is found in the "*Ts'in Han wa tang wên tzŭ*" 秦漢瓦當文字 (Book 2, p. 12 b) and in the "*Kin shih so*," section *shih so*, Vol. 6 [title No. 46]. FORKE (Plate 5, No. 85) gives another variant with this character in a different form, and graved in; while in our tile, as in all others, it is in low relief. Four pairs of spirals with an anchor-like continuation at the side are grouped around the square.

In Fig. 4 (diameter of disk, 16.3–16.5 cm) the knob (diameter, 2.1 cm) and the inscription are the same as in Fig. 3, Plate LXV. The varnish is dark brown, and there is an ink-slab of slate.

The circumference of the next disk (Fig. 5, Plate LXVI) forms a very irregular circle 16.4–16.8 cm in diameter. It is varnished brown, with an ink-slab on the back also varnished brown, without slate. There are two characters on it, 上林 *Shang lin*, the name of the great park established by *Ts'in shih Huang ti* in 200 B.C., and extended by the Han Emperor *Wu* (140–86 B.C.). It is said to have measured 400 *li* in circumference, and to have teemed with castles and palaces. The tiles are presumed to come from the surrounding wall, or from the gates and watch-towers. Ten variations of this tile are known (see FORKE, l.c., p. 79).

The specimen represented in Fig. 6, Plate LXVI, measures 17.4–17.8 cm in diameter. It has a round flat knob (2.3 cm in diameter) in the middle, surrounded by twelve small "nails," as the Chinese expression is, and it bears the inscription

20

長生無極 *ch'ang shêng wu chi*, "Long life without end!" (see FORKE, l.c., p. 64). This tile is known in fifteen variants. The surface is not varnished, but is of the natural gray clay color. The grooves are still partially covered with yellow loess, and there is a slate ink-slab on the back.

Plate LXVII, Fig. 1, shows another disk (diameter 15.5–15.8 cm) varnished a dark brown, and with the usual slate ink-slab on the back. The characters on this are written in scrolls or flourishes called "crow-feet," and are explained as 永受嘉福 *yong shou chia fu*, "May you always have good luck!" (see FORKE, l.c., p. 74).

The specimen represented in Fig. 2 of this plate (diameter, 15.5–15.7 cm) has a large flat knob 3.8 cm to 4 cm in diameter. The inscription reads 長生未央 *ch'ang shêng wei yang*, "Long life without end!" (see FORKE, l.c., p. 67), — the words *wei yang* implying an allusion to the *Wei yang* Palace, west from *Ch'ang an*. Over thirty different kinds of this tile have been found.

In Fig. 3, Plate LXVII, we see a disk whose diameter is 15.8–15.9 cm, and which has a brown varnish and an ink-slab of slate. There is a large flat central knob (2.3 cm in diameter) surrounded by twelve small "nails." The inscription reads 高安萬世 *kao an wan shih*, "Sublime peace to the numerous generations!" According to another but rather artificial explanation, this tile is said to come from the palace of *Tung hsien*, a favorite of the Emperor *Ai* (6–1 B.C.), who conferred upon him the rank of Marquis of *Kao an* 高安侯, so that the above characters would require the translation "Many generations to [the Marquis of] *Kao an*" (see FORKE, l.c., p. 82).

The last disk on this plate (Fig. 4) measures 14.8–15 cm in diameter. It is of a light-yellow clay color outside, but

PLATE LXVII.

Han Pottery Tile Disks.

inside of a bluish-gray color, similar to that of modern Pekingese tiles and bricks. It is not varnished, and has no ink-pallet. Some fragments of the cylindrical part still adhere to the disk. The design on this tile represents a conventionalized· bird in low relief. A similar one, reproduced by FORKE (Plate 12, No. 144), is found in the "*Ts'in Han wa tang wên tzŭ*" (Book 2, p. 14 a). The form of the bird on our tile differs from that one in some respects: the posture of head, neck, tail-feathers, and feet, for instance, are here much better drawn; and moreover the feet of our bird have three front-claws and one hind-claw, while these are wanting in the design mentioned. The explanation of CH'ÊNG TUN, author of the above work, that this bird is the *chu niao* 朱鳥,[1] and his identification of it with the hawk on the military flags of the *Chou*, are certainly of little or no value, as they represent no tradition of the Han time, but merely the personal view of this modern author. Likewise FORKE'S interpretation (l.c., p. 99), "a kind of bird-of-paradise or lyre-bird," is subjective. What the people of the Han time, or the artist who sketched the design for this tile, meant to represent by it, we do not know; and all we can conscientiously say is, that it is a conventionalized bird, whose style of conventionalization is due to a considerable extent to the rough clay material, and comes very near to the numerous and manifold representations of the cock and pheasant.

The decorative forms of the characters on the Han tiles could not fail to impress the ornament-loving mind of Chinese artists. They found ready imitators and an appreciative public. The forms of Han tile-disks with characters are frequently

---

[1] Literally, "the red bird." SCHLEGEL (Uranographie chinoise, Vol. I, p. 69) claimed that this term is identical with the *fêng huang* ("phœnix").

employed as weaving-patterns in silk,[1] and are printed on paper-hangings, which are manufactured in *Hsi an*, naturally, for it was there that the Han tiles first made their appearance, and appealed to lovers of antiquity. From there they were traded and transported to other parts of the country. Specimens of such paper-hangings, which I brought from *Hsi an*, are printed in red and blue by means of copper-plates, and show the four characters *yen nien i shou* (see Plate LXV, Fig. 1) in the style of the tile script. The characters are surrounded by five bats, the symbol of good luck. The walls and ceilings of the palace[2] in *Hsi an*, in which the imperial family took refuge in 1900 after the capture of Peking by the allies, are entirely decorated with paper of the tile-pattern; and I found it quite frequently also in Yamens, *kung kuan*, and inns in the province of Shensi.

Still more curious is the fact that a new form of vessel, based on these tile-disks, has been invented, shaped like the half of such a disk; as seen, for example, in Fig. 1, Plate LXVIII, — a modern terra-cotta teapot from *Yi hsing hsien*, with the two characters in low relief reading from right to left 延年 *yen nien*, "May thy years be prolonged!" (see Plate LXV, Fig. 1). This teapot is 9.2 cm high; the length at the base being 13.5 cm, and the width at the base 5.6 cm. The inner side of the cover bears the seal-mark *Yi hsing*. Indeed, also half-disks (半瓦) have been found, among them one with the same characters, *yen nien* (illustrated in

---

[1] Also in Japan the patterns on the front sides of roofing-tiles are applied to textiles. Compare JUSTUS BRINCKMANN, Japanische Flächenornamente (Aarau, 1892), pp. 33, 35, and Plates 23 and 41. — On the other hand, it is notable that the Han tiles are not imitated in the modern Chinese roofing-tiles, on which I find only the following patterns: the character *shou* ("longevity") in various ornamental forms, dragons, flowers in a flower-pot, plum-blossoms.

[2] For a description of this palace, see F. H. NICHOLS, Through Hidden Shensi (New York, 1902), p. 209 et seq.

PLATE LXVIII.

Modern Teapot and Metal Ink-Box Decorated with Han Tile Inscriptions.

PLATE LXIX.

Modern Flower-Pots Decorated with Han Tile Inscriptions.

the "*Ts'in Han wa tang wên tzǔ*," Appendix, p. 24 b; and
FORKE, l.c., Plate 8, No. 134).[1] The above characters on
the teapot are not identical with those of the tiles, but seem
to be derived from the "*Shuo wên*," with fanciful alterations
tending to attain superior gracefulness of form. It seems
that 延 is confounded with 建.

The application of Han tiles is illustrated further in four
modern flower-pots on Plate LXIX, made in a village outside
of the northeastern gate of Peking. The shape of these
pots is likewise derived from the half of a tile-disk, and
adorned with characters in imitation of the Han-tile writing.
The bases are 25 cm long, 15 cm wide; the height, 16.2 cm.
The pattern is graved in deep and broad grooves, which
are colored, — the border-line of the two upper ones, blue;
the double border-lines of the two lower ones, green and
blue; the characters, green; the dividing-lines and the knobs
and half-circles in the centre, red, green, and yellow. The
characters are not exact facsimiles, but rather free imitations
of the Han style. Neither does the succession of the characters
correspond to that on the tiles, where *wei yang*, *ch'ang lo*,
*wan sui*, are written one below the other; while in this case,
where only two characters were selected, they had necessarily to be arranged in juxtaposition. The mere copying
of the upper or under half of the tile-inscription would naturally have been senseless. The characters on Fig. 1 are
未央 *Wei yang*, the name of a palace west from *Ch'ang
an;* on Fig. 2, 長樂 *Ch'ang lo* ("lasting joy"), the name
of a palace built 200 B.C.; on Fig. 3, 萬歲 *Wan sui*, "ten
thousand years" (see FORKE, l.c., p. 86); on Fig. 4, 千秋
*Ch'ien ch'iu*, "thousand autumns" (see FORKE, l.c., p. 96).

---

[1] For other half-disks, see Ibid., Plate 6, Figs. 94, 95.

Fig. 2, Plate LXVIII, represents a modern rectangular ink-box of white copper (白銅), with a lining of red copper, and a layer of slate inside of the lid. It is 8.9 cm long, 5.9 cm wide, and 3.5 cm high. The surface of the cover is decorated with an etching consisting of seven Han tile-disks, one overlapping the other. This design is surrounded by twigs of the plum-tree, with leaves and blossoms.

### EXPLANATION OF THE TILE CHARACTERS.

*a* = 金 (see *Ts'in Han wa tang wên tzŭ*, Appendix, pp. 7 b and 8 a; FORKE, l.c., p. 87).

*b* = 甘泉上林 (see FORKE, l.c., p. 88, and Plate 8, No. 122).

*c* does not exactly correspond to any tile known to me. The two upper characters read 延宮.

*d* = 黃山 (see *Ts'in Han wa tang wên tzŭ*, Book 下, p. 18 b; FORKE, l.c., p. 77; and Plate 5, No. 87). The original tile is fragmentary, the upper part of the character *huang* being missing: this part is here supplemented.

*e* = 延年 (see *supra ad*, Plate LXV, Fig. 1).

*f* = 宜當 (only half of tile and its inscription). See *Ts'in Han wa tang wên tzŭ*, Appendix, p. 13 b; FORKE, l.c., p. 89, No. 124.

*g*, identification doubtful.

From the standpoint of art-history, the chief interest in these tile-disks centres around their composition in concentric zones, which, in my opinion, must be derived from that of coeval metal mirrors. The flat knob in the centre of the disk, surrounded by a circle of dots, is certainly an imitation of the same design on mirrors. The mirror-knob is of essential importance, and testifies to its *raison d'être* through its perforation for the passage of a cord by which to hold the mirror in the hand when in use. There is no utilitarian view-point imaginable for the same knob on the tile-disks, where it appears to be merely ornamental. There is one tile (see FORKE, l.c., p. 95; and Plate 9, No. 136) on which four animals used in divination are represented. Exactly the same subject occurs on metal mirrors of the Han period. This is further proof for my statement that the mode of composition and decoration of these tiles has been suggested and influenced by that of the metal mirrors.

# APPENDIX II.

I WISH to make known here a series of specimens of mortuary pottery of the *Sung* dynasty (A.D. 960–1278) which were exhumed in 1903 on the premises of the American Presbyterian Mission, one mile from *Wei hsien*, Shantung, by Mr. Frank H. Chalfant, and kindly given to me by that gentleman in China for presentation to the American Museum of Natural History (cf. p. 12). Two reasons induce me to give here a brief description of these finds: first of all, they represent the first intelligent excavation of Chinese tombs undertaken by a foreigner, and thus afford an opportunity to judge how mortuary offerings were arranged in the grave; secondly, they show us what kind of objects these were during the Chinese middle ages.

On the grounds excavated by Mr. Chalfant there were four groups of graves, — (1) in the northwestern corner a group of three domed graves of the time of the *Ming* dynasty (A.D. 1368–1628), yielding no finds, and an isolated oblong grave of the same period in the northeast portion; (2) on the southeastern side two *Ming* graves with no finds of any value; (3) in the southwestern corner a group of three graves of the *Sung* dynasty; and (4) in the centre, southeast from Group 1, a cluster of five *Sung* dynasty graves.

312

### FINDS FROM THE GRAVES OF GROUP 3.

The accompanying sketch (Fig. 55), furnished by Mr. Chalfant, shows a sectional view of the interior of these graves. Mr. Chalfant comments on this as follows: "Interment was through an arched door at south. A mock-door (Fig. 55, *c*) was usually found on north side, built of carved brick in relief, with appearance of door-leaves ajar, so that the spirit might have exit and ingress. Some graves had a square well or hole in floor, evidently for drainage. The domed grave seems to have appeared first in the early *Han* dynasty (*circa* 200 B.C). The *Ming* dynasty graves here were all domed, but varied as to shape of bottom from round to square, some being octagonal, and one elliptical or oblong. Adjacent graves of the first emperor's time of the present dynasty were found as now made, — an arched parallelogram. The reason for this sudden change I cannot give." The apex

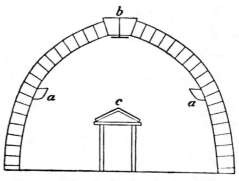

Fig. 55.   Sketch showing Sectional View of the Interior of a Grave.

of the graves rises from 60 cm to 90 cm below the surface; the circular bottom is from 1.5 m to 2.5 m in diameter. In the apex of the grave (Fig. 55), *b* denotes a brick plug, to which a bronze mirror, face downward, was attached by means of an iron wire, and set in lime-mortar. The mirror, thus reflecting downwards, was meant to light the grave.[1] On both sides (*a*) of the grave there were two brackets or

---

[1] Compare DE GROOT, The Religious System of China, Vol. II, p. 399.

projecting shelves, on which oil lamps were placed, at a height of three or four feet from the bottom of the grave.

The following set, consisting of eight pieces (Plates LXX, LXXI), was found in one of these graves: —

A whitish water-jar of light-reddish clay (Fig. 1, Plate LXX), decorated with a wave-line around the neck and scroll-work around the shoulders, both ornaments in brown. The interior is coated with a dark-brown glaze. The exterior white glaze has lost its brilliancy. When found, it was coated with alkali, so that the glaze seems to have been injured by the alkaline fumes in the grave. The jar measures 25.6 cm in height; the diameter of the mouth being 15.8 cm, and that of the bottom 12.1 cm.

A bowl (Fig. 1, Plate LXXI) 6.4 cm high, with a diameter at the opening of 16.9 cm, made of a coarse gray clay, covered inside with a dark-gray crackled glaze.

Another bowl (Fig. 2, Plate LXXI) 6.6 cm high and 18 cm maximum diameter, of very irregular shape, but of similar description to that of Fig. 1.

A small saucer (Fig. 3, Plate LXXI) 2.4 cm high and 11.2 cm in diameter, the glaze of which is almost destroyed. The three last-named vessels held food for the dead.

A sacrificial vessel (Fig. 4, Plate LXXI) of the *tou* 豆 type, 8.4 cm high and 10.2 cm in diameter, with short hollow standard. Its interior has a light-yellowish glaze.

A water-bottle (Fig. 2, Plate LXX) 18.9 cm high, the diameter of the mouth being 6.4 cm. The lower part of the surface is glazed a dark brown, with yellow flecks scattered here and there; the upper part, yellowish white, with bamboo-leaf ornamentation painted in black over the glaze.

The above six pieces of pottery are described in the same succession as they were arranged on the bottom of the grave.

Mortuary Pottery of the Sung Period.

PLATE LXX

Finds from a Grave of the Sung Period (Figs. 1-5. Pottery; Fig. 6. Bronze Mirror).

PLATE LXXII.

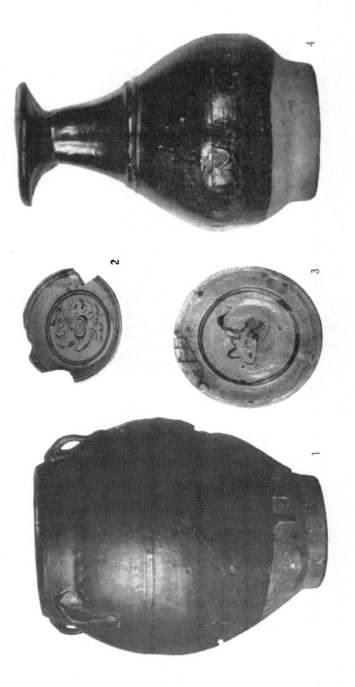

Mortuary Pottery of the Sung Period.

The earthenware lamp (Fig. 5, Plate LXXI) comes from the same grave, where it was placed on a bracket at the side (see above). It is 4 cm high, with a diameter of 8.4 cm. The thickness of the clay is 1 cm. Inside it is coated with a fine chocolate-brown glaze. Its outside shows a dark-red burnt clay. The edge of the rim for a short distance is serrated, apparently for receiving the wicks.

Fig. 6, Plate LXXI, shows the bronze mirror mentioned above as being in the apex of the grave. It has a diameter of 8.2 cm, and the rim is 0.5 cm thick. The surface is covered with a green-and-blue coating of patina; the back is unornamented, and layers of lime-mortar (see above) still adhere to it; it has the well-known perforated knob in the centre.

The contents of another grave of the same group is a set of four ceramic objects (Plate LXXII) and five copper coins, very thin and much worn, oxidized with green patina. The legends on these coins run as follows: (1) 咸平元寶, "Coin of the period *Hsien p'ing*, A.D. 998–1003" (under the Emperor *Chên Tsung*); (2) 祥符元寶, "Coin of the period *Hsiang fu*, A.D. 1008–1016" (under the same emperor); (3) 天聖元寶, "Coin of the period *T'ien shêng*, A.D. 1023–1031" (under the Emperor *Jên Tsung*); (4 and 5) 熙寧元寶, "Coin of the period *Hsi ning*, A.D. 1068–1077" (under the Emperor *Shên Tsung*). Several other coins buried in this grave were so brittle that they crumbled on handling, but all proved to be of the *Northern Sung* dynasty, prior to the *Hsi ning* period. It therefore follows that this grave must have been made shortly after the year 1077, and that the pottery in it must be ascribed also to the end of the eleventh century. The mirror belonging to this grave could not be found: it had dropped into the grave long ago, and was lost in the great deposit of earth that had gradually half filled the grave.

Mr. Chalfant emphasizes the fact that a curious feature of this pottery is, that each piece is of a different color, which was surely intentional. One small black bottle of the same size as that in Fig. 2, Plate LXIX, was shattered to fragments by a workman. The four pieces are: —

A water-jar (Fig. 1, Plate LXXII) 17.9 cm high, with a diameter of 8.8 cm at the mouth, and of 7.1 cm at the bottom, glazed a chocolate-brown, and having a row of small knobs around the lower end of the neck. It is furnished with four ears for the passage of a cord. Such jars with ears are still made, and of exactly the same shape and style, in *Po shan hsien;* and from a comparison with the modern *Po shan* ware, of which I made a collection on the spot, I am inclined to think that the whole of this *Sung* pottery described here was turned out of the famous kilns of that place.

A broken plate (Fig. 2, Plate LXXII) found in fragments. No shards belonging to it, however, were discovered in the grave. It is 3.7 cm high, with a diameter of 17.4 cm at the opening, and of 7.6 cm at the bottom. The clay, as shown by the bottom, is of a grayish color. A yellowish-white glaze covers the interior, and also the outside of the rim. The bottom of the plate inside is decorated with a painting in brown, a spiral in the centre. It is a work of the same glaze and style as those represented in Figs. 1 and 2, Plate LXX.

A small dish (Fig. 3, Plate LXXII) 2.4 cm high and 12.7 cm in diameter, with a beautiful gobelin-blue crackled glaze. Artistically it is the most exquisite of these *Sung* pieces. It is made of a finer quality of clay than the others. A circle is painted in black around the bottom, on the inside, and within it appear the head and neck of a bird in black, emerging from the character 福 *fu* ("happiness") in cursive

PLATE LXXIV.

Mortuary Pottery of the Sung Period.

Finds from a Grave of the Sung Period; (Fig. 1. Clay Cash; Figs. 2, 5. Water-Jars; Fig. 3. Lamp; Fig. 4. Strike-a-Light; Fig. 6. Saucer; Fig. 7. Bronze Mirror).

script. The background from which the character stands out has assumed a light sea-green tint, probably due to some chemical action on the original blue. Only traces of the character are now visible: the color in which it was executed — presumably also black — has disappeared, and the glaze is unfortunately exfoliated in several places.

The fourth piece (Fig. 4, Plate LXXII) is a mottled brownish-drab bottle with funnel-like rim, being 12 cm high, with a diameter of 5 cm at the mouth. It resembles in shape the modern oil-bottles used by teamsters.

A third grave in the same group contained two skeletons, male and female, presumably husband and wife. A large number of clay cash, some of which are shown in Fig. 1, Plate LXXIII, were found with the female skeleton. They are made in various irregular sizes, on an average 1.7 cm in diameter and 0.8 cm thick, with small central round holes. It would seem that clay money served well enough for the wife, while the man was supplied with forty real copper cash, thirty-eight of which were secured. The dates of these coins range from A.D. 758 (*T'ang* dynasty) to the period *Chêng lung* (A.D. 1156–1160), which is the latest represented on the legends in this lot. This is a year period of the *Kin* 金 dynasty (1115–1234). The coins with the legend "*Chêng lung*" were issued by the *Kin* from 1158 to 1170; so that, supposing this to be the current coin at time of interment, the grave in question must be about that date, — say, about 1170 or a few years later.[1] The crown of the grave was

---

[1] In a letter dated Dec. 13, 1905, Mr. CHALFANT makes the following additional remarks regarding the finds of coins in graves: "Copper coins were found with finger-bones of hand attached. The Chinese seemed familiar with the old customs, and said that coins were put into the hands of the dead. This I actually found in the other graves whose skeletons had not been disturbed. No coins were found in the ears or mouths of the remains we unearthed, but I saw a skeleton once exhumed from an old

early broken, and the grave rifled, and filled with earth. The bones of the skeletons, and a few remaining vessels, were scattered through the hard clay. The water-jar (Fig. 2, Plate LXXIII) was discovered in eight separate fragments. Several bits could not be found. It is 19.6 cm high, 12.3 cm in diameter at the mouth, and 8.2 cm at the bottom. It is covered with a fine brilliant jet-black glaze, which is much superior to similar glazes in modern ware. A lamp from the same grave (Fig. 3, Plate LXXIII), 2.5 cm high and 6.7 cm in diameter, has the same smooth, brilliant deep-black glaze. It is exceedingly well made. The material is a dark-gray clay. The bottom is decorated on the outside with a svastika in low relief. The oil-bowl proper is surrounded by a narrow flat rim ornamented with a row of tiny raised dots. The inside edge of the rim, which is left unglazed, forms a somewhat raised ring, open for a short space on one side to admit the wick. On the side opposite the opening, the rim broadens out; and on this part there is a raised leaf-shaped porous portion, unglazed, but blackened and rough. This was intended for the thumb in holding the lamp, with the idea of preventing the fingers from being smeared with oil, which could not flow over this spot. Just below it, on the outside, is a small oblong aperture in the clay for inserting a handle. A bit of cast iron (Fig. 4, Plate LXXIII) 6.1 cm long and 2.5 cm wide, very rusty, for striking fire from flint, came out of the same grave. Such strike-a-lights are also nowadays frequently placed in graves.

---

grave at *Ch'ing chou fu*, Shantung Province, which had copper coins in ears and mouth. These I identified, and found the oldest to be of the *Yüan* dynasty (about A.D. 1300). The coins, both copper and clay, were evidently strung, for the copper coins were always found rusted together in the shape of a cylinder. The faces of two of the clay coins were still adhering, which shows they had once been strung."

FINDS FROM THE GRAVES OF GROUP 4.

Among these are the following: —

A water or wine jar (Fig. 5, Plate LXXIII) 22.7 cm high; diameter of opening, 13.4 cm; diameter of bottom, 10 cm. The upper part of the surface is glazed a yellowish brown; the shoulders are decorated with scroll-work, and the neck with a wave-line in black under the glaze. The glaze is crackled. The interior is glazed a black. The lower part of the outer surface is unglazed, and exhibits a light-reddish clay with grooves in parallel circles.

A saucer (Fig. 6, Plate LXXIII) 3.5 cm high, with a diameter of 16.5 cm, and of 5.7 cm on the under side. The glaze is the same as that of the preceding, except a ring along the edge (glazed black) and a ring on the bottom (left unglazed).

A bronze mirror (Fig. 7, Plate LXXIII) 12.3 cm in diameter, and 0.4 cm thick over the rim, covered on both sides with green patina. On the back, on either side of the knob, is a fish in low relief, the one with head turned upward, the other with head downward. This was a favorite pattern on mirrors of the *Sung* dynasties,[1] this mirror being styled *shuang yü ching* ("double fish mirror").

A wine-jar (Fig. 1, Plate LXXIV) in mottled brownish-drab, 19.8 cm high, 11.1 cm in diameter at the opening, and 8 cm in diameter at the bottom. The glaze extends alike over both interior and exterior, and even over the bottom, except the raised ring surrounding it, on which the vessel stood in the furnace.

An oval-shaped wine-bottle with constricted neck and funnel-like mouth (Fig. 2, Plate LXXIV), 22.3 cm high, the diameter

---

[1] One is illustrated in the *K'in shih so*, section *kin so*, Vol. 6. There is one with the double fish, of highly elaborate workmanship, in our collections.

of the mouth being 5 cm, and that of the bottom 7.2 cm. There is a yellowish-brown glaze over the upper part of the exterior surface, separated from the unglazed portion around the base by a thickened black circle. A curious *craquelé* running in diagonal lines, and forming large irregular squares and rhomboids, appears under the glaze, where are also two designs in dark brown, one being a spiral enclosed by fanciful wave-lines.

The specimen illustrated in Fig 3, Plate LXXIV, is inserted here only as forming part of the Chalfant collection. It did not come from the *Sung* graves, nor is it *Sung* pottery. "It is said," remarks Mr. CHALFANT, "to have come from a grave exhumed on the Shantung Railway, but there is no definite evidence of this. It is a curious sort of ware not now made (so the Chinese say). It is not very old, perhaps *circa* A.D. 1800, possibly of *K'ang hsi* period (1662–1723)." The piece in question is 14.7 cm high, its mouth having a diameter of 4.5 cm. Its bottom, which is concave, measures 14 cm in diameter. It represents a shape well known in porcelain and ordinary pottery. It has a light-green glaze resembling jade in color, with very coarse crackles, like that turned out nowadays in *Yi hsing hsien;* and from a comparison with the modern pottery of this locality, of which there is a large variety in our collections, I should judge that this piece, coming from Shantung, is a *Yi hsing* production, although not modern, but one about a century or so old, which, comparatively speaking, might be styled old *Yi hsing* ware.

# APPENDIX III.

## MORTUARY POTTERY OF THE PRESENT TIME.

PROFESSOR DE GROOT says, in his book "The Religious System of China" (Vol. II, p. 387), "The dynasty which now bears sway in China has abolished the burial of victuals as an official rite, at least the *Ta Ts'ing t'ung li* does not give any precepts on this head. This work orders, however, that at burials of members of the Imperial family, offerings and libations shall, with the accompaniment of the wailing voices of all those attending, be made upon the tomb by the principal mourner; while at those of the nobility, the official classes, and the common people, such sacrifices shall be set out on the spot in front of the soul tablet. The actual state of matters seems to be in conformity with these precepts, for we have never seen or heard anything of a still prevailing custom of placing food in the graves; while, on the other hand, offerings upon the tombs, both at the burial and afterwards, are very general."

In such a generalization this statement is not quite correct. It may doubtless hold good for the province of Fuhkien and the south of China, where the observations of Professor de Groot have been chiefly made; but in the kilns of Peking, of *Po shan* in Shantung, and of *Yi hsing* in Kiangsu, mortuary vessels for storing up food and placing in the grave are still turned out, from which it follows that this custom prevails nowadays in the northern and eastern portions of the country. It has been observed and described by Professor

321

W. Grube.[1] It is known under the name "the filling of the urn," 夾礶兒 *chien-* [not to be read *chia* in this case] *kuan'rh*. *Chien* means "to seize something with tongs or chopsticks." The term *chien-kuan'rh* means, accordingly, "to put food into an urn by means of chopsticks." The procedure connected with this observance is, according to Grube, as follows. The urn is covered with a *lao-ping*, a kind of pancake, the edge of which must overlap a little the rim around the opening of the urn. The oldest son takes it off. Thereupon the daughters of the deceased (first the married ones, then the unmarried), after them the oldest son, then his wife, then the second son and his wife, and so on, each places in the urn, by means of chopsticks, a morsel of the offerings on the sacrificial table. Great care must be taken that the sticks pass from hand to hand without interruption. He who, by mistake, lays them down on the table after using them, interrupts the continuity of the turns, which may cause a death in his family. Should the urn not be full after each of the participants has contributed his share, it is the duty of the oldest son to fill it up to the brim; and as soon as this is done, he covers the urn with the pancake, from which he has nibbled off enough to make it fit the mouth of the vessel. This custom accounts for the aversion that mothers have to seeing their children nibble around the edge of a pancake, and they even regard it as an evil omen. Finally a red silk cloth is thrown over the mouth of the urn, and wound around with a variegated cord. The sacrificial cups are then removed, and the urn put in their place

---

[1] W. Grube, Pekinger Totenbräuche (Journal of the Peking Oriental Society, Vol. IV, Peking, 1898, p. 110). As this journal is now very scarce, and presumably in the hands of only a few readers, almost all remaining copies having been burned during the Boxer uprising at Peking, I have translated above the passage in question.

PLATE LXXV.

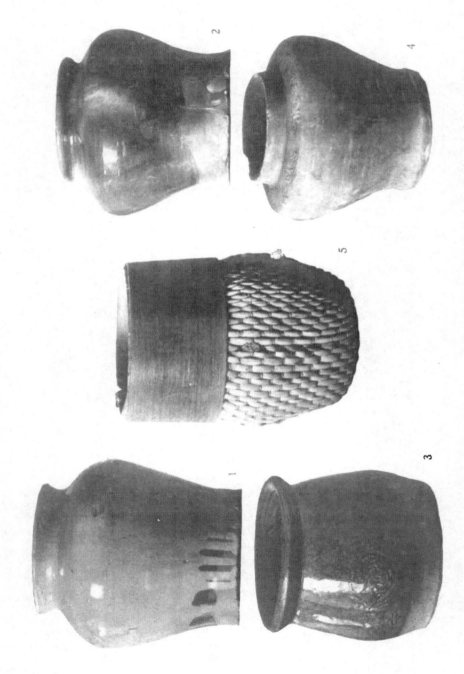

Figs. 1-4. Modern Pottery Funeral Urns; Fig. 5. Mortuary Basket Urn.

in the middle of the sacrificial table.    Afterwards it is interred
in the grave with the coffin, on the step on its front side.
I have not yet succeeded, remarks GRUBE, in interpreting
this strange observance, but I suppose it should be looked
upon as a food-sacrifice for the soul-substance tarrying in the
grave, with the idea of keeping it from starvation.

On Plate LXXV, Figs. 1–4, four such modern funeral urns
are illustrated.    They bear no resemblance in shape to any
of the types of mortuary pottery of the *Han* or the *Sung*
dynasty.    That shown in Fig. 1 is 14 cm high, and its mouth
measures 7.6 cm in diameter.    It is of light-reddish clay,
covered with a very thin transparent yellowish-white glaze
exhibiting greenish and yellow spots.    The next one (Fig 2)
is 12 cm high, with the same diameter and the same glaze,
but with more prominent shoulders.    Both are Peking ware.
The one represented in Fig. 4 was manufactured in the same
kiln, northeast from Peking, but is unglazed, and has a band
of leaf-ornaments around the shoulders, effected by the stamp-
wheel (see Plate XXXVI, Fig. 1).    It is 11.3 cm in height, and
the mouth 6.7 cm in diameter.    In shape it is similar to
Fig. 2, except that it has a straight rim.    The third urn
(Fig. 3 of this plate) is a product of *Yi hsing hsien*, and is
made of a coarse grayish clay coated with the well-known
bluish-green jade-like glaze, under which designs of blossoms
and spirals, ending in a bird's head, stand out in the clay.
Its height is 10.9 cm, and the diameter of the mouth 8.4 cm.
In *Po shan hsien*, I was told at the time of my visit, similar
funeral urns are manufactured; but unfortunately none were
then ready that I could take along.    In Peking the colloquial
designation for them is *yeh ssŭ kuan*, which was understood
also in *Po shan*, and for which the characters 耶死鑵
were given to me.    No Chinese dictionary that I have ever

seen contains this term. I cannot make out what *yeh ssŭ* means; and I doubt whether the characters given for it, if there are any, be correct. Some people, continues GRUBE in the passage quoted above, avail themselves, instead of the more usual clay urn, of a vessel plaited of willow-twigs, having the shape of a *shêng* 升 and serving as a grain-measure, because victuals buried in a clay vessel in the ground for a year or longer are believed to change into a liquid by fermentation. This fluid is regarded as an efficient remedy for a disease of the stomach called 噎膈. For this reason it is feared lest such a clay urn might be pilfered, and the grave thus be defiled.

A basket of the kind just described, made in the environs of Peking, is shown in Fig. 5, Plate LXXV. It is 16.5 cm high, with a circumference at the opening of 36.8 cm. These oval-shaped baskets serve the double purpose of a grain-measure (holding one small *shêng*) and a mortuary vessel (in the latter case, the wooden hoop around the rim is removed). I was informed that it is utilized as a mortuary vessel, particularly by the peasantry in the neighborhood of the capital, from which it might follow that it bears a certain relation to the rural occupation of the deceased. The modern reasoning given for its introduction, as set forth by Professor Grube, is certainly a subsequent and secondary explanation, because baskets were employed for this purpose in the days of antiquity, and are mentioned as early as in the *I li*.[1] This is one of the interesting cases where basketry alternates with pottery, one of the most striking examples in modern China being the large waterproof wicker-ware baskets in which wine and oil are transported.

---

[1] See DE GROOT, The Religious System of China, Vol. II, p. 382.

# INDEX.

Abel, Clarke, 18.

Agricultural work, representation of scenes of, on pottery plaque, 36, 44.

Albrecht, Oskar, 250, 258, 278.

Alchemy in China, beginnings of, 191.

Anderson, John, 259.

Andree, R., 215, 223.

Animals, plastic figures of, 246.

Annals of *Ch'êng tê fu*, 265.

Annals of the Later *Han* Dynasty, 6, 170.

Annals of the *Liang* Dynasty, 228.

Annals of Shantung Province, 267, 279.

Annals of *Ssŭ-ma Ch'ien*, 252.

Annals of the *T'ang* Dynasty, 241.

Annam Baroos, importations of camphor and cloves from, 180.

Anon, first European print on falconry by, 234.

Appendix I, Han roofing-tiles, 300–311.

Appendix II, mortuary pottery of the *Sung* dynasty, 312–320.

Appendix III, mortuary pottery of the present time, 321–324.

Archer, in flying gallop, in Japanese art, 217; mounted, in ancient Siberian art, 213, 214, 217, 245; mounted, in Persian art, 219, 220; on courser, with head *en face*, 218, 219; on horseback, on Han pottery reliefs, 160, 162, 166, 205 (see also *rider*); on hydra, on Han pottery reliefs, 167.

Architecture, Han roofing-tiles, 53, 55, 63, 300–311; model of farm-shed, 40; model of grain-tower, 52, 53; model of house, 42; roofs of granary urns, 54–58.

Arrow, Scythian manner of shooting, 218.

Artemisia, use of, for censers during *Chou* period, 180.

Asia, domesticators of the dog in Eastern, 252; four centres of bronze culture in, 259.

Aspelin, I. R., 217, 218, 219, 220, 223, 232.

Assurbanipal, Assyrian sculptures in Palace of, 248.

Aston, W. G., 35.

Aubazac, L., 20.

Automobile, anticipation by Chinese author of seventeenth century of idea of, 19.

"Axe" pattern, 123.

Baber, E. C., 275.

Ball, J. Dyer, 26, 266, 277, 280.

Baltzer, F., 65.

Bamboo, teeth in grain-mills made from, 24, 27; vessels made of, 123, 124.

Bamboo poles, well-sweep made from, 67.

Bamboo-drawing, 284.

Barbou, Alfred, 271.

Basket, imitation of, in bronze and pottery, 116.

325

# AMERICAN ANTHROPOLOGIST

## NEW SERIES

ORGAN OF THE AMERICAN ANTHROPOLOGICAL ASSOCIATION,
THE ANTHROPOLOGICAL SOCIETY OF WASHINGTON,
AND THE AMERICAN ETHNOLOGICAL
SOCIETY OF NEW YORK

### PUBLICATION COMMITTEE

F. W. HODGE, *Editor*, WASHINGTON, D. C.

J. R. SWANTON, *Associate Editor*, WASHINGTON, D. C.

## VOLUME 12

LANCASTER, PA., U. S. A.
PUBLISHED FOR
THE AMERICAN ANTHROPOLOGICAL ASSOCIATION
1910

to deal with the higher castes in future volumes. This book is free from many of the flaws which I have found in Thurston's work. It does not contain any anthropometric data, as the author has planned to confine himself to descriptions of the customs, manner, traditions, etc., of the various castes. The information contained in the book is well digested and is presented in an agreeable form. The superiority of Iyer's work over Thurston's is partially due to the fact that Mr Iyer is a native of southern India, and partially to the greater care and patience which he has shown in performing his task. The illustrations in this volume are also superior to those of Thurston.

Prefacing it are two introductions, one by John Beddoe and the other by A. H. Keane. Keane's introduction materially increases the value of the work. He takes this opportunity to present his theory of the racial composition of India, and to criticize those of Ripley and others. Though it would be too much to say that Keane has established his thesis beyond doubt, yet in the present condition of our knowledge regarding the ethnography of southern India, his view appears to be more probable than those which he opposes. This theory is as follows:

In the present general amalgam are represented five primary stocks: a submerged *Negrito*, probably from Malaysia; *Kolarian, Dravidian*, and *Aryan* who arrived in the order named from beyond the Hindu Kush and the Himalayas; lastly the *Mongul*, mainly confined to the Himalayan slopes. To the Kolarian, Dravidian, and Aryan ethnical stocks, correspond three distinct linguistic stocks, Kolarian being radically different from Dravidian, and both from Aryan. There is therefore no "Dravido-Kolarian" or "Dravido-Munda" mother-tongue, and these and the other compound terms like Indo-Aryan, Scytho-Dravidian, etc., are for the most part meaningless if not actually misleading.

If Dr Keane has any more evidence than what is published in this introduction to support the above mentioned theory, that evidence deserves to be published. Its publication will facilitate further research on the subject.

SHRIDHAR V. KETKAR.

*Chinese Pottery of the Han Dynasty.* BY BERTHOLD LAUFER. Leyden: E. G. Brill, 1909. Roy. 8°, pp. 339.

This is a publication of the East Asiatic Committee of the American Museum of Natural History—the Jacob H. Schiff Chinese Expedition as stated on the title page. The committee organized to direct the endowment of Mr Schiff appointed Dr Berthold Laufer, explorer and collector, as eminently qualified to take charge of the work, and this gentleman spent nearly three years in exploration in China and in studying native books on the subject of his investigations. The large number of objects figured were secured by Dr Laufer in Hsi an fu, province of Shensi.

The title of the book conveys little meaning of its contents. The author not only describes the pieces of Han pottery brought home by him from China, but examples of Han pottery in the possession of public museums and private collectors. The description of the pieces, however, is but a prelude to an exhaustive discussion of the culture of the Han dynasty as illustrated by its pottery, and this last sentence as a title would have conveyed a better idea of the remarkable work done by Dr Laufer in actual research and excavation and in the study of many Chinese works relative to the subject than the title used. Indeed the author confesses the broader character of his work when he says (p. 9), "In the following study it has been the aim of the author to furnish contributions, not so much to Chinese ceramics as to the archeology or culture of the period in question, and to illustrate ancient Chinese culture by means of these finds." Only one versed in the written language can profit by the numerous foot-notes in which are long quotations in Chinese, and titles of books, and even in the text are many quotations not translated. In fact this feature of the book compelled its publication in Holland, where, through the labors of De Groot, Schegel, and other sinalogues, the proper fonts and the men to set them have been brought together.

The reader may wonder how the Han pottery can give one a clue to the culture of that period, but an examination of the excellent figures, of which there are many, will convince him that many texts are presented for the discussions that follow. A parallel example is offered in the study of early Egyptian culture derived from the representation of trades, customs, etc., in wood, stone, and pottery, supplemented, and often confirmed, by the mural decorations in ancient tombs. In like manner, then, the mortuary vessels of the Han period present to us many important features in the life of the people. We may remark in passing that the general reader will have to turn over many pages before he finds the date of the Han dynasty in our chronology, which is B.C. 206 to A.D. 25. (Eastern Han lasted from A.D. 25 to A.D. 220.)

In these objects are found models of animals, houses and household utensils, agricultural devices, hunting scenes, and a variety of material which reveals to us in a fragmentary way the culture of that remote period. An enumeration of the many features exhaustively discussed would exceed the limits of this brief review, but the student interested in the relations of China with neighboring countries will find in this work a rich mine of information of a most varied character and with ample references.

A pottery hand mill leads to an interesting discussion, with many illustrations from Chinese sources, of various grinding devices, such as earth mills, hulling and winnowing mills, worked by water, animal, and man power. The student of the Chinese written language will be interested in the discussion as to the best rendering of the characters. The author criticizes Giles as to the translation of two characters which certainly read "wind" and "wheel" and which Giles naturally translates as wind-mill. By interpolating the character for fan, in another rendering we get the idea of winnowing. Dr Laufer shows that the wind-mill did not originate or exist in China prior to its introduction from occidental nations and is referred to in Chinese books, which he cites, as a foreign device. The peculiar cross-hatching on certain pieces he compares to similar markings on the mortuary pottery of Korea and Japan, which is supposed to be 1,200 years old or more. Judging from figures 1 and 5 and the color of the clay it is highly probable that the unglazed, lathe-turned mortuary vessels of Korea were derived from China, as a similar class of pottery in Japan was certainly derived from Korea. The rice pounder is a very ancient device judging from a Han example. The rice pounder to-day consists of a large mortar with a pestle on the end of a long lever. A man standing at the farther end of the lever presses it down by bearing his weight upon it, and, then, lifting his foot, the pestle, which is often weighted with stone, drops heavily into the mortar.

Vessels representing granaries with tiled roofs are shown, and, judging from the ridges, the tile consisted of imbrex and tegula, representing the oldest form of roofing tile known and the universal type to-day.

The dragon design leads to an interesting discussion as to whether it was pre-Buddhistic. Attention might be called to the *Institutes of the Chou Dynasty* (Gingell's translation), in which flag poles are figured with dragon-head terminations, also nobles' flags with two dragons intertwined. If the *Chou li* was really written in the Han period, which Dr Laufer says is suspected, then these allusions would have no weight in the argument. The old New England well-sweep was known in China 2000 years ago. I have seen the typical form in Satsuma and this simple device probably had many centers of origin. The forms of kitchen ranges discovered by Dr Laufer have survived until to-day. The pottery stoves on four legs are very curious. The slight observations I was able to make in the country about Shanghai and Canton revealed only the solid kitchen range, of a variety of designs, however, which I figured in my *Glimpses of China*. The projecting collar about the hole of the Han stove is also seen in the kitchen range of Canton. Portable cooking braziers, so common in China to-day, are also shown in the Han material. A primitive form of kitchen ladle made by splitting a long-necked gourd longitudinally is represented in the Han pottery by devices of the same form. A tazza made of rough clay and unglazed, figured on page 122, is like a typical form found in ancient graves in Korea and Japan. A form of censer with perforated cover, known as *po shan lu*, or hill-censer is discussed at length. It originated in the Han period. A vessel, not remotely unlike the *po shan lu* in size and perforated cover, without the saucer however, is found in old Korean pottery, white and brown glazed.

Is it not possible that some of these objects were used as mosquito smokers. known in Japan as *kaibushi?* Dr Laufer speaks of them in one place as fumigators. Some of them might have been hand-warmers or *teburo*.

The figure of an oval sanitary vessel and its survival today in a square form is in an accordance with the persistence of many other devices in China and emphasizes the unchangeableness of the masses.[1]

The triangular device of clay to support pottery in the furnace while baking is found in its simplest form among the Han objects. This simple device had a wide distribution in the eastern hemisphere in ancient times and survives today in the potteries of Japan, China, and Europe. It is a device of necessity and does not indicate community of origin.

Representations of dogs in figure and in relief among these mortuary objects lead to certain considerations regarding the breeds of dogs in China and their origin. One will find in the *Institutes of the Chou Dynasty* allusions to dogs as food. I quote the following: "If the dogs had their posteriors bare and red, and they were fleet of foot, the meat was deemed rank and of bad smell." The fat of the dog is often alluded to in the *Institutes*.

In discussing the influence of Siberian art and culture on ancient China Dr Laufer finds in the design of a horse in "flying gallop" evidence of its derivation from Turkish-Siberian sources; particularly does he mention the attitude of the mounted archer in shooting backward. In this connection it is interesting to note that in the fêtes given in honor of General Grant's visit to Japan, and which I was fortunate enough to witness, a tournament of mounted archers was the most unique. Three small targets were set up on poles, thirty or forty feet apart and perhaps fifty feet from the side of the road down which the archers in turn galloped at a furious rate, literally a "flying gallop." The archer shot at right angles to the direction of his motion. It

---

[1] Dr Laufer quotes from my paper on the *Latrines of the East* the statement that a certain pottery device sometimes served as a headrest or pillow. He says the statement in some respects seems erroneous. Some years ago Prof. F. W. Williams of Yale wrote to me as follows : " I am particularly pleased to find in your *Latrines* a reference (never before printed I am sure) to the portable stoneware urinals of China, which recalls my early childhood when an old groom of ours in Peking showed me one and declared that it made a mighty cozy pillow on a cold night."

seemed as if the archer could hardly have time to draw an arrow from his quiver which hung on his back before he would get by his target, yet one archer, having hit two targets in succession, waved the third arrow in bravado at the vast audience as he passed the third target, and then, quickly adjusting the arrow, shot backward and hit the target! Had he missed after this display he would have been justified in committing *harakiri*, and in feudal days would probably have done so. Dr Laufer observes the heavy stirrup with broad base of the Han period. The Japanese stirrup, though of an entirely different type, has a broad base in which the entire foot is supported, and this greatly assists the standing attitude of the archer in hunting or in warfare. Evidences are given which indicate that falconry was derived from Persia, while the game of polo originated in Turkestan.

Dr Laufer believes that the horse was brought to China by the Turks. In the *Institutes of the Chou Dynasty*, already alluded to, I find that six kinds of horses were distinguished as follows: thoroughbreds, chargers, horses of a color, roadsters, hunters, and commonbred. Allusions are also made to grooming, breeding, castrating, etc.

The absence of chairs, tables, and bedsteads has also been a marked characteristic of Asiatic nations with the exception of the Chinese, and in China one finds all these articles of furniture. The Chinese sit in chairs, have their meals at tables, though using chop sticks and eating out of communal dishes. May not this primitive method be a survival of the time when the family gathered about a common receptacle for food while sitting on the floor or ground?

Bas reliefs of the Han period show the absence of household furniture, and this leads Dr Laufer to an interesting discussion as to the origin of chairs, tables, etc., in China, and here again the author gives reasons for believing that these objects have been derived from the Turks. In the '*Institutes*' I find no reference to chairs, tables, or bedsteads, but there are references to five kinds of stools to lean upon—arm rests, literally. These arm rests are often represented in ancient Chinese pictures and are still surviving in Japan today. The five kinds of arm rests described are "(1) gems; (2) inlaid with imitation gems; (3) red-colored stool; (4) black-colored stool, varnished; (5) white plain unvarnished stools." Also makers of these objects are named in which the wood radical appears.

The ancient inscriptions on pottery and bronze will interest those who are familiar with Chalfant's valuable memoir on *Early Chinese Writing* published by the Carnegie Museum, Pittsburg.

Dr Laufer's comments are quite fair concerning the making of imitations in China. It is unjust in the foreigner to assert that there is a counterfeit industry in China for the purpose of deceiving foreigners. In most cases these imitations are made exclusively for the Chinese, who readily know their character and buy them as such, just as we make furniture and other objects after old models—of pottery scarabs as paper weights, bronze objects with the green patina imitated, etc. These and other common examples are familiar to all and are not intended to deceive. The Japanese have two words for these kinds of objects, one indicating an imitation for the purpose of deceiving and another an honorable imitation in which the fabricator signs his name.

The numerous and beautiful examples of Han pottery figured in this book will certainly surprise students and collectors of early Chinese pottery who know their rarity. In Japan thirty years ago an attaché of the Chinese legation showed me a single example of Han pottery which was esteemed a rare and valuable object. I made a careful study of its clay and of the few remaining patches of its thin crackled glaze. When a few years ago there suddenly appeared in our market a number of large vases and other pieces purporting to be Han I could not believe such rare objects in such numbers could be genuine, and though a critical examination failed to reveal a fraudulent make, I wondered at the rare skill of the counterfeiter, despite the fact that the Chinese are past masters in this art. A paragraph in Dr Laufer's book explains for the first time the reason for the sudden appearance of these remarkable pieces. In his Introductory the author says:

"Specimens of this pottery first came to light, towards the end of the seventies, in Hsi an fu, according to the statements of the dealers in antiquities living there, and consequently no mention of it is made in the archaeological or ceramic literature of the Chinese. In the beginning, nobody cared for these pieces, and they were indifferently thrown away for a few *cash*, until, after some years, larger cargoes of them reached Peking, where they brought enormous prices among Chinese lovers of antiquity. Nowadays it is the unanimous opinion of all Chinese judges, that this pottery represents genuine and most precious relics of the Han time."[1]

In summing up Dr Laufer expresses his conviction that the pottery of the Han dynasty originated in that period. He says (p. 212):

---

[1] The Museum of Fine Arts, Boston, has lately acquired through the efforts of Dr Okakura a remarkable collection of figurines and other objects of the Han dynasty. The curious head coverings, peculiar dresses, and, in one case, a broad-nosed figure with curly hair will interest the student of Chinese antiquities. In this collection is a hut urn and a cylindrical vessel made in imitation of a granary, in which is represented minutely the character of the tiled roof. The tegula is represented as a flat piece like the ancient Greek tile, though the curved tegula is represented in other examples. The collection also has a large number of the terminal or eaves imbrices with circular disc. It is hoped that these objects will soon be arranged for public exhibition.

"The pottery of the Han dynasty, from the view-point of technique and form, is doubtless a genuine and original production of Chinese culture. So far as this pottery represents imitations of actual objects, we have seen that these objects are true constituents of Chinese civilization, and existed in the period to which the pottery is to be ascribed. So far as it embodies vessels of a great variety of shapes, we have noticed that a great number of these can be traced back to forms in bronze and nephrite which were developed either during the age of the Han, or, for the most part and to the greatest extent, in ages much earlier, on Chinese soil. Of the remaining pieces which do not betray an affinity to forms in other substances, it may be asserted with perfect safety that they do not display any traces suggestive of foreign influence, but appear as thoroughly Chinese ceramic productions. Indeed, this entire pottery forms a well-defined group, which, in its character as pottery, bears no resemblance to any other known groups of ancient pottery in Asia, nor to that of Siberia or of Turkistan."

In Appendix One there is given a description of the roofing tiles of the Han. The circular discs which terminate the eaves imbrices are very interesting. The decoration in strong relief consists in most instances of characters quartering the circle and these have been translated as reading "Infinite like heaven," "Thousands and thousands of years without end," "Sublime peace to the numerous generations," "Long life without end," and similar sentiments. One is amazed at the possibility of translating the more obscure ones, particularly the disc represented on Plate LXVII, Fig. 1, a complicated mass of scrolls which are, however, rendered "May you always have good luck!" Dr Laufer expresses the opinion that the decorative composition in concentric zones on these discs must have been derived from coeval metal mirrors. While this supposition may be correct I would point out that a circular area for decoration would naturally lead to concentric treatment. That the occurrence of four animals used in divination should be found on a metal mirror as well as on a tile does not seem strong evidence in support of his suggestion, as such animal symbols as well as sentiments and mottoes are widely used as decorative features on the most dissimilar objects.

In Appendix Two, Dr Laufer describes a number of pottery objects which Mr Frank H. Chalfant exhumed from Sung tombs in Shantung. Most of the forms closely resemble those made to-day. The oviform bottle figured on plate LXXIV, in form and glaze is identical with the wine bottle which one might find cast away in the ash barrel of a Chinese laundryman in this country. The Sung period takes us back eight hundred years or more and the persistence of these forms is one of the amazing characteristics of the Chinese. The two pieces figured on plate LXX resemble pottery with a light-cream colored glaze and rude decoration of brush strokes in brown that one meets with in Japan and which the Japanese antiquarian identifies as Korean.  Dr Frederick Hirth assured me that the pieces were made in China and were probably shipped in numbers to Korea from which country they found their way to Japan.

For those who believe that the culture of Middle America owes its origin to Asiatic contact the picture so graphically given by Dr Laufer of the culture of the Han period affords no evidence.  Neither in the methods of pottery making, glazing of various colors; the triangular device for supporting pottery in the furnace, old-world-wide in its distribution, but singularly absent in the western hemisphere; nor in the potters' wheel and wheels of other kinds, even the pulley, can any resemblance be found in Middle America.  So we might enumerate other features none of which do we find in the ancient culture of Middle America, such as roofing tiles, grain towers, grinding mills, water mills, mortars of various kinds, irrigating devices, well sweeps, decorative motives, knife money and other forms of money, cooking stoves, cross-bows, falconry, furniture, the forms of bronze vessels, and characters used in inscriptions.

In conclusion, I think the author will agree with me when I suggest that it would have added greatly to the usefulness of this unique work if there had been appended a bibliography of the subject.  Even an alphabetical list of the numerous works he refers to in the text would have been welcomed.  The importance of this addition would most likely have occurred to Dr Laufer had he been at home during the making of the book, but the last half of the volume was sent to press during his second visit to China.  In such a bibliography he might well have included for the general reader Alfred E Hippisley's article entitled "A Sketch of the History of the Ceramic Art in China," published in the *Smithsonian Annual Report* for 1887–88, and again published in the Report of 1900.  The more so as in this article the various Chinese dynasties are given with the chronology of the successive productions of pottery and porcelain, and for the added reason that these reports were freely distributed and are easily accessible to every reader.  In such a bibliography also might have been included William Henry Gingell's *Translation of the Institutes of the Chou Dynasty*, B.C. 1121, a rare book but of surpassing interest as showing the ceremonial usages of the Chinese nearly a thousand years before the Han dynasty.  If the *Chou li* was really written in the Han period Dr Laufer would have added greatly to the interesting assemblage of facts he has brought together regarding the customs of the period.

EDWARD S. MORSE.

# BULLETIN

DE

# l'Ecole Française

## D'EXTRÊME - ORIENT

---

### TOME X.  1910

HANOI

IMPRIMERIE  D'EXTRÊME-ORIENT

—

1910

# Chine

Berthold LAUFER. — *Chinese Pottery of the Han Dynasty*. (Publication of the East Asiatic Committee of the American Museum of Natural History. — The Jacob H. SCHIFF Chinese Expedition). — Leiden, Brill, 1909; 1 vol. in-8°, XVI-339 pp., fig.

En 1901, M. Jacob H. SCHIFF fit don à l'American Museum of Natural History d'une somme destinée à rechercher et à réunir des collections en Chine. L'administration de ces fonds fut confiée par M. Schiff à un comité organisé sous les auspices de l'American Museum. Le Docteur Berthold Laufer, chargé de mener à bien l'entreprise patronnée par M. Schiff, passa près de trois années en Chine (de 1901 à 1904) et recueillit de nombreux objets, spécialement à Si-ngan fou 西安府 (¹). Les résultats de ces longues recherches sont exposés dans le beau livre de M. L. Disons tout de suite qu'ils sont d'une importance considérable pour l'histoire de la Chine antique. — M. L. étudie successivement la poterie antérieure aux Han, la poterie des Han, les objets d'ornementation, les inscriptions. En appendice sont traités les sujets suivants: tuiles des Han; poterie mortuaire des Song; poterie mortuaire actuelle.

Dans les quelques pages qu'il consacre à la poterie antérieure aux Han, M. L. étudie surtout la poterie de l'époque des Tcheou. Quelques exemplaires (²) accusent des motifs d'ornementation rudimentaires sous la forme de lignes obliques ou horizontales irrégulièrement disposées et dans lesquelles M. L. voit une caractéristique de la poterie des Tcheou. Evidemment ce motif peut aider à dater une pièce; pourtant il n'est pas suffisant. Les poteries de fabrication et de décoration grossières sont de toutes les époques. Dans beaucoup de pays, entre autres en Egypte, en Kabylie, on emploie encore des vases d'argile crue ornés de dessins rudimentaires et qui n'en sont pas moins de fabrication très récente.

Les objets de l'époque des Han étudiés par M. L. ont été trouvés dans des tombes. Ils sont de deux sortes. Les uns sont des vases, des bols, des plats, etc., tels qu'on les utilisait dans la vie quotidienne; les autres sont des modèles ou des reproductions, à échelle réduite, de maisons, de moulins, de bergeries, de puits et d'appareils usuels d'assez grandes dimensions. Toute cette poterie est spécialement fabriquée en vue d'une destination funéraire. La coutume qui consiste à ensevelir avec le défunt des objets qu'il utilisait de son vivant est très ancienne en Chine, et l'on voit de suite quel intérêt peut offrir l'étude de ce mobilier funéraire. Nous sommes en présence d'une sorte de microcosme, d'une réduction fidèle et durable de la vie chinoise telle qu'elle était à l'époque où l'on a enfoui ces objets.

Que tous ces objets datent effectivement de l'époque des Han, on a le droit d'en douter. M. L. essaie pourtant de l'établir dans son introduction (pp. 5 et ss). Son

---

(¹) C'est aussi à Si-ngan-fou que M. PELLIOT, lors de sa dernière mission, acquit des bronzes archaïques, des miroirs métalliques et plusieurs types de poterie des Han. La comparaison est intéressante à faire entre ces derniers et ceux qui sont reproduits dans le livre de M. L. Cf. aussi la collection rapportée de la Mandchourie méridionale par M. TORII Ryūzō 鳥居龍藏 et décrite dans la *Kokka*, 1909, n° 235; 1910, nᵒˢ 237, 239, 241, 243, 245, sous le titre: *Relics of the Earlier Han Dynasty in South Manchuria*.

(²) Voyez notamment pl. I, fig. 2, et pl. III, fig. 1.

argumentation peut se résumer ainsi qu'il suit: les inscriptions sur ces poteries sont analogues en tous points aux inscriptions sur les bronzes des Han; toutes les dates déchiffrées se réfèrent à cette époque; le style et les sujets d'ornementation correspondent à ceux des bas-reliefs bien connus des Han; d'ailleurs le *Heou Han chou* 後漢書 énumère les différents objets de terre qu'on avait accoutumé de placer dans les tombes, et ce texte est une base historique de première importance. Ce sont là d'excellents arguments, mais la conclusion qu'en tire M. L. est peut-être un peu étroite et rigide. En tout cas ce qu'on peut dire, c'est que si ces objets ne sont pas tous de l'époque des Han, ceux qui peuvent s'en éloigner le font de très peu; il ne me semble pas possible d'en situer aucun plus tard qu'au commencement du ve siècle de notre ère, et aucun ne présente une trace quelconque d'influence gréco-bouddhique.

Toutes ces poteries paraissent, dit M. L., avoir été faites au tour. Le tour du potier est évidemment très ancien et consistait d'abord en un simple plateau auquel la main de l'ouvrier imprimait un mouvement de rotation (¹). Mais les poteries de M. L. ont des formes si régulières que le tour des Han devait être déjà assez perfectionné.

M. L. étudie des moulins à main (pl. IV). Il les compare aux moulins actuels, à main également, employés dans le nord de la Chine, et il est le premier à distinguer le *long* 磑 (ou 礱) « for husking grain » du *mo* 磨 « for pounding hulled grain into flour ».

Tou Yu 杜預 est l'inventeur des moulins à eau, en Chine (²). Ce Tou Yu vivait au IIIᵉ siècle de l'ère chrétienne (222-284). Comme les moulins à eau apparaissent en Europe dès le Iᵉʳ siècle avant l'ère chrétienne, on est en droit de se demander s'ils n'ont pas été introduits en Chine par des étrangers. M. L. pense que « l'invention a dû

---

(¹) C'est ce tour primitif qui est cité dans Homère (*Iliade*, chant VII) et qu'on peut voir représenté sur les peintures murales des hypogées égyptiens de Thèbes. Cf. aussi *Che ki* 史記, k. 83, 6 b, et le commentaire, qui renvoie à l'histoire des Han.

(²) La biographie de Tou Yu se trouve dans le *Tsin chou* 晉書 (éd. de Chang-hai, k. 34, 6 a et ss). On en trouvera un extrait, enrichi de quelques notes originales, dans le récent ouvrage japonais de [Sōshō] IWATARE Kentoku [蒼松] 岩 垂 憲 德, le *Jugaku taikwan* 儒 學 大 觀, p. 664. — Cf. aussi GILES, *Biogr. Dict.*, nº 2072. — CHAVANNES, *Mémoires historiques...*, t. V, Append. I, pp. 449 et ss. — *San kouo tche* 三 國 志, *Wei chou* (éd. Ki-kou ko, k. 16, 4 a; éd. Chang-hai, k. 16, 8 a). Ce Tou Yu (*tseu*: Kong-k'ai 公 凱) est surtout célèbre par ses études sur le *Tch'ouen ts'ieou* 春 秋 et le *Tso tchouan* 左 傳. Parmi ses ouvrages, je citerai: le *Tch'ouen ts'ieou king tchouan tsi kiai* 春 秋 經 傳 集 解, en 30 *kiuan*, incorporé à divers *ts'ong-chou* (相 臺 岳 氏 本 五 經; 正 誼 齋 叢 書; etc.); le *Tch'ouen ts'ieou che li* 春 秋 釋 例, en 15 k. (entré dans le 古 經 解 彙 函, dans le 岱 南 閣 叢 書, etc.); deux courtes œuvres, d'un chapitre chacune, le *Tch'ouen ts'ieou t'ou ti ming* 春 秋 土 地 名 et le *Tch'ouen ts'ieou tch'ang li* 春 秋 長 曆, incorporées toutes deux au *Wei p'o sie yi chou* 微 波 榭 遺 書, etc. — Kou Yen-wou a consacré tout un ouvrage à l'étude d'un commentaire de Tou Yu: c'est le *Tso tchouan tou kiai pou tcheng* 左 傳 杜 解 補 正, en 3 k. (voyez 指 海, VI, 1; 經 學 叢 書, 乙 集, 3; 皇 清 經 解, kk. 1 à 3; 頣 志 齋 叢 書.) — Le célèbre auteur diplomate Li Chou-tch'ang 黎 庶 昌 a écrit, lui aussi, une étude critique du commentaire de Tou Yu sur le *Tso tchouan*: c'est le *Tch'ouen ts'ieou tso tchouan tou tchou kiao k'an ki* 春 秋 左 傳 杜 注 校 勘 記, paru en 1894 dans l'admirable petite collection qu'est le *Ling fong ts'ao t'ang ts'ong chou* 靈 峯 草 堂 叢 書 de Tch'en Kiu 陳 榘 (cf. PELLIOT, *Notes de Bibl. ch.*, le *Kou yi ts'ong chou*, BEFEO, II, 1902, p. 315 et 339-340.)

être faite antérieurement dans une région intermédiaire entre l'Empire romain et la Chine, et qu'elle se répandit ensuite également vers l'Est et vers l'Ouest ». Au Japon, les moulins à eau n'apparaissent qu'au VIIe siècle (670) : c'est du moins le *Nihongi* 日本紀 (1) qui l'avance, et c'est là la seule mention que nous en ayons, car le *Kojiki* 古事記, antérieur de quelques années seulement au *Nihongi*, est muet à ce sujet. Le moulin à eau fut introduit de Chine au Tibet à peu près à la même époque, en 641 (2), lors du mariage du roi tibétain Sroṅ btsan sgam po avec la princesse chinoise Wen-tch'eng 文成.

Dissertant sur les mortiers à grains, M. L. a un excellent raisonnement à propos d'une figure (p. 39, fig. 8) extraite du *Cheou che t'ong k'ao* 授時通考 et illustrant un *kang-touei* 壙碓, appareil composé d'un pilon de pierre (碓) fixé à l'extrémité d'un levier formant bascule. Ce pilon retombe dans une sorte de jarre (壙) à l'intérieur de laquelle sont placées les graines à écraser.

Aux planches VI et VII, M. L. donne trois photographies extrêmement importantes pour l'histoire de l'architecture chinoise. Ces illustrations sont celles de maisons reproduites à petite échelle. Nous en avons des vues de l'intérieur et de l'extérieur. La première maison (pl. VI) présente un toit à tuiles plates réunies entre elles par d'autres tuiles semi-cylindriques; les extrémités de la poutre faîtière et des quatre arbalétriers se relèvent comme actuellement, mais, chose remarquable, le toit n'est pas incurvé. Je ne crois pas que la théorie dite « de la tente » (3) puisse être à nouveau sérieusement

---

(1) *Nihonshoki* 日本書紀 (ou *Nihongi* 日本紀) (édition du Kokushi taikei 國史大系, 1897), vol. I, p. 484. On peut même se demander s'il ne s'agit pas de moulins à eau dans la première citation de l'année 610 (3e mois de la 18e année de Suiko 推古 (593-628) : v. *Nihonshoki*, ibid, pp. 384-386) ; ASTON, *Nihongi*, II, p. 140 et note, suppose qu'il est ici question de moulins à main, mais dans l'édition dont je me sers l'expression *tengai* 碾磑 est expliquée par *miẕuusu* ミヅウス, ce qui signifierait « mortier (ou moulin) à eau ». Les moulins à eau auraient donc été connus au Japon, déjà en 610. — Dans ce sens voyez encore le *Nihonshoki tsūshaku* 日本書紀通釋, par IIDA Takesato 飯田武鄉, vol. IV, chap. 54, p. 299, col. 10.

(2) « In about A. D. 635 » (p. 35) est une imprécision et une erreur. M. L. ne fait d'ailleurs que suivre ROCKHILL (*Notes on the Ethnology of Tibet*. Report of the National Museum, 1893, p. 672) qu'il cite. M. Rockhill a confondu la date de la demande en mariage (634) avec celle du mariage lui-même (641). Voyez le *Tseu tche t'ong kien pou tcheng* 資治通鑑補正, éd. lith. (太宗貞觀八年，甲午), k. 194, 4 b. — Le roi du Tibet dépêcha à l'empereur de Chine des ambassadeurs chargés d'apporter un tribut et de demander la main d'une princesse chinoise pour le roi. Sroṅ btsan sgam po essuya un refus qui déchaîna même une guerre où le roi tibétain eut le désavantage. Lorsque la paix survint, Sroṅ btsan sgam po renouvela sa demande, et cette fois T'ai Tsong lui accorda la main de la princesse Wen tch'eng. Voy. *Tseu tche t'ong kien pou tcheng*, id. (貞觀十五年，辛丑), k. 196, 1 a. Dans un précédent travail (*Tibet. A geographical, ethnographical and historical sketch, derived from chinese sources*. J. R. A. S., N. S., XXIII, p. 190), M. Rockhill, encore que ne donnant pas la date du mariage, n'avait pas fait cette confusion. Dès 1880, BUSHELL avait indiqué la date exacte (641) d'après le *T'ang chou* (*The Early History of Tibet from chinese sources*. J. R. A. S., N. S., XII, p. 443 et p. 444). Cf. aussi WIEGER, *Textes historiques*, vol. III, p. 1569 et 1575-1576.

(3) Cf. EDKINS, *Chinese Architecture*, J. C. B. R. A. S., N. S., XXIV, 1889-90, no 3, pp. 253-288. — S. Ritter von FRIES, *The Tent Theory in Chinese Architecture*, Ibid.; ibid., pp. 303-306. — BUSHELL, *Chinese Art*, trad. fr., pp. 53 et ss. — FERGUSSON, *History of Indian and Eastern Architecture*, 2e éd, vol. II, pp. 451 et ss. — S. LÉVI, *Le Népal*, vol. II, pp. 10-11.

soutenue. Cette interprétation, théoriquement possible, ne tient pas devant les faits. Les maisonnettes que M. L. nous fait connaître prouvent que l'incurvation du toit chinois est postérieure aux Han. Elle est d'ailleurs plutôt due à l'influence des conceptions bouddhiques et de l'art hindou qu'à d'autres causes. La seconde maison (pl. VII), dont le toit est de forme intéressante, a ceci de particulier qu'à l'intérieur se trouve un lit de briques analogue au *k'ang* 炕 actuel. Ce dernier serait donc très ancien : ceci est une notion nouvelle.

Nous pouvons voir (fig. 10, p. 45) la représentation d'une des plus curieuses pièces de la poterie des Han : c'est un bercail avec ses moutons. Après avoir décrit les petits animaux de terre et leur habitat, M. L. essaie d'expliquer leur raison d'être dans une tombe. Reprenant un épisode bien connu de la vie de Confucius, rapporté d'ailleurs par Sseu-ma Ts'ien dans sa biographie du Maître (¹), M. L. fait une longue dissertation sur le terme *fen yang* 墳羊 (écrit quelquefois 羵羊 et aussi 羵羘), sans pourtant obtenir de résultat bien précis. M. L. cite Sseu-ma Ts'ien, mais seulement à travers les intermédiaires. Si M. L. s'était reporté au texte même du *Che ki* 史記, il aurait eu le mot de l'énigme (²). Enfin il aurait pu se renseigner de façon parfaite dans le Sseu-ma Ts'ien de M. CHAVANNES (vol. V, pp. 310 et ss), où il eût trouvé la traduction du passage en question, l'indication des sources et de copieuses notes. M. L. n'ignore pas l'important ouvrage de M. CHAVANNES (cf. p. 195, lignes 21 et ss., et p. 214, lignes 11 et ss.), et c'est sans doute par inadvertance que ce passage lui a échappé.

Les greniers publics existent en Chine depuis l'antiquité la plus reculée. Ils sont de deux sortes. Les uns, *k'iuan* 囷, sont ronds, les autres, *ts'ang* 倉, sont carrés. S'inspirant de leurs formes, les potiers fabriquèrent ce que le chinois moderne appelle communément les « *wou kou kouan* 五穀罐, bocaux aux cinq sortes de graines » (³). Les *wou kou kouan* qui nous sont représentés pl. IX, sont aux morts ce que les greniers sont aux vivants. M. L. a d'excellentes pages sur divers types de ces urnes à grains (⁴), sur

---

(¹) Ki Houan-tseu 季桓子 creusant le sol, déterra un vase de terre dans lequel se trouvait un mouton. Il apporta sa trouvaille à Confucius et pria le Maître de l'éclairer. Confucius lui répondit entre autres choses : « 土之怪曰墳羊. Les manifestations surnaturelles du sol s'appellent *fen yang*. » Voyez Sseu-ma Ts'ien, *Che ki* (édition de Chang-hai, k. 47, 3 a ; éd. du Ki-kou ko. k. 47, 1 b).

(²) Il eût pu même faire quelques remarques intéressantes sur certaines divergences de texte. C'est ainsi que M. L. eût rencontré la mention : « 得狗. J'ai trouvé un chien, » parole que prononce Ki Houan-tseu au moment où il présente le mouton de pierre à Confucius. Le commentaire de ce petit membre de phrase est assez intéressant : « 獲羊而言狗者以孔子博物測之. (Ki Houan-tseu) avait trouvé un mouton et disait que c'était un chien pour éprouver la pénétration de Confucius. » (*Che ki*, loc. cit.). — Comme le fait d'ailleurs remarquer M. CHAVANNES (*Mémoires historiques*, t. V. Note add., pp. 437-438), il est étrange de voir Confucius discourir sur un pareil sujet ; il y a là une contradiction formelle avec ce qu'avance le *Louen-yu* (VII, 20) : « Le Maître ne discourait pas sur les prodiges... et les êtres surnaturels. » Cf. encore CHAVANNES, ibid., pp. 412 et ss.

(³) Cf. CHAVANNES, op. laud., t. I, p. 28 ; BRETSCHNEIDER, *Botanicon Sinicum*, n° 335. (J. C. B. R. A. S., N. S., XXV, 1893, pp. 137 et ss.).

(⁴) Voyez, pl. X, l'illustration d'une de ces urnes dont les trois pieds portent chacun très distinctement une figure d'ours. C'est le décor ordinaire de ce genre de poteries. M. L. (p. 57, note) cherche l'explication de ce motif d'ornementation. Pourquoi un ours ? se demande-t-il. Et, citant un passage du *Po kou t'ou lou*, M. L. en conclut que « l'ours

des puits à contre-poids, à poulie et sur des fourneaux de cuisine. La présence de fourneaux de cuisine dans des tombes peut étonner au premier abord, mais le *Heou Han chou* mentionne expressément les *wa tsao* 瓦竈 dans l'énumération des objets qu'on enfouissait avec le mort. Les fourneaux sont quelquefois rectangulaires, ils épousent le plus souvent la forme d'un fer à cheval (pl. XVII et XVIII). Ces derniers, qu'il sera intéressant de comparer avec quelques spécimens de la collection PELLIOT, présentent à leur surface des motifs d'ornementation en relief. Les différents objets utiles à la préparation des mets et des repas : cuillères, petits plats de toutes formes, couteaux, crochets à feu, etc., y sont reproduits; on y voit même des aliments tout préparés (¹). — Notons aussi le motif de décor en losanges si fréquent dans l'art de cette époque (²).

Des terrines à feu et des ustensiles de cuisine je dirai peu de chose. M. L. nous montre surtout des reproductions de ces objets en bronze, et citant un passage du *Tong t'ien ts'ing lou* 洞天清錄, il distingue nettement deux sortes de « cooking vessels » un peu confondues jusqu'alors : le *tiao teou* 刁斗, vase *sans pied*, à longue poignée (voyez pl. XXI. fig. 1), et le *tsiao teou* 鐎斗, (pl. XXI, fig. 4), vase à trois pieds que M. L. croit être le prototype de la théière actuelle. Après une étude sur les cuillères, étude qui nous fait mieux saisir l'expression *p'ao teou* 匏斗 (³), « récipient en forme de calebasse » (voyez pl. XXII, et cf. *Heou Han chou*, loc. cit.), M. L. passe

---

était un vœu de progéniture mâle, à cause de son endurance et de sa force. (熊男子之祥取其有所堪能故也). » Dans le *Che king* 詩經, *Siao ya* [LEGGE, C. C., vol. IV, p. 306], on lit: « ... 維熊維羆男子之祥.... The bears and grisly bears are the auspicious intimations of sons. » Il reste à savoir pourquoi, plutôt qu'un autre animal, l'ours est considéré comme le symbole de la force. La similitude graphique, et peut-être phonétique, qui existe entre 能 et 熊, n'est probablement pas étrangère à la formation de cette conception. Le texte ci-dessus tendrait un peu à l'établir: l'auteur, en employant le mot 能 dans son explication, semble avoir voulu attirer l'attention sur ce point. On trouvera des interprétations analogues dans le travail de M. CHAVANNES : *De l'expression des vœux dans l'art populaire chinois* (J. A., 2e semestre 1901, pp. 193 et ss.). Le *Si ts'ing kou kien* 西清古鑑, (k. 38, 28 a, de la grande éd. imp.; k. 38, 38 a, de la petite éd. de 1888), donne une illustration d'un ours identique, ornant un bronze des T'ang. L'animal est appelé cette fois *fei hiong* 飛熊, « ours volant ». Cette dernière expression est l'équivalent de 非熊.

(¹) Par exemple un plat de poissons prêt à être servi (pl. XVII).

(²) Voyez fig. 23. Ce motif apparaît plusieurs fois sur les piliers et sur les bas-reliefs des chambrettes de Wou-leang ts'eu 武梁祠. Cf. CHAVANNES, *Mission archéologique dans la Chine septentrionale*, pl. XXXIV, 58; XXXV, 59; XXXVII, 61; XXXIX, 65 et 66; XL, 67 et 68; XLI, 69-70; XLIII, 73-74; XLIV; XLV; XLVI; L; LIII; LVIII; LXI; LXXI, 135 et 137; LXXIV; LXXV, etc. Le décor est composé d'une suite de losanges, le plus souvent doubles. A noter une particularité sur la dalle verticale à l'Est et en dehors de la chambrette du Hiao-t'ang chan (CHAVANNES, id., pl. XXX, 54). Au sommet des angles de chaque losange est représenté un moule à sapèques. Les côtés du parallélogramme représentent probablement les canaux minuscules servant à amener le métal en fusion jusqu'aux moules. — Cf. l'expression 棗核 *tsao ho*, noyaux de jujube (*P'ei wen yun fou* 佩文韻府, k. 100 下; rime 陌; s. v. 核).

(³) 夸斗 (p. 107) est une faute d'impression pour 匏斗. Cf. *Po kou t'ou lou*, k. 16, 4 a.

rapidement en revue des bols, des plats (pl. XXIII), une table (pl. XXIV), des vases rectangulaires (pl. XXV) identiques aux vases de bronze du type *fou* 盙 (ou 簠) de l'époque des Ming, des urinaux (pl. XXVI), à propos desquels l'auteur admire « this evidence of refinement and progress in hygienic matters among the people of the Han time ».

La partie la plus captivante du livre est certainement celle qui concerne les vases et les jarres. M. L. a eu la bonne idée de multiplier les illustrations, et c'est un véritable plaisir d'avoir à lire ces pages pleines de notes intéressantes et bourrées d'aperçus, un peu osés parfois, mais toujours ingénieux.

Le livre de M. L. sera indispensable pour les recherches sur les origines des formes de la porcelaine chinoise. Ces formes dérivent de celles des poteries que M. L. étudie. La poterie est évidemment antérieure aux bronzes ; mais plus tard, les formes de la poterie s'inspirèrent des formes des vases de bronze lorsque ces derniers furent devenus d'un usage courant. C'est ainsi qu'on arriva à fabriquer des poteries avec de pseudo-anses, simples saillies décoratives, rompant la sécheresse des formes, mais rappelant malgré tout la destination des anses analogues de vases de bronze.

Parmi les bronzes des Han il faut noter les *po chan lou* 博山鑪 (ou 博山香爐 *po chan hiang lou*), « brûle-parfums en forme de montagnes ». M. L. date ces « hill censers » du commencement du 1er siècle avant l'ère chrétienne. Le couvercle de ces bronzes figure une montagne au milieu des vagues [1]: M. L. donne du moins cette explication qu'il tire du *K'ao kou t'ou* 考古圖. M. Chavannes (*T'oung pao*, mai 1910, pp. 301 et ss.) serait disposé à y voir les pics des quatre points cardinaux entourant le pic du centre. Je ne suis pas éloigné de me rallier à l'opinion de M. Chavannes, encore que sur certains spécimens n'apparaisse qu'un seul pic et que sur presque tous les autres exemplaires l'aspect des vagues se confonde étrangement avec celui des soi-disant montagnes. Le monticule central reste immuable dans tous les cas. Il y a là une intéressante question à résoudre.

D'après M. L. la théorie donnant la Corée comme ayant inventé la poterie doit être rejetée. La poterie aurait été introduite de Chine en Corée. M. L. a probablement raison, mais nous aurions voulu quelques preuves plus convaincantes que celles qu'il nous donne ; la question est très complexe et ne peut être résolue en quelques pages.

La plupart des sujets exécutés en relief sur les vases représentent des scènes de chasse ou des animaux isolés. Les différents sujets sont identifiés de très heureuse façon par M. L., dont la perspicacité est pourtant mise en échec par une forme de figure quasi-humaine « with animal like demoniacal grimace and wide open mouth ». Ce démon ne serait-il pas tout simplement un ours ?

Les décors en reliefs de deux vases différents (pl. XLVIII et XLIX) reproduisent un archer monté qui décoche une flèche en se retournant. Le cheval est à l'allure de « galop volant », pour employer l'expression désormais célèbre de M. Salomon Reinach [2].

---

[1] Le même décor apparaît aussi sur le couvercle de jarres en terre.

[2] Chavannes, *Mission archéologique dans la Chine Septentrionale* (I, xix, n° 35), a donné une intéressante scène de chasse illustrant un bas-relief d'un des piliers du Chao-che 少室 (Teng-fong hien). Là aussi nous trouvons un archer monté et se détournant pour décocher une flèche à un animal alors que son cheval est à l'allure du galop volant. — Cf. S. Reinach. *La représentation du galop dans l'art ancien et moderne* (Revue archéologique, 1900-1901).

Dans ses notes sur l'ornementation, M. L. recherche l'origine des motifs artistiques qui apparaissent sur les poteries. Le style est dominé par le dessin conventionnel du galop volant et présente une ressemblance si frappante avec le style de l'art ancien scytho-sibérien que des rapports doivent être admis presque *a priori* (¹). Le motif du lion, animal inconnu des Chinois, ne peut être considéré que comme un emprunt. Le motif de l'archer-cavalier se retournant pour décocher une flèche est absolument étranger à la pensée chinoise et est dérivé de l'art turc. Les Chinois avaient en effet, vers le IVᵉ siècle avant l'ère chrétienne, pris au peuple turc la tactique des archers montés, et comme les Turcs eux-mêmes avaient trouvé là l'expression d'une représentation artistique, il faut en conclure que les Chinois reçurent des Turcs le motif artistique en même temps qu'ils recevaient la tactique.

Signalons, avant de terminer, l'étude très poussée de M. L. sur les races de chiens dans l'ancienne Chine (pp. 247-281) et enfin le chapitre sur les inscriptions (²) (pp. 287 et ss.).

P. 5. M. L. cite le *Kouei sin tsa tche* 癸辛雜識 de Tcheou Mi 周密, et oppose l'opinion de M. Hirth à celle de Wylie sur la date probable de la rédaction de l'ouvrage. La vérité est que Tcheou Mi (*tseu* : Kong-kin 公謹) a vécu successivement sous les Song et sous les Yuan, puisqu'il est dit dans sa notice biographique qu'à la chute des Song, il ne demanda point d'emploi et se retira chez lui où il mourut. Le *Kouei sin tsa tche* a probablement été rédigé vers la fin des Song, mais nous ne savons rien de précis à cet égard et l'on ne peut se baser sur la note des bibliographes de K'ien-long indiquant que l'auteur vivait sous les Song. L'immense *ts'ong chou* qu'est le *Chouo feou* 說郛 renferme plus d'une dizaine d'œuvres de Tcheou Mi. Il faut noter que pour le *Tche ya t'ang tsa tch'ao* 志雅堂雜抄, il donne l'indication suivante : « composé par Tcheou Mi des Yuan 元 ». Ce même *Tche ya t'ang tsa tch'ao* est aussi incorporé au *Tö yue tch'e ts'ong chou* 得月簃叢書, qui indique : « composé par Tcheou Mi des Song ». Il n'y a aucune base sérieuse de discussion dans ces indications de tradition (³).

P. 23. M. L. donne un extrait du *Cheou che t'ong k'ao* à propos du *mo* 磨, et le traduit pp. 20, 21 et 22. Le passage « 方言或謂之㾦 » ne doit pas être traduit comme le fait M. L. par : « In local dialects it is sometimes called ch'i ». Le sens est : « le *Fang yen* 方言 dit qu'il est parfois appelé *ts'i* 㾦 ». Grammaticalement l'interprétation de M. L. est exacte, mais M. L. aurait dû être frappé par ce fait que dans le passage du *Cheou che t'ong k'ao* étaient cités deux autres dictionnaires, le *T'ang yun* 唐韻 et

---

(¹) M. L. ne fait d'ailleurs que reprendre et développer la théorie de M. Salomon Reinach (op. laud.), qui fait remonter l'origine de ce motif jusqu'à l'art mycénien.

(²) M. Chavannes (*T'oung pao*, mai 1910, p. 302, note) a donné la lecture exacte de la date inscrite sur un *fang hou* 方壺, vase de forme carrée (voyez Laufer, pp. 140-141 et p. 290). Il faut lire 始建國四年柒月. Le doute portait sur le caractère 柒.

(³) Cf. *Nan Song wen fan* 南宋文範, k. 9, 13*b*. ; *Sseu k'ou ts'iuan chou tsong mou*, k. 140, 42 *b* ; k. 141, 34 et ss. — Tcheou Mi est l'auteur d'un ouvrage peu connu, où sont décrits des peintures, des bronzes et des jades, le *Yun yen kouo yen lou* 雲烟過眼錄, en 2 k., dont T'ang Yun-mo 湯允謨, des Yuan, a écrit un *Siu lou* 續錄, incorporé avec l'ouvrage principal au *Che wan kiuan leou ts'ong chou* 十萬卷樓叢書, III, 45. Voyez Pelliot. *L'Œuvre de Lou Sin-yuan*, BEFEO, IX, 1909, p. 246.

le *Chouo wen* 説文. J'ai pourtant voulu en avoir le cœur net et je me suis adressé au *Fang yen* lui-même (éd. du 漢魏叢書; k. 5, 3 b, col. 2.) à l'expression 碓機 qu'il glose ainsi : « 陳魏宋楚自關而東謂之㮛碾或謂之硬. Dans les pays de Tch'en, de Wei, de Song et de Tch'ou, et à l'Est des passes, on l'appelle *yen-wei*; il est parfois appelé *ts'i*. »

— 甬 est une inadvertance pour 通 dans 四川通志 (p. 50) et dans 山東通志 (p. 267, note 4).

P. 88. M. L. cite un passage de l'encyclopédie *San ts'ai t'ou houei* 三才圖繪 à propos du mot *tsao* 竈. Ce passage est le suivant: « 淮南子曰炎帝王於火死而爲竈. » M. L. traduit ainsi : « Huai Nan tzŭ (philosopher of the second century before Christ) says, Yen-ti (i. e., the Emperor Shên-nung 2838-2698 B. C.) ruled by virtue of the fire, and when dying made kitchen ranges. » Le sens est : « Houai-nan Tseu dit : Yen-ti régna par la vertu du feu et devint après sa mort le dieu du foyer. » Le mot *tsao* désigne les « fourneaux de cuisine », mais a aussi par extension le sens de « foyer » et de « dieu du foyer ». L'expression *tsao-chen* 竈神 (quelquefois 竈王 et aussi 竈君), le dieu ou l'esprit du foyer, est très connue et désigne souvent Yen-ti Chen-nong (¹). Le commentaire de la phrase de Houai-nan Tseu (²) ne peut laisser aucun doute : 炎帝神農以火德王天下死託 (une autre édition écrit 托) 祀子竈神. Le *K'ang-hi tseu tien* (s. v. 竈), citant aussi Houai-nan Tseu, est plus clair encore : « 炎帝作火官死而爲竈神. Yen-ti institua les *mandarins-flammes* (³) et à sa mort devint le dieu de foyer. » D'ailleurs l'interprétation de M. L. n'est guère possible, car, en dehors de toute question de langue chinoise, j'ai peine à me représenter l'empereur Chen-nong se mettant à fabriquer des fourneaux de cuisine au moment de sa mort, et il y aurait de plus contradiction avec la suite de la citation du *San ts'ai t'ou houei* (LAUFER, p. 88), qui indique Houang-ti, postérieur à Chen-nong, comme étant l'inventeur des fourneaux de cuisine.

Léonard AUROUSSEAU.

---

(¹) Cette expression désigne aussi parfois 黃帝 Houang-ti (Cf. DE GROOT, *Les Fêtes annuellement célébrées à Emoui*, Ann. du Musée Guimet, II, pp. 449 et ss.), parfois Tchou Yong 祝融. (Cf. *Li-ki*, trad. COUVREUR, passim).

Noter (DE GROOT, op. laud. p. 452, dernier paragraphe et note 6) un autre extrait de Houai-nan Tseu que M. DE GROOT tire du *Ko tche king yuan* (k. 19, s. v. 竈): « 黃帝作竈死爲竈神. Houang-ti inventa le foyer et devint après sa mort dieu du foyer. » Je ne connais aucune édition de Houai-nan Tseu donnant cette phrase et j'ignore d'où le *Ko tche king yuan* a pu l'extraire. Cette question du dieu du foyer est très obscure, et des divergences aussi marquées dans un même auteur ne sont pas faites pour l'éclaircir.

(²) Voyez Houai-nan Tseu, k. 13, 氾論訓, trois derniers folios.

(³) Voyez *Tso tchouan* (LEGGE, Chinese Classics, vol. V, II, 17ᵉ année du duc 昭, pp. 665 et 667 a): « 炎帝氏以火紀故爲火師而火名. Yen-te (Shen-nung) came to his (rule) with the (omen of) fire and therefore he had fire officers, naming them after fire... » COUVREUR (*Dictionnaire classique de la langue chinoise*, 2ᵉ éd., 1904, p. 663, col. 3) traduit ainsi la citation de Houai-nan Tseu insérée dans le *K'ang-hi tseu tien* : « L'officier qui fut préposé au feu par Chen-nong devint après sa mort le dieu du foyer. » Cette interprétation ne peut pas se soutenir : 官 n'est certainement pas sujet dans la phrase ci-dessus. De plus COUVREUR n'a pas eu connaissance du passage du *Tso tchouan* indispensable pour comprendre l'expression 火官.

# T'OUNG PAO

通 報

OU

ARCHIVES

*CONCERNANT L'HISTOIRE, LES LANGUES,*
*LA GÉOGRAPHIE ET L'ETHNOGRAPHIE*
*DE*
*L'ASIE ORIENTALE*

Revue dirigée par

**Henri CORDIER**
Membre de l'Institut
Professeur à l'Ecole spéciale des Langues orientales vivantes

ET

**Edouard CHAVANNES**
Membre de l'Institut, Professeur au Collège de France.

**VOL. XI**

LIBRAIRIE ET IMPRIMERIE
CI-DEVANT
E. J. BRILL
LEIDE — 1910.

qu'elle est énoncée par les Chinois et telle qu'elle fut connue en Grèce, soit qu'il explique le sens de la réforme de 1596 qui accorde les tubes sonores suivant le principe de notre tempérament égal, soit qu'il étudie les cinq notes de la gamme ancienne ou les sept notes de la gamme complétée, ou les neuf notes de la gamme importée par la dynastie mongole, M. Laloy fait œuvre originale et le lecteur tirera beaucoup de profit de ses remarques. Les chapitres suivants sont consacrés à la description des divers instruments de musique, aux modes de notation, à la musique religieuse, à la musique de chambre, à la musique populaire et enfin à la musique de théâtre. Dans le chapitre sur la musique religieuse, M. Laloy a traduit l'hymne qui est chanté lors des sacrifices de printemps et d'automne dans les temples de Confucius; il ne paraît pas avoir connu la traduction qui en a été donnée par G. E. Moule [1]) et qui est supérieure à celle de van Aalst.

Ed. Chavannes.

Berthold Laufer, *Chinese pottery of the Han dynasty*. — 1 vol. in-8° de 339 p.; Leyde, Brill, 1909. Publication of the East Asiatic Committee of the american museum of Natural History. The Jacob H. Schiff Chinese Expedition. — Cf. Berthold Laufer, *Kunst und Kultur Chinas im Zeitalter der Han* (Globus, vol. xcvi, n° 1 et 2; 8 et 15 Juillet 1909).

La porcelaine Chinoise a depuis longtemps attiré l'attention des Européens par sa beauté sans rivale; mais la poterie qui l'a précédée avait été presque entièrement négligée; M. Laufer est le premier qui ait eu l'idée de la prendre pour sujet d'étude et les

---

1) *Notes on the ting-chi, or half-yearly sacrifice to Confucius* (Journal of the China Branch of the Roy. as. Soc., vol. XXXIII, n° 2, p. 37—72.

résultats qu'il a obtenus sont considérables. Les objets dont il a fait une ample collection proviennent pour la plupart de la province de *Chàn-si*; c'est aussi à *Si-ngan fou*, soit dit en passant, que M. Pelliot a pu acquérir un grand nombre de pièces analogues dont on verra les principaux spécimens dans la salle de la mission Pelliot au Louvre. Ces objets appartiennent tous à des mobiliers funéraires; déposés dans des tombes à l'usage des morts, ils étaient des modèles réduits de ce qui servait aux vivants; ce sont des moulins, des maisons, des bercails avec leurs moutons, des urnes en forme de greniers, des puits où la poulie sur laquelle passait la corde soutenant le seau est parfaitement visible, des fourneaux à la surface desquels sont représentés les principaux ustensiles culinaires, des aiguières, des vases et des plats de diverses formes, voire même des urinoirs. Tous ces types divers ont été parfaitement décrits par M. Laufer qui a pu, par leur moyen, reconstituer la civilisation dont ils sont l'image.

Certains vases ont fourni à M. Laufer l'occasion de remarques du plus haut intérêt. Les uns affectent exactement la même forme que nous observons dans les bronzes; ils ont sur leur panse deux anneaux qui, étant en terre, sont collés sur la paroi du récipient, mais qui simulent évidemment les anneaux mobiles en métal dont on se servait pour soulever les vases de bronze; d'où résulte la conclusion que ces vases en terre sont la réplique à bon marché des vases de bronze et leur sont chronologiquement postérieurs. D'autres vases en forme de jarre présentent des bandes circulaires en relief sur lesquelles on voit des tigres galopant ou trottant, des hydres, des oiseaux, des lions, des démons, enfin des cavaliers qui se retournent pour tirer de l'arc derrière eux sur quelque animal; ce dernier motif, remarque M. Laufer, est proprement scythique. Enfin d'autres vases ont une forme cylindrique et sont munis d'un couvercle qui figure, d'après M. Laufer, les îles des immortels entourées des flots de la mer, mais qui pourrait aussi bien figurer les pics des quatre points cardinaux entourant le pic du centre.

M. Laufer estime que les objets qu'il étudie remontent tous à l'époque des *Han* et sans doute a-t-il raison dans la majorité des cas; il y aura lieu cependant, lorsque le matériel soumis à nos investigations, sera plus abondant, de faire certaines distinctions chronologiques et je serais, pour ma part, assez incliné à croire que plusieurs des poteries classées par les Chinois comme poteries des *Han* appartiennent en réalité à l'époque des *Tsin*, peut-être même à celle des *Wei* du Nord. Ce qui est du moins évident, c'est que les objets dont M. Laufer nous donne la reproduction, sont entièrement étrangers à l'art bouddhique; on n'y remarquera donc pas le moindre vestige de cette influence hellénistique qui se laisse encore entrevoir dans les monuments dont l'inspiration est bouddhique. Reprenant et développant l'hypothèse qui a été proposée pour la première fois par M. Salomon Reinach dans ses remarquables articles sur le galop volant, [1]) M. Laufer croit que c'est dans l'art dit mycénien qu'il convient de rechercher l'origine de quelques uns des motifs qui apparaissent dans l'art des *Han*. [2])

ED. CHAVANNES.

---

1) S. Reinach, *La représentation du galop dans l'art ancien et moderne* (Revue archéologique, 1900—1901).

2) A la p. 291, M. Laufer a cité une lettre que je lui avais écrite il y a cinq ans et que je ne croyais point destinée à la publicité; la date dont il est question dans cette lettre doit être lue 始建國四年七月. «Le septième mois de la quatrième année *che-kien-kouo* (12 p. c.)». Le caractère 七 est écrit ici avec une graphie qui se retrouve fréquemment sur les fiches de l'époque des *Han* découvertes par M. Stein à l'Ouest de *Touen-houang*; cette graphie a donné naissance à la forme moderne 柒.

# ARTIBUS ASIAE

## INSTITUTE OF FINE ARTS · NEW YORK UNIVERSITY

## VOL. XXX

*This publication has been assisted by a grant from the*
*J. D. R. 3rd Fund*

MCMLXVIII

ARTIBUS ASIAE — PUBLISHERS · ASCONA · SWITZERLAND

# BIBLIOGRAPHIA

*Berthold Laufer, Chinese Pottery of the Han Dynasty. Photographic reprint of the 1909 first edition. Charles E. Tuttle Company, Rutland, Vermont, 1962. xx & 339 pp., 55 figs., 75 pls.*

Review of this volume presents peculiar difficulties. Any Sinologist must be grateful for the recent increase in publication of reprinted books which have been long out of print and impossible to acquire. One hesitates, therefore, to question the value of any such reprint. Then, too, Berthold Laufer remains one of the greatest among Western Sinologists. His "pioneering work" embraces an amazing variety of studies. In this case, his introduction of Han pottery to the West in the first decade of this century is both a "classic work" and a "landmark" as the reprint's dust jacket claims. But, was it "undeservedly" out of print?

Thanks in large part to the publication of its first edition, Dr. Laufer's introductory remarks on examples of Han pottery in Western collections and publications, and his listing of those with dated inscriptions, have—ironically enough—been rendered obsolete. Now, every major museum and many private collections contain notable specimens. Interest in early Chinese ceramic art has been aroused to the extent that it would be possible to compile an extensive bibliography of studies on this subject from the last half century.

Dr. Laufer's survey is based upon a collection made by purchase in Hsi-an between 1901 and 1904. He gave to it the meticulous examination of a conscientious scholar, and sought further clarification of its manufacture in Chinese literature and comparative archaeology, the only sources for research available at that time. Now, in the second half of the 20th century, we are not so limited. For the Han Dynasty alone, hundreds of excavated sites have been reported in considerable detail. Han pottery can be discussed in its total archaeological context. Deductions as to its derivation, evolution, function, etc. can be made from a mass of data whose numbers and distribution may even permit some statistical evaluation. It is true that material gathered within the last three decades of archaeological activity remains scattered amongst many excavation reports and a few specialized analyses, most of them in the Chinese language. Dr. Laufer's more general presentation in English has not been superceded by another such, of similar scope but with more complete information. Is this, however, a sufficient reason for perpetuating its misconceptions and limitations?

Any reader of this second reprint edition will recognize the inadequacy of Dr. Laufer's section on pre-Han pottery. It was written before the discovery of the sites at Yang-shao, Ch'eng-tzu-yai and An-yang. Today's students of Chinese art and archaeology know, not only of Neolithic "painted" and "black" pottery, Bronze Age "white" or "grey" pottery, but also of many refinements within these large groups. Excavation of dwelling sites, as well as burials, has revealed the relationship between domestic and ceremonial or mortuary wares. Their wide geographical distribution has permitted the distinction of local variations—eastern (Honan, Shansi) from western (Shensi, Kansu) painted pottery, northern (Shantung) from southern (Anhui, Chekiang) black pottery, and many minor types which developed in the isolation of peripheral regions during the later Neolithic "Lungshan" phase and Bronze Age Shang Dynasty of the central Yellow River valley. Attention to finds in stratified areas has yielded closer and more reliable dating, so that technical advances in the development of ceramics may be observed and charted. We know now, for example, that ash glazes were in use by the mid Chou Dynasty.

Han Dynasty ceramic development is less known. The problem of its study is not due to insufficient data. Rather, a profusion of sites and finds presents formidable tasks of arranging, grouping and analyzing. A review is not the medium for surveying such a complex and rich field. It may, however, prove illuminating to compare Dr. Laufer's material with similar examples from one excavated Han site.

The Loyang Shao-kou site was excavated in 1953. It is on the south slope of Mang Shan, long known to be the location of a Han necropolis. 1,040 burials were found there, mostly distributed in an orderly manner, with those of any given period confined to a single locale. The mode of burial, including its architectural construction, and the contents of each

347

clearly indicate a chronological progression from mid-West Han (mid-2nd century B.C.) through late East Han (early 3rd century A.D.). The Shao-kou site, therefore, provides a more or less complete record of changes during the entire dynasty. It is, to be sure, only one area and a different location from the place of origin for Dr. Laufer's collection. Still, it remains within the focal range of early Chinese culture and succeeded Hsi-an (Ch'ang-an) as the capital for the East Han Dynasty. Reference to the excavation report *Loyang Shao-kou Han mu* (Peking, 1959) will be made, hereafter, with the abbreviation—SK.

Dr. Laufer begins his description of Han pottery, in section III, with "imitative forms".

Grist-mills Pls. IV & V

SK Pl. 62.2 (also fig. 90.1) and 4

Both of Laufer's clay examples are glazed and comparable in size to Shao-kou's single grey clay specimen, plain and carved out rather than wheel thrown or moulded, and 3 stone models. These emerged from burials of the later 1st century B.C., and 1st through mid-2nd centuries A.D. Their forms differ from Laufer's examples, and also from modern Chinese hand-mills, but relate to his second type which is, however, more elaborately constructed. In the Shao-kou ceramic finds, brown glazed wares emerged from burials of the mid-2nd century B.C. through the first decade of the 1st century A.D.; painted wares began in the early 1st century A.D. and continued to the early 3rd century; while green glazed wares were also introduced ca. 100 A.D. but increased during the mid-2nd century and retained popularity through the end of the dynasty. It seems likely that Laufer's examples are from the later 2nd century A.D. In any case, his statement (p. 8) that the 1st century A.D. "may be considered as the terminus ad quem" for the pottery discussed is untenable.

Mortars and rice-pounders fig. 7

SK Pl. 62.1 of stone

Animal pens figs. 10 & 12    SK Pl. 37.1–3

Shao-kou yielded 26 pig pens from burials after 100 A.D. through the mid-2nd century; of grey or reddish clay, some with painted ornament on walls, and most with buildings attached. Tall structures of rectangular plan may represent storehouses. One example of a low hut with a hole through its floor is identified as a latrine.

Granary-urns Pls. IX, X, XII, XIII

SK Pls. 23–25, fig. 54

983 pieces were found at Shao-kou, usually in groups

of 5 or 10 to a burial. Many contained remnants of millet, rice, maize, soy beans, sesame or barley. Some bear inscriptions; either specifying the grain within, or noting the volume measure. The report divides this vessel form into 8 types on the bases of form and decoration. These divisions fall into progressive chronological units. Laufer's selection may be correlated with these to form the following sequence—

Pl. XIII  is related to Type 1:1 (SK fig. 54.1)—mid-2nd century B.C. to ca. 100 A.D.

XII.1  is of Type 1:2 (SK fig. 54.2)—same time span as Type 1:1.

XII.2  is of Type 2:2 (SK fig. 54.5)—ca. 100 A.D. through mid-2nd century.

IX.1  is related to Variant Type 1 (SK fig. 54.7)—ca. mid-2nd century A.D.

IX.2  is related to Variant Type 2 (SK fig. 54.8)—later 2nd to early 3rd century A.D.

Laufer's granary-urns in Pls. IX and X have ribbed roof-like shoulders, but none of the Shao-kou types do. This imitation of an architectural element may have been a local Shensi stylistic feature. At Shao-kou, the forms are straightforwardly designed for vessels. Note that Pl. XI illustrates a completely different vessel form; that of the *lien*, a cosmetic box or food container which will be discussed later.

Draw-well jars Pls. XIV & XVI

SK Pls. 29.3–6, 30.1–5, 31; figs. 59–61

There are 97 Shao-kou examples, distinguished into 6 types. Laufer's first reproduction is related to Type 3:1 (SK fig. 60.3), produced from the 1st century A.D. throughout the East Han Dynasty. Three more on Pl. XVI are related to Type 3:2 (SK fig. 60.4) which appeared in the mid-1st century A.D. and continued through the dynasty. Laufer shows no earlier types.

Cooking-ranges fig. 19, Pls. XVII & XVIII

SK Pls. 28, 29.1–2; figs. 57, 58

About ½ of the 155 examples from Shao-kou are coated with a silvery slip. None are glazed. All of Laufer's selection bears green glaze, except Pl. XVIII which is unglazed. The Shao-kou finds are divided into 8 types. Laufer's fig. 19 is related to Type 3:2 (SK fig. 58.2) and his Pl. XVIII is a more developed version of the same. The horse-shoe shaped examples on Pl. XVII do not find a parallel among Shao-kou forms, but their surface details resemble those of Type 4:1 (SK fig. 58.3). Thus, again, Laufer is confined to the East Han, and probably to the later half of that period. Objects moulded upon the cooking

348

surfaces certainly represent the common equipment of kitchens; cauldrons, steamers, spoons, ladles, knives, bowls, ear-handled cups, platters, meats and fish. Borders with geometric ornamentation may have been confined to mortuary models, but can we be so sure of their specialized appearance as to state (p. 88)—"Certainly, the kitchen ranges in actual use during the Han time were not ornamented at all, but were plain..."

Section IV describes vessels.

Braziers Pls. XIX, XX.1        SK Pl. 33.4, fig. 63.1–3
There seems to be nothing about the example on Pl. XIX to distinguish it as of Han manufacture. Pl. XX.1 resembles one of the 14 finds made at Shao-kou in general shape (SK fig. 63.2), and another in the prongs protruding from the bowl's rim (SK fig. 63.3 & Pl. 33.4). Both of the excavated specimens came from burials of the 1st century A.D.

Dr. Laufer begins his discussion of cooking-vessels with a distinction between *chiao tou* as a tripod cooking-vessel and *t'iao tou* as a cooking-pan without feet. His bronze example of the former (Pl. XX.3) certainly looks to be possibly of the Han Dynasty, but reproductions taken from the Ch'ing Dynasty antiquaries catalogue *Hsi Ch'ing ku chien* (figs. 25–30) cannot be regarded as reliable illustrations of Han works. The bronze *t'iao tou* on Pl. XXI.1 is described as being "of very thin dark-red copper, as is peculiar to the work of the Han time" (p. 101). If such a material is distinctive for that period, this reviewer is ignorant of the "fact". No vessel of similar shape has emerged among excavated finds.

*Tou* Pl. XX.2        SK Pl. 35.8, fig. 65.1
10 examples of this vessel form emerged at Shao-kou; all placed in the category of food servers, such as ear-handled cups, bowls and plates. It seems, therefore, to have been used to present food, rather than to cook it. Laufer's example resembles the Shao-kou one from a burial of the later 2nd century A.D., except that its dragon-head handle is more sharply bent down.

Laufer's Pl. XXI.2 is related to a Shao-kou ladle form (SK fig. 64.11) from a burial of the later 2nd century A.D. His Pl. XXI.3 represents, of course, a *ho* which suggests a slightly pre-Han dating.

Ladles Pl. XXII        SK Pl. 35.6–7, fig. 64.10
The sizes of these two groups of ladles are roughly comparable. Laufer's examples are photographed from above, so it is difficult to judge their curvature in profile. They appear cruder in manufacture than the excavated pieces. Of the latter, the report notes that all are hand-made in imitation of carved wood, and some are ornamented in red and/or black pigment on their interiors or exteriors.

Laufer correctly identifies the ear-handled cups on Pl. XXXIII.3–4 as being for wine. The first is described as having handles in horizontal position, the second slightly turned upward. At Shao-kou, examples with up-turned handles came from burials of the 1st century A.D., those with horizontal handles from burials of the later 2nd through early 3rd centuries A.D. (SK fig. 64.5–6).

The sacrificial table on Pl. XXIV may be compared to one from Shao-kou (SK fig. 64.2). 46 rectangular or round low tables were excavated at that site. The first type is reported as being from burials from the 1st century A.D. on, and increasing in numbers about the early 2nd century.

In Laufer's discussion of tazzas or sacrificial vessels, the nomenclature of two unrelated vessel forms, is confused.

*Tui* Pl. XXVII    SK Pls. 26.2 & 4, 27.2,4 & 6; fig. 56
103 examples came from Shao-kou burials and were usually found with *ting* tripods. Some bear inscriptions; either simply stating the sacrificial function, or noting the kind of grain contained. The report also mentions discovery of animal bones within a few, so these covered bowls presumably were used for meat sacrifices as well. Their formal development paralleled that of the accompanying *ting*, and is divided into 3 types or phases. Laufer's examples belong to Type 1, and are closest to one from a Shao-kou burial of the later 1st century B.C. (SK fig. 56.4).

Laufer's Pl. XXVIII.2 is more correctly named a *tou*. At least, it is like the later vessel form called *tou*. Interestingly, his discussion cites a passage from the *Li Chi* (Book of Rites) to affirm that *tou* existed in the Hsia Dynasty and its original material was wood. He suggests that the pottery form was introduced in the Han Dynasty. Archaeological evidence shows that this vessel appears in the ceramics of the later Neolithic "Lung-shan" culture, but impressions from wood *tou* remained in one or two of the large tombs at An-yang. 3 objects of similar shape were uncovered at Shao-kou (SK Pl. 30.6). The report calls these lamps, but does not explain the reason for doing so. Again, in Laufer's survey of jars and vases, the nomenclature of variant forms requires clarification. They may be divided according to the Shao-kou report.

349

*Kuan* Pl. XXIX.3        SK Pls. 16 & 17, fig.49
*Wen* Pls. XXIX.1–2, XXX, XXXII, XXXIV
       SK Pls. 21.4, 22; fig. 53

Essentially, both *kuan* and *wen* are simple jars, such as have been made since the invention of ceramics. We need not search for their prototypes in other materials. Laufer's belief that such jars were made in imitation of a bronze form (p. 129) should be reversed. Bronze followed ceramic forms. The Shao-kou report separates jars with a clear distinction between body, neck and mouth *(kuan)* from those less well defined *(wen)*. This division is useful for descriptive purposes only. Shao-kou yielded 598 *kuan* which the report classifies in 9 types. Some examples bear inscriptions indicating that they had contained liquids. 337 Shao-kou *wen* fall into 6 types. Several of these are also inscribed with the graphs for wine and for the volume measure. One contained remains of a grain. Laufer's *kuan* resembles one of Type 1 from a burial of the later 2nd century B.C. (SK fig.49.1). All his examples of *wen* conform to Type 2.1 which was, however, in use throughout the Han Dynasty. His arrangement of them on the plates does offer a roughly accurate sequence for their evolution. Brown glazed examples (Pl. XXIX.1–2) precede green glazed ones (Pl. XXX) and green glazed jars with moulded belt of decoration (Pl. XXXII).

*Hu* Pls. XXXVII–XLV

       SK Pls. 19A–20.3, figs. 51 & 52

Since more examples of this vase form were available for study, Dr. Laufer could attempt a typological analysis. His criteria in doing so are similar to those of the Shao-kou report; vessel shape and decor. We can conveniently compare their results. The Shao-kou examples are divided into one group of over 32 cm. height, with 1183 finds, and another shorter group with 369 pieces. Some of the first group is inscribed with the graphs for various grains or pickles. 8 of the second group are specified as being for salt or other flavouring matter.

Pl. XLIV.3 (Laufer Type 3) is related to SK fig. 51.1 —SK Type 1.1, mid-2nd century B.C. to ca. 100 A.D.

Pl. XLI.1 (Laufer Type 2) is related to SK fig. 51.2 —SK Type 1.2, the same time span as Type 1.1.

Pl. XLIV.2 (Laufer Type 3) is related to SK fig. 51.3 —SK Type 2.1, 1st century A.D. through mid-2nd century.

Pl. XLV.1–2 (Laufer Type 3) is related to SK fig. 51.4—SK Type 2.2, same time span as Type 2.1.

Pls. XXXVIII.3–4, XXXIX.2 (Laufer Type 1a) are related to SK fig. 51.5—SK Type 3.1, same time span as Type 2.1.

Pls. XLII.2, XLIII.1, XLIV.1 (Laufer Type 2); XLI.3, XLIII.2–3 (Laufer Type 2a); and XLI.4 (Laufer Type 4) are variants of SK fig. 51.6—SK Type 3.2, same time span as Type 2.1.

Pl. XXXIX.1 (Laufer Type 2) is related to SK fig. 51.7—SK Type 4, from the 1st century through the early 3rd century A.D.

Pls. XLI.2 and XLII.1 (Laufer Type 2b) are related to SK fig. 51.8—SK Type 5, from the later 1st through the early 3rd centuries A.D.

Pls. XXVII.1 & 3, XXVIII.1, XXXIX.3 and XL.1 (Laufer Type 1) can be compared to SK fig. 52.9—SK Group 2 Variant 4 from a burial of the 1st century A.D.

Pls. XXXVII.2, XXXVIII.2 and XL.2 (Laufer Type 1) can be compared to SK fig. 52.10—SK Group 2 Variant 5 from a tomb of the late 2nd to early 3rd centuries A.D.

Note that most of Laufer's specimens are glazed and many bear decorative relief bands, suggesting that these should be placed well within the East Han period and probably after the 2nd century A.D.

Pl. XXXV shows a vase with moulded medallions on its shoulder. Both its form and decoration would place it in the T'ang Dynasty.

The hill-censer of Pl. LV.1 is similar to one from a Shao-kou burial of the later 2nd to early 3rd centuries A.D. (SK Pl. 34.6, fig. 64.1). Only 3 examples of this vessel form were discovered at the Shao-kou site.

*Lien* Pls. LVI, LVII, LIX & LX

       SK Pls. 32 A & B.1, figs. 62.1–6

Dr. Laufer calls these hill-jars, though not all of them have lids in the hill shape. It has become common practice to designate the vessel form by this term in Western literature, but it is not accurately descriptive. 161 examples were found at Shao-kou; some containing fragments of red powder, lacquer and bronze, others a ladle or animal bones. They are judged to have been used either to hold toilette equipment or to serve food. The report groups them in 7 types, of which one displays certain formal

350

characteristics descended from lacquer *lien* of the Warring States period. Laufer's Pls. LVI and LVII are related to the Shao-kou Type 1.3 from burials of the later 1st through mid-2nd centuries A.D. (SK Pl. 32 A.3, fig. 62.2). Pl. LIX.1 is similar to Shao-kou Type 1.1 which was introduced in the 1st century A.D. and continued through the East Han Dynasty (SK Pl. 32 A.1). Pl. LIX.2 can be compared to Shao-kou Type 1.5 popular about the mid-2nd century A.D. (SK fig. 62.4). Again, only the Hsi-an examples bear relief bands. Similarly, only those have lids moulded into hill shapes, though the Shao-kou Types 1.2–3 show lids rising from rim to central knob in concentric tiers.

All the evidence suggests that glazes were more commonly used in ceramics of north-western China during the Han Dynasty, while painted wares were prevalent in north-eastern China. Also, relief decoration with mythological or landscape motifs was distinctive to Shensi pottery.

Many parts of Dr. Laufer's work are interesting to read; in particular, his descriptions of how tools and vessels were used. His Section V on ornamentation of Han pottery remains useful to those who are curious about the lion motif, races of dogs in China, etc.

His appendices on Han roofing-tiles, Sung mortuary pottery (all Tz'u-chou ware vessels) and modern mortuary pottery are based upon such scanty material that they can be of little use to us now.

It is to be hoped that some scholar will be moved to investigate Han pottery in the light of available archaeological materials, and that some publisher will choose to publish such a study instead of the less expensively produced reprint of an out-dated work.

*University of Toronto*                    Hsio-Yen Shih

*Eli Lancman, Chinese Portraiture. 188 pp., 65 monochrome plates, 8 color plates. Rutland, Vermont and Tōkyō, Japan, Charles E. Tuttle Company, 1966, $ 10.00.*
This is a book on portraiture in the widest sense of the word: the representation of an individual living or dead, real or imaginary, by depicting his physical form or his moral traits, or both. The author indicates that Chinese portraiture "includes what western art calls figure painting" (p. 34). This is only to be expected because the Chinese, except for ancestor paintings, rarely made portraits in the strict sense of the term.
The first chapter is a fifteen page, capsule résumé of

Chinese history, culture, and the general outlines of the development of major artistic trends.

In the second chapter, "Understanding Chinese Portraiture," the author outlines his approach to the study of Chinese portraiture and suggests the significance of portrait-figure painting to the Chinese. The conclusion reached by some critics of Chinese art that depictions of human figures occupy an insignificant place in the scope of Chinese painting he believes to be incorrect. He feels not only that "figure-painting has been one of the highest art forms in all periods of Chinese art" (p. 34) but also that "like figure-and-landscape painting, the art of portraiture had its periods of rise and decline" (p. 34). As for the meaning of portraiture, the Chinese, he says, looked upon it "as a composite art, an amalgamation of picture-making and biography" and "held that literal reproduction of the features never fully revealed the character of the subject, but that the painter must use what today would be called 'psychological clairvoyance' in order to reveal his subject's soul... Consequently, Chinese portraits were never merely a transcription of the features of the subject but rather a composite of what the painter thought the essence of the man to be, founded on his knowledge of his life and lit by his imaginative insight" (pp. 34–35).
From these statements, the reader would expect, in the remainder of the book, to learn of the history of portrait-figure painting, and to gain some insight into the methods used by the artists to convey the characters of those whom they portrayed and indeed something about those portrayed themselves. The reader will be disappointed, for Mr. Lancman at best has given a superficial account of these aspects of figure painting. His grasp and understanding of Chinese painting is limited, the value of his comments is often dubious, and there is a definite bias toward the anecdotal and purely literary part of the history of Chinese art.
The main chapters of the book are the "Early Stages", from legendary times through the Han Dynasty (Chapter 3); "The Classical Period", the Chin through the Five Dynasties (Chapter 4); a period of "Development and Stagnation", Sung and Yüan (Chapter 5); "Decline and Revival", Ming and Ch'ing (Chapter 6), including a section on imperial portraits. The last chapter (7) treats Western influence in Chinese portraiture.
The "early stages" of Chinese portraiture are of necessity presented as a collection of literary refer-

351